Mechanical and Thermophysical Properties of Polymer Liquid Crystals

Polymer Liquid Crystals Series

Series editors: D. Acierno, Department of Chemical Engineering, University of Salerno, Italy
Witold Brostow, Department of Materials Science, University of North Texas, Denton, USA
A.A. Collyer, formerly of the Division of Applied Physics, Sheffield Hallam University, UK

The series is devoted to an increasingly important class of polymer-based materials. As discussed in some detail in Chapter 1 of Volume 1, polymer liquid crystals (PLCs) have better mechanical performances, higher thermal stabilities and better physical properties than flexible polymers. They are more easily processable than reinforced plastics and can be used in small quantities to lower dramatically the viscosities of flexible polymer melts. PLCs can be oriented easily in shearing, electric and magnetic fields. They also have interesting optical properties, making applications in optical data storage possible, such as light valves and as erasable holograms. Current applications of PLCs include automobile parts. The areas of *potential* applications include the electrical, electronic, chemical, aircraft, petroleum and other industries.

One measure of the rapidly increasing interest in PLCs is the number of names that are used for them: *in-situ* composites, molecular composites, liquid crystalline polymers (LCPs), self-reinforcing plastics, etc. The very rapidly growing literature on the subject makes it more and more difficult for researchers, engineers, faculty members and students to keep up with the new developments. Newcomers to the field are typically overwhelmed by the complexity of these materials in comparison to traditional engineering plastics – as reflected in sometimes mutually conflicting conclusions in publications, phrased moreover in difficult terminology. The present book series solves these problems for people already in the field as well as for the novices. Experts in the field from all over the world have been called upon to clarify the situation in their respective areas. Thus, conflicting evidence is sorted out and general features are stressed – which becomes achievable after a uniform picture of the structure of these materials is provided.

Volume 1 gives an introduction to liquid crystallinity, describes characterization of LC phases including NMR studies, discusses lyotropic (produced in solution) as well as thermotropic (produced by manipulating the temperature) PLC phases. Volume 2 deals with rheology and processing. Volume 3 deals with mechanical and thermophysical properties of PLCs and PLC-containing blends, including inorganic PLCs, formation of PLC phases – also in non-covalently bonded systems – memory effects, phase diagrams, relaxation of orientations, creep and stress relaxation, thermoreversible gels, acoustic properties and computer simulations of PLCs. Volume 4 deals with electrical, magnetic and optical properties, including a discussion of displays and also of optical storage. Overall, the book series constitutes the *only truly comprehensive source* of knowledge on these exciting materials.

Titles in the series

1. **Liquid Crystal Polymers: From structure to applications**
   Edited by A.A. Collyer
2. **Rheology and Processing of Liquid Crystal Polymers**
   Edited by D. Acierno and A.A. Collyer
3. **Mechanical and Thermophysical Properties of Polymer Liquid Crystals**
   Edited by Witold Brostow
4. **Electrical, Magnetic and Optical Effects on Polymer Liquid Crystals**
   Edited by Witold Brostow and A.A. Collyer (forthcoming)

Mechanical and Thermophysical Properties of Polymer Liquid Crystals

Edited by

Witold Brostow

Department of Materials Science
University of North Texas, Denton
USA

CHAPMAN & HALL
London · Weinheim · New York · Tokyo · Melbourne · Madras

Published by Chapman & Hall, an imprint of Thomson Science, 2–6 Boundary Row, London SE1 8HN, UK

Thomson Science, 2–6 Boundary Row, London SE1 8HN, UK

Thomson Science, Pappelallee 3, 69469 Weinheim, Germany

Thomson Science, 115 Fifth Avenue, New York, NY 10003, USA

Thomson Science, Suite 750, 400 Market Street, Philadelphia, PA19106, USA

First edition 1998

Thomson Science is a division of International Thomson Publishing

Typeset in 10/12 Palatino by AFS Image Setters, Glasgow

Printed in Great Britain by the University Press, Cambridge

ISBN 0 412 60900 2

A catalogue record for this book is available from the British Library

Library of Congress Catalog Card Number: 96-72034

Contents

Contributors

C. Géraldine Bazuin
Centre de Recherche en Sciences et Ingénerie des Macromolécules (CERSIM), Faculté des sciences et de génie, Université Laval, Cité Universitaire, Québec, Province of Quebec, G1K 7P4, Canada

Rita B. Blumstein
Department of Chemistry, University of Massachusetts Lowell, One University Avenue, Lowell, MA 01854, USA

Witold Brostow
Department of Materials Science, University of North Texas, Denton, TX 76203-5310, USA

C.L. Choy
Faculty of Applied Sciences and Textiles, Hong Kong Polytechnic University, Hung Hom, Kowloon, Hong Kong

Cameron G. Cofer
Composites Innovation Laboratory, Owens-Corning Science & Technology, 2790 Columbus Road, Route 16, Granville, OH 43023, USA

James Economy
Department of Materials Science and Engineering, University of Illinois at Urbana-Champaign, 1304 West Green Street, Urbana, IL 61801, USA

Lydia Fritz
Institut für Chemie, GKSS-Forschungszentrum Geeschacht GmbH, Kantstraße 55, D-14513 Teltow, Germany

Ulf W. Gedde
Department of Polymer Technology, The Royal Institute of Technology, 100-44 Stockholm, Sweden

Andreas Greiner
Department of Physical Chemistry, Polymer Institute, University of Marburg, Hans-Meerwin-Straße/Geb. 23, 35032 Marburg, Germany

Michael Hess
Department of Physical Chemistry, Gerhard Mercator University, 47048 Duisburg, Germany

Gabor Kiss
Bellcore 2F-125, 445 South Street, Morristown, NJ 07960, USA

Josef Kubát
Department of Polymeric Materials, Chalmers University of Technology, 412-96 Gothenburg, Sweden

Betty L. López
Department of Chemistry, University of Antioquia, Apartado Aereo 1226, Medellin, Columbia

Robert Maksimov
Institute of Polymer Mechanics, Latvian Academy of Sciences, 23 Aizkraukles Street, 1006 Riga, Latvia

Roger S. Porter
Polymer Science and Engineering Department, University of Massachusetts, Amherst, MA 01003, USA

Willie E. (Skip) Rochefort
Chemical Engineering Department, Gleeson Hall 103, Oregon State University, Corvallis, OR 97-331-2702, USA

Joachim Rübner
Macromolecular Chemistry, Technical University of Berlin, Straße des 17. Juni 135, 10623 Berlin, Germany

George P. Simon
Department of Materials Engineering, Monash University, Clayton, Melbourne, Victoria, Australia 3168

Ram Prakash Singh
Materials Science Centre, Indian Institute of Technology, Kharagpur 721302, India

Jürgen Springer
Macromolecular Chemistry, Technical University of Berlin, Straße des 17. Juni 135, 10623 Berlin, Germany

Richard G. Weiss
Department of Chemistry, Georgetown University, Washington, DC 20057-2222, USA

Göran Wiberg
Dietmar Wolff
Macromolecular Chemistry, Technical University of Berlin, Straße des 17. Juni 135, 10623 Berlin, Germany

Yang Zhong
Beijing Institute of Aeronautics and Astronautics, 37 Xueyuan Road, Haidan District, Beijing, P.R. China and Mechanical Engineering Department, Prairie View A & M University, Prairie View, TX 77446, USA

Part One
Formation of polymer liquid crystals

1

Creation of liquid crystalline phases: a comparative view emphasizing structure and shape of monomer liquid crystals

Richard G. Weiss

1.1 INTRODUCTION

A crystal may be defined microscopically as a condensed phase in which molecules possess both orientational and three-dimensional long-range order and are not able to change conformations; macroscopically, the shape of a crystal depends upon (microscopic) molecular positions within a lattice, but not upon the shape of the vessel in which it is held. A liquid is a condensed phase in which there is no long-range order and individual molecules are able to alter their conformations; the shape of a liquid conforms to the vessel in which it is held. Condensed phases with properties intermediate between these two extremes are called *mesophases*. It is now recognized that this designation encompasses plastic crystals (that is, condensed phases in which molecules are ordered with respect to their positions in a lattice, but are disordered with respect to their orientations), condis crystals (that is, condensed phases whose molecules possess a great deal of both positional and orientational order, but are conformationally disordered [1; and Wunderlich, B. and

Mechanical and Thermophysical Properties of Polymer Liquid Crystals
Edited by W. Brostow
Published in 1998 by Chapman & Hall, London.
ISBN 0 412 60900 2

Chen, W. (1996) in *Liquid-Crystalline Polymer Systems* (ed. A.I. Isayev, T. Kyu and S.Z.D. Cheng), American Chemical Society, Washington, chapter 15], and liquid crystals (that is, condensed phases in which molecules exhibit orientational order and varying degrees of positional order [2]). The distinctions among these three mesophase types will be maintained throughout this chapter; liquid crystalline systems which are thermotropic (that is, induced by changes in the temperature or pressure of a sample) rather than lyotropic (that is, induced by addition of an isotropic liquid as 'solvent') will be emphasized.

Although the focus of interest of this series is polymer liquid crystals (**PLC**s), the factors responsible for their mesomorphism are the same as those leading to monomer (low molecular weight) liquid crystals (**MLC**s). Since **MLC**s have been investigated much more extensively and are intrinsically easier to characterize than are **PLC**s, we will focus on the former in this chapter. Some structural classifications and reviews of **PLC** structural properties [3–7] and an extensive review of the consequences of polymerizing **MLC**s [8, 9] have appeared. These are in addition to information found in this and other volumes of the *Polymer Liquid Crystals* series. It should be kept in mind that the concepts of molecular design and shape of **MLC**s which have been exploited so elegantly during the last few decades can be and, in many cases, have been extrapolated to **PLC**s [10].

However, some important caveats are necessary. An **MLC** phase consists of no more than a few different molecules, each of which can be identified completely by its discrete molecular weight, atomic connectivities, stereochemistry, etc. Most **PLC** phases consist of molecules that are related by their (main chain or side chain) repeat units, but differ in their molecular weights [9], tacticities, and/or sequence of repeat units. On a more macroscopic level, all **MLC**s are homogeneous when viewed as an average over a short time period. Local inhomogeneities, due to orientations of molecules in different domains and molecular packing near domain interfaces, can diffuse on 'short' time scales. **PLC**s may exist with domains that are virtually time-invariant or with quasi-crystalline and quasi-isotropic regions. Morphology has different meanings when applied to **MLC**s and **PLC**s (see Chapter 5).

Furthermore, not all polymerizable **MLC**s are precursors of **PLC**s. A polymer derived from an **MLC** may not be mesomorphic due to packing constraints introduced by the backbone of the chains and the change in entropy that attends linking the monomer units by covalent bonds. On the other hand, a non-liquid crystalline monomer may yield a **PLC** (due, in this case, to favorable entropic changes; see Chapter 15).

The manner in which **PLC**s are synthesized and processed can have an enormous influence on their phase properties. Unlike **MLC**s, **PLC**s can have long memories of their prior treatment (see Chapter 5) and may never overcome the effects of hysteresis and stress (see Chapters 6 and 12).

The first sentence of the reference work, *Handbook of Liquid Crystals*, reads: 'The terms liquid crystals, crystalline liquid, mesophase, and mesomorphous state are used synonymously to describe a state of aggregation that exhibits a molecular order in a size range similar to that of a crystal but acts more or less as a viscous liquid.' [2] In other words, molecules within a liquid crystalline phase possess some orientational order and lack positional order; furthermore, the shape of a liquid crystalline sample is determined by the vessel in which it is contained rather than by the orientational order of its aggregated molecules. The authors recognized the limitations and imprecision of this definition but, like others preceding them, could not devise a simple and generally applicable one that is better. Regardless, the terms 'liquid crystal' and 'mesophase' should *not* be used interchangeably. As mentioned above, all liquid crystals are mesophases, but all mesophases are not liquid crystals.

Recent studies, employing elaborate and sophisticated analytical techniques, have permitted finer distinctions between classical crystals and mesophases. At the same time, they have made definitions like that from the *Handbook of Liquid Crystals* somewhat obsolete for reasons other than terminology. One part of the problem arises from the use of a combination of bulk properties (like flow) and microscopic properties (like molecular ordering) within the same definition. Another is that both the bulk and microscopic properties are expressed in terms that may be difficult to relate from one system to another; this is especially true of the bulk properties, but it can also lead to uncertainties in interpretation of the microscopic ones.

In spite of these potential ambiguities, an enormous number of neat molecular systems (estimated to be 50 000 at the end of 1991 [11]) have been identified as forming one of the three main types of liquid crystalline states – nematic (including twisted nematic or cholesteric), smectic and discotic – and many **MLC**s have been classified further as belonging to one of the subtypes of smectic or discotic phases [11, 12]. Unfortunately, there appear to be no extensive compendia of thermotropic or lyotropic **MLC** mixtures or of single substance **PLC**s. Undoubtedly there are many more single substances and mixtures which are mesomorphic than those which have been found thus far. In principle, almost any isotropic liquid phase which is supercooled sufficiently without becoming a glass will become a mesophase; in practice, crystallization or glass formation occurs before many liquids do.

Experimentally, the distinction between an isotropic liquid, a classic crystal and a mesophase usually relies upon one or a combination of observables from techniques such as polarized optical microscopy, rheology and X-ray diffraction. Even the most 'microscopic' of these is then related to the structures of the constituent molecules within the phase in order to propose a packing model. This conceptual leap of faith is a key step in our ability to design new mesomorphic molecules and to devise theoretical paradigms of mesomorphism [13–17]. To understand the nature of a mesomorphic state, it is necessary to recognize the structural and electronic features of the molecules that comprise it and the important intermolecular interactions that lead to partial ordering without crystallization. For instance, it is generally accepted that existence of the least ordered thermotropic liquid crystalline phase, the nematic, can be explained by the anisotropic application of hard-core repulsive forces and attractive dispersive forces among rod-shaped (calamitic) constituent molecules [18, 19]. **Mesomorphism may thus be described as schizophrenia at the level of molecular interactions**.

It is a misnomer to ascribe mesomorphism, an aggregate phenomenon, to individual molecules; it is also incorrect to describe aggregates as mesomorphic without providing the conditions of their existence (such as pressure [20] and temperature) and, in some cases, their history (such as rates of heating or cooling and periods of annealing). Unless noted otherwise, all phase behavior described here is referenced to atmospheric pressure.

Possible distinctions among liquid, plastic, condis and fully organized crystals [1] and among the various types of liquid crystalline phases [2] based on thermodynamic considerations have been discussed. The consequences of those distinctions will be emphasized here using anecdotal but related examples. Selected molecules, grouped by shape, functionality, etc., will be examined (and when possible compared) to discern the salient structural and electronic features that may be key to whether they form thermotropic mesophases and, in a few cases, lyotropic mesophases. Since the common structural features of molecules which become plastic crystalline [21–27] and liquid crystalline [12] are well known, an emphasis will be placed upon more exotic characteristics. Questions to be addressed include the following.

1. Which molecules are beyond the scope of 'normal' mesomorphic structures and yet provide mesophases?
2. What subtle nuances of structure are responsible for one molecule forming only crystalline and liquid condensed phases while related molecules yield mesophases?
3. Why do seemingly similar molecules form different types of mesophases?

We shall see that available information does not always make the answers to these questions obvious. However, there are abundant hints about how to proceed to find the answers. A general description of

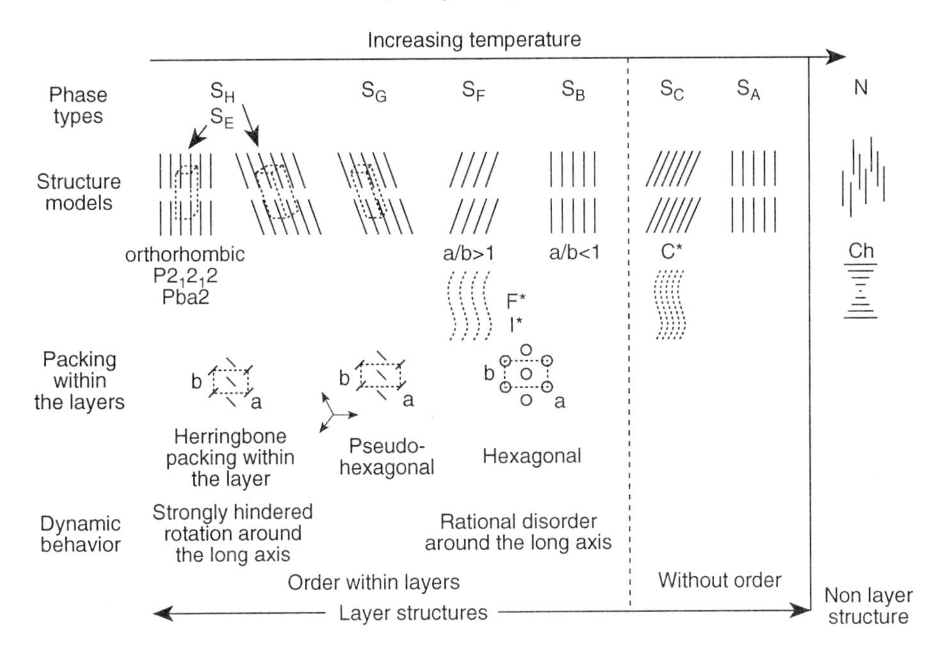

Figure 1.1 Cartoon representations of molecular packing in various smectic (**S**), nematic (**N**) and cholesteric (**Ch**) liquid crystalline phases comprised of rod-like molecules [26].

the molecular packing models of all of the mesophase types discovered thus far will not be presented; due to their frequent use here, a representation of the molecular organization of various smectic phases is included in Figure 1.1 [26]. Other pertinent information may be found in several of the references.

1.2 PHASES OF SOME SIMPLE ACYCLIC (CALAMITIC) ALKANES AND THEIR FUNCTIONALIZED ANALOGS

Figure 1.2 provides a conceptual flow chart that links many of the molecules discussed in this section by their structural similarities and functional groups. The consequences of the structural changes indicated among molecules linked by arrows are discussed in the text.

1.2.1 Rod-like molecules as single substances

(a) n-Alkanes

There is no evidence for either a plastic or a condis phase of short *n*-alkanes. For instance, hexane exhibits a triclinic solid phase that melts

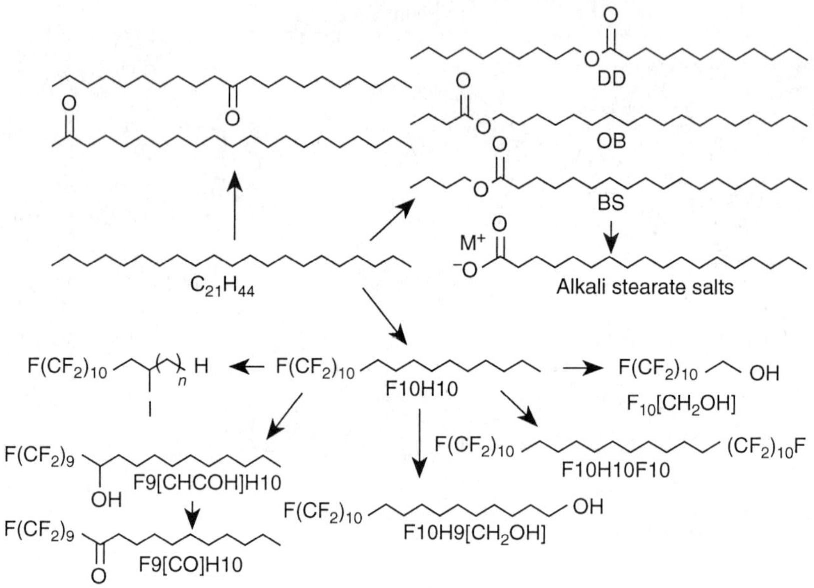

Figure 1.2 Conceptual links among many of the molecules discussed in section 1.2.

directly to a liquid at 177.8 K [28]. Many longer *n*-alkanes exist as condis crystals within a temperature range between the melting transition and the transition to a lower temperature solid phase. They have low temperature solid phases in which molecules in extended, well-ordered conformations are packed triclinically, monoclinically or orthorhombically in layers [28–30]. Below their melting temperatures, many longer (especially odd) *n*-alkanes from C_9H_{20} to $C_{43}H_{88}$ also exhibit intermediate 'rotator' phases with hexagonally packed layers [28–30]. Individual molecules within rotator phases have their long axes orthogonal to the layer planes and are able to rotate about this axis. Within the rotator phases, the incidence of gauche conformational defects along a chain increases from a molecule's center to its ends [31, 32]. These structural and dynamic characteristics show that the rotator phases of *n*-alkanes are condis crystals. They are also analogous to smectic B liquid crystalline phases [33] in the way their molecules are organized within the crystalline lattices.

An example is heneicosane ($C_{21}H_{44}$). The probability of gauche defects along the molecular chains in the rotator phase at two different temperatures near the transition temperatures (305.7–313.4 K [29]) is shown in Figure 1.3 [31]. Below 305.7 K, the hexagonal packing within layers collapses to an orthorhombic arrangement (like that of a smectic E liquid crystalline phase [33]) due to dampened rotations about the

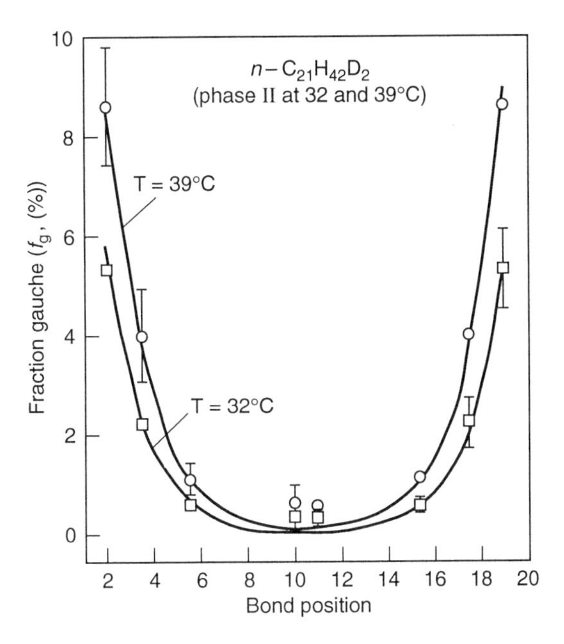

Figure 1.3 Fraction of C—C bonds that are gauche plotted as a function of bond position of heneicosane chains for the rotator phase (phase II) at 32°C (305 K) and 39°C (312 K) [31].

long molecular axes [28] (Figure 1.4). By contrast, eicosane ($C_{20}H_{42}$) forms only one solid phase below its melting temperature (326.5 K), a rigid triclinic in which the long molecular axes are fully extended and inclined at a non-normal angle with respect to the layer planes (as in smectic G and H liquid crystalline phases [33]) [28]. Molecules in layers of even-numbered *n*-alkanes are usually packed more closely than those of the odd-numbered ones [35]; methylene groups near the center of a triclinic layer are constrained even more than those in an orthorhombic phase [36].

(b) Alkanones

Insertion of a carbonyl between two methylene groups transforms an *n*-alkane into an alkanone. Since the group volume and bond angle to neighboring carbon atoms of a carbonyl are only *c.* 15% and *c.* 4° larger, respectively, than those of a methylene [37–40], the intrinsic shapes and sizes of alkanes and alkanones with the same number of carbon atoms are very similar. On that basis, they should form similar solid phase structures; differences may be attributable to the opposite

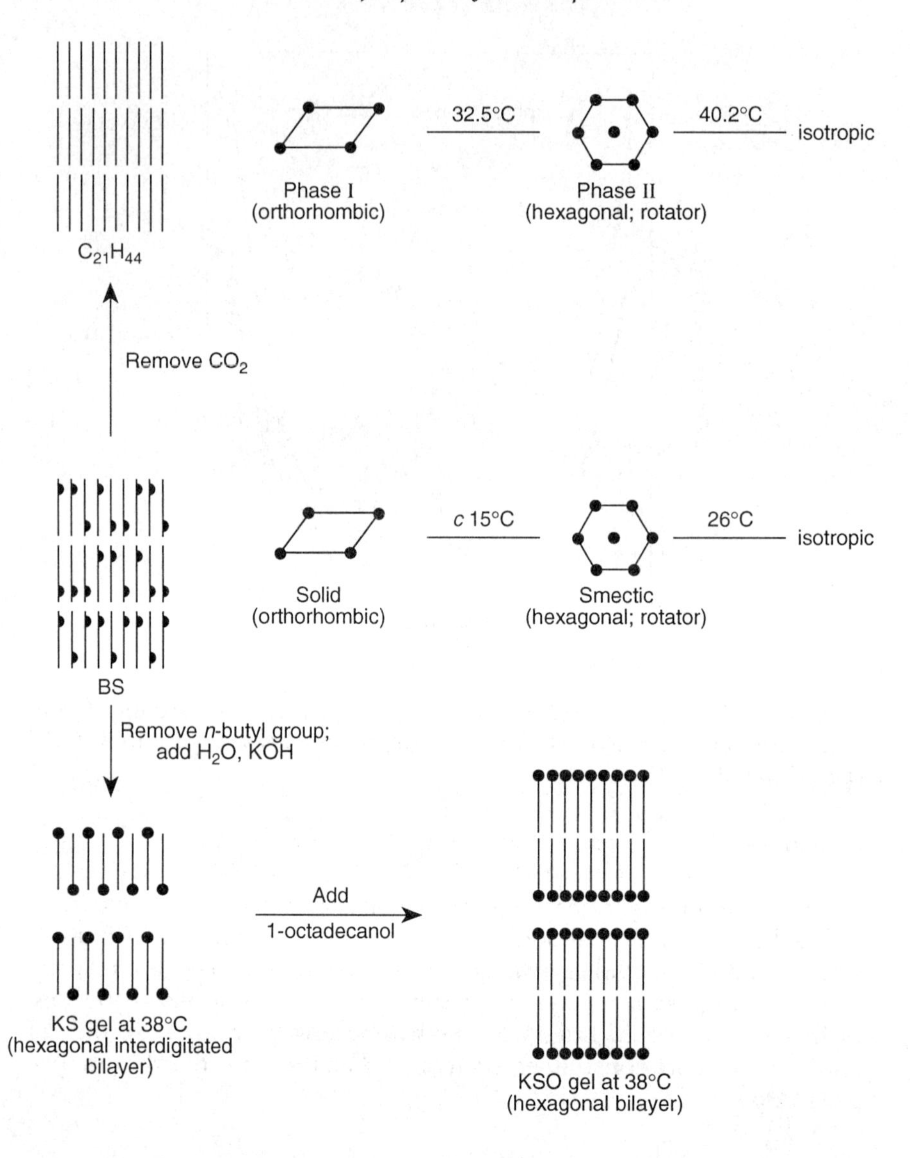

Figure 1.4 Cartoon representation of molecular packing of layers of heneicosane and related molecules demonstrating the similarities of the organization of their condensed phases.

directions and disparate magnitudes of the group dipoles of a methylene (0.35 D [41]) and a carbonyl (2.4 D [41]). An experimental manifestation of the structural similarities is that up to 15% of 2-eicosanone, 2-heneicosanone or 11-heneicosanone can be added to heneicosane without observing significant changes in thermograms of the latter that encompass both of its solid phase transitions [42]; the alkanones must be incorporated nearly isomorphously into the heneicosane lattice. Calorimetric studies on a limited number of neat 2- and symmetrical alkanones with carbon chain lengths comparable to that of heneicosane provide no evidence for solid–solid transitions [21]. Like many of the *n*-alkane solid phases, those of the alkanones consist of layers of molecules in extended conformations and with their long axes perpendicular to the layers planes [43–46]. The 2-alkanones appear to pack in pseudo bilayers, so that the carbonyl groups from molecules in adjacent layers are near each other [45, 46]. Limited NMR spectroscopic evidence from solid samples indicates that the probability of gauche defects along an alkanone chain follows the same trend detected with *n*-alkanes [42]. These results and others from attempts to force alkanones to undergo photochemical reactions which require moderate conformational changes [47–49] indicate that the solid phases of the alkanones examined are not condis crystals. Although additional experiments are required to confirm this conclusion, it appears that the introduction of a carbonyl group along an alkyl chain creates inadequate disruption to change the general mode of molecular packing within the solid lattices. If anything, attractive dipolar interactions between neighboring carbonyl groups raise the solid–liquid transition temperatures (with respect to those of analogous alkanes) and may preclude the formation of condis phases.

(c) Alkyl alkanoates

In principle, insertion of a carboxy group instead of a carbonyl group along an alkyl chain induces a larger steric perturbation and greater ease of chain deformation: the barrier to internal rotation in dimethyl ether, 13.0 kJ mol^{-1}, is 1.0 kJ mol^{-1} smaller than the barrier in propane [50]; although the *s-trans* conformer of a carboxy unit permits an ester to fit better into a lattice in which molecules are in fully extended conformations than does the gauche-like *s-cis* conformer (Figure 1.5), the group dipole of the former, *c.* 1.5 D [50], is 2 Debye units smaller than the latter [50] and, therefore, less capable of stabilizing intermolecular interactions through van der Waals forces.

n-Butyl stearate (**BS**) is heneicosane into which a carboxy group has been inserted between C4 and C5. It is known to form two solid and one mesophase [43, 51–53]. Transitions are observed at 299.3 K (liquid $\leftrightarrow \alpha_1$), 288 K ($\alpha_1 \leftrightarrow \alpha_2$) and 284.3 K ($\alpha_2 \leftrightarrow \alpha_3$) [51–53]. In all of the α phases, **BS** molecules are in extended conformations and arranged in layers ($d = 31.7$ Å) with their long axes normal to the layer planes (Figure 1.4). When **BS** is annealed for very long periods or obtained directly from crystallization, a solid phase whose layer thickness is

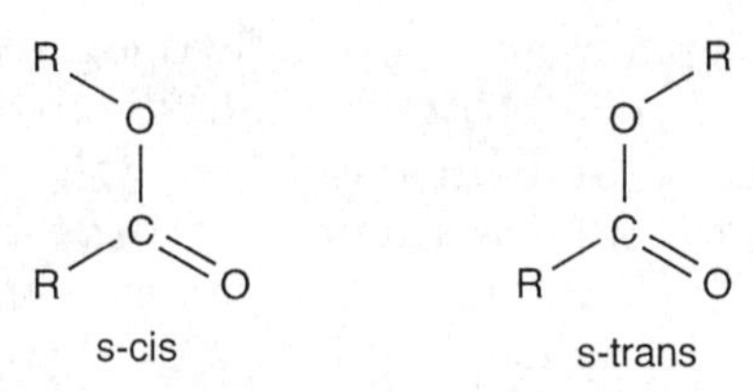

Figure 1.5 Representation of the *s-cis* and *s-trans* conformations of an ester linkage.

27.6 Å can be obtained [43]. The exact nature of the α_3 phase is not known, but it must be very similar in lattice structure to the orthorhombic arrangement of the α_2 phase; recent X-ray diffraction data are consistent with molecular rotations in the α_3 phase being frozen and those in the α_2 phase being slower than in the α_1 phase [54; and Baldvins, J.E. and Weiss, R.G. (1997) *Liq. Cryst.*, submitted]. Thus the organizations of the lower temperature solid phases of **BS** and heneicosane are very similar. The α_1 phase, a waxy solid that flows very slowly, is known to have hexagonal packing and rotational disorder along the long molecular axes of **BS** molecules [51–53]. The question arises whether it is a condis crystal (like the higher temperature solid phase of heneicosane), a true smectic B liquid crystalline phase as initially indicated [52, 53], or a hexatic smectic B phase in which there are some interlayer molecular correlations in addition to the intralayer ones. Subsequent optical microscopic [55] and X-ray diffraction [54, 56, 57] experiments provide support for the latter attribution. Diffractograms from the rotator phase of heneicosane, which shows at least three orders of intralayer diffractions, and from the α_1 phase of **BS** are compared in Figure 1.6: the α_1 phase of **BS** is more ordered than a normal smectic B phase, but less ordered than a condis phase; it is a hexatic B liquid crystalline phase.

By inverting the direction of the carboxy group along the chain of heneicosane, one can imagine transforming **BS** to an isomeric ester, octadecyl butyrate (**OB**). **OB** also has two solid phases and one mesophase [54]. The transition temperatures and heats of transition are very similar for the two esters, and they are miscible at all compositions of their mixtures. Furthermore, diffractograms of **OB**, **BS** and their mixtures at 291 K (within the α_1 phase) are very similar in peak appearance and diffraction angle. Thus, **OB** appears to form a hexatic smectic liquid crystalline phase [54].

Interestingly, a liquid crystalline phase is not formed upon heating or cooling another positional isomer of **OB** and **BS**, decyl dodecanoate (**DD**), in which the carboxy group is inserted near the middle of heneicosane [54]. Unlike **OB** and **BS**, **DD** does not flow under force and it exhibits only one (melting) transition by DSC when heated or cooled in the temperature range when the other two isomers are polymorphic. Its X-ray powder diffraction pattern indicates that **DD**

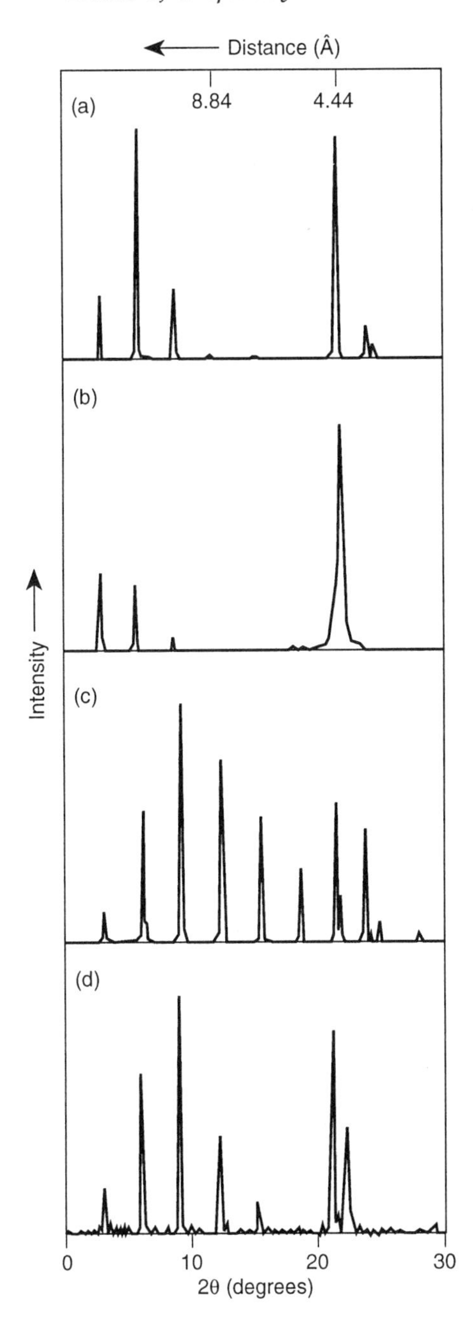

Figure 1.6 X-ray powder diffraction patterns of **BS** in its α_3 (278 K) and α_1 (291 K) phases (a and b, respectively) and of $\mathbf{C_{21}H_{44}}$ in its orthorhombic (299 K) and hexagonal (311 K) solid phases (c and d, respectively) [54].

molecules are in extended conformations and packed normal to the layer planes ($d = 29.4$ Å [54]).

Evidence for polymorphism and possible mesomorphic phases, but not for liquid crystallinity, has been found in studies on methyl, ethyl and propyl stearate (i.e. molecules in which the length of the 'acid' part is maintained at 18 carbon atoms and the length of the 'alcohol' part is shorter than in **BS**) [43, 52–55]. We hypothesize that enhanced rotation of an alkoxy group of moderate length in these carboxy-inserted alkanes (perhaps in combination with dipolar effects) includes sufficient disorder within molecular layers to transform the phases from condis to liquid crystals; embedding the carboxy group near the middle of an alkane (as with **DD**) serves to attenuate alkyl chain motions near a layer interface and thus precludes the formation of liquid crystalline phases. These qualitative assertions can and should be tested experimentally using a wider variety of molecules than have been examined thus far.

(d) Alkanoic acids and their salts

Insertion of the carboxy group in a terminal C—H bond of an *n*-alkane, creating an alkanoic acid, is an additional and rather drastic modification of an alkane structure. The mesomorphic behavior of long-chain carboxylic acids and other surfactants as neat soaps and lyotropic systems (usually with water) has been investigated extensively [58–60] and simple theoretical models of the aggregate structures based on simple packing and electrostatic considerations have been advanced [61]. Only selected phases involving stearic acid will be discussed here. The neat (anhydrous) alkali metal salts of stearic acid and other fatty acids are known to exist in several mesomorphic states, depending upon temperature [58], in which the molecules, with differing degrees of deviation from all-*trans* chain conformations, are organized into various layered forms [62].

Detailed analyses of the temperature dependence of X-ray diffraction patterns of alkali stearate salts in water have allowed rather complex phase diagrams to be constructed [63]. An example involving potassium stearate (**KS**), based on samples cooled from 373 K but not annealed at subambient temperatures, is shown in Figure 1.7. The lowest temperature dotted line represents the separation between the 'coagel' phase (solid soap particles (C) dispersed in water (E)) and temperatures/compositions at which various gel phases (G) can exist or coexist with a solid. Between 303 and *c.* 316 K and at $\geqslant$30 wt% of **KS**, a stable transparent gel (G) is observed. The L (neat) phase consists of bilayer sheets (lamellae) of **KS** molecules whose alkyl chains are 'melted' and separated by aqueous layers whose average thickness depends upon the phase composition. The M (middle) phase is composed of

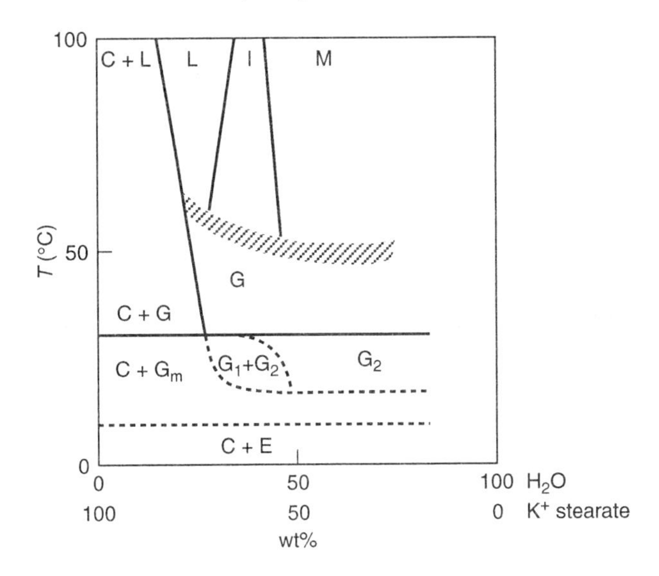

Figure 1.7 Phase diagram for **KS** gels [63].

rod-shaped micelles whose long axes are oriented parallel to each other on average (as in a thermotropic nematic phase). The I (intermediate) phase is probably a progression of different **KS** organizates which are related somewhat to the L and M phases [58].

We will concentrate on the structure of the gel phases [33]. Presumably due to the reluctance to separate from their carboxylate counterions, the lithium and sodium salts of stearic acid do not form gel phases with water [63]. However, the heavier alkali metal salts do. In each case, the stearate anions are arranged in nearly fully interdigitated bilayers (Figure 1.4) which vary in thickness from 25.2 Å (**KS** at 318 K) to 26.8 Å (resium stearate at 298 K); the average carboxylate head group area, *c.* 40 Å^2, is twice the value of that in a closely packed non-interdigitated bilayer. These distances require the stearate chains to be fully extended and normal to the layer planes. As expected from the model, the thickness of a layer is virtually unchanged as the water content is altered within a gel phase. Furthermore, the thicknesses of layers in gel phases of other potassium alkanoates increase or decrease by the distances predicted when methylene groups are added or subtracted from a fully extended alkyl chain [63].

Within a layer of a **KS** gel, stearate molecules are hexagonally packed so that each layer resembles the organization of a smectic B liquid crystalline phase. Rubidium stearate can have gel phases in which stearate chains are packed either in a hexagonal (G_H) or in a rectangular (orthorhombic-like; G_R) array; they resemble the arrangements of molecules in smectic B and E phases, respectively. Since the interiors of the stearate layers are extremely lipophilic, they can be swelled by hydrocarbons [58]. For example, anhydrous sodium stearate can accept large weight fractions of *n*-hexadecane (cetane), leading to the formation of new phases (e.g. a middle phase) not found in either pure component [64]. 1-Octadecanol, an alcohol very close in size and shape to the stearate anion, can be added in equimolar amounts to **KS** in water to create a new set of phases [63]. Among these are hexagonal and rectangular gel phases (Figure 1.8) in which the surfactants are arranged in non-interdigitated bilayers with an alternation between hydroxy and carboxylate head groups at an aqueous interface (Figure 1.4) [63]: the layer thickness is twice that of a stearate molecule and the head group area is again *c.* 40 Å^2.

Stearate chains within a hexagonally packed layer of a gel phase rotate relatively freely as in the smectic B and condis phases of rod-like molecules mentioned above. The similarities in the structure and dynamics among these 'rotator' phases are unmistakable, as are the similarities among their lower temperature orthorhombically packed counterparts. However, there are equally striking differences: NMR spectroscopic studies of deuterated stearate chains in both **KS** [65] and **KS**/1-octadecanol (**KSO**) [66] hexagonal gel phases demonstrate that the methylene groups at C2 (i.e. nearest a carboxylate head group and therefore near a layer end) are the most ordered; constraints to motion

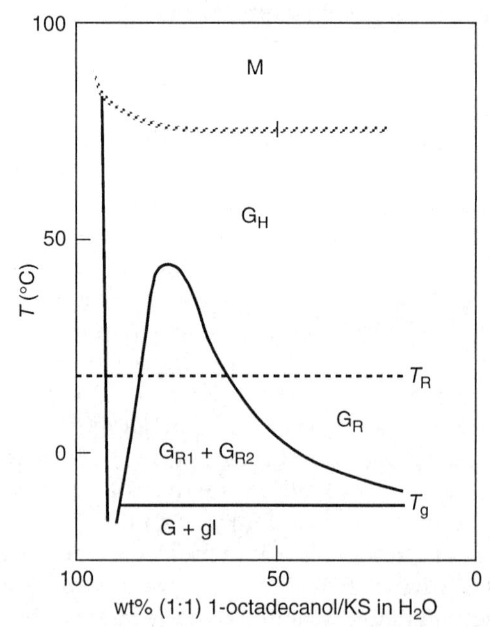

Figure 1.8 Phase diagram for 1:1 1-octadecanol/**KS** (**KSO**) gels [63].

are attenuated as the distance from the head group increases. The values of the quadrupolar splitting parameter ($\Delta\nu_{90}$) for the deuterium atoms at C2 are near the theoretical limit for a rigid, rotating all *trans* chain; in general, the values of $\Delta\nu_{90}$ in the **KS** gel are smaller along the chain than those in the **KS**/1-octadecanol gel. Stiffening of lipophilic chains upon addition of 1-octadecanol is consistent with what is known about the addition of alcohols to bilayer (lamellar) phases [67].

(e) Partially fluorinated alkanes and related molecules

A very different kind of surfactant molecule obtains if the carboxylate head group of an alkanoic acid is replaced by one that is both hydrophobic and lipophobic, like a perfluoroalkyl group. Such 'diblock' molecules, $\mathbf{F(CF_2)_m(CH_2)_nH(F}m\mathbf{H}n\mathbf{)}$, have been shown to form normal and reverse micelles in perfluoroalkanes and alkanes, respectively [68, 69]. On the bases of the well-known antipathy between hydrocarbons and perfluorocarbons, the disparity between cross-sectional areas and volumes of CH_2 (18.5 Å^2 and 10.22 Å^3) [70, 71] and CF_2 (28.3 Å^2 and 16.04 Å^3) [70, 71] groups, and the tendency of long perfluoroalkyl chains to adopt helical conformations, it is expected and found that organized assemblies of **F*m*H*n*** molecules exhibit some strange properties. For instance, the alkyl portions of many of these molecules 'melt' before their perfluoroalkyl portions [72], and the allowed motions resemble closely those of *n*-alkanes in their rotator phases [73, 74].

Detailed studies of the homolog, **F10H10**, have provided strong evidence for formation of two smectic phases below the melting temperature [75, 76]. Viney *et al.* [76] have proposed that they are of the highly ordered G or J type which have positional order between layers, pseudohexagonal ordering within layers, and extended molecules whose long axes are tilted with respect to the layer planes. In the higher temperature phase (SmII), the angle of tilt is *c.* 5° and the molecules are thought to be fully interdigitated; in the lower temperature phase (SmI), the angle of tilt is *c.* 30° and the model requires interdigitation of only the alkyl portions of the molecules (Figure 1.9). We have expressed doubts about the correctness of the model for SmII [71], and Höpken and Möller [70] have found very complex forms of layered packing in mesophases of molecules like **F10H10**. Regardless of the details of their molecular packing, the mesophases of **F10H10** are smectic or smectic-like and have many of the lattice characteristics of the hydrocarbons from which they derive.

The effects on mesomorphic behavior of placing additional substituents at or near the α (molecular middle) or ω (terminal methyl) carbon atom of the alkyl portion of **F10H10** and related molecules have been explored. Iodination at the β carbon of the alkyl chain leads to molecules

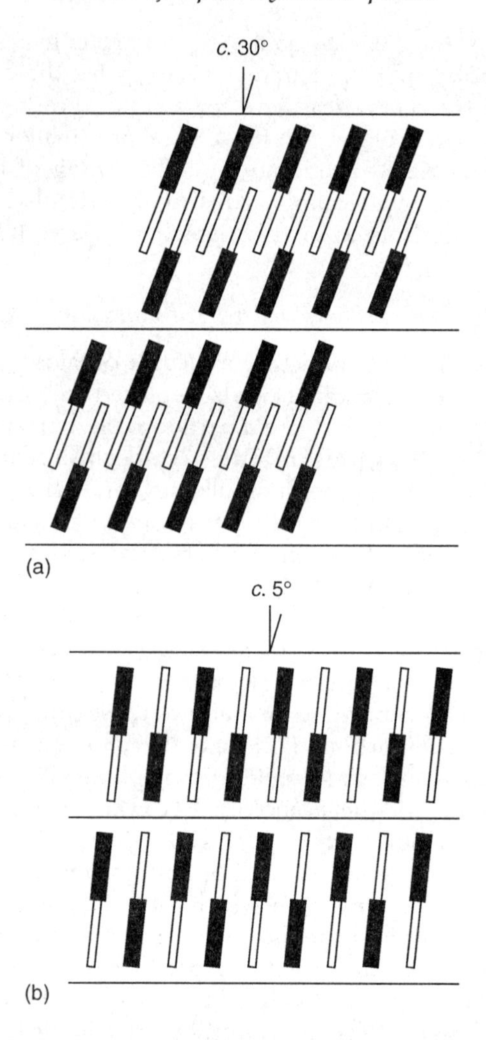

Figure 1.9 Representations of SmI (a) and SmII (b) packing arrangements for **F10H10** according to Viney *et al.* [76]. Filled and unfilled blocks represent perfluorodecyl and decyl groups, respectively, in their elongated conformations.

F10[CH$_2$CHI]H*n* ($4 \leqslant n \leqslant 10$) which form both an enantiotropic smectic B and a smectic A phase [77]. The related alcohol **F9[CHOH]H10** and a shorter homolog, **F7[CHOH]H8** are also liquid crystalline; two monotropic smectic phases and a lower temperature (probably solid) phase have been observed for each [78, 79]. By contrast, cooled melts of the corresponding ketones **F9[CO]H10** and **F7[CO]H8** transform directly to solid phases [78–80]. Since a carbonyl group is intermediate in polarity between a –**CHOH**– and a –**CHI**– group and is smaller than either of them (but larger than a methylene group), it is not apparent why the keto compounds are not mesogens also. Examination of more examples with greater diversity of group polarity and size is needed.

A hydroxy group placed on the terminal carbon atom of **F10H10** provides **F10H9[CH$_2$OH]** which again forms a smectic phase [81, 82]. In contrast to the methacrylate esters of the two above mentioned alcohols **F*m*[CHOH]H*n*** which are not mesogenic, the methacrylates of several homologs of **F10H*n*[CH$_2$OH]** ($n = 3, 5$) do form smectic phases [83], as do the allyl ethers ([81, 82, 84]; J. Höpken, personal communication) of several **F*m*H*n*[CH$_2$OH]** homologs. Previously we reported that the methacrylate of **F7[CHOH]H8** also appears to form a smectic phase [77]. Subsequent experiments have demonstrated that it and the methacrylate of **F9[CHOH]H10** do not (K.J. McGrath, A. Fiseha and R.G. Weiss, unpublished results). Of the two vinyl ethers of the alcohols **F*m*H1[CH$_2$OH]** reported thus far, the $m = 6$ molecule does not form a liquid crystalline phase, but the $m = 8$ one does [85]. Finally, smectic B-like phases have been found for the ethers derived from **F10[CH$_2$OH]** and alkyl α-(hydroxymethyl)acrylate when alkyl is ethyl, decyl and 2-[2-(ethoxy)ethoxy]ethyl [86, 87]. When alkyl is a bulky *tert*-butyl group, no mesophase is formed by the ether [86, 87] due to the introduction of severe deviations from a rod-like molecular shape.

Through these anecdotal examples, a clear pattern emerges in which smectic-like phases can form when the 'crystallinity' of the laterally associated perfluoroalkyl groups can be retained at temperatures where the (terminally modified) alkyl portions are melted. However, when a substituent is near the middle of a molecule, a much more complicated interplay between energies of association and steric and electronic interactions determines whether mesomorphism can occur.

'Triblock' partially fluorinated hydrocarbons **F*m*H*n*F*m*** ($m = 12, n = 8$; $m = 10, n = 10$; $m = 12, n = 12$) have also been examined and all are found to exhibit very narrow ($<1°$) range of thermotropic smectic B phases between their crystalline and isotropic liquid phases [88]. When the polymethylene portions of such molecules are replaced by a much stiffer *para*-substituted aromatic ring, the mesogenic character is lost [89].

1.3 PHASES OF SOME SIMPLE CYCLIC ALKANES AND THEIR FUNCTIONALIZED ANALOGS

Figure 1.10 is a flow chart that includes the structures of many of the monocyclic molecules discussed in this section, with arrows representing a progression of functionalization. The development is subjective and intended to emphasize the progression of phase behavior that accompanies increased molecular complexity.

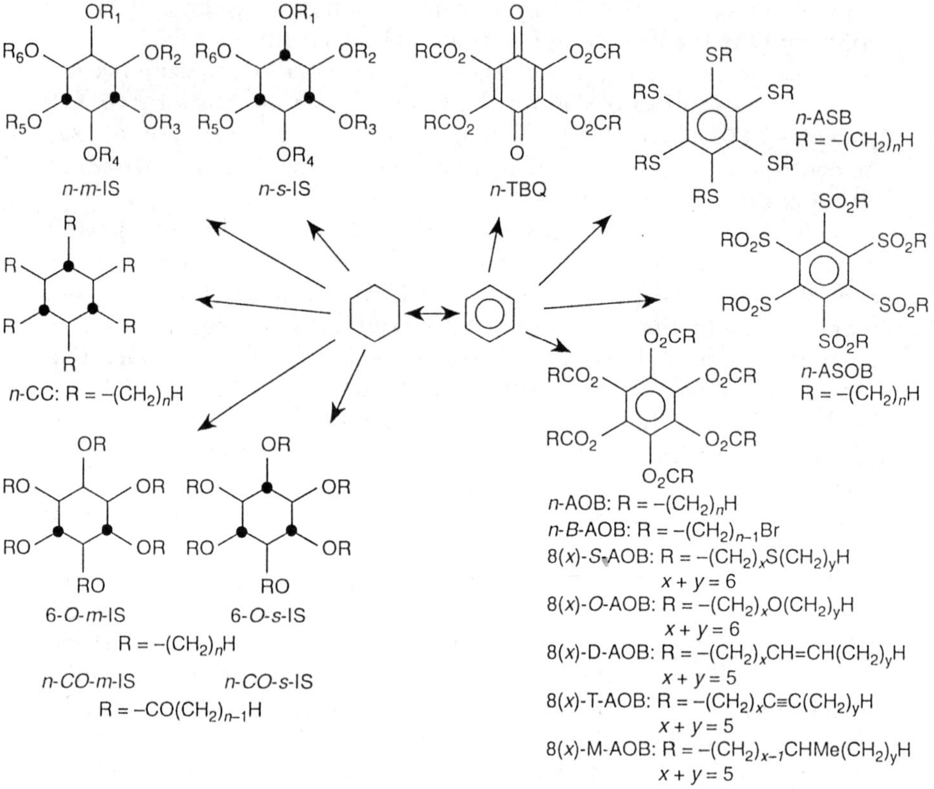

Figure 1.10 Conceptual links among many of the molecules discussed in section 1.3.

1.3.1 Cyclic hydrocarbons

There is ample evidence that small cycloalkanes and many of their derivatives with similar sphere-like shapes can exist as plastic crystals [21, 22]. They are not birefringent and exhibit fewer X-ray diffraction lines in the plastic phases than in the corresponding (lower temperature) normal crystalline phases. Many plastic phases have cubic lattices like those of the closely related smectic D phases [33] which are comprised of aggregates of rod-like molecules. In both cases, more than one molecule can constitute the unit cell or fundamental lattice unit so that several molecules may tumble in a correlated motion. A similar microscopic model explains the lack of optical birefringence in the related 'cholesteric blue' liquid crystalline phases from some chiral rod-like molecules [90]. The distinction between a plastic phase and a smectic D phase seems to be related phenomenologically more to the nature of the transitions to other phases within a sample than to physical or mechanical differences between samples. Plastic phases have no liquid crystalline phases in adjacent temperature ranges but smectic D phases do [15].

A well studied but not fully understood example of plastic crystallinity is from cyclohexane [21, 22, 91, 92], the closed ring analog of hexane, a non-mesomorphic molecule. Its monoclinic solid structure can be transformed to a face-centered cubic plastic phase by heating to above 186 K. This phase persists to the melting temperature, 280 K. Although some intermolecular correlations continue in the plastic phase, they are severely reduced with respect to those in the monoclinic solid. Since cyclohexane molecules are able to diffuse relatively rapidly through the lattice ([21, 22, 93–95]; K.J. McGrath and R.G. Weiss, *Langmuir*, submitted) and undergo chair–chair conformational interconversions (rate $= 51\,s^{-1}$ at 222 K [96]) in the plastic phase, there is ample evidence that it is both structurally disordered and dynamically mobile. Larger cycloalkanes, due to their plate-like shapes having higher aspect ratios (defined as the ratio between a molecule's van der Waals diameter and thickness) and greater conformational flexibility, do not form plastic crystalline phases. Instead, they can exist as condis crystals [97, 98].

Benzene, a molecule whose aspect ratio is similar to that of cyclohexane but does not possess its conformational mobility, forms neither a plastic phase nor any other type of mesophase. It crystallizes from the melt at 279 K in an edge-to-face herringbone-like arrangement (orthorhombic) [99] that is the antithesis of the preferred packing for disk-like molecules to exhibit mesomorphism [100–104]. Some motion within the crystal is possible: individual molecules perform 60° jumps in the plane defined by the atoms at a rate of *c.* $10^5\,s^{-1}$ at 100 K [93]; they do not undergo the tumbling motions that are characteristic of a plastic phase.

1.3.2 Substituted rod-like cyclic hydrocarbons

The perturbations to rod-like molecules created by functional groups can operate also on disk-like ones. When substituted with at least one flexible chain and at least one other (frequently polar) group, preferably in a *para* orientation, many molecules with benzene or cyclohexane as cores provide nematic and smectic phases [11, 12]. Additional substituents can also lead to other mesomorphic aggregation schemes [105]. The shapes of these molecules can no longer be considered disk-like, and they owe their ability to be liquid crystalline to shape anisotropy and electronic factors alluded to previously [18, 19].

1.3.3 Substituted disk-like cyclic hydrocarbons

Cyclic hydrocarbons with six equivalent substituent chains emanating from several carbon atoms about the ring can again adopt conformations in which the molecular shape is disk-like, but the aspect ratio is much larger. If the chains are somewhat flexible and the molecules possess an appropriate balance between order and disorder and attractive and repulsive intermolecular forces, the creation of new mesophase types can be envisioned – and several have been found [102, 104]. The two general classes of aggregation in liquid crystalline phases of discotic molecules are lenticular nematic (N_D) and columnar discotic (D) [100–104]. Carbonaceous pitch mesophases, discussed in section 1.4.5, resemble N_D phases. Only those discotic mesophases with benzene, cyclohexane and shape-related cores having primarily alkyl chains as substituents will be discussed here.

(a) Benzene derivatives

Of the three tetrakis(alkanoyloxy)benzoquinone homologs (***n*-TBQ**) that have been reported, the hexanoyloxy homolog (**6-TBQ**) is not liquid crystalline, but the heptanoyloxy and octanoyloxy (**7-** and **8-TBQ**) form enantiotropic **D** phases over small (2–3°) temperature ranges [106]. These molecules have shapes that are intermediate between the rods of 1,4-disubstituted cores and the disks of hexasubstituted ones, and yet they are able to stack in mesomorphic columns.

Of the several functional group types that have been used to attach six alkyl chains to benzene cores and, thereby, to increase the aspect ratios of the molecules, only three – oxycarbonyl (–$\mathbf{O_2CR}$), thia (–**SR**), and sulfono (–$\mathbf{SO_2R}$) – will be mentioned here. They provide provocative insights into the factors that determine the ease of liquid crystal formation in discotic molecules with small rigid cores. Although there is some discrepancy between the transition temperatures of hexakis(alkanoyloxy)benzene (***n*-AOB**) samples prepared in different laboratories [101, 107–109], the trends in the dependence between chain length and columnar phase formation are in agreement; we report the data of Collard and Lillya [108]. Whereas **5-**, **6-** and **10-AOB** are not liquid crystalline, the heptanoyloxy and octanoyloxy homologs form enantiotropic columnar phases over $<6°$ temperature ranges) and the nonanoyloxy forms a monotropic one. All of the molecules become isotropic below 386 K. By contrast, the heptyl through the pentadecyl homologs of hexakis(alkylsulfono)benzene (***n*-ASOB**) form enantiotropic columnar discotic phases over a broader temperature range and at somewhat higher temperatures than the analogous ***n*-AOB** [110, 111]. The shorter and longer members of the series melt directly to the isotropic state and do not form even monotropic mesophases (Figure 1.11)

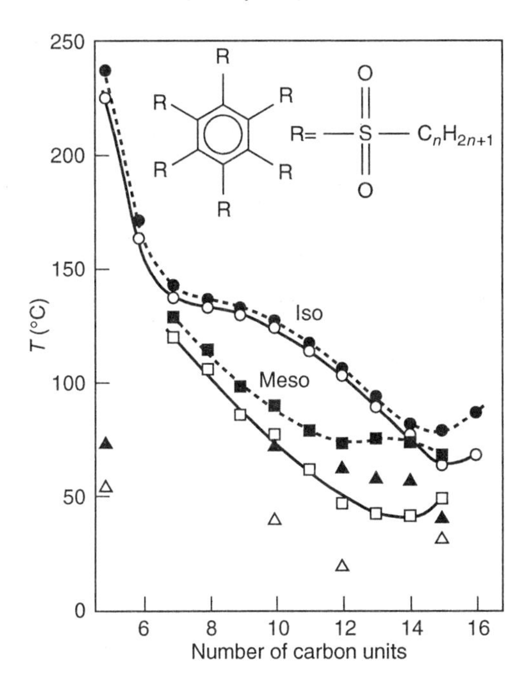

Figure 1.11 Transition temperatures for ***n*-ASOB** from DSC measurements [111]. Open symbols are for transitions approached from higher temperatures and solid symbols are for transitions approached from lower temperatures. Circles: isotropic–mesophase; squares: mesophase–solid; triangles: solid–solid.

[111]. Thus the sulfono group has the effect of making a more rigid core (with perhaps more attractive intermolecular interactions) than does the oxycarbonyl group. As a result, columnar stacking of ***n*-ASOB** molecules persists to higher temperatures, resisting better the influence of the melted alkyl chains [108, 109] to disorder completely the aggregates.

By contrast, none of the hexakis(alkylthia)benzenes (***n*-ASB**) with $n = 5-16$ forms a liquid crystalline phase [111]. The homologs with $n = 5-10$ are viscous oils at room temperature; the longer ones are crystalline solids with melting points from 302–303 K ($n = 11$) to 330–332 K ($n = 16$). A possible reason for the lack of mesomorphic character of the **ASB** is that the thia ethers are conformationally more labile than the sulfonoalkyl groups, making it more difficult for the core groups to retain their columnar stacking once the polymethylene chains have 'melted'.

The consequences of modifying the chains of ***n*-AOB** molecules on their mesomorphic behavior have been investigated in some detail. Perhaps the least disruptive modifications place a group at the end of or within an alkyl chain. For instance, ω-substitution of one bromine atom on each of the chains of **7-AOB** produces **7-*B*-AOB** [112] whose shape and size are very similar to those of the mesogen, **8-AOB**. However, neither **7-*B*-AOB** nor **8-*B*-AOB** appears to form liquid crystalline **D** phases [112]. Exchange of a methylene group by a sulfur atom in each of the alkyl chains in **8-AOB** results in isomeric **8(χ)-*S*-AOB** thia ethers which differ only in the location of the substitution. Replacement of C3 or C4 by sulfur ($x = 2$ or 3, respectively) does not yield liquid crystalline molecules, but **8(6)-*S*-AOB** is discotic columnar between 304.7 and 310.6 K [113]. This behavior is consistent with that reported for the ***n*-ASB** molecules.

The analogous oxa series, **8(*x*)-*O*-AOB**, is liquid crystalline for a larger range of isomers [114]. Thus, although the carbonate (formally from replacement of C2 for an oxygen atom; $x = 0$) remains isotropic to below 200 K, the isomers from oxygen substitution for $x = 2$, 3, 4 and 7 form columnar discotic phases and over temperature ranges of 26, 2, 1 and 9° (289–298 K for **8(6)-*O*-AOB**; compare with **8(6)-*S*-AOB**), respectively [114]. Clearly, the presence of the smaller oxygen (perhaps by virtue of lower crystal ↔ mesophase transition temperatures caused by enhanced ease of chain bending [50]) has a much greater effect than sulfur on the ability to induce liquid crystallinity in **AOB**-type molecules.

Chains of **8-AOB** can be both stiffened and/or projected in off-angle directions when unsaturated groups are introduced. For example, a double bond places a stiff, bent segment along an alkyl chain. The liquid crystalline phase of **8-AOB** is lost when an E or Z configuration double bond is placed near a chain middle (between C4 and C5; **8(3)-D-AOB**), but the clearing temperature – 356.8 K for the E isomer and 354 K for the Z isomer, compared to 356.5 K for the saturated (parent) molecule – is affected only slightly [108]. By contrast, placement of the double bond at the ends of the alkanoyloxy chains (**8(6)-D-AOB**), stabilizes the columnar discotic phase which persists from 306 to 343 K [108].

A triple bond introduces a stiff, linear segment along an alkyl chain. When a triple bond is placed between C4 and C5 or between C7 and C8 of the chains, the effects are qualitatively the same as those from double bond insertion, but quantitatively different: the isomer with the C4–C5 yne group (**8(3)-T-AOB**) has a higher clearing temperature, 420 K, than **8-AOB** but forms no discotic phase; **8(6)-T-AOB** does form a D phase whose range, 2.3°, is comparable to that of **8-AOB** but whose clearing temperature, 321.6 K, is much lower [108]. These results demonstrate the importance that unsaturation along an alkyl chain can have on mesophase transitions, but they are not sufficiently extensive to permit many conclusions. The influence of unsaturation on molecules forming other mesophase types is well documented [2, 11, 12].

The influence of methylation along the alkyl chains of ***n*-AOB** molecules has also been investigated [107, 108]. For example, the consequences to **D** phase formation caused by introducing a methyl group at various positions of the six octanoyloxy groups at **8-AOB**, as in **8(*x*)-M-AOB**, are enormous. **8(2)-M-AOB** is isotropic at room temperature and **8(3)-M-AOB** exhibits only a crystal isotropic transition at 369 K. However, **8(4)-** and **8(5)-M-AOB** are in a columnar discotic phase at room temperature and remain so to 369 and 375 K, respectively. **8(6)-** and **8(7)-M-AOB** are discotic over more limited temperature ranges: 334–360 K for $x = 6$; 358–360 K for $x = 7$. As in the case of oxygen insertion, the largest disturbing influence to phase order occurs when methylation is closest to the aromatic core, and it decreases in a sometimes irregular fashion as the substituent is moved toward the chain ends. At the intermediate positions, the disturbance is large enough to allow a stabilization of D phases (decreased melting points and wider mesophase temperature ranges); near the chain ends, methylation may serve primarily to fill space within the diameter of the molecular disk.

(b) Cyclohexane derivatives

Unlike benzene, cyclohexane can be polysubstituted in configurations that either encourage or discourage conformations in which the molecular shapes are rod-like or disk-like. When a cyclohexane molecule adopts its most favored chair conformation, substituents may be directed axially or equatorially. Only in the latter does the molecular shape approximate that of classical mesogens. The added complexity associated with the directionality of substituents with respect to a plane defined roughly by the cyclohexane carbon atoms permits the design and comparison of the mesomorphic properties of configurational isomers with differing degrees of linearity and planarity and varied aspect ratios. Molecules with no more than one substituent at each ring carbon of cyclohexane will be discussed here.

If all of six *n*-alkyl chains, one on each carbon atom, are equatorial when the core cyclohexane is in its preferred chair conformation, the resulting molecule (***n*-CC**) is more disk-shaped and less spherical than cyclohexane itself. As a result, it should be incapable of forming a plastic phase, but may exhibit other forms of mesomorphism. In fact, **6-CC** has been shown to melt directly from its crystalline phase at 339.4 K [115]. Additionally, no liquid crystalline phases are found in molecules like ***n*-CC** whose chains are 2-hydroxyethyl, 2-(butanoyloxy)ethyl, (butyloxycarbonyl)methyl or other substituted linear chains [115]. The possibility that the crystalline phase possesses condis properties was not mentioned. It would be instructive to examine other homologs of ***n*-CC** and isomers in which one (or more) alkyl chain is forced into an axial orientation; both variations would provide important information concerning the relationships between modified disk-like shapes of molecules and the various phases that they can form.

Since the structural and substituent requirements to form nematic and smectic phases of 1, 4-disubstituted cyclohexanes, especially those with *trans* configurations, have been explored exhaustively [2, 11, 12], only one rather unconventional example, the *n*-alkyl ethers of *myo*- and *scyllo*-inositols (***m*-** and ***s*-IS**, respectively) [116, 117], will be discussed here. The mono dodecyl ether of ***s*-IS** (**1-*s*-IS**) and three (racemic) mono dodecyl ethers of ***m*-IS** (**1-**, **2-** and **3-*s*-IS**) shown can adopt roughly rod-like shapes. However, hydrogen bonding interactions of the inositol head groups are expected to (and do) mediate molecular packing very strongly: the highly symmetrical compound **1-*s*-IS** is not liquid crystalline; compounds **2-** and **3-*m*-IS**, with an equatorially projected dodecyloxy group and one axially projected hydroxy group, form enantiotropic smectic A phases over *c.* 70° temperature ranges; compound **1-*m*-IS**, whose dodecyloxy group is projected axially, becomes liquid crystalline (probably smectic) only on cooling and over a small temperature range [116, 117].

As found for the hexakis(alkanoyloxy)benzenes with oxa or thia atoms inserted along the chains [113, 114], thia ethers of the analogous **IS** have smaller liquid crystalline temperature ranges and higher clearing points than the corresponding **IS** oxa ethers [118].

The 1, 4-di-*n*-hexyl ethers of ***m*-** and ***s*-IS** (**4-*m*-IS** and **2-*s*-IS**, respectively) also are rod-like in shape. Again, the less symmetrical *myo* isomer is able to form an enantiotropic smectic A phase (417.6–449.3 K), albeit a high temperature one, but the more symmetrical *scyllo* isomer is not (m.p. 531.5 K) [116, 117]. At least some *myo* and *scyllo* 1, 2-di-*n*-hexyl ethers of **IS** can form **D** phases in spite of their not being disk-like in shape [116, 117,119]. The key to their success is in the formation of hydrogen-bonded aggregates of *c.* 5 molecules that, as a result, possess disk shapes; the five **IS** rings behave as a core and the alkyl chains are projected radially from the center of the aggregate. Dimeric hydrogen bonded units lead also to columnar discotic phases of 1, 2, 3, 4-tetra-*n*-hexyl **5-*m*-IS** and **3-*s*-IS** [119, 120]. This phase behavior should be compared to that of the ***n*-TBQ** [106], discussed in section 1.3.3(a), whose shape and lack of hydrogen bonds allow their columnar phases to have only one molecule per disk unit.

The vast majority of discotic mesogens with cyclohexane cores and six substituent chains are based on ***m*-** and ***s*-IS**. Pertinent to the discussion above concerning the relationship between the direction of projection of chains on 1, 2, 3, 4, 5, 6-hexasubstituted cyclohexanes, the hexa(*n*-hexyl) ether of ***m*-IS** with one axial hexyloxy (**6-*O*-*m*-IS**) has been shown not to be liquid crystalline; the corresponding ***s*-IS** hexaether with a more disk-like shape (**6-*O*-*s*-IS**) forms an enantiotropic columnar discotic phase from 291.6 to 364.0 K [121]. The ability of the all-equatorial hexaether to exist as a stable liquid crystal, while the hexaalkylated cyclohexane **6-CC** does not, demonstrates again the importance of enhanced rotational mobility of substituents to the induction of mesomorphism.

Carboxy groups have also been inserted between alkyl chains and cyclohexane cores in hexa-substituted, all-equatorial disk-shaped molecules. Generally, 1, 2, 3, 4, 5, 6-hexakis(alkanoyl) derivatives of *scyllo* inositol have wider discotic temperature ranges and higher clearing temperatures than the ethers with the same number of carbon atoms in the chains. For instance, **6-CO-*s*-IS** exhibits a columnar discotic ($\mathbf{D_{ho}}$) range of 130° and a clearing temperature of 471.2 K (versus 72° and 364 K, respectively, for **6-*O*-*s*-IS**) [121]. When ***n*-CO-*s*-IS** has $n \leqslant 5$, a discotic phase type is formed which is more highly ordered than $\mathbf{D_{ho}}$, has a much narrower range and clears at higher temperatures [122]. The ***n*-CO-*s*-IS** with $n = 7-11$ (the longest homolog studied) behave anologously to the $n = 6$ ester but with progressively lower clearing points (Table 1.1) [122].

Detailed substitution studies along the chains of **8-CO-*s*-IS**, like those with **AOB**, have been conducted [105–107]. They demonstrate that when the core group is very amenable to and stabilizing toward discotic phases, substitutions along the chains have a much smaller influence on the mesophase properties. Thus methylation, oxygen and sulfur insertions and ω-bromination lead to relatively small changes of the parent molecule's transition temperatures.

Molecules related to the esterified ***m*-** and ***s*-IS** mesogens, but with

Table 1.1 Transition temperatures of hexaalkanoyl *scyllo*-inositols [122]

***n*-CO-*s*-IS** $n =$	Transition temperatures (K) $K \rightarrow \mathbf{D_1}$ [a]	$K \rightarrow \mathbf{D_{ho}}$	$\mathbf{D} \rightarrow i$
1	561		566
2	*c.* 485		549
3	*c.* 486		533
4 [b]	458		481
5		342	473
6		343	475
7		353	473
8		354	469
9		356	462
10		361	456
11		365	450

[a] $\mathbf{D_1}$ is a very organized discotic phase. [b] Forms a possible cubic mesophase at temperatures above the transition to the $\mathbf{D_1}$ phase.

optically active cores, have also been investigated. Most are saccharides containing one chain [123–126]. Strong hydrogen bonding interactions among the saccharide head groups of neighboring molecules lead to bilayer packing arrangements that favor thermotropic smectic phases [126].

Of the peresterified monosaccharides, the α- and β-anomers of *D*-glucopyranose (***α*- and *β*-*n*-GP**; Figure 1.12) have received the closest scrutiny [127; and Mukkamala, R. Burns, C.L., Jr, Cathings, R.M. and Weiss, R.G. (1996) *J. Am. Chem. Soc.* **118**, 9498], although the peresters of other monosaccharide and some disaccharide cores have been examined [128–131]. Like ***m*-IS**, α-glucopyranose has one axial hydroxy group and the rest equatorial; like ***s*-IS**, β-glucopyranose has all of its hydroxy groups axially disposed. Both glucopyranose anomers adopt chair conformations preferentially. Besides chirality, a significant difference between the **GP** and **IS** is the presence of an oxa group (and therefore the absence of a hydroxy group) in the six-membered ring of the monosaccharides. As a result, the peresters of the **GP** have shapes like a pie with a slice missing.

In both the α and β series of **GP** anomers, alkanoyl chain lengths of 10–16 and 18 carbon atoms have been explored [127]. All members of both series form at least one **D** phase. There is evidence that the shorter chained α-anomers, especially, can be supercooled since slow heating of the rapidly cooled mesophases gives rise to reentrant solid phases. These same molecules pack helically in columns, with six molecules completing a full twist. The axially oriented ester chain is hypothesized to fit into the space near the oxygen of the core of a neighboring molecule (i.e. the missing slice) [127]. Consistent with this model, even the shorter homologs of the β-anomer stack in columns which may become twisted. The longer homologs of both anomers appear to lack rotational ordering within the columns. Due to the tendency of the longer chained **GP** to become glassy, it was not possible to discern whether their **D** phases are enantiotropic; those of the shorter chained **GP** are clearly monotropic [127]. Clearing temperatures, ranging from 306 K (for ***α*-10-GP**) to 353 K (for ***β*-18-GP**), are much lower than those of the analogous inositol peresters. The decreased symmetry of the core groups of the **GP** apparently creates a significant disturbance to molecular packing which can be overcome at much lower temperatures.

1.4 INDUCED MESOPHASES IN MIXTURES

The examples of section 1.2.1 show that mesomorphism depends, in an abstract sense, upon ordering and disordering factors which apply themselves simultaneously but in spatially different parts of a molecule. We have seen the consequences of such influences on the ability of single component samples of rod-like molecules to be mesomorphic.

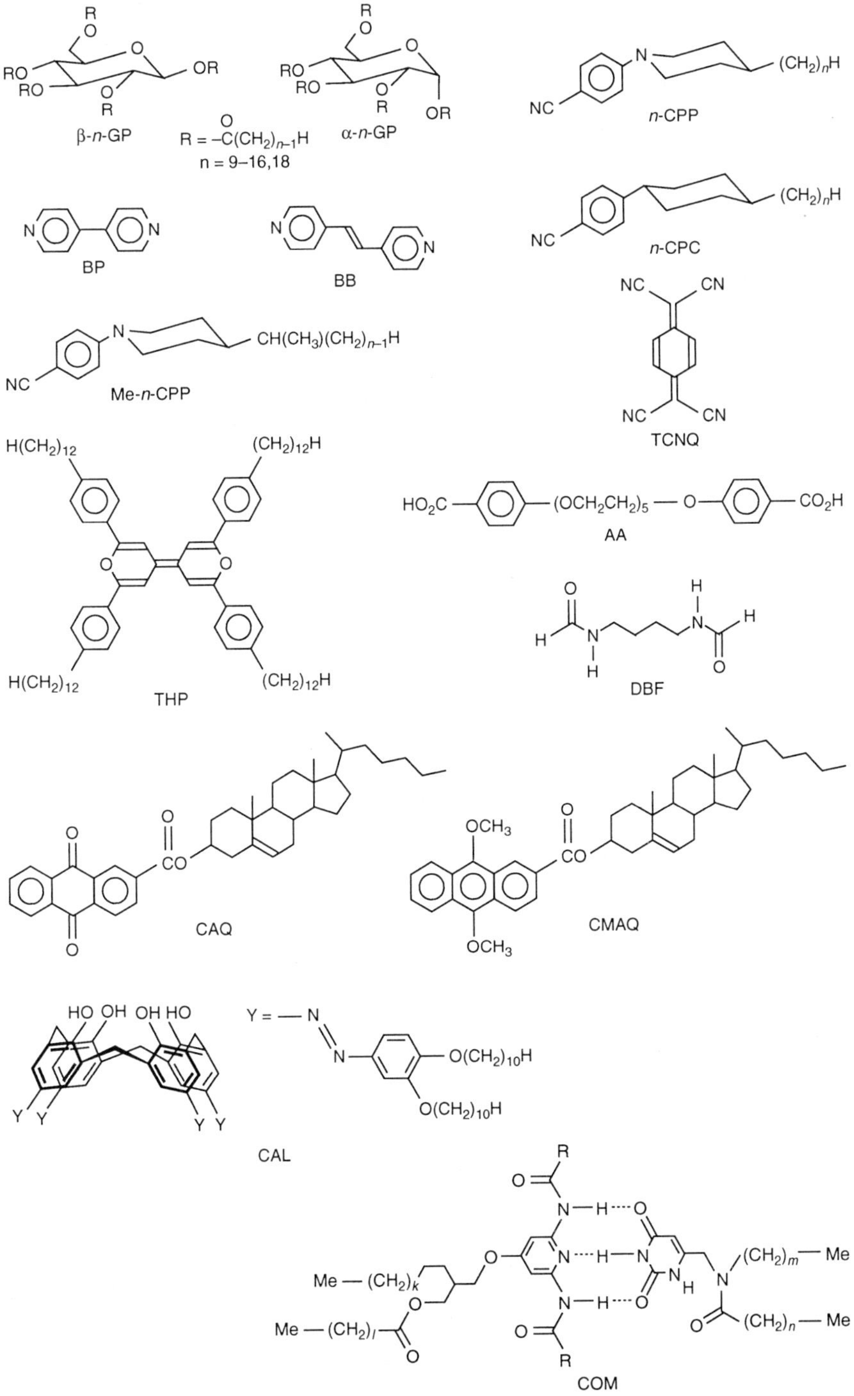
β-n-GP
R = –C(O)(CH2)n–1H
n = 9–16,18
α-n-GP
n-CPP
BP
BB
n-CPC
TCNQ
Me-n-CPP
CH(CH3)(CH2)n–1H
H(CH2)12
(CH2)12H
AA
HO2C
(OCH2CH2)5
CO2H
THP
DBF
CAQ
CMAQ
OCH3
Y =
O(CH2)10H
CAL
COM
Me — (CH2)k
Me — (CH2)l
(CH2)m — Me
(CH2)n — Me

Figure 1.12.

An alternative, less exploited methodology for attaining the necessary order–disorder balance is to blend two (or more) non-mesogenic molecules. Several early attempts to supercool such mixtures sufficiently to effect liquid crystallinity met with limited (and sometimes erroneous) success [132]. Greater success was found when one of the components was liquid crystalline [132, 133]. In both types of experiments, the virtual melting temperatures of the mixtures can be estimated using the Schröder–van Laar equation and assuming ideal mixing behavior [80, 134, 135].

1.4.1 *n*-Alkanes

There have been a few examples in which mesophases are induced when two homologs of simple, rod-like molecules, neither of which is liquid crystalline, are mixed. For example, Sirota *et al.* observed that mixtures of the *n*-alkanes, tricosane ($\mathbf{C_{23}H_{48}}$) and *c.* 25–50 wt% octacosane ($\mathbf{C_{28}H_{58}}$), form a new rotator phase, which they believe is a (perhaps hexatic) smectic phase, between the hexagonal rotator and orthorhombic solid phases found for the individual components [136]. Other mixtures of *n*-alkanes which vary in length from eicosane to tritriacontane ($\mathbf{C_{33}H_{68}}$) also form new phases not observed for the individual components, but the authors do not use the word 'smectic' in describing them [137, 138; and Jouti, B., Provost, E., Petitjcan, D., Bouroukba, M. and Dirand, M. (1996) *J. Mol. Struct.* **382**, 49]. A characteristic of the mixtures is that they show increased ranges of stability for rotator phases [138]. Dorset has interpreted the microscopic nature of solid solutions of *n*-alkane homologs somewhat differently [139], preferring to consider them as aggregates of inhomogeneous microcrystals. Consistent with the view of Dorset, mixtures of the shorter *n*-alkanes, nonadecane and heneicosane, form 'disturbed' phases of the types found for the individual components, but no new phases [140].

1.4.2 Alkyl alkanoates

Insertion of carboxy groups between C2 and C3 of the molecules in mixtures of even shorter *n*-alkanes, hexadecane and octadecane, does allow a new hexatic B liquid crystalline phase to form [52]. Thus, between 10 and 75 wt% of methyl stearate ($\mathbf{H(CH_2)_{17}CO_2CH_3}$) in methyl palmitate ($\mathbf{H(CH_2)_{15}CO_2CH_3}$), an α_1 phase whose birefringent pattern is indistinguishable from that of **BS** can be detected. From X-ray diffraction measurements, the layer spacing of a 1:1 mixture in its α_1 phase, 25.7 Å, is 2.5 Å longer than that found in the lower temperature (and more densely packed) solid phase.

1.4.3 More complex rod-like molecules

In the same way that these non-mesogenic molecules can be cajoled to form liquid crystalline mixtures, so can mesogenic ones be made non-mesogenic through small structural changes; this is a common and well-known phenomenon [2, 23–27]. For instance, many homologs of *trans*-1-alkyl-4-(4-cyanophenyl)cyclohexane (***n*-CPC**; Figure 1.12) are known to form enantiotropic nematic and monotropic smectic phases [12, 141]. However, except for the hexyl homolog (which forms an extremely unstable monotropic nematic phase [142]), the corresponding 4-alkyl-*N*-(4-cyanophenyl)piperidines (***n*-CPP**; Figure 1.12), in which a methyne group of the cyclohexyl ring has been exchanged for an **N** atom of almost equal size, do not [143–146]. Detailed single crystal X-ray diffraction studies on ***n*-CPP** ($n = 4-8$) [146] show that there are minor differences between the molecular packing arrangements of the two series of molecules that may have important consequences with regard to mesophase formation: in general, the angle made by the pseudo plane of the saturated ring and the phenyl ring of the **CPC** are much larger (*c.* 65° [147]) than those of the **CPP** (15–20° [12, 141]). We conjecture that the loss of liquid crystallinity of the ***n*-CPP** is related both to differences in the projections of the alkyl chains that these angles require and to larger intermolecular electronic interactions between **CPP** molecules. By contrast, several two and three component mixtures of ***n*-CPP** homologs do form relatively stable monotropic nematic phases [143–146]. Mixtures of 1:1 composition were maintained in their nematic state for more than 1 day when one-half the sum of the number of carbon atoms in the alkyl chains of the two components was near six (i.e. the number of carbon atoms in the one homolog of **CPP** that is liquid crystalline as a single substance [142]); phase stability decreased with increasing deviation from the optimal value of six for one-half of the sum of the chain carbon atoms or for the number of the carbon atoms in the chain of an individual component [144]. On this basis, we conjecture that the induced mesogenicity of the mixtures is related to the delicate balance between ordering and disordering of molecular packing.

Support for this contention is found in the observation that although the liquid phases of **Me-*n*-CPP** (**CPP** methylated at the 1-position of the alkyl chain; Figure 1.12) can be supercooled significantly below their melting points, they do not form mesophases as single substances [143–146]. In fact, 1:1 mixtures of **Me-6-CPP** and other **Me-CPP** homologs can be cooled to 220 K without crystallizing or becoming birefringent. An even more impressive demonstration of the influence that methylation of alkyl chains can exert on the phase behavior of liquid crystalline molecules has been provided by examples included in section 1.3.3(a) where the discotic temperature range of the isomeric **8(*x*)-M-AOB** was suppressed completely or expanded by moving the six methyl groups of each molecule to different positions ($x = 2-8$) along the octanoyloxy chains [108].

There are many other examples of the dramatic effects that small groups and their positions along the 'main chain' of an **MLC** can have on phase behavior. Demonstrations of both induction and destruction of liquid crystallinity have been documented when even small groups like methyl are added [148–150]; they can affect packing arrangements by influencing conformations of single molecules and electrostatic interactions between molecules. Unfortunately, the manifestations of these perturbations on **MLCs** and **PLCs** are not easily related in many cases.

1.4.4 Hydrogen bonding systems

Some binary mixtures of protic molecules have also been reported to form mesophases that are not present in either pure component. For instance, Lawrence observed mesophase formation in 1:2 molar ratios of cholesterol and 1-alkanols containing 12, 13, 14, 16 and 18 carbon atoms [151]. Mixtures of 1-hexadecanol and dodecanoic acid exhibits a lamellar liquid crystalline (smectic) phase [152]. Intermolecular hydrogen bonding must be an important contributing factor to maintain the necessary aggregation for mesomorphism in these cases. For applications of this concept to **PLCs**, see Chapter 3.

Many ring-substituted aromatic carboxylic acids form thermotropic liquid crystalline phases [11, 12] as a result of their ability to exist as stable hydrogen-bonded dimers. Several carboxylates and β-diketonates as square planar complexes with Cu(II) and other metal ions produce thermotropic columnar discotic phases [153–158]. Recently, several examples of liquid crystalline behavior by mixtures of non-mesogenic carboxylic acids and non-mesogenic pyridine-based hydrogen bond acceptors have been reported [159, 160]. Aggregated mesogenic units are present when monocarboxylic acids and simple pyridines (dimeric units), either monocarboxylic acids and dipyridines or dicarboxylic acids and simple pyridines (trimeric units), or dicarboxylic acids and dipyridines (polymeric units), are employed. For example, both nematic and smectic A liquid crystalline phases consisting of trimeric units were detected in mixtures of 4, 4′-bipyridyl (**BP**; Figure 1.12) and the nematogenic single substance, 4-*n*-hexylbenzoic acid [159]. **BP** and heptanedioic acid, neither of which is liquid crystalline as a single substance, are not mesogenic as mixtures, also [159]. However, a 1:1 (mol/mol) mixture of the two non-mesogenic molecules **AA** and **BB** (Figure 1.12) [160] forms a thermotropic smectic A phase after the first heating to the isotropic phase where the two components apparently have the mobility to polymerize in hydrogen-bonded, alternating **...(.AA...BB.)$_N$...** chains [160]. More specific complementary hydrogen bonding between non-mesogenic molecules, as 1:1 molar mixtures, has been shown to induce metastable liquid crystallinity via structures thought to be like **COM** (Figure 1.12) [161]. Recently, the diformamide **DFB** has been shown to induce a hexagonal columnar bowlic phase (B_h) in the otherwise non-liquid crystalline calix[4]arene **CAL** ([162]; T.M. Swager, personal communication). The nature of the interactions between the two components must be rather specific since simpler formamides like *N*, *N*-dimethylformamide and *N*-butylformamide are ineffective with **CAL**.

1.4.5 Carbonaceous phases

A final example of induced mesogenicity in a multicomponent system is the well studied, but less well understood, carbonaceous mesophases which are comprised of a myriad of unidentified molecules which are created *in situ* as petroleum pitches are heated to temperatures where chemical transformations occur [163]. The processes leading to a mesophase involve decreases in both the elemental weight fraction of hydrogen and the group fraction of aliphatic carbon atoms [164]. Model studies have demonstrated that the component molecules of these phases are fused, polycyclic aromatic molecules with disk-like shapes; the exact structures of the components depend upon the natures of the precursor molecules which are heated [164–167]. All of the carbonaceous mesophases somewhat resemble discotic nematic phases [168]. At least some of them probably represent another example of liquid crystallinity induced by mixing molecular components which, when separated, are not mesogenic.

1.4.6 Charge transfer interactions

Liquid crystalline phases can also be stabilized via charge transfer interactions. Exploitation of charge transfer interactions in **PLC**s is described in Chapter 3. Saeva *et al.* [169] demonstrated that the temperature of the crystal ↔ **D** transition of the electron-rich bi-4*H*-pyran **THP** (Figure 1.12) can be depressed from 363 K to 284 K and the clearing temperature can be increased from 422 K to 616 K (decomposition) when an equal molar amount of the (non-mesogenic) efficient electron acceptor, 7, 7, 8, 8-tetracyano-*p*-quinodimethane (**TCNQ**; Figure 1.12) is added. The liquid crystalline phase remains discotic and, in fact, a second type of **D** phase appears at 307 K. Cation radicals of **THP** with BF_4^- and ClO_4^- also form two discotic phases. The crystal ↔ D transition temperatures of the salts are somewhat higher than for **THP** alone, but the clearing temperatures are at the decomposition point, above 610 K, making the discotic ranges for the salts wider than that for the parent **THP** [169].

In another application of the same principle, an electron donor/electron acceptor pair of disk-shaped groups were attached through a flexible atom chain [170]. When incorporated into columns of discotic molecules containing only the electron donor group, the tethered donor/acceptor molecule was able to influence the motions of columns of donor molecules in their **D** phases [170].

Charge transfer interactions have been shown to depress the crystal ↔ cholesteric transition temperatures, also, while increasing the temperature range of the liquid crystalline phase (E. Ostuni and R.G. Weiss, unpublished results). Thus, a 1:2 wt/wt mixture of **CMAQ/CAQ** forms a glassy cholesteric phase on cooling instead of crystallizing like each of the single substances do. Upon heating, the cholesteric phase is retained until the molecules decompose (>570 K). As single substances, **CAQ** and **CMAQ** (Figure 1.12) exhibit enantiotropic cholesteric phases in the temperature range *c.* 433–508 K and 345–470 K, respectively.

There are many apparent advantages to using strong attractive charge transfer interactions to stabilize core group aggregates in liquid crystalline systems. The disturbing influences of melted chains can be withstood by the stacked core (aromatic) groups to higher temperatures, allowing the phases to maintain their packing arrangements over very large temperature ranges. This is a strategy that should be exploitable to effect liquid crystallinity in systems and at temperatures where it is not usually found.

1.4.7 Other related phenomena

There are several related phenomena that will not be treated in this chapter. They include the well documented transformation of nematic phases into cholesteric (twisted nematic) phases by adding small amounts of optically active molecules to a nematogen [171], creation of smectic phases from mixtures of molecules which alone form only nematic phases [172, 173], and the presence of reentrant phases [174] due to molecular reorganizations based upon the relative importances of various short- and long-range intermolecular interactions in different temperature regimes (as mediated by the interplay of entropy and enthalpy terms) [175–177]. Each has been exploited to create interesting and novel systems and devices based upon mesomorphism.

ACKNOWLEDGMENT

The National Science Foundation is thanked for its support of the cited research from the laboratory of the author.

REFERENCES

1. Wunderlich, B. and Grebowicz, J. (1984) in *Liquid Crystal Polymers II/III* (ed. N.A. Platé), Springer-Verlag, Berlin, p. 1.
2. Kelker, H. and Hatz, R. (1980) *Handbook of Liquid Crystals*, Verlag Chemie, Weinheim, p. 2.
3. Brostow, W. (1990) *Polymer*, **31**, 979.
4. Percec, V., Heck, J., Johansson, G. *et al.* (1994) *J. Macromol. Sci.*, **A31**, 1031.
5. Finkelmann, H., Meier, W. and Scheuermann, H. (1993) *Liquid Crystals. Applications and Uses* (ed. B. Bahadur), Vol. 3, World Scientific, Singapore, p. 345 and references cited therein.
6. Weiss, R.A. and Ober, C.K. (eds) (1990) *Liquid Crystal Polymers*, American Chemical Society, Washington, DC.
7. Ciferri, A. (ed.) (1991) *Liquid Crystallinity in Polymers*, VCH, New York.
8. Percec, V., Jonsson, H. and Tomasos, D. (1992) in *Polymerization in Organized Media* (ed. C.M. Paleos), Gordon and Breach, Philadelphia, Ch. 1.
9. Percec, V. and Lee, M. (1991) *Macromolecules*, **114**, 1269 and references cited therein.
10. Kosaka, Y. and Uryu, T. (1994) *Macromolecules*, **27**, 6286.
11. Vill, V. (1992–1993) *Liquid Crystals* (ed. J. Thiem), Vol. 7, Parts a–c (Landolt-Börnstein, Group IV) Springer-Verlag, Berlin.
12. Demus, D. and Zaschke, H. (1974, Vol. I; 1984, Vol. II) *Flüssige Kristalle in Tabellen*, VEB Deutscher Verlag für Grundstoffindustrie, Leipzig.
13. Kelker, H. and Hatz, R. (1980) *Handbook of Liquid Crystals*, Verlag Chemie, Weinheim, Chapter 3 and references cited therein.
14. de Gennes, P.G. (1974) *The Physics of Liquid Crystals*, Clarendon Press, Oxford.
15. Vertogen, G. and de Jeu, W.H. (1988) *Thermotropic Liquid Crystals Fundamentals*, Springer-Verlag, Berlin.
16. Chandrasekhar, S. (1977) *Liquid Crystals*, Cambridge University Press, Cambridge.
17. Madhusudana, N.V. (1990) in *Liquid Crystals. Applications and Uses* (ed. B. Bahadur), Vol. 1, World Scientific, Singapore, p. 37.
18. Cotter, M.A. (1983) *Phil. Trans. Roy. Soc. London*, **A309**, 217 and references cited therein.
19. Cotter, M.A. (1983) *Mol. Cryst. Liq. Cryst.*, **97**, 29.
20. Chandrasekhar, S. and Shashidhar, R. (1979) in *Advances in Liquid Crystals* (ed. G.H. Brown), Vol. 4, Academic Press, New York, p. 83.
21. Sherwood, J.N. (ed.) (1979) *The Plastically Crystalline State*, Wiley, New York.
22. Smith, G.W. (1975) in *Advances in Liquid Crystals* (ed. G.H. Brown), Vol. 1, Academic Press, New York, p. 189.
23. Gray, G.W. and Winsor, P.A. (eds) (1974) *Liquid Crystals and Plastic Crystals*, Vol. 1, Ellis Horwood, Chichester.
24. Gray, G.W. (1962) *Molecular Structure and the Properties of Liquid Crystals*, Academic Press, London.
25. Demus, D. (1990) *Liquid Crystals. Applications and Uses* (ed. B. Bahadur), Vol. 1, World Scientific, Singapore, p. 1.
26. Demus, D., Diele, S., Grande, S. and Sackmann, H. (1983) *Advances in Liquid Crystals* (ed. G.H. Brown), Vol. 6, Academic Press, New York, p. 1.
27. Sackmann, H. (1984) in *Progress in Colloid and Polymer Science* (ed. A. Weiss), Vol. 69, Steinkopff Verlag, Darmstadt, p. 73.
28. Broadhurst, M.G. (1962) *J. Res. NBS*, **66A**, 241.

29. Schaerer, A.A., Busso, C.J., Smith, A.E. and Skinner, L.B. (1955) *J. Am. Chem. Soc.*, **77**, 2017.
30. Ewen, B., Strobl, G.R. and Richter, D. (1980) *Faraday Soc. Disc.*, **69**, 19.
31. Maroncelli, M., Strauss, H.L. and Snyder, R.G. (1984) *J. Chem. Phys.*, **82**, 2811.
32. Kelusky, E.C., Smith, I.C.P., Ellinger, C.A. and Cameron, D.G. (1984) *J. Am. Chem. Soc.*, **106**, 2267.
33. Gray, G.W. and Goodby, J.W.G. (1984) *Smectic Liquid Crystals. Textures and Structures*, Leonard-Hill, Glasgow, Ch. 4.
34. Kutsumizu, S., Yamada, M. and Yano, S. (1994) *Liq. Cryst.*, **16**, 1109 and references cited therein.
35. Craivich, A.F., Doucet, J. and Denicolo, I. (1985) *Phys. Rev. B*, **32**, 4164.
36. Okazaki, M. and Toriyama, K. (1989) *J. Phys. Chem.*, **93**, 2883.
37. Herzberg, G. (1967) *Electronic Spectra of Polyatomic Molecules*, Van Nostrand, Princeton, NJ.
38. Sutton, L.E. (ed.) (1965) *Tables of Interatomic Distances and Configurations in Molecules and Ions*, The Chemical Society, London.
39. Sutton, L.E. (ed.) (1958) *Tables of Interatomic Distances, Special Publication No. 11*, The Chemical Society, London.
40. Kitaigorodsky, A.I. (1973) *Molecular Crystals and Molecules*, Academic Press, New York, p. 21.
41. Smyth, C.P. (1937) *J. Chem. Phys.*, **4**, 209.
42. Nunez, A., Hammond, G.S. and Weiss, R.G. (1992) *J. Am. Chem. Soc.*, **114**, 10258.
43. Sullivan, P.K. (1974) *J. Res. NBS*, **78A**, 129.
44. Vilalta, P.M., Hammond, G.S. and Weiss, R.G. (1993) *Langmuir*, **9**, 1910.
45. Saville, W.B. and Shearer, G. (1925) *J. Chem. Soc.*, **127**, 591.
46. Bailey, A.V., Mitcham, D. and Skau, E.L. (1970) *Chem. Eng. Data*, **15**, 542.
47. Weiss, R.G., Chandrasekhar, S. and Vilalta, P.M. (1993) *Collect. Czech. Chem. Commun.*, **58**, 142.
48. Treanor, R.L. and Weiss, R.G. (1987) *Tetrahedron*, **43**, 1371.
49. Slivinskas, J.A. and Guillet, J.E. (1973) *J. Polym. Sci. Chem.*, **11**, 3043.
50. Ferguson, L.N. (1975) *Organic Molecular Structure*, Willard Grant Press, Boston, Ch. 18.
51. Dryden, J.S. (1957) *J. Chem. Phys.*, **26**, 604.
52. Krishnamurti, D., Kristmamurthy, K.S. and Shashidhar, R. (1969) *Mol. Cryst. Liq. Cryst.*, **8**, 339.
53. Krishnamurthy, K.S. and Krishnamurti, D. (1970) *Mol. Cryst. Liq. Cryst.*, **6**, 407.
54. Baldvins, J. (1996) PhD Thesis, Georgetown University, Washington, DC.
55. Krishnamurthy, K.S. (1986) *Mol. Cryst. Liq. Cryst.*, **132**, 255.
56. Moncton, D.E. and Pindak, R. (1979) *Phys. Rev. Lett.*, **43**, 701.
57. Leadbetter, A.J., Mazid, M.A., Kelley, B.A. *et al.* (1979) *Phys. Rev. Lett.*, **43**, 630.
58. Ekwall, P. (1975) in *Advances in Liquid Crystals* (ed. G.H. Brown), Vol. 1, Academic Press, New York, p. 1.
59. Winsor, P.A. (1968) *Chem. Rev.*, **68**, 1.
60. Friberg, S.E. (1991) *Liquid Crystals. Applications and Uses* (ed. B. Bahadur), Vol. 2, World Scientific, Singapore, p. 157 and references cited therein.
61. Mitchell, D.J. and Ninham, B.W. (1981) *Faraday Trans. 2*, **77**, 601.
62. Luzzati, V. (1968) in *Biological Membranes* (ed. D. Chapman), Academic Press, New York, p. 71.

63. Vincent, J.M. and Skoulios, A. (1966) *Acta Crystallogr.*, **20**, 432, 441, 447.
64. Doscher, T.M. and Vold, R.D. (1946) *J. Colloid Sci.*, **1**, 299.
65. Mely, B. and Charvolin, J. (1977) *Chem. Phys. Lipids*, **19**, 43.
66. Treanor, R.L. and Weiss, R.G. (1988) *J. Am. Chem. Soc.*, **110**, 2170.
67. Boden, J. and Jones, S.A. (1985) in *Nuclear Magnetic Resonance of Liquid Crystals* (ed. J.W. Emsley), D. Reidel, Dordrecht, Ch. 22.
68. Turberg, M.P. and Brady, J.E. (1988) *J. Am. Chem. Soc.*, **110**, 7797.
69. Pugh, C., Höpken, J. and Möller, M. (1988) *Polym. Prepr. Am. Chem. Soc.*, **29**, 460.
70. Höpken, J. and Möller, M. (1992) *Macromolecules*, **25**, 2482.
71. Vilalta, P.M., Hammond, G.S. and Weiss, R.G. (1993) *Langmuir*, **9**, 1910.
72. Twieg, R. and Rabolt, J.F. (1983) *J. Polym. Sci. Lett.*, **21**, 901.
73. Rabolt, J.F., Russell, T.P. and Twieg, R.J. (1984) *Macromolecules*, **17**, 2786.
74. Russell, T.P., Rabolt, J.F., Twieg, R.J. *et al.* (1986) *Macromolecules*, **19**, 1135.
75. Mahler, W., Guillon, D. and Skoulios, A. (1985) *Mol. Cryst. Liq. Cryst. Lett.*, **2**, 111.
76. Viney, C., Russell, T.P., Depero, L.P. and Twieg, R.J. (1989) *Mol. Cryst. Liq. Cryst.*, **168**, 63.
77. Viney, C., Twieg, R.J. and Russell, T.P. (1990) *Mol. Cryst. Liq. Cryst.*, **182B**, 291.
78. Vilalta, P.M. and Weiss, R.G. (1992) *Liq. Cryst.*, **12**, 531.
79. McGrath, K.J. and Weiss, R.G. (1993) *J. Phys. Chem.*, **97**, 11115.
80. Vilalta, P.M., Hammond, G.S. and Weiss, R.G. (1991) *Photochem. Photobiol.*, **54**, 563.
81. Höpken, J. (1991) PhD Thesis, University of Twente, CIP-DATA, Koninklijke Bibliotheek.
82. Höpken, J. and Möller, M. (1990) *Polym. Prep. Am. Chem. Soc.*, **31**, 324.
83. Höpken, J., Faulstich, S. and Möller, M. (1992) *Mol. Cryst. Liq. Cryst.*, **210**, 59.
84. Höpken, J., Möller, M. and Boileau, S. (1991) *New Polymeric Mater.*, **2**, 339.
85. Höpken, J., Möller, M., Lee, M. and Percec, V. (1992) *Makromol. Chem.*, **193**, 275.
86. Jariwala, C.P., Sundell, P.-E.G., Hoyle, C.E. and Mathias, L.J. (1991) *Macromolecules*, **24**, 6352.
87. Jariwala, C.P. and Mathias, L.J. (1993) *Macromolecules*, **26**, 5129.
88. Viney, C., Twieg, R.J., Gordon, B.R. and Rabolt, J.F. (1991) *Mol. Cryst. Liq. Cryst.*, **198**, 285.
89. Schulte, A., Hallmark, V.M., Twieg, R., Song, K. and Rabolt, J.F. (1991) *Macromolecules*, **24**, 3901.
90. Stegemeyer, H., Blumel, T., Hiltrop, K., Onusseit, H. and Prosch, F. (1986) *Liq. Cryst.*, **1**, 3.
91. Wurflinger, A. (1991) *Ber. Bunsenges. Phys. Chem.*, **95**, 186 and references cited therein.
92. Farnam, H., O'Mard, L., Dore, J.C. and Bellissent-Funel, M.-C. (1991) *Mol. Phys.*, **73**, 855.
93. McGrath, K.J. (1994) PhD Thesis, Georgetown University, Washington, DC.
94. Andrew, E.R. and Eades, R.G. (1953) *Proc. Royal Soc.*, **A215**, 398.
95. Hood, G.M. and Sherwood, J.N. (1966) *Mol. Cryst.*, **1**, 97.
96. McGrath, K.J. and Weiss, R.G. (1993) *J. Phys. Chem.*, **97**, 2497.
97. Grossman, H.-P. (1981) *Polymer Bull.*, **5**, 137.
98. Shannon, V.L., Strauss, H.L., Snyder, R.G. and Elliger, C.A. (1989) *J. Am. Chem. Soc.*, **111**, 1947.

99. Cox, E.G., Cruickshank, D.W.J. and Smith, J.A.S. (1958) *Proc. Royal Soc.*, **A247**, 1.
100. Singh, S. (1976) Annual Convention of Chemists, Indian Chemical Society, Bangalore, India.
101. Chandrasekhar, S. (1982) in *Advances in Liquid Crystals* (ed. G.H. Brown), Vol. 5, Plenum Press, New York, p. 47.
102. Destrade, C., Foucher, P., Gasparoux, H. *et al.* (1984) *Mol. Cryst. Liq. Cryst.*, **106**, 121.
103. Destrade, C., Tinh, N.H., Gasparoux, H. *et al.* (1981) *Mol. Cryst. Liq. Cryst.*, **71**, 111.
104. Lei, L. (1987) *Mol. Cryst. Liq. Cryst.*, **146**, 41 and references cited therein.
105. Johansson, G., Percec, V., Ungar, G. and Abramic, D. (1994) *Perkin Trans. 1*, 447.
106. Lillya, C.P. and Thakur, R. (1989) *Mol. Cryst. Liq. Cryst.*, **170**, 179.
107. Collard, D.M. and Lillya, C.P. (1989) *J. Am. Chem. Soc.*, **111**, 1829.
108. Collard, D.M. and Lillya, C.P. (1991) *J. Am. Chem. Soc.*, **113**, 8577 and references cited therein.
109. Fontes, E., Heiney, P.A., Ohaba, M. *et al.* (1988) *Phys. Rev.*, **A37**, 1329.
110. Sarkar, M., Spielberg, N., Praefcke, K. and Zimmermann, H. (1991) *Mol. Cryst. Liq. Cryst.*, **203**, 159.
111. Praefcke, K., Poules, W., Scheuble, B. *et al.* (1984) *Z. Naturforsch.*, **B39**, 950.
112. Lillya, C.P. and Collard, D.M. (1990) *Mol. Cryst. Liq. Cryst.*, **182B**, 201.
113. Collard, D.M. and Lillya, C.P. (1991) *J. Org. Chem.*, **56**, 6064.
114. Tabushi, I., Yamamura, K. and Okada, Y. (1987) *J. Org. Chem.*, **52**, 2502.
115. Praefcke, K., Psaras, P. and Kohne, B. (1991) *Chem. Ber.*, **124**, 2523.
116. Praefcke, K. and Blunk, D. (1993) *Liq. Cryst.*, **14**, 1181.
117. Praefcke, K., Blunk, D. and Hempel, J. (1994) *Mol. Cryst. Liq. Cryst.*, **243**, 323.
118. Praefcke, K., Kohne, B., Psaras, P. and Hampel, J. (1991) *J. Carbohydr. Chem.*, **10**, 539.
119. Praefcke, K., Marquardt, P., Kohne, B. *et al.* (1991) *Mol. Cryst. Liq. Cryst.*, **203**, 149.
120. Praefcke, K., Marquardt, P., Kohne, B. *et al.* (1991) *Liq. Cryst.*, **9**, 711.
121. Praefcke, K., Kohne, B., Stephan, W. and Marquardt, P. (1989) *Chimia*, **43**, 380.
122. Kohne, B., Praefcke, K. and Billard, J. (1986) *Z. Naturforsch.*, **41B**, 1036.
123. Jeffrey, G.A. (1986) *Acc. Chem. Res.*, **19**, 168.
124. Pfannemüller, B., Welte, W., Chin, E. and Goodby, J.W. (1986) *Liq. Cryst.*, **1**, 357.
125. Jeffrey, G.A. and Wingert, L.A. (1992) *Liq. Cryst.*, **12**, 179.
126. Goodby, J.W. (1984) *Mol. Cryst. Liq. Cryst.*, **110**, 205.
127. Morris, N.L., Zimmermann, R.G., Jameson, G.B. *et al.* (1988) *J. Am. Chem. Soc.*, **110**, 2177.
128. Zimmermann, R.G., Jameson, G.B., Weiss, R.G. and Demailly, G. (1985) *Mol. Cryst. Liq. Cryst. Lett.*, 183.
129. Sugiura, M., Minoda, M., Fukuda, T. *et al.* (1992) *Liq. Cryst.*, **12**, 603.
130. Itoh, T., Takada, A., Fukuda, T. *et al.* (1991) *Liq. Cryst.*, **9**, 221.
131. Vill, V. and Thiem, J. (1991) *Liq. Cryst.*, **9**, 451.
132. Dave, J.S. and Dewar, M.J.S. (1955) *J. Chem. Soc.*, 4305 and references cited therein.
133. Dave, J.S. and Dewar, M.J.S. (1954) *J. Chem. Soc.*, 4616.
134. Dorset, D.L. (1990) *Macromolecules*, **23**, 894.
135. Hsu, E.C.-H. and Johnson, J.F. (1974) *Mol. Cryst. Liq. Cryst.*, **27**, 95.

136. Sirota, E.B., King, H.E., Jr, Hughes, G.J. and Wan, W.K. (1992) *Phys. Rev. Lett.*, **68**, 492.
137. Sirota, E.B., King, H.E., Jr, Singer, D.M. and Shao, H.H. (1993) *J. Chem. Phys.*, **98**, 5809.
138. Sirota, E.B., King, H.E., Jr, Shao, H.H. and Singer, D.M. (1995) *J. Phys. Chem.*, **99**, 798.
139. Dorset, D.L. (1993) *Acta Chim. Hung.*, **130**, 389.
140. Maroncelli, M., Strauss, H.L. and Snyder, R.G. (1985) *J. Phys. Chem.*, **89**, 5260.
141. Eidenschink, R., Erdman, D., Krause, G. and Pohl, L. (1977) *Angew. Chem. Int. Ed. Engl.*, **16**, 100.
142. Karamysheva, L.A., Kovshev, E.I., Pavluchenko, A.I. *et al.* (1981) *Mol. Cryst. Liq. Cryst.*, **67**, 241.
143. Sheikh-Ali, B.M. and Weiss, R.G. (1991) *Liq. Cryst.*, **10**, 575.
144. Sheikh-Ali, B.M. and Weiss, R.G. (1994) *Liq. Cryst.*, **17**, 605.
145. Sheikh-Ali, B.M., Khetrapal, C.L. and Weiss, R.G. (1994) *J. Phys. Chem.*, **98**, 1213.
146. Sheikh-Ali, B.M., Rapta, M., Jameson, G.B. and Weiss, R.G. (1995) *Acta Cryst. B.*, **B51**, 823.
147. Paulus, H. and Haase, W. (1983) *Mol. Cryst. Liq. Cryst.*, **92**, 237.
148. Gray, G.W. and Kelly, S.M. (1984) *Mol. Cryst. Liq. Cryst.*, **104**, 335.
149. Ohta, K., Morizumi, Y., Fujimoto, T. and Yamamoto, I. (1991) *Mol. Cryst. Liq. Cryst.*, **208**, 55.
150. Menger, F.M., Wood, M.G., Jr, Zhou, Q.Z. *et al.* (1988) *J. Am. Chem. Soc.*, **110**, 6804.
151. Lawrence, A.S.C. (1970) *Liquid Crystals and Ordered Fluids* (eds J.F. Johnson and R.S. Porter), Plenum Press, New York, p. 289.
152. Maryanamurthy, K., Revannasiddaiah, D., Madhova, D.S. and Somakeshar, R. (1992) *Abstracts 14th Annual Liquid Crystal Conference*, Pisa, Italy, I-P46 (p. 738).
153. Giroud-Godquin, A.M. and Billard, J. (1981) *Mol. Cryst. Liq. Cryst.*, **66**, 147.
154. Cayton, R.H., Chisolm, M.H. and Darrington, F.D. (1990) *Angew. Chem. Int. Ed. Engl.*, **29**, 1481.
155. Burrows, H.D. (1990) *The Structure, Dynamics and Equilibrium Properties of Colloidal Systems* (eds D.M. Bloor and E. Wyn-Jones), Kluwer, Dordrecht, p. 415.
156. Sadashiva, B.K., Ghode, A. and Rao, P.R. (1991) *Mol. Cryst. Liq. Cryst.*, **200**, 187.
157. Ohta, K. *et al.* (1991) *Mol. Cryst. Liq. Cryst.*, **208**, 21, 33, 43, 55.
158. Giroud-Godquin, A.-M. and Maitlis, P.M. (1991) *Angew. Chem. Int. Ed. Engl.*, **30**, 375.
159. Kresse, H., Szulzewsky, I., Diele, S. and Paschke, R. (1994) *Mol. Cryst. Liq. Cryst.*, **238**, 13.
160. Lee, C.-M., Jariwala, C.P. and Griffin, A.C. (1994) *Polymer*, **35**, 4550.
161. Brienne, M.-J., Gabard, J., Lehn, J.-M. and Stibor, I. (1989) *J. Chem. Soc. Commun.*, 1868.
162. Xu, B. and Swager, T.M. (1995) *J. Am. Chem. Soc.*, **117**, 5011.
163. Zimmer, J.E. and White, J.L. (1982) in *Advances in Liquid Crystals* (ed. G.H. Brown), Vol. 5, Academic Press, New York, p. 157.
164. Azami, K., Yamamoto, S. and Sanada, Y. (1994) *Carbon*, **32**, 947.
165. Azami, K., Kato, O., Takashima, H. *et al.* (1993) *J. Mater. Sci.*, **28**, 885.
166. Fortin, F., Yoon, S.-H., Korai, Y. and Mochida, I. (1994) *Carbon*, **32**, 979.
167. Azami, K., Yamamoto, S., Yokono, T. and Sanada, Y. (1991) *Carbon*, **29**, 943.

168. Chandrasekhar, S. (1982) in *Advances in Liquid Crystals* (ed. G.H. Brown), Vol. 5, Academic Press, New York, p. 47.
169. Saeva, F.D., Reynolds, G.A. and Kaszczuk, L. (1982) *J. Am. Chem. Soc.*, **104**, 3524.
170. Möller, M., Wendorff, J.H., Werth, M. *et al.* (1994) *Liq. Cryst.*, **17**, 381.
171. Gottarelli, G., Mariani, P., Spada, G.P. *et al.* (1983) *Tetrahedron*, **39**, 1337.
172. Haddawi, S., Diele, S., Kresse, H. *et al.* (1994) *Liq. Cryst.*, **17**, 191 and references cited therein.
173. Dave, J.S., Patel, P.R. and Vasanth, K.I. (1966) *Ind. J. Chem.*, **4**, 505.
174. Cladis, P.E. (1975) *Phys. Rev. Lett.*, **35**, 48.
175. Netz, R.R. and Berker, A.N. (1992) in *Phase Transitions in Liquid Crystals* (eds S. Martinellucci and A.N. Chester), Plenum Press, New York, Ch. 7.
176. Berker, A.N. and Indekeu, J.O. (1987) in *Incommensurate Crystals, Liquid Crystals, and Quasi-Crystals* (eds J.F. Scott and N.A. Clark), Plenum Press, New York, p. 205.
177. Moore, J.S., Zhang, J., Wu, Z. *et al.* (1994) *Macromol. Symp.*, **77**, 295.

2

Inorganic polymer liquid crystals

Cameron G. Cofer and James Economy

2.1 INTRODUCTION

This chapter is concerned with a discussion of completely inorganic polymers that display liquid crystalline character. Until recently there was no description of such material in the literature. Part of the problem lies in the fact that most inorganic materials when prepared in the form of polymeric chains, tend to be very water sensitive which tends to inhibit further study. Furthermore, to prepare such materials in the form necessary to achieve liquid crystalline character greatly complicates the process for designing synthetic methodologies necessary to produce rod-like or discotic shapes.

There have been reports of rod-like inorganic structures such as $(SiS_2)_x$ which is usually depicted as an extended chain polymer [1]. Also, polythiazyl, $(SN)_x$, has been observed to display highly anisotropic character exhibiting pseudo-metallic behavior [2]. However, these materials do not melt, and the close proximity of the rods suggests that they are better classified as three-dimensional inorganic structures than as polymers.

There are some interesting allusions in earlier literature to the potential for liquid crystalline character in inorganic materials. Bragg demonstrated in 1960 that he could stretch graphite rod at temperatures

Mechanical and Thermophysical Properties of Polymer Liquid Crystals
Edited by W. Brostow
Published in 1998 by Chapman & Hall, London.
ISBN 0 412 60900 2

in excess of 2700°C to achieve a highly oriented graphite structure with a very high modulus [3]. This work was followed by a number of studies in the 1960s that demonstrated that carbon fibers derived from rayon [4–6], polyacrylonitrile [7–9] or pitch [10–13] could also be oriented by applying tension at elevated temperatures. This resulted in orientation of the layered carbon structures parallel to the fiber axis accompanied by very high mechanical properties in the axial direction. In only one of these cases was liquid crystallinity demonstrated, namely in the pitch-based carbon [11–13]. It was shown that, by designing pitch materials that exhibited substantial liquid crystalline character, this greatly facilitated the orientation of the pitch chains during melt spinning. Subsequent carbonization of the pitch fiber resulted in a very high modulus carbon fiber. While the pitch-based liquid crystalline phase is generally not considered to be an inorganic material, its close similarity to the inorganic liquid crystalline system described in this chapter, provides a useful analogy.

Numerous close parallels can be identified between the carbon system and boron nitride (BN). As shown in Figure 2.1, both materials in their graphitic forms display a stacked layer structure where the bond distances and the interlayer spacing are almost identical [14]. Both materials can be made in a diamond structure with extremely high hardness. They have also been prepared as highly disordered glassy materials, i.e. structures with a mixture of sp^2 and sp^3 bonding. However, glassy BN finds little use because of its very high sensitivity to water [15]. Around 1970, it was reported that BN fibers could be stretched in a manner similar to carbon to achieve a major increase in strength and modulus [16]. This experiment, which was carried out at 2000°C, permitted longitudinal stretching of the fiber up to 15% and a corresponding orientation of the BN layered structure parallel to the fiber axis.

Unquestionably the first report of a liquid crystalline inorganic polymer was that made in 1994 in a borazine oligomer which was designed as a melt processible precursor to BN [17]. More specifically, in using this oligomer as a matrix with pitch-based carbon fibers, it was observed that excellent mechanical properties were realized after heating to 1200°C [18]. In fact, these properties increased dramatically on heating to 1500°C. One would have anticipated relatively poor mechanical properties because of the large mismatch in coefficient of thermal expansion (CTE) between the highly anisotropic carbon fibers [19,20] and the relatively isotropic BN matrix. Hence, the only explanation that seemed plausible was that the borazine oligomer was able to orient at the fiber surface and in effect match the CTE of the fibers. Subsequently, we were able to develop convincing evidence that the borazine oligomer could indeed form a liquid crystalline state [17].

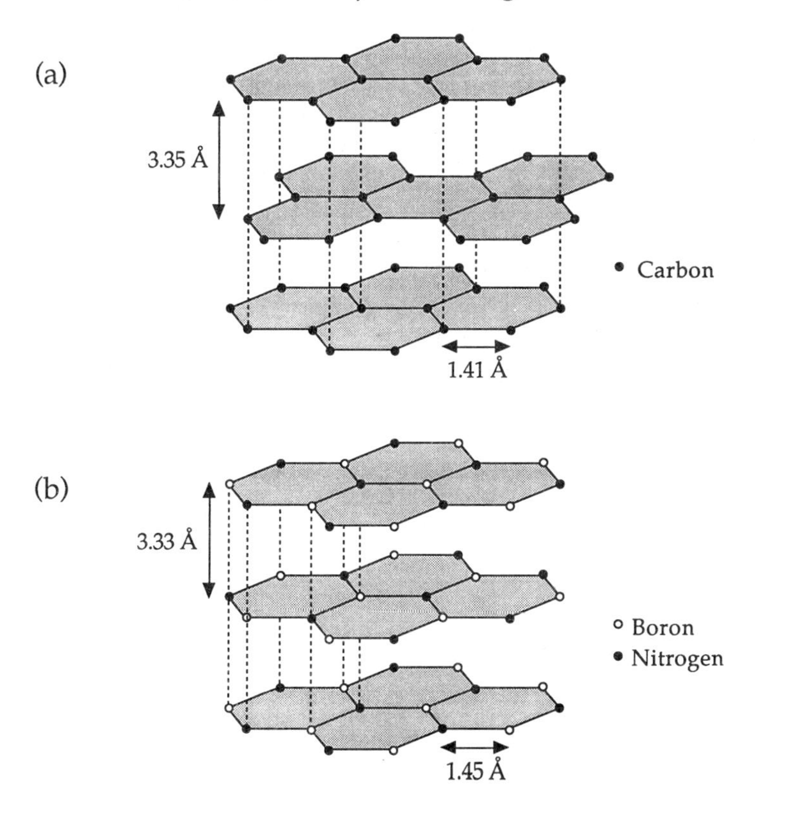

Figure 2.1 Crystal structures of (a) graphitic carbon, and (b) hexagonal BN.

Furthermore, we now believe that this kind of intermediate is critical to achieving many of the unique features associated with BN prepared from this precursor.

In this chapter, the key characteristics of the oligomer necessary for successful processing of composites is described. Specific data confirming the liquid crystalline character of the oligomer are presented. Finally, some of the key advantages of BN formed via this route are described.

2.2 CHARACTERISTICS OF BORAZINE OLIGOMERS

Borazine oligomers are formed by means of condensation reactions of the 6-membered ring compound borazine, $B_3N_3H_6$. These reactions occur very readily upon heating in a sealed vessel at temperatures between 60 and 80°C [21]. Both molecular weight and viscosity increase as a function of polymerization time, allowing the flexibility of tailoring the melt viscosity to meet the needs of the application.

As polymerization of borazine proceeds, a wide range of condensed polynuclear species are observed. Figure 2.2 shows a mass spectrum of the borazine oligomer with a weight average molecular weight of 5000 prepared by heating at 70°C for 35 h. The mass spectrum indicates the presence of a series of condensed polynuclear molecules corresponding to discrete planar hexagonal units.

While borazine is analogous in shape and structure to benzene, the behavior observed here differs sharply from that of benzene. During the pyrolysis of benzene at temperatures in excess of 500°C, linear biphenyl, terphenyl and polyphenyl units are formed which do not exhibit liquid crystallinity [22]. In contrast, when naphthalene or anthracene are heated, they will enter a liquid crystalline 'mesophase' and will produce graphitizable carbon upon continued pyrolysis (Figure 2.3). This mesophase consists of a wide range of planar, polynuclear species which have a tendency to stack, thereby forming a discotic phase. The mesophase is also observed to form when a 50:50 mixture of benzene and naphthalene is heated [23].

As shown in Figure 2.2, the condensation of borazine results in a mixture of biphenyl and naphthalene analogs in addition to larger planar units. Like graphitizable pitch, these appear to have the potential to stack in a discotic phase allowing for formation of a graphitic boron

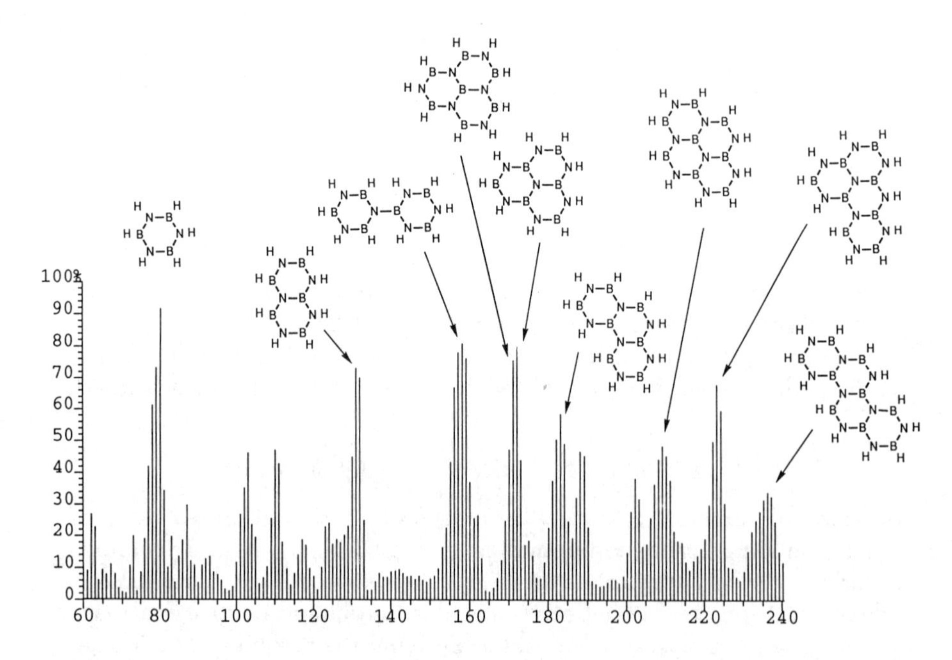

Figure 2.2 Mass spectrum of borazine oligomer with average M_w of 5000.

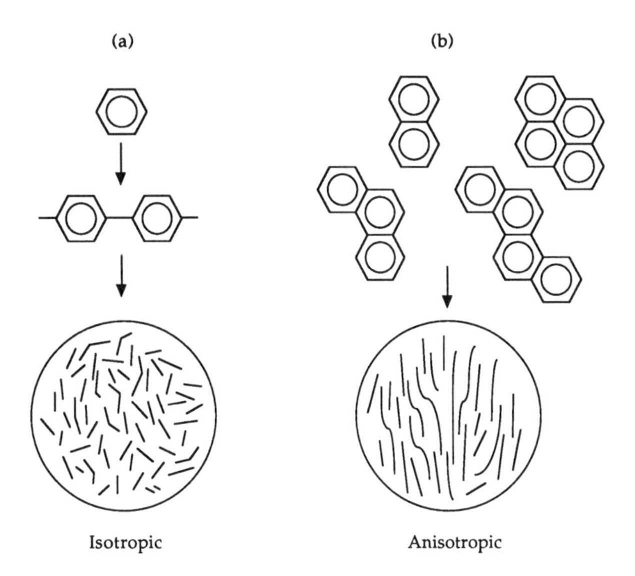

Figure 2.3 Schematic contrasting (a) non-graphitizable carbons versus (b) graphitizable carbons.

nitride at reasonably low temperatures. However, unlike graphitizable pitch which must be heated above 450°C to form a liquid crystalline phase, the borazine oligomers can exhibit liquid crystallinity after heating at temperatures below 100°C.

Optical texture has been observed in the borazine oligomer when viewed through cross-polarizers (Figure 2.4). This behavior can be enhanced either by cooling at 5°C or by applying a low shear stress between glass slides [17]. Large anisotropic patches are detected with layers lying either parallel or perpendicular to the plane of polarization of the incident light. Depending on the molecular orientation, the isochromatic areas of the anisotropic domains change to blue, yellow or magenta.

X-ray diffraction (XRD) of a borazine oligomer melt at 25°C is shown in Figure 2.5. This specimen was formed by polymerization at 70°C and then cooled to 5°C for 2 weeks. At 25°C , the oligomer shows a characteristic (002) peak at 3.72 Å. A reversible transition to a more isotropic phase occurs upon heating to 70°C. Following heating to 100°C, this transition becomes irreversible and the peak is broadened. Presumably disorder becomes permanent in the isotropic melt as additional polymerization proceeds at the elevated temperatures.

It is noteworthy that the borazine oligomer does not show a transition enthalpy in differential scanning calorimetry (DSC) measurements. This is also comparable to the carbonaceous mesophase in which a nematic to isotropic transition enthalpy has not been observed [24]. In both the

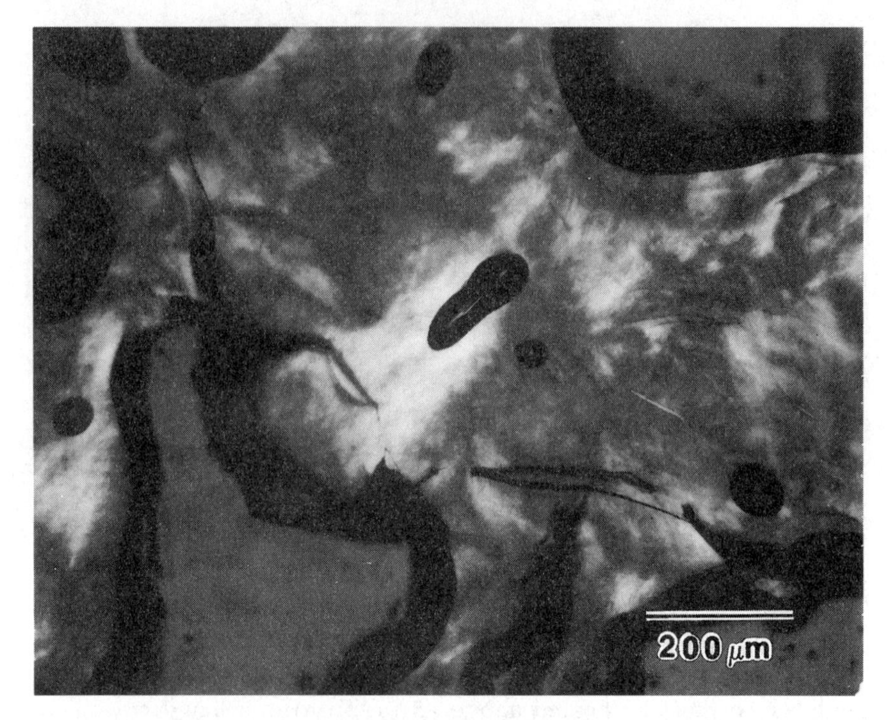

Figure 2.4 Optical texture of the borazine oligomer.

borazine and the carbonaceous systems, transitions may occur over a broader temperature range resulting in the absence of an observable transition enthalpy.

Both composition and physical treatments affect the degree of crystallinity in the borazine oligomer melt. For example, the propensity to exhibit anisotropic behavior is enhanced by cooling the oligomer at 5°C for extended periods of time [17]. This allows segregation and molecular orientation at very slow polymerization rates.

The application of pressure to the melt has also been observed to enhance orientation in the borazine oligomer [25]. Figure 2.6 shows X-ray diffraction (XRD) scans of specimens which have undergone various low temperature processing conditions. Figure 2.6a, shows an oligomer with an average molecular weight of 2400. Here a broad amorphous peak is observed, characteristic of a rather isotropic melt. Upon heating in the absence of pressure, the diffuse peak shifts from 4.10 to 3.56 Å indicating modest ordering (Figure 2.6b). When pressure is applied and the specimen is heated to 100°C, a sharp peak at 3.37 Å is observed (Figure 2.6c). This process may be related to the presence of a small fraction of anisotropic phase which can then convert to a BN structure with a much higher degree of order.

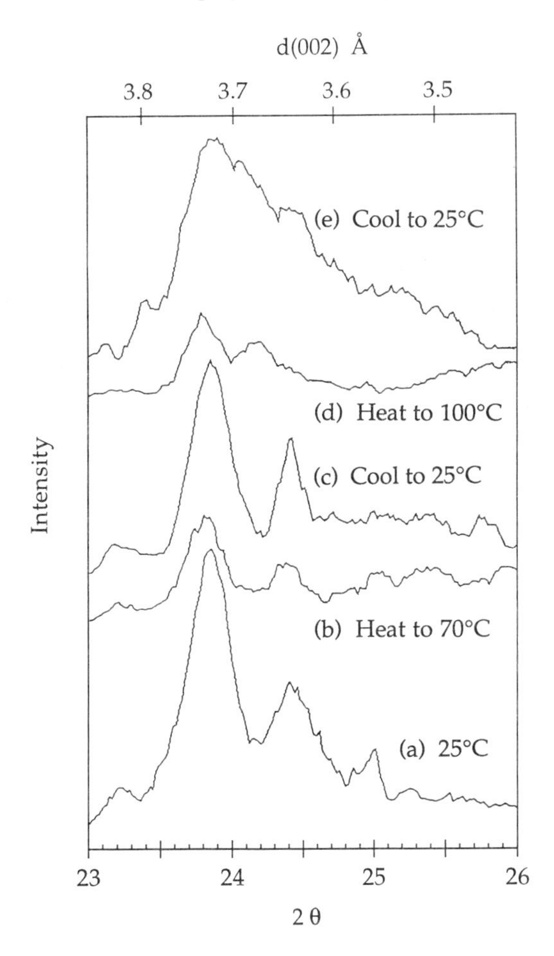

Figure 2.5 Hot stage XRD demonstrating reversible transition in the melt.

2.3 PROCESSING OF BORAZINE OLIGOMERS

By heating the monomer at 70°C in a pressure reactor vessel, self-condensation occurs and hydrogen is evolved. This reaction proceeds in a very easy, controlled manner without the need for additional catalysis.

During polymerization, the weight average molecular weight (M_w) is observed to increase exponentially (Figure 2.7). The kinetics depend on both temperature and pressure. Figure 2.7a demonstrates the molecular weight evolution when the pressure is held slightly above 1 atmosphere. A portion of the evolved hydrogen in addition to some borazine gas should be periodically eliminated from the reactor vessel.

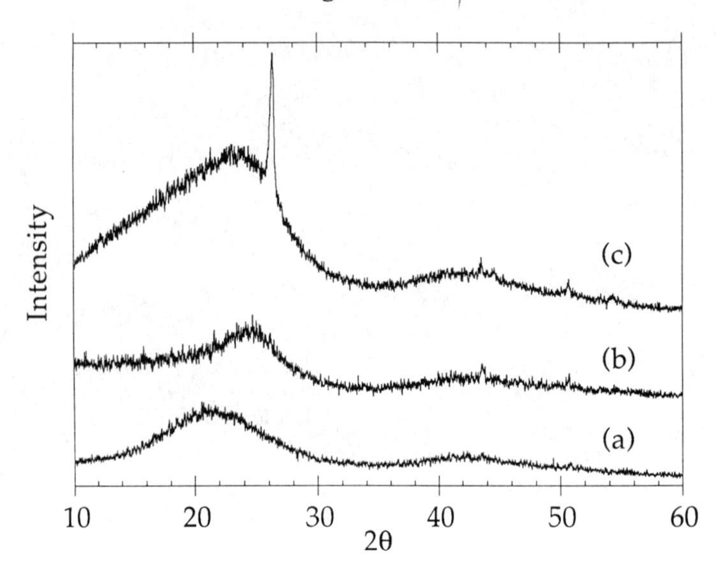

Figure 2.6 XRD of borazine oligomer (a) as produced, (b) after heating to 70°C and (c) after applying pressure and heating to 100°C.

In contrast, when the gaseous products are not continuously removed and internal pressure continues to increase, the reaction proceeds more slowly (Figure 2.7b). Mass spectrometry of the product indicates that the structural evolution of the borazine oligomer is not affected by these changes.

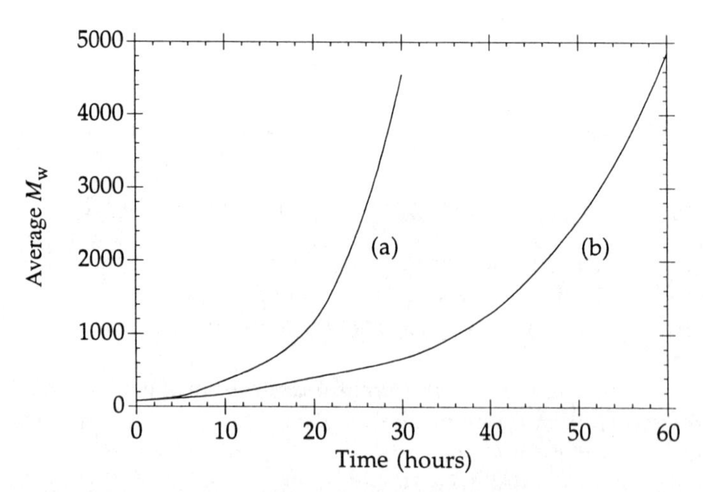

Figure 2.7 Average M_w of borazine oligomer as a function of polymerization time for (a) pressure slightly above 1 atmosphere and (b) without degassing.

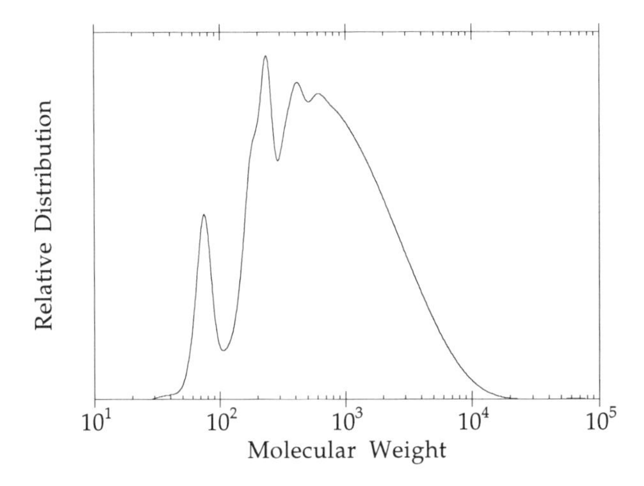

Figure 2.8 Molecular weight distribution after 20 h polymerization.

Figure 2.8 displays a typical molecular weight distribution of intermediate oligomer with an average molecular weight of 1130. A very broad molecular weight distribution is observed, including the borazine monomer with M_w of 80 to large molecules with M_w of several thousand.

Mass spectrometry can also be used in order to gain insight into the mechanism of borazine polymerization. Of particular interest is the means by which polynuclear species such as the naphthalene analog are formed. Examination of the gaseous products reveals that hydrogen is the primary byproduct of the reaction with a small amount of ammonia.

The mass spectrum of borazine, $B_3N_3H_6$, shows that it has a propensity to decompose into several stable fragments, including $B_3N_2H_3$, $B_2N_2H_4$ and BN_2H (Figure 2.9). Presumably the ring fragments such as $B_2N_2H_4$ could react with borazine rings to form the naphthalene-like species. Addition of the BN_2H fragment to the naphthalene analog could account for the presence of the perinaphthalene-like species. The formation of the various borazine fragments is also accompanied by ammonia evolution and by the addition of pendant amine groups on both borazine and larger oligomeric species.

Infrared spectra for the borazine oligomer reveals strong absorptions at 3440 and 2490 cm^{-1} which are assigned to the N–H and B–H groups, respectively (Figure 2.10) [26]. While the B–N absorption of 1430 cm^{-1} is very sharp in borazine, it begins to broaden when borazine oligomerizes. As the specimen is heated, further dehydrogenation occurs and it crosslinks to form an insoluble material. During this process, the N–H and B–H peaks gradually disappear and the B–N peak continues to broaden until it is nearly identical to that observed in fully crystalline BN.

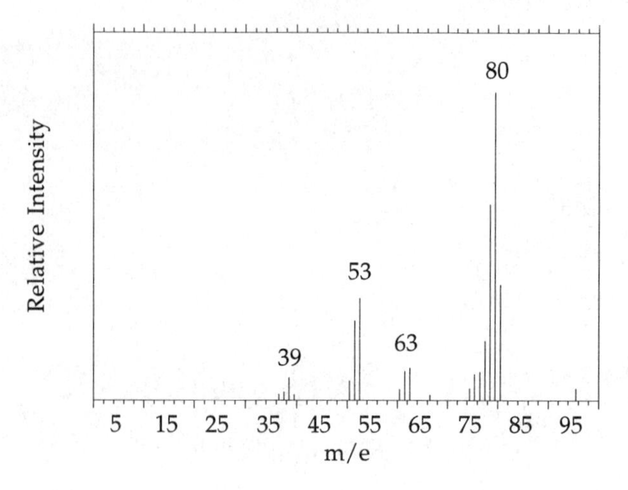

Figure 2.9 Mass spectrum of borazine monomer.

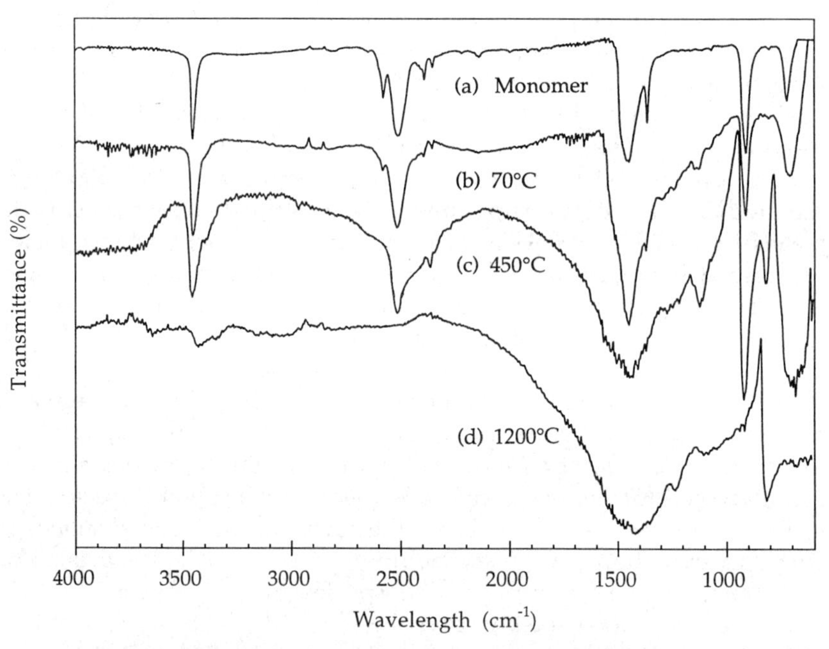

Figure 2.10 FTIR of products during conversion from borazine to BN.

2.4 SYNTHESIS OF BORON NITRIDE VIA LC BORAZINE OLIGOMERS

Boron nitride is formed when the borazine oligomer is heated above 1000°C. This is a particularly novel processing route because it offers the ability of forming highly crystalline BN thin films or composites at rather low temperatures. In contrast, temperatures of 1800°C are required to form a stable BN via carbothermal reduction. The BN phase is quite similar to graphitic carbon, except that BN has a significantly higher oxidation resistance and is electrically insulating.

One unique characteristic of the borazine oligomer is that it has a very high mass yield observed upon conversion to BN. Figure 2.11 shows a thermogravimetric analysis (TGA) plot of the borazine oligomer. This data reveals that the mass yield upon conversion to BN is approximately 85% which is very high for a ceramic precursor [27]. Additional examination has revealed evidence for viscoplastic flow in the crosslinked oligomer under the application of pressure. This may increase the effective BN mass yield even further for a bulk structure such that the volume yield is between 60 and 70%.

The crystallinity and morphology of the BN has been shown to be dependent on various processing parameters including heating rate, maximum temperature, soak time and atmosphere. One primary factor controlling the structure of the ultimate BN, however, is the development of structure in the LC oligomer at low temperatures.

To illustrate this point, Figures 2.12 and 2.13 compare the crystallinity as a function of processing temperature for two different specimens [25]. The first specimen (Figure 2.12), consisted of the borazine oligomer from the melt to which no external pressure has been applied. A turbostratic BN phase with an intermediate *d*(002) value of 3.41 Å is formed only upon heating to 1200°C. Specimens like that shown in Figure 2.6c were also investigated. In these specimens, pressure was

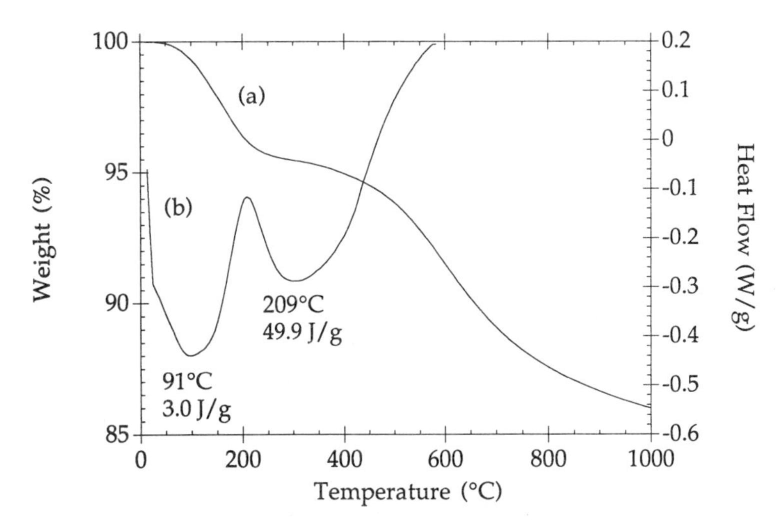

Figure 2.11 (a) TGA and (b) DSC of borazine oligomer during heating to 1000°C.

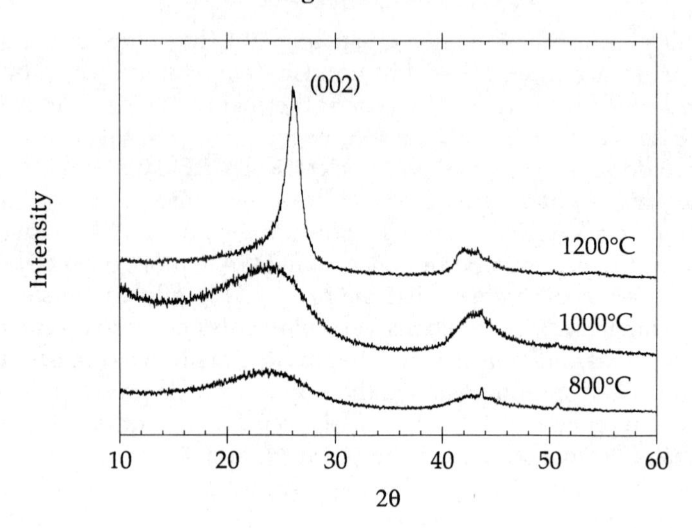

Figure 2.12 XRD of specimens with no prior pressure which were heated to 800, 1000 and 1200°C.

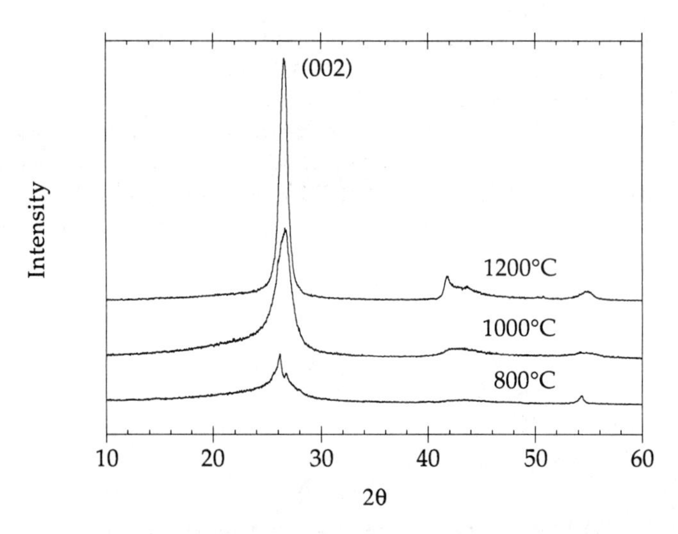

Figure 2.13 XRD of specimens with 20 MPa pressure applied from 70 to 400°C and subsequently heated to 800, 1000 and 1200°C.

applied only up to 400°C and then released. Subsequent heat treatments resulted in a much more ordered BN at lower temperatures (Figure 2.13). Upon heating to 1200°C, the resulting phase had a near theoretical $d(002)$ of 3.34 Å with an L_c crystallite size of 200 Å. This suggests that the processing history of the oligomeric phase has a profound impact on the ordering of the resulting BN.

The crystallinity of the BN is an important consideration because the interlayer spacing has been shown to control the stability of BN to both oxidation and hydrolysis [15]. This is demonstrated with several BN specimens which have been produced with *d*(002) spacings ranging from 3.67 to 3.33 Å. When exposed to dry oxygen at 950°C, the rate of oxidation is significantly slower in specimens with interlayer spacings closer to the theoretical 3.33 Å (Figure 2.14). The stability of carbon has been observed to have a similar dependence on its *d*(002) spacing [15]. However, while BN oxidizes to B_2O_3 with an accompanying weight gain, carbon incurs rapid weight loss as it volatilizes to CO and CO_2. Further, oxidation of carbon is observed at temperatures as low as 500°C [28].

The stability of BN against hydrolysis has also been observed to depend on the interlaying spacing. As shown in Figure 2.15, the smaller *d*(002) specimens are significantly more stable against reaction with moisture [15].

Boron nitride derived from the borazine oligomer has been successfully used to replace the carbonaceous matrix in carbon fiber/carbon matrix composites (C/C) [15,18]. The carbon fiber/boron nitride matrix composites (C/BN) exhibit improved oxidation resistance over the C/C. Excellent mechanical strength and toughness have also been observed in the C/BN. As shown in Figure 2.16, both the flexural strength and modulus of pitch carbon fiber/BN specimens actually increase with extended heat treatment [18].

The increase in the properties of the C/BN suggests a structure more similar to that observed in C/C with a mesophase pitch-derived matrix.

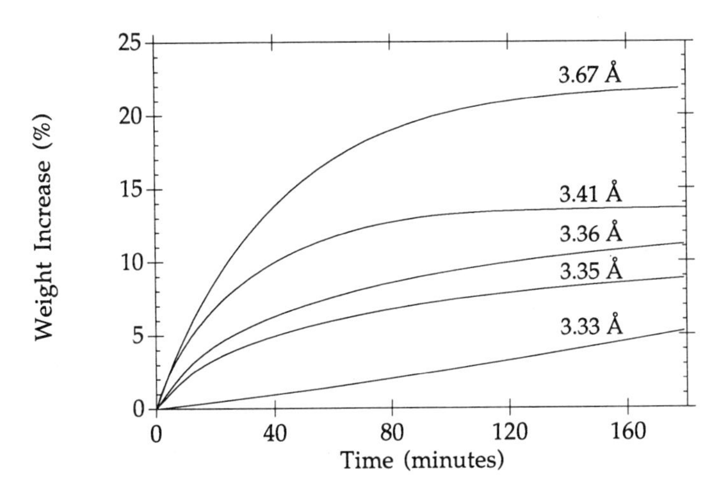

Figure 2.14 Isothermal weight gain in air at 950°C for BN specimens with *d*(002) from 3.33 Å to 3.67 Å.

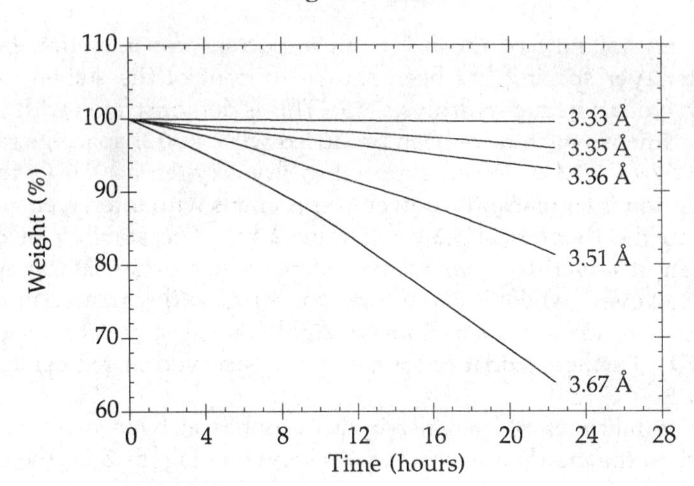

Figure 2.15 Isothermal weight loss in H_2O/air at 700°C for BN specimens with *d*(002) from 3.33 Å to 3.67 Å.

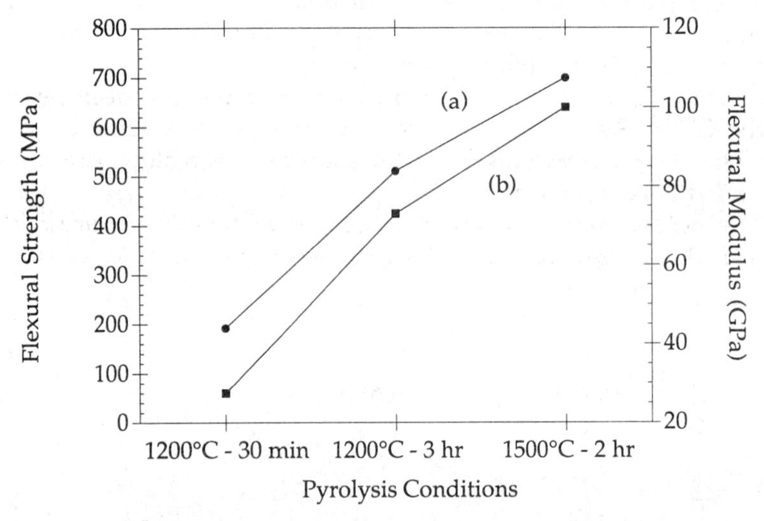

Figure 2.16 Dependence of (a) flexural strength and (b) modulus on heat treatment for pitch C/BN.

In both composites, the planar molecules facilitate orientation of the crystalline structure parallel to the fiber–matrix interface.

Transmission electron microscopy (TEM) of ceramic fiber/BN matrix composite has verified the preferential orientation of BN parallel to the interface (Figure 2.17). In this figure, the diffraction pattern in the BN matrix reveals a very well aligned crystalline structure. This evidence suggests that the liquid crystallinity of the borazine oligomer provides

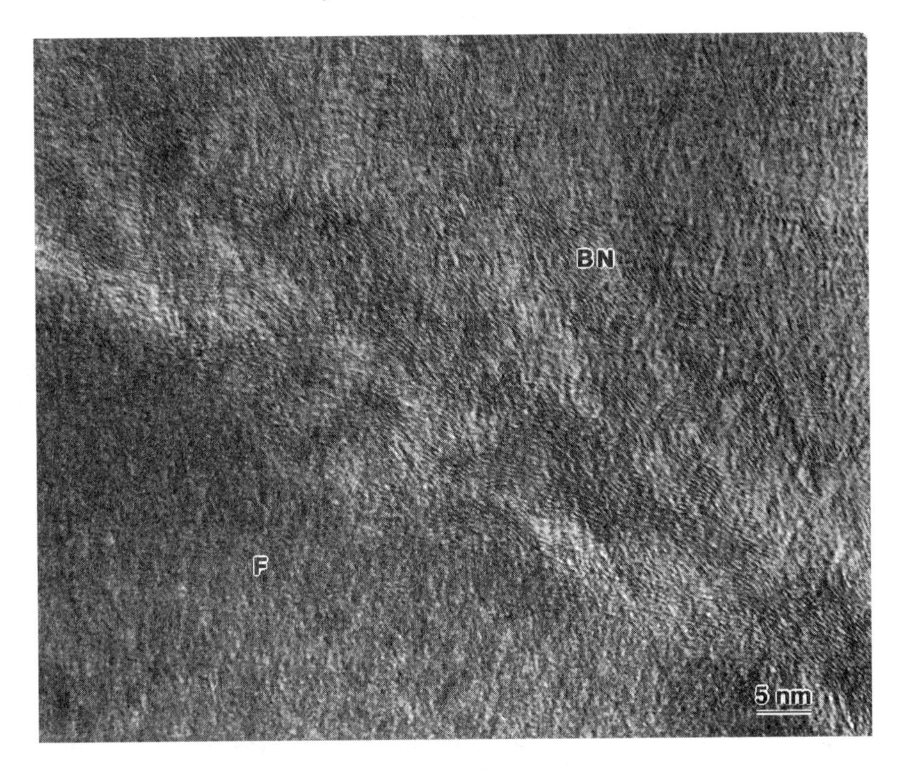

Figure 2.17 TEM micrograph of the fiber (F)/matrix (BN) interface in a Nicalon/BN composite demonstrating preferential orientation of the BN matrix.

a pathway for alignment of the structure parallel to the fiber surfaces during melt processing. In the case of C/BN, ordering persists and continues to match the CTE of the fiber, thereby minimizing the stress state at the interface.

Some unexpected results have been observed in the wear and friction studies of C/BN versus C/C [29]. C/C composites are used as brake pads in most commercial aircraft because they display good wear resistance at high temperatures. Of considerable significance is the observation that under sliding contact, the surface of the C/C can actually deform plastically producing an oriented graphitic surface [30]. When a shear force is applied at the surface, the carbon can flow and orient in much the same manner as Bragg had observed 30 years ago but at much lower temperatures [3]. Similar behavior has been observed in the C/BN in which an oriented surface film of BN will develop during sliding contact.

For both the C/C and C/BN, the coefficient of friction is observed to increase as a function of temperature (Figure 2.18). Recent results

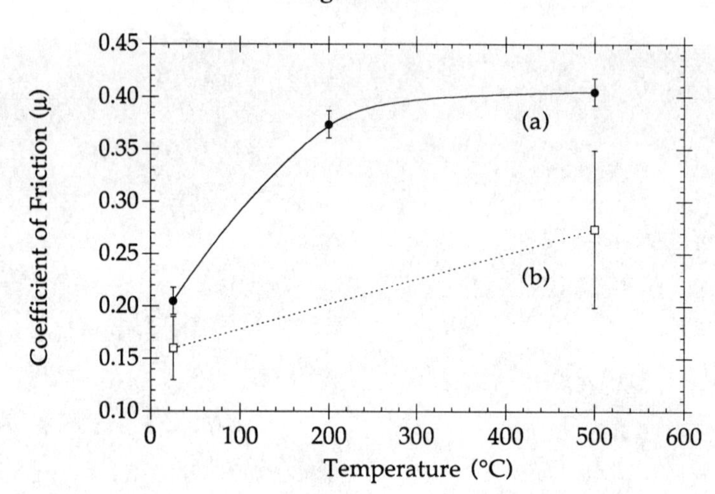

Figure 2.18 Coefficient of friction measured as a function of temperature for (a) C/C and (b) C/BN.

appear to confirm that this increase occurs because of the desorption of moisture from the surface of either carbon or BN as temperature increases. While the wear rate of C/C increases rather dramatically at temperatures of 100–500°C, preliminary results indicate that the wear rate of C/BN is far lower than C/C at these temperatures. Since aircraft brakes reach high temperatures during landing, the lower wear rates of C/BN suggest a very attractive alternative to C/C.

The borazine oligomers have also been used as a solution-based route to form BN films and coatings. By either spin or dip coating, films ranging in thickness from 0.2 to 9 μm have been prepared. Figure 2.19 shows a typical BN film which was dip coated onto the surface of a silicon wafer. The dielectric constant of dip coated films was observed to vary between 4.0 and 5.5 depending on both the measurement frequency and the film thickness (Figure 2.20). The dielectric loss (tan ∂) was typically between 0.001 and 0.1. Generally, the dielectric constant was observed to decrease in the thinner films, suggesting the possibility of an orientational effect. This is supported by work of Westphal *et al.* [29], who showed that the dielectric constant of highly aligned BN was 2.9–3.3 in the direction perpendicular to its crystallographic planes and 5.0–5.2 in the parallel direction.

In conclusion, the liquid crystalline character which has been observed in the inorganic borazine oligomers provides a mechanism for the development of an oriented structure in BN. It is now understood that the interlayer spacing must be close to that of theoretical BN to produce a stable structure. The borazine oligomer offers a convenient

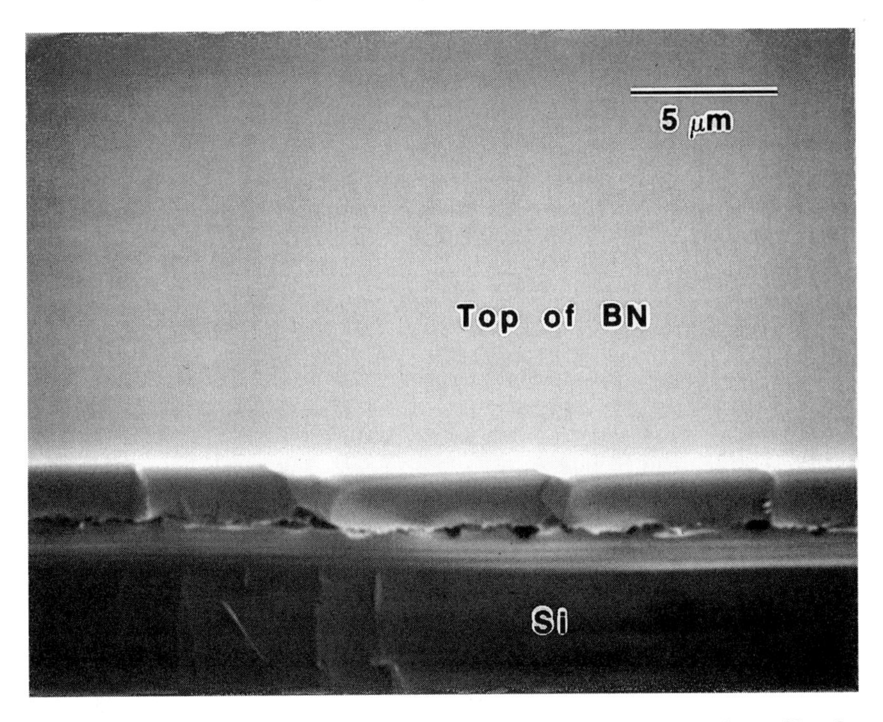

Figure 2.19 SEM micrograph showing cleaved cross-section of BN film dip coated onto Si substrate under nitrogen.

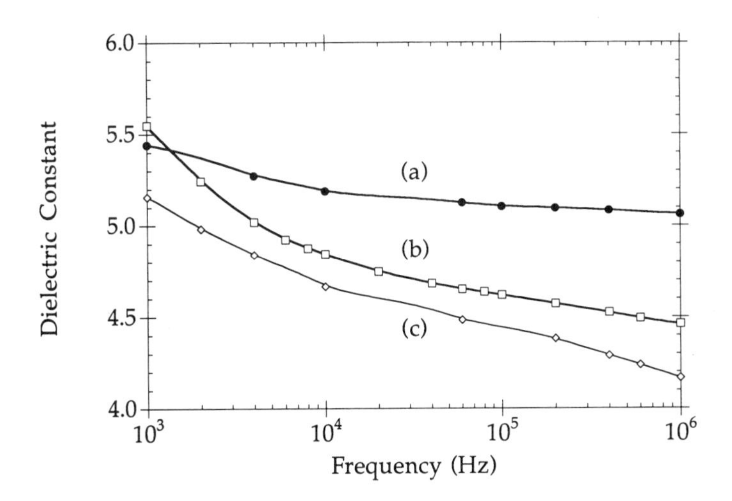

Figure 2.20 Dielectric constant for BN films with thickness of (a) 9.5 μm, (b) 3.6 μm and (c) 2.3 μm.

route to prepare stable BN either in two-phase composites or in thin films at reasonably low temperatures.

REFERENCES

1. Flory, P.J. (1953) *Principles of Polymer Chemistry*, Cornell University Press, Ithaca.
2. MacDiarmid, A.G., Mikulski, C.M., Saran, M.S. *et al.* (1976) *Adv. Chem. Ser.*, **150**, 63.
3. Bragg, R.H., Crooks, D.D., Fenn, R.W., Jr and Hammond, M.L. (1964) *Carbon*, **1**, 171.
4. Ford, C. and Mitchell, C. (1960) US Patent 3107152 (Union Carbide Corp.).
5. Bacon, R. *et al.* (1967) US Patent 3305315.
6. Schalamon, W. *et al.* (1973) US Patent 3716331 (Union Carbide Corp.).
7. Shindo, A. (1961) *J. Ceram. Assoc. Jap.*, **69**, C195.
8. Prescott, R. and Standege, A. (1966) *Nature*, **211**, 169.
9. Watt, W., Phillips, L. and Johnson, W. (1966) *Engineering* (London), **221**, 815.
10. Otani, S. (1965) *Carbon*, **3**, 213.
11. Brooks, J.D. and Tayor, G.H. (1965) *Carbon*, **3**, 185.
12. Brooks, J.D. and Tayor, G.H. (1966) *Adv. Chem. Ser.*, **55**, 549.
13. Hawthorne, H.M. (1971) *Proceedings of the First International Conference on Carbon Fibres*, Plastics Institute, London, 81.
14. Pease, R.S. (1952) *Acta Cryst.*, **5**, 356.
15. Cofer, C.G. and Economy, J. (1995) *Carbon*, **33**, 4.
16. Economy, J. and Anderson, R.V. (1967) *J. Polymer Sci. C*, **19**, 283.
17. Kim, D.P. and Economy, J. (1994) *Chem. Mater.*, **6**, 395.
18. Kim, D.P. and Economy, J. (1993) *Chem. Mater.*, **6**, 1216.
19. Yasuda, E., Tanabe, Y., Machino, H. and Takaku, A. (1987) *Proceedings of the 18th Biennial Conference on Carbon*, 30.
20. Sheaffer, P.M. (1987) *Proceedings of the 18th Biennial Conference on Carbon*, 20.
21. Fazen, P.J., Beck, J.S., Lynch, A.T. *et al.* (1990) *Chem. Mater.*, **2**, 96.
22. Fitzer, E., Mueller, K. and Schaefer, W. (1971) *Chemistry and Physics of Carbon*, Marcel Dekker, New York, **7**, 237.
23. Marsh, H., Dachille, F., Melvin, J. and Walker, P.L., Jr (1971) *Carbon*, **9**, 159.
24. Marsh, H. and Walker, P.L. (1979) *Chemistry and Physics of Carbon*, Marcel Dekker, New York, **15**, 229.
25. Cofer, C.G., Kim, D.P. and Economy, J. (1995). *Cer. Trans.*, **46**, 189.
26. Paine, R.T. and Narula, C.K. (1990) *Chem. Rev.*, **90**, 73.
27. Rice, R.W. (1983) *Cer. Bull.*, **62**, 889.
28. Rodriguez-Mirasol, J., Thrower, P.A. and Radovic, L.R. (1993) *Carbon*, **31**, 789.
29. Westphal, W.B. and Sils, A. (1972) Massachusetts Institute of Technology Technical Report AFML-TR-72-39.
30. Cofer, C.G., Saak, A.W. and Economy, J. (1995) *Cer. Eng. Sci. Proc.*, **16**, 663.

3

Design of polymer liquid crystals with non-covalent bonds

C. Géraldine Bazuin

3.1 GENERAL CONCEPTS

Polymer liquid crystals (PLCs) are traditionally designed with covalent bonds as connectors of the various structural subunits. Recently, it has been demonstrated that non-covalent bonds can also be used to advantage in the molecular construction of PLCs. They can involve hydrogen bonds, ionic interactions, coordination complexes, charge transfer interactions or other donor–acceptor effects. A variety of molecular designs that incorporate non-covalent bonds are schematically illustrated in Figure 3.1 for longitudinal and comb-like PLCs possessing thermotropic mesogens.

In order to obtain effective non-covalent bonds, complementary functional groups are usually necessary. Strong hydrogen bonds, for example, can be generated by the use of carboxylic acid and pyridine moeities. Ionic bonds can be created by proton transfer from sulfonic acid to basic amine moieties, or by ion exchange involving sulfonate and ammonium groups.

Longitudinal or main chain-like architecture can be generated from two bifunctional (or ditopic) low molar mass constituents that possess the necessary complementary moieties. As shown in Figure 3.1, the

Mechanical and Thermophysical Properties of Polymer Liquid Crystals
Edited by W. Brostow
Published in 1998 by Chapman & Hall, London.
ISBN 0 412 60900 2

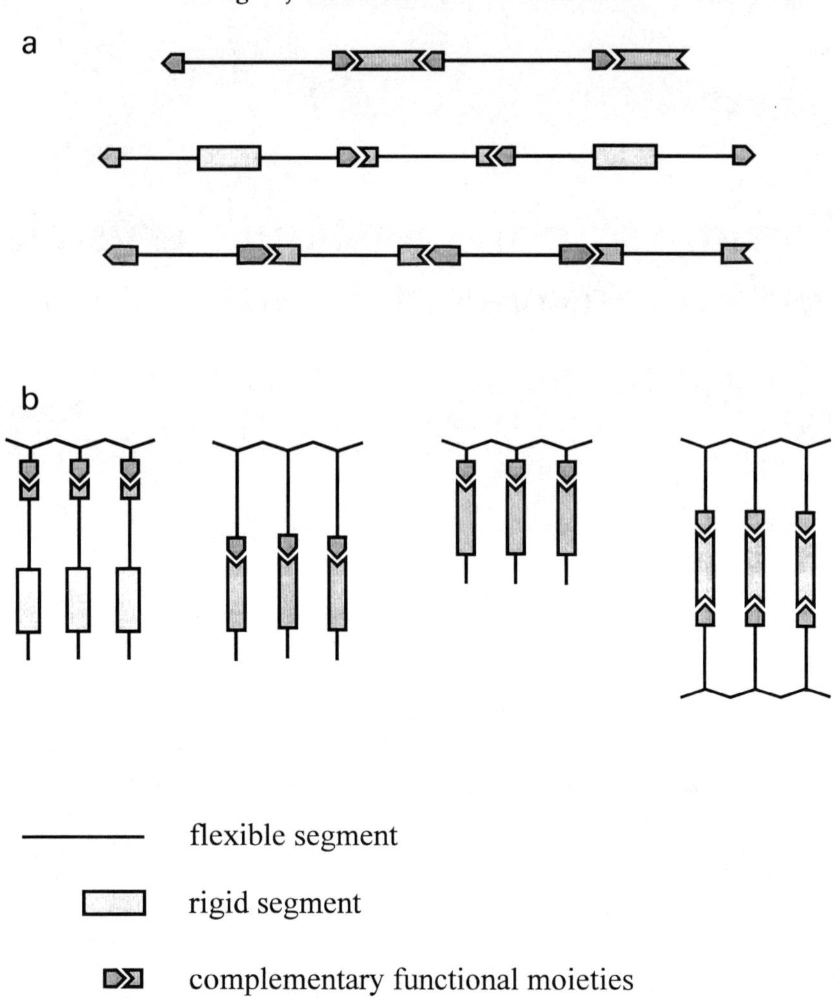

Figure 3.1 Schematic designs of (a) longitudinal and (b) comb-like polymer liquid crystals with non-covalent bonds.

non-covalent bond can, in principle, be found at the interface between the flexible and rigid segments, within the flexible segment or within the rigid segment. In the first two cases at least one of the constituents possesses a rigid mesogenic moiety. In the third case, the non-covalent interaction can itself be responsible for creating the rigid segment.

To obtain comb-like or side chain-like architecture, a monofunctional, low molar mass mesogen (or monomer liquid crystal, MLC) can be complexed to a polymer possessing complementary groups. The non-covalent bond may then be located within or near the polymer backbone, or within or near the rigid mesogen. In the first case, the flexible spacer is part of the low molar mass mesogen; in the second, it constitutes the polymer side chain. Coupling of functionalized MLCs and appropriate polymers may also be envisaged without the presence of a flexible spacer, as shown in Figure 3.1. If the MLC is fixed with a functional group at both of its extremities, a non-covalently cross-linked structure, or network may be generated. Various copolymeric structures may also be envisaged; they may involve less than equimolar proportions of MLC to polymer repeat unit, or mixtures composed of more than one MLC, or parent copolymers.

The creation of PLCs with noncovalent bonds has been termed 'supramolecular chemistry' or 'molecular recognition-directed' 'self-assembly'. The application of these concepts to a variety of systems, including PLCs, has been reviewed by Lehn [1–3]. A first general overview of non-covalently bonded PLCs is given in reference [4].

In this chapter the focus will be principally, although not exlusively, on longitudinal and comb-like systems, since these compose the majority of the non-covalently bonded PLCs published to date. However, it is no doubt only a question of time before the concepts of non-covalent bonding will be applied to other molecular PLC structures such as those classified by Brostow [5]. Furthermore, this chapter will consider only thermotropic systems, and only systems where the building up of a larger polymeric molecule through non-covalent bonds involves at least two dissimilar components, both of which possess segments which are not directly part of the noncovalent bond and at least one of which is a low molar mass substance. The latter criteria exclude, for instance, ionic PLCs with simple counterions, single component PLCs whose mesogenic character nevertheless depends on intersegmental interactions, or PLC blends whose intermolecular interactions involve covalently bonded mesogens. Finally, no attention will be given to low molar mass LC systems involving non-covalent bonds, often considered as models for the polymeric or polymeric-like systems.

3.2 HYDROGEN BONDED PLC STRUCTURES

3.2.1 Longitudinal or main chain architecture

A simple application of the self-assembly of complementary bifunctional small molecules to form extended-chain structures, or longitudinal PLCs, is given in references 6–8. Griffin and coworkers report that hydrogen bond-driven association between bispyridyl-terminated and bisbenzoic acid-terminated species can lead to liquid crystalline materials with polymeric characteristics. Typical compounds employed are shown in Figure 3.2. None of the starting components are liquid crystalline.

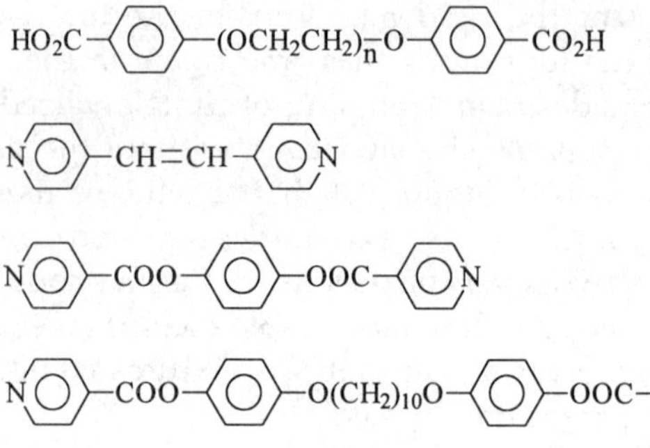

$HO_2C-C_6H_4-(OCH_2CH_2)_n-O-C_6H_4-CO_2H$ n = 4 **1a**; n = 5 **1b**

$NC_5H_4-CH=CH-C_5H_4N$ **2a**

$NC_5H_4-COO-C_6H_4-OOC-C_5H_4N$ **2b**

$NC_5H_4-COO-C_6H_4-O(CH_2)_{10}O-C_6H_4-OOC-C_5H_4N$ **2c**

Figure 3.2 Bifunctional molecular components for longitudinal PLCs assembled with hydrogen bonds.

However, equimolar mixtures of the complementary components, which associate as shown in Figure 3.3, produce nematic and/or smectic A mesophases. The temperature ranges of the mesophases are generally small (in the region of 160–180°C) in heating cycles, although they tend to be considerably extended in cooling cycles due to significant supercooling of the mesophase–crystalline transition. It was shown for the ester-containing species, specifically (1b) + (2b), that prolonged exposure to melt temperatures induces ester exchange reactions, which affect the mesophases observed, in particular by increasing the liquid crystalline stability (diminishing the crystallinity) [8]. The hydrogen bonding that takes place between the pyridyl and acid moieties is easily detected by infrared spectroscopy.

Away from equimolar stoichiometry, the transition temperatures decrease, as shown in Figure 3.4 for the (1a) + (2a) mixture. This can be understood by considering the association as an A_2B_2 step growth polymerization [6–7]. From this point of view, it is obvious that maximum effective chain length is achieved for the equimolar complex, and this would be consistent with its displaying the highest temperature transitions. A theoretical analysis of the extent of hydrogen bonding and mesophase ordering in these systems confirms that there is

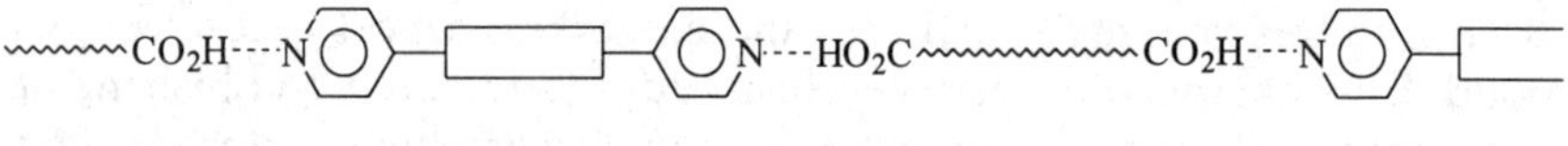

$\sim CO_2H\cdots NC_5H_4-\square-C_5H_4N\cdots HO_2C\sim\sim CO_2H\cdots NC_5H_4-\square$

Figure 3.3 Representation of the self-assembly of longitudinal PLCs through single hydrogen bonds, using the complementary components of Figure 3.2.

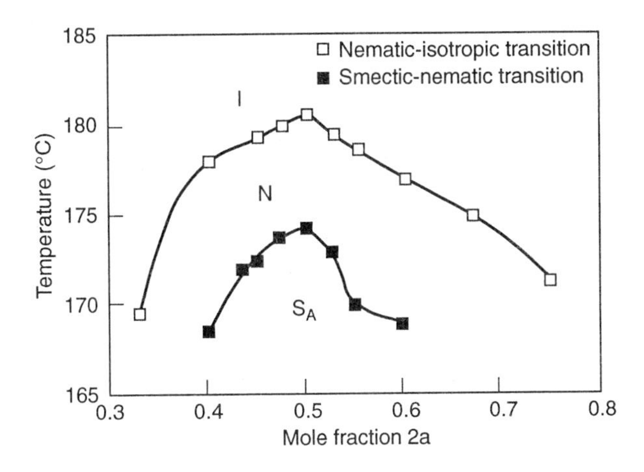

Figure 3.4 Partial phase diagram for complex (1a) + (2a) (permission being sought from [7]).

significant coupling between the two effects [9]. These systems can be expected to have a temperature-dependent viscosity due to a temperature-dependent effective chain length. In some cases, it is possible to pull fibers (albeit brittle) of these materials and observe glass transition behavior; see a note given in reference 31 of [10].

Variants of the above complexes were prepared by using tetrapyridyl-functionalized molecules (shown in Figure 3.5) with the diacid [10–12], which can be compared to A_2B_4 step growth polymerization [10]. Doubling the amount of binding sites in the pyridyl component allows easier pulling of fibers and easier observation of a glass transition, thus demonstrating the greater polymeric character of these complexes [10–12], although at high temperatures the viscosity is typical of low molar mass molecules rather than of polymers [10]. Again, none of the constituents alone exhibit thermotropic mesophases; however, the 1:1 complexes of the diacid (1a) with pyridyls (3b) [10] and (3c) [11, 12] are liquid crystalline, the latter appearing to possess a smectic A phase. The core of the tetrafunctional molecules may be expected to have a tetrahedral conformation and give rise to a kind of network structure that could hardly be liquid crystalline (Figure 3.6a); however, molecular modeling indicates that the flexible nature of the core allows these molecules to take on an essentially rod-like or calamitic shape, which gives rise to an extended-chain ladder-like structure (Figure 3.6b) [10–12]. On the other hand, complex (1a) + (3b) appears to lose its mesophase after repeated thermal cycling, interpreted as reflecting a

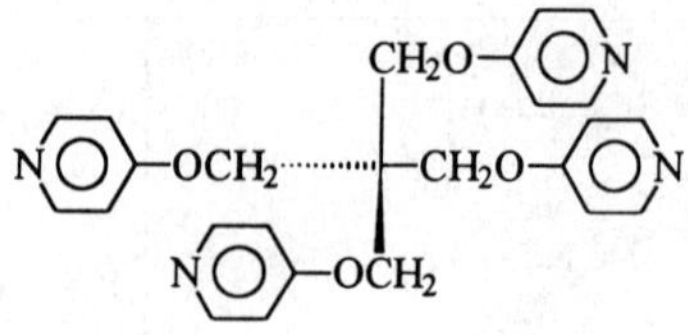

CH_2O–(pyridyl N)
N(pyridyl)–OCH_2 ········ CH_2O–(pyridyl N)
N(pyridyl)–OCH_2

(3a)

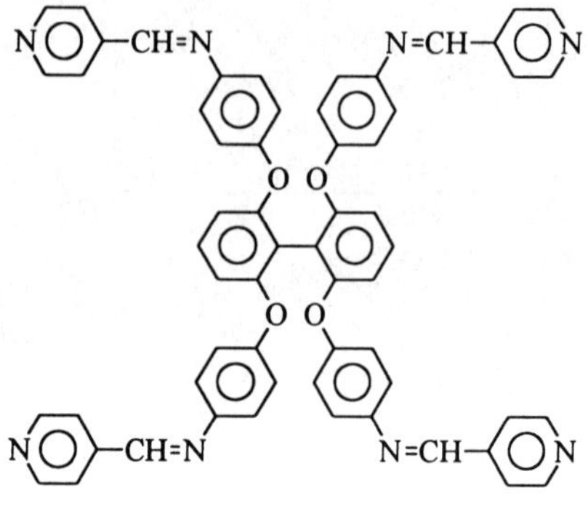

N–CH=N   N=CH–N
O O
O O
N–CH=N   N=CH–N

(3b)

CH_2O–(phenyl)–N=CH–(pyridyl N)
N(pyridyl)–CH=N–(phenyl)–OCH_2 ········ CH_2O–(phenyl)–N=CH–(pyridyl N)
N(pyridyl)–CH=N–(phenyl)–OCH_2

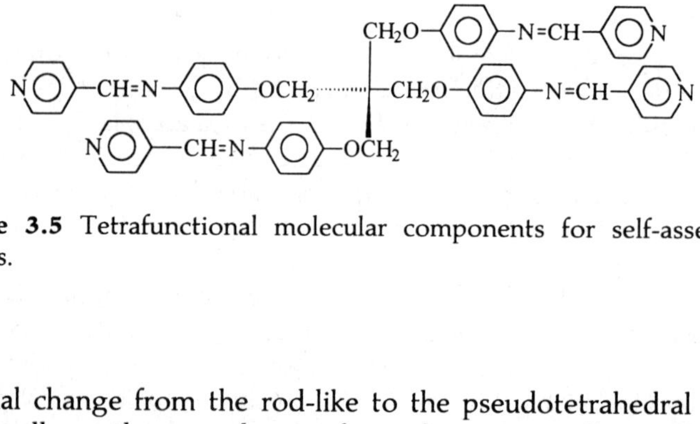

(3c)

Figure 3.5 Tetrafunctional molecular components for self-assembly with diacids.

gradual change from the rod-like to the pseudotetrahedral geometry, which allows the complex to form the entropically more favorable network structure [10]. The flexible arms and short, aromatic segments of pyridyl (3a) apparently favor the three-dimensional network structure for complex (1a) + (3a) under all conditions [10].

Lehn and coworkers [1–3, 13] have demonstrated that the simple self-assembly of two complementary ditopic components can generate hexagonal columnar mesophases. This was accomplished by mixing long-chain derivatives of (chiral) tartaric acid bifunctionalized by the complementary groups, 2, 6-diaminopyridine (P_2) and uracil (U_2) (Figure 3.7a). The two functional groups spontaneously associate via triple hydrogen bonds. The resulting material appears as a highly birefringent glue that forms fibers on spreading [13]. Whereas the

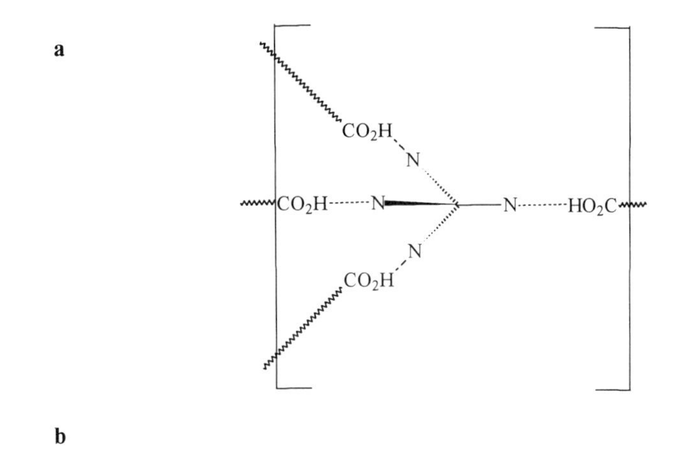

Figure 3.6 Schematic of (a) a network-like structure and (b) a ladder-like structure of self-assembled diacid and tetrapyridyl components (redrawn from [10]).

pure components exhibit a complex polymorphism, the mixture displays a liquid crystalline mesophase from below room temperature to above 200°C. The specific chirality of the tartaric acid was observed to influence profoundly the superstructure formed. Derivatives of L-tartaric acid ($LP_2 + LU_2$) form columns consisting of three polymeric strands in a triple helix superstructure, whereas those based on meso-tartaric acid ($MP_2 + MU_2$) form columns built of three strands in a zigzag conformation (Figure 3.7b). A mixture based on L- and D-tartaric acid produces yet another columnar arrangement.

It is notable that, in the cases presented above, the liquid crystalline mesophase is generated from components that do *not* have such mesophases, and that the molecular recognition process involves creation of the rigid mesogenic core. In the first cases presented, a single hydrogen bond per site connects the components; in the later cases, triple hydrogen bonding between the dissimilar components takes place. To our knowledge, there are not yet any published examples of longitudinal PLCs where hydrogen bonds are located within the flexible segment.

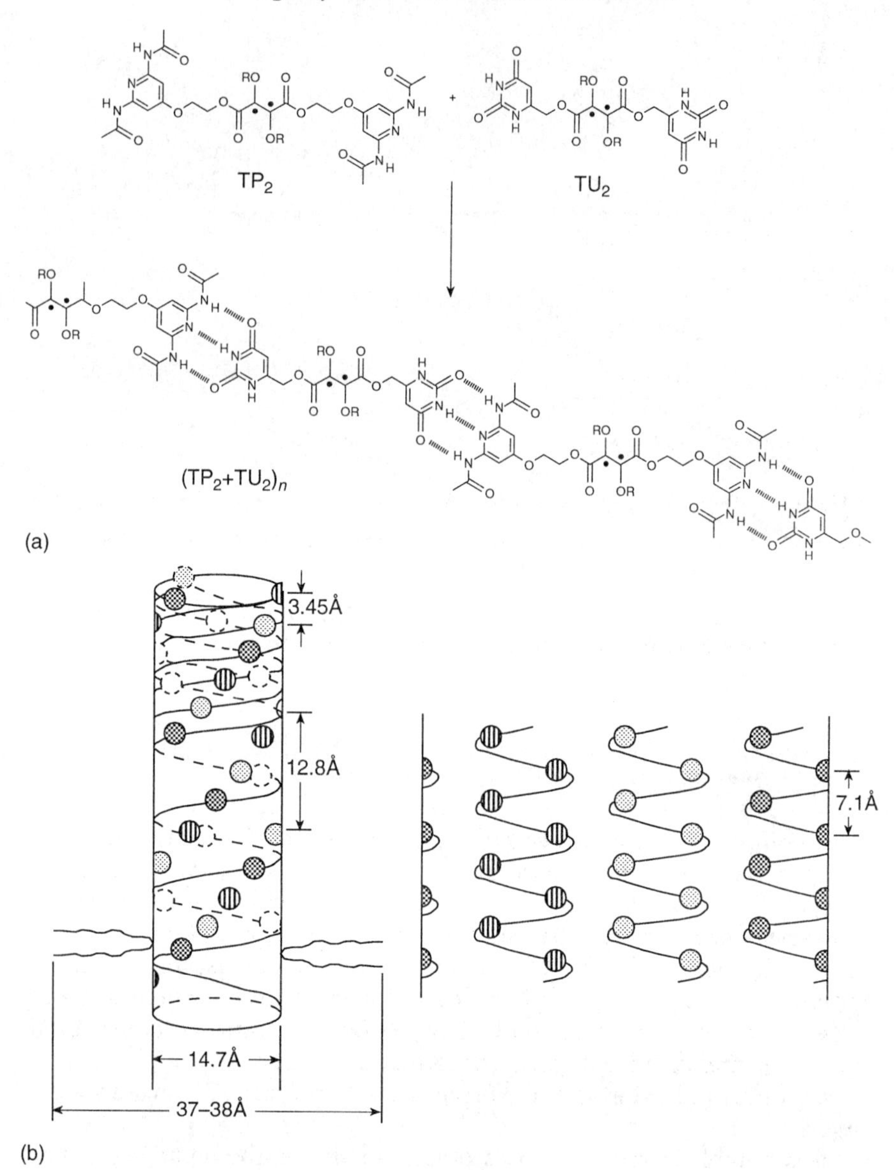

Figure 3.7 (a) Self-assembly of the supramolecular species $(TP_2 + TU_2)_n$ from the complementary chiral components, TP_2 and TU_2, via triple hydrogen bonding; T represents L-, D- or M-tartaric acid, $R = C_{12}H_{25}$. (b) Schematic representation of the columnar superstructures for $(LP_2 + LU_2)_n$ (left) and $(MP_2 + MU_2)_n$ (right); each spot represents a PU or UP base pair, with spots of the same type belonging to the same supramolecular strand; the aliphatic chains emerge from the cylinder more or less perpendicularly to its axis; for $(LP_2 + LU_2)_n$, a single helical strand and the full triple helix are represented at the bottom and the top, respectively, of the column; for $(MP_2 + MU_2)_n$, the representation corresponds to the column cut parallel to its axis and flattened out. (Permission being sought from [13].)

3.2.2 Comb-like or side chain architecture

In one of the first published reports of comb-like PLCs incorporating non-covalent bonds, Kato and Fréchet [14] showed that it is possible to stabilize an already existing liquid crystalline phase through such interactions. The polymer ((4a) in Figure 3.8), a polyacrylate with pentamethylene side chains terminated with a 4-oxybenzoic acid unit itself displays a nematic phase; this is attributed to the acid group preferring to exist as hydrogen-bonded dimers, thereby creating a core of sufficient length and rigidity to allow formation of the mesophase. The complementary small molecule ((6a) in Figure 3.8), a stilbazole derivative with a methoxy tail, also possesses a nematic mesophase. When the two components are mixed, the temperature range of the nematic phase is greatly extended at both its lower and upper boundaries, but especially the upper, as shown in Figure 3.9. The extent of stabilization depends on the molar ratio of the dissimilar components, and is maximized when they are equimolar. This mesophase enhancement is attributed to the formation of an extended rigid core (as shown in Figure 3.9) composed of the 4′-pyridyl-terminated stilbazole unit hydrogen-bonded to the benzoic acid unit, an attribution which is supported by infrared spectroscopy. In other words, the single, directed

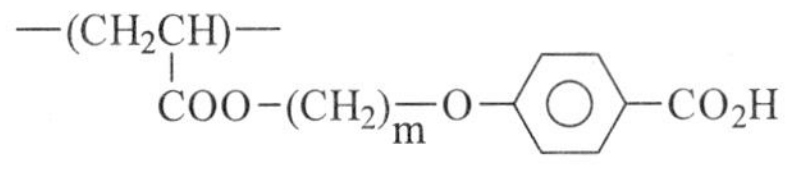

$-(CH_2CH)-$ with side group $COO-(CH_2)_m-O-C_6H_4-CO_2H$ — m = 5 **4a**; m = 6 **4b**

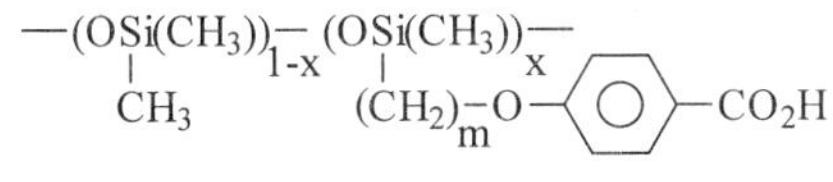

$-(OSi(CH_3))_{1-x}-(OSi(CH_3))_x-$ with side groups CH_3 and $(CH_2)_m-O-C_6H_4-CO_2H$ — x = 0.3, 0.4, 1.0; m = 5, 8, 10 **5**

$NC_5H_4-CH{=}CH-C_6H_4-O_2C-C_6H_4-OCH_3$ **6a**

$NC_5H_4-CH{=}CH-C_6H_4-O(CH_2)_{n-1}CH_3$ — n = 1-8, 10 **6b**

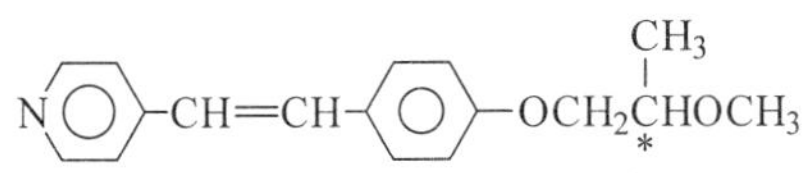

$NC_5H_4-CH{=}CH-C_6H_4-OCH_2C^*H(CH_3)OCH_3$ **6c**

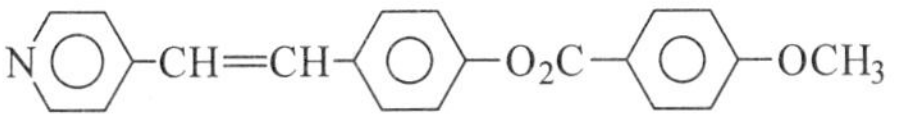

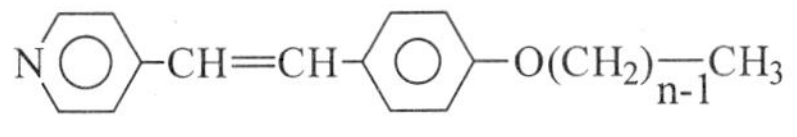

$NC_5H_4-C_5H_4N$ **6d**

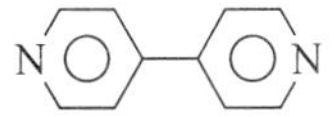

Figure 3.8 Molecular components used in the hydrogen-bonded comb-like PLCs of Kato and Fréchet and coworkers.

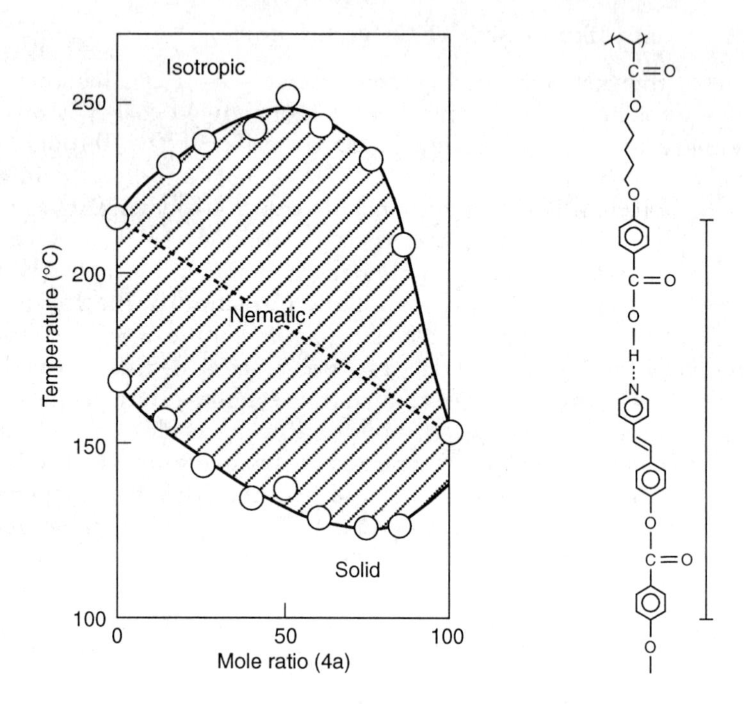

Figure 3.9 Phase diagram for the binary mixtures of polymer (4a) and stilbazole derivative (6a), and representation of the hydrogen-bonded complex with the extended rigid core indicated (permission being sought from [14]).

hydrogen bond connecting the two components increases the length-to-diameter ratio of the effective rigid core, thus thermally stabilizing the mesophase.

Since their first report, Kato, Fréchet and coworkers have investigated a number of variations on the theme [15–19], illustrating the versatility with which non-covalently assembled PLCs can be adapted for specific purposes. All cases involve essentially the same hydrogen bonding interaction between pyridyl and benzoic acid-terminated moieties, and this interaction always forms part of the resulting mesogenic core. First, the same polyacrylate derivative as above, but this time with a hexamethylene spacer (4b), was mixed in stoichiometric proportions with a series of *trans*-4-alkoxy-4′-stilbazoles having varying tail lengths ((6b) in Figure 3.8) [15]. No liquid crystalline behavior was observed for the pure stilbazoles with $n = 1–5$; however, those with $n = 6–8$ and 10 exhibit smectic mesophases. The 1/1 complexes with the polymer all display smectic A phases, the temperature range of the phase increasing with decreasing tail length (Figure 3.10), due principally to the decreasing solid–S_A transition temperature. The isotropization

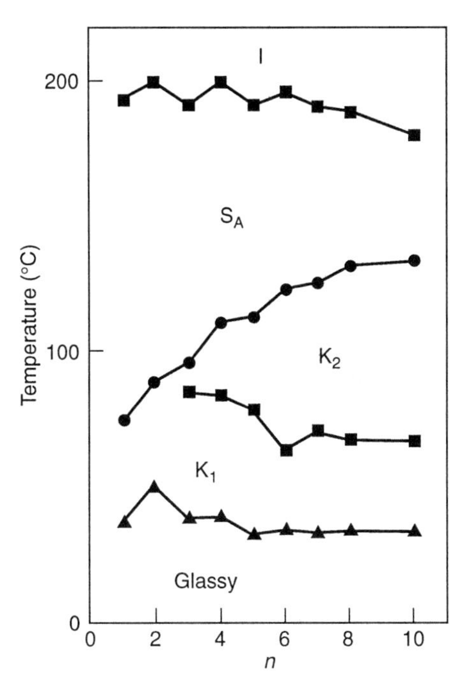

Figure 3.10 Transition temperatures as a function of alkyl chain length for the 1:1 complexes of polymer (4b) and stillbazole derivatives (6b) (permission being sought from [15]).

temperature varies little with tail length, although both it and the melting temperature show an odd–even effect. The enthalpies of isotropization are high, probably reflecting the breaking of hydrogen bonds during this transition. The solid phase below the S_A mesophase for $n \geqslant 3$ was identified as a smectic B phase in a later study [19]. A glass transition was observed near 35–40°C. Homogeneous copolymer-like complexes were obtained by mixing the polymer with binary mixtures of stilbazoles, such as shown in Figure 3.11a for polymer (4b) + stilbazoles (6b) with $n = 1$ and $n = 6$. Furthermore, it was shown that blends of a polymeric acid–stilbazole complex appear miscible with its homologous monomeric acid–stilbazole complex (Figure 3.11b).

Polysiloxane-based complexes were also prepared by Fréchet and coworkers [16]. The siloxane polymers ((5) in Figure 3.8) themselves, containing variable levels of substitution (0.3, 0.4 and 1.0 molar fraction) by 4-alkoxybenzoic acid side chains of variable lengths (5, 8 and 10 methylene units), generally possess a lamellar mesophase, identified as smectic C. Equimolar complexes of these polymers with stilbazole (6a) (which, as noted above, possesses a nematic phase) also give rise to a smectic C phase, with the clearing temperatures being higher than

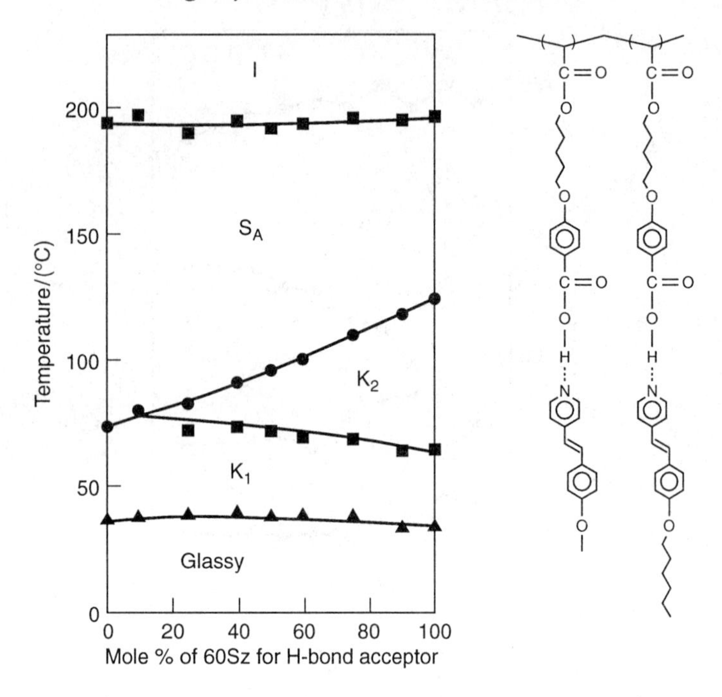

Figure 3.11 (a) Binary phase diagram of 1:1 (acid:pyridyl) copolymer-like complexes of polymer (4b) with varying proportions of 2 stilbazole derivatives (6b) ($n = 1$ and $n = 6$), as shown in the schematic representation; the abscissa represents the variation in terms of mole percent (6b) with $n = 6$ (permission being sought from [15]).

those of the individual components. Lower side chain substitution levels for a given spacer length tend to produce lower smectic–isotropic transition temperatures (probably reflecting a dilution effect) but higher crystalline–smectic transition temperatures. An increase in spacer length, for a given substitution level, tends to increase the transition temperatures, especially for the crystalline–smectic transition. It was noted that the transition temperatures are dependent on thermal history, and they tend to be broad (this broadness was attributed to the polydispersity of the polymers). Glass transition temperatures were observed between 7 and 25°C, in some cases higher and in some cases lower than those of the polymer alone (−20 to 60°C, where detectable).

Stilbazole (6b) with $n = 2$, which has a shorter axial length than (6a) and possesses no mesophase; it also displays a smectic mesophase (A or C) when complexed with the siloxane copolymers, although with lower clearing temperatures than the complexes with stilbazole (6a) (and with no dependence on thermal history) but higher than the

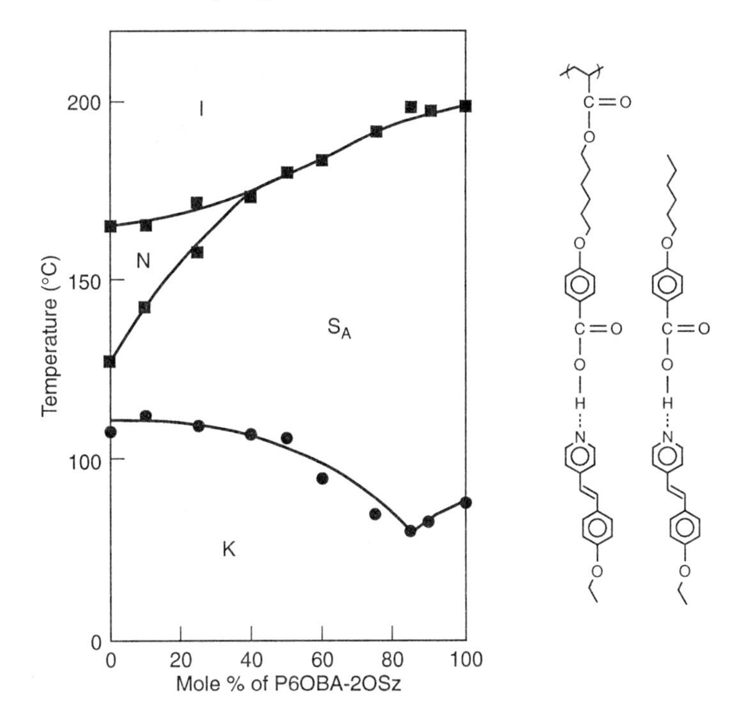

Figure 3.11 (b) Binary phase diagram of mixtures of the equimolar polymeric complex of (4b) + (6b) ($n = 2$) and the equimolar monomeric analog, as shown in the schematic representation; the abscissa is in terms of the mole percent polymeric complex (permission being sought from [15]).

individual components [16]. It is notable that a mesophase was observed as well for the complex with the single polymer which is not liquid crystalline; in other words, a mesophase was generated in this case from two constituents which individually possess none. No glass transition was observed for these complexes. It was also indicated [17] that the temperature range of mesophases can be adjusted with a polar lateral substituent located in the small molecule component and that it is not necessary for the complexes to be coplanar in order for extensive hydrogen bonding to take place and liquid crystalline behavior to be observed.

In still another variation, a chiral or optically active stilbazole (6c), which by itself displays no mesophase, was complexed with the polysiloxane-based polymers [18]. Again, mesomorphicity was obtained, with the usual effects on the transition temperatures (which are, once more, broad and dependent on thermal history). No glass transitions were detected. X-ray data indicate that the mesophase is smectic C,

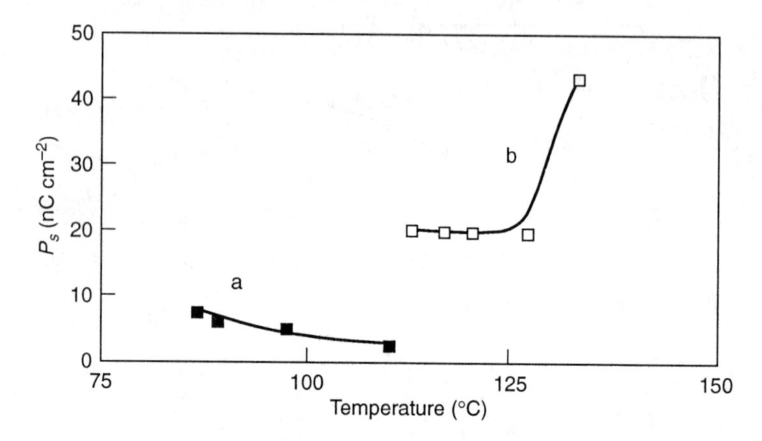

Figure 3.12 Temperature dependence of the spontaneous polarization, P_s, for equimolar complexes of (a) polymer (5) ($x = 0.4$, $m = 8$) and (b) polymer (5) ($x = 0.4$, $m = 5$) with chiral stilbazole (6c) (permission being sought from [18]).

possibly accompanied by a higher temperature smectic A phase (not observed by differential scanning calorimetry, DSC). Figure 3.12 illustrates that spontaneous polarization (P_s) of these complexes is possible, one complex showing the usual decrease in P_s with increase in temperature, another showing an unusual increase in P_s with temperature (thought to be related to the shorter spacer length which hinders rotation of the hydrogen-bonded mesogen about its molecular axis [18]). This demonstrates that hydrogen-bonded complexes are potentially useful for the preparation of ferroelectric liquid crystals.

In addition, it was shown that liquid crystalline network-like structures, in which the polymer side chain is non-covalently cross-linked, are possible [19]. A bifunctional mesogenic molecule, 4,4′-bipyridine ((6d) of Figure 3.8), mixed with an acid polymer (polyacrylate (4b)) was shown to exhibit smectic A mesomorphicity up to 205°C (accompanied by a large enthalpy change); it is frozen into the glassy state at about 85°C. Ternary mixtures of the bipyridine and one of the monofunctional stilbazoles ((6b) with $n = 6$), blended with the polymer such that an equimolar pyridyl/acid ratio is maintained, are miscible over the entire composition range of bi- vs. monofunctional components. These blends display a continuous change in the phase transition curves, with only the isotropization temperatures remaining approximately constant near 200°C (Figure 3.13). The crosslinking effect of the bipyridyl component is expressed by the increase in the glass transition temperature that is observed as the proportion of this component in the mixture is increased.

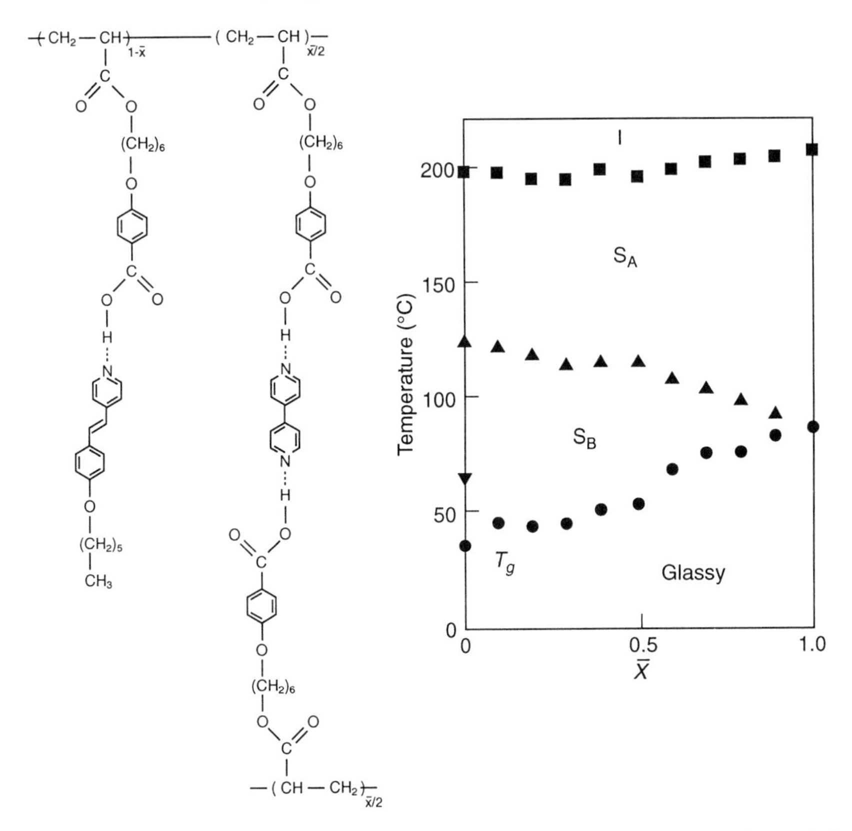

Figure 3.13 Schematic representation and phase diagram of 1:1 (acid/pyridyl) complexes of polymer (4b) with varying proportions of mono- and bifunctional monomers (6b) ($n = 6$) + (6d) (permission being sought from [19]).

In a new twist on the theme, double hydrogen bonds were utilized for the construction of the structure shown in Figure 3.14 [20]. The pyridine components by themselves are non-mesogenic. A monotropic mesophase is obtained in the equimolar complexes. The temperature range of the mesophase can be increased by preparing equimolar copolymeric complexes combining both pyridine components ($m = 3$ and $m = 6$) with the polymer.

In all of the non-covalently assembled comb-like PLCs described above, the spacer group is covalently part of the polymer constituent and the hydrogen bond is structurally a part of the (extended) mesogenic core. In a study by Brandys and Bazuin [21, 22], a mesogenic small molecule ((7a) in Figure 3.15) was functionalized by a carboxylic acid group at the extremity of an alkyl chain spacer, the rigid core being located at the other end of the spacer, and this was mixed to varying proportions with a simple complementary polymer, poly(4-vinyl

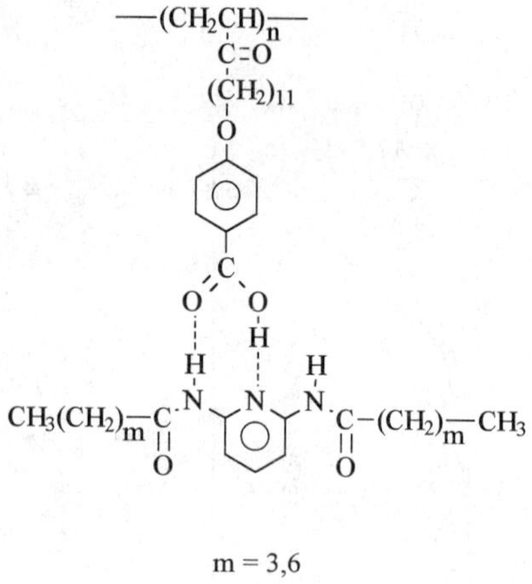

$-(CH_2CH)_n-$
C=O
$(CH_2)_{11}$
O
C
O O
H
H H
N N N
$CH_3(CH_2)_m-C$ $C-(CH_2)_m-CH_3$
O O

m = 3,6

Figure 3.14 Structure proposed for a doubly hydrogen-bonded comb-like PLC complex [20].

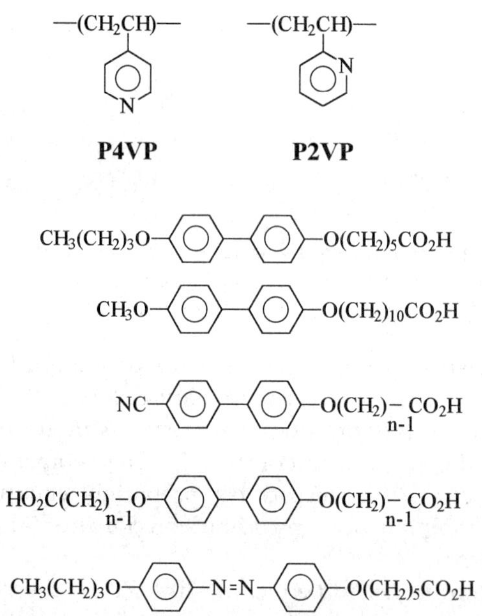

$-(CH_2CH)-$ N **P4VP**

$-(CH_2CH)-$ N **P2VP**

$CH_3(CH_2)_3O-$⟨⟩⟨⟩$-O(CH_2)_5CO_2H$ **7a**

CH_3O-⟨⟩⟨⟩$-O(CH_2)_{10}CO_2H$ **7b**

$NC-$⟨⟩⟨⟩$-O(CH_2)_{n-1}-CO_2H$ n = 6,11 **7c**

$HO_2C(CH_2)_{n-1}-O-$⟨⟩⟨⟩$-O(CH_2)_{n-1}-CO_2H$ n = 6,11 **8**

$CH_3(CH_2)_3O-$⟨⟩$-N=N-$⟨⟩$-O(CH_2)_5CO_2H$ **9**

Figure 3.15 Molecular components in the mixtures of poly(vinyl pyridine)s (P4VP and P2VP) and acid-functionalized (mono- and bi-) mesogens.

pyridine) (P4VP). In this case, although hydrogen bonding between the two constituents takes place, sufficient to strongly plasticize the polymer (depressing its T_g by some 70°C), it is limited to about 20%

of the pyridine groups (where maximum plasticization is achieved). Beyond this solubility limit, the excess small molecules coalesce, apparently in microdomains since no evidence of phase separation is observed through optical microscopy. It is noteworthy that the functionalized mesogen itself possesses a monotropic smectic A phase, which is suppressed in the mixtures for all of the compositions studied (excess pyridine up to equimolar ratios).

A partial phase diagram for the system is given in Figure 3.16. A glass transition is detectable up to a molar ratio (acid/pyridine) of 0.5. Below molar ratios of about 0.1–0.2, the system forms a miscible, amorphous blend. The ordered phase, detected by both DSC and X-ray techniques for molar ratios of about 0.2 and above, is thought to be smectic E (as in the small molecule alone) and is made up essentially of acid-dimerized small molecules. A possible picture representing the system at the higher molar ratios is given in Figure 3.17, where it is suggested that some (probably not all) of the hydrogen-bonded mesogens may at the same time participate in the solid phase. In the melt at these molar ratios, infrared and X-ray data showed that a significantly greater amount of acid–pyridine hydrogen bonding occurs (this may well be dynamic in nature) and it can be frozen in by quenching. When quenched, the molecular organization within layers still resembles that of the neat mesogen, except that the lamellar spacing now corresponds to a single molecular length including a vinyl pyridine unit rather than to that of an acid dimer. The system reverts to a microphase-separated one when reheated to the (shearable) intermediate

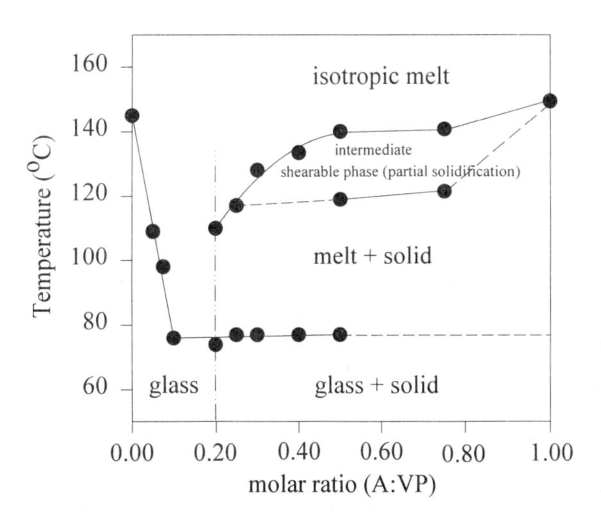

Figure 3.16 Partial phase diagram of the mixtures of P4VP and (7a), in terms of the molar ratio of acid to pyridine moieties.

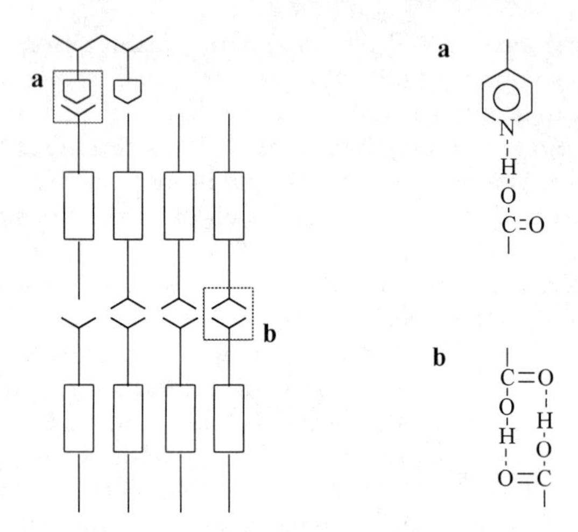

Figure 3.17 Schematic representation of an interfacial portion of an acid–dimer microdomain within which a polymer hydrogen-bonded mesogen is incorporated.

region indicated in Figure 3.16. This system, then, is an example where acid–pyridine interactions are not thermodynamically favored over acid dimerization (except at low small molecule content), as is the case for the Kato–Fréchet systems.

The picture represented by Figure 3.16 was confirmed for a similar small molecule (7b), with a longer spacer chain (10 methylenes) and shorter tail (methoxy), mixed with P4VP, where the tendency to crystallize in separate microdomains is, if anything, more pronounced [23]. Analogous behavior was observed for di-alkoxy biphenyl mesogens ((8) in Figure 3.15) functionalized at both extremities by carboxylic acid and mixed with P4VP [24]; these mixtures potentially resemble (dynamically) crosslinked systems. Again, they are amorphous at low acid/pyridine molar ratios (less than 0.25 for $n = 6$ and 0.1 for $n = 11$, noting that these values must be reduced by half in terms of mesogen/pyridine molar ratios). Glass transition temperatures for these systems were not determined. Quantitative infrared analysis of the system with $n = 11$ indicates that acid–pyridine interactions are significantly less than 100% of that theoretically possible for molar ratios above about 0.25; indeed, the number of interactions increases very little above 0.25, and are not more than 30% of the theoretical maximum at equimolar acid/pyridine ratios.

Similar behavior was observed for another acid-functionalized liquid crystal molecule ((9) in Figure 3.15 and possessing a monotropic nematic phase) blended with P4VP and also P2VP [25]. It was pointed out that partial complexation results in what can be considered as a supramolecular copolymer that acts as an isotropic solute for the acid dimers, and is thereby responsible for the reduction in transition temperatures observed for the latter. The similarity observed between the P4VP and P2VP systems was attributed to the random coil conformation of the polymers. In contrast, when the same acid-functionalized mesogen is mixed with polystyrene, the system is almost completely immiscible and gives optical textures like those observed for polymer-dispersed liquid crystals.

Preliminary data for mixtures of cyano-terminated biphenyl mesogens (7c) with P4VP (studied at 0.4 acid/pyridine molar ratio) indicate that more acid–pyridine hydrogen bonding occurs in these systems than in the systems with alkoxy-terminated mesogens (F.A. Brandys and C.G. Bazuin, unpublished). Indeed, a disordered phase is obtained at about 90°C on cooling from the isotropic melt for the mixture involving the mesogen with the longer spacer (the mesogens themselves possess a short-range nematic phase, monotropic for $n = 6$ and enantiotropic for $n = 11$); this phase appears to be stable at ambient temperature.

One example has been published of mixtures of functional vinyl polymers and rigid aromatic derivatives (shown in Figure 3.18), characterized by complementary hydrogen bonding interactions, where there is no flexible spacer (or only a very short one) present. These mixtures were reported to generate liquid crystalline behavior, whereas none of the components are mesogenic [26]. The mesophase, whose temperature range increases with increasing concentration of small molecule although the isotropization temperature remains approximately constant, was claimed to be nematic.

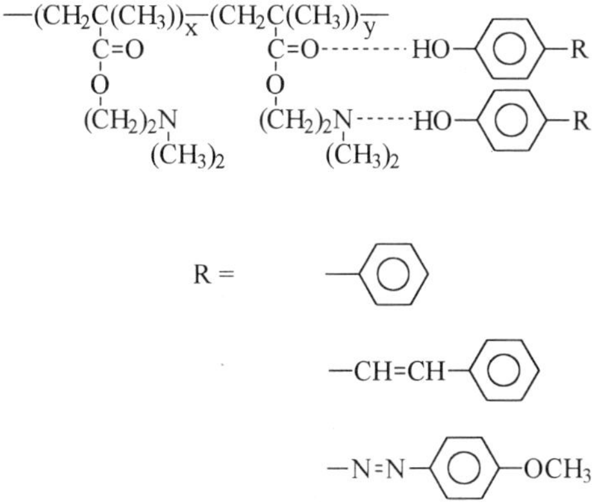

—(CH₂C(CH₃))ₓ—(CH₂C(CH₃))ᵧ— with side groups C=O–O–(CH₂)₂N(CH₃)₂; on the y units, C=O ······ HO–C₆H₄–R and (CH₂)₂N ······ HO–C₆H₄–R

R = –C₆H₅; –CH=CH–C₆H₅; –N=N–C₆H₄–OCH₃

Figure 3.18 Representation of a hydrogen-bonded PLC with no (or a very short) flexible spacer [26].

3.3 IONICALLY BONDED PLC COMPLEXES

The use of ionic bonds to couple dissimilar components has been shown to be a viable alternative to hydrogen-bonded assemblies in constructing PLCs. The ionic bond concept has been applied almost exclusively to comb-like PLCs to date. In general, the architecture of the examples published resembles that of the (potentially) hydrogen-bonded system in Figure 3.15, except that the hydrogen bond is replaced by an ion pair; that is, the flexible spacer is part of the small molecule constituent.

This family of noncovalently bonded PLCs may also include polyelectrolyte + surfactant complexes which, as will be seen, can also give rise to liquid crystalline mesophases. In these complexes there is only a flexible alkyl chain attached to the ionic head group in the small molecule constituent, with no rigid aromatic core present. Since surfactants themselves are frequently thermotropic liquid crystals, it is not surprising that their complexes with polyelectrolytes may produce PLCs, in both cases driven by the incompatibility between the ionic and aliphatic parts leading to amphitropic systems [27].

Since the ionic units play a role similar to that of rigid mesogens, and since the ionic interactions in most polyelectrolyte + surfactant systems take place very near the polymer backbone, the resulting architecture can be considered to resemble that of inverse comb PLCs [5]. From this point of view the question may be posed, for the complexes where there is also a rigid core, whether the rigid core and the ionic groups are competitive or synergistic in determining the mesomorphic behavior or if one simply dominates the other. Another aspect that must be kept in mind is related to the well known fact that ionic groups attract H_2O molecules and that when bound they are not easy to remove; if present in the complexes, they can be expected to have some effect on the behavior observed.

The first published report, by Ujiie and Iimura [28], of an ionically bonded PLC possessing a rigid core involves complexes of poly(vinyl sulfonate) (PVSA) and an ammonium-capped mesogenic molecule ((10a) in Figure 3.19), as shown in Figure 3.20. Individually, both constituents are ionic and, presumably but not necessarily, their counterions were removed as 'micro-ions' in the precipitation step of the complex preparation. The ionic small molecule (as well as a non-ionic precursor, tertiary amine-capped) is itself liquid crystalline, displaying a smectic A phase over a 130 or 150°C range (40–170°C for quaternization by iodoethane and 57–207°C for quaternization by HBr) compared to a 60°C range (61–120°C) for the non-ionic precursor. The complex was constructed using the I^--neutralized molecule, and it displays a smectic A phase over a 185°C range (50–235°C), frozen in below the glass transition at 50°C. Because the layer spacing obtained from X-ray data

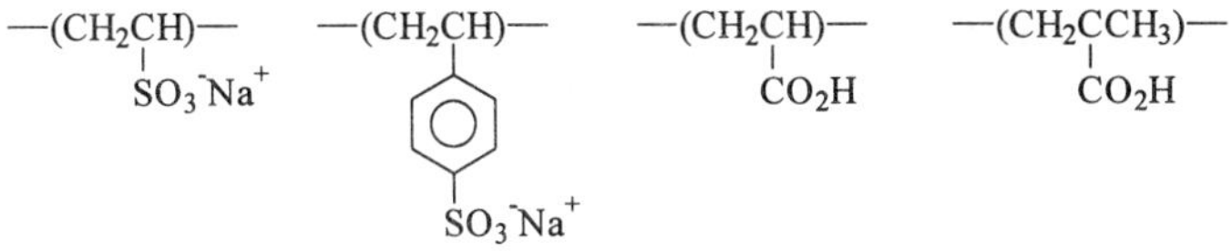

—(CH_2CH)— with $SO_3^-Na^+$ (PVSA)

—(CH_2CH)— with phenyl–$SO_3^-Na^+$ (PSSA)

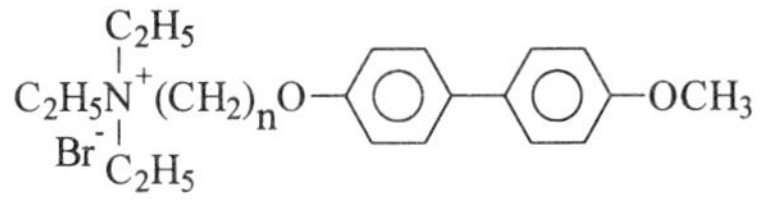

—(CH_2CH)— with CO_2H (PAA)

—(CH_2CCH_3)— with CO_2H (PMAA)

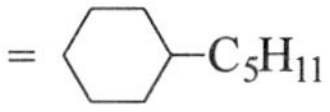

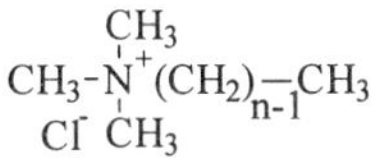

PVSA **PSSA** **PAA** **PMAA**

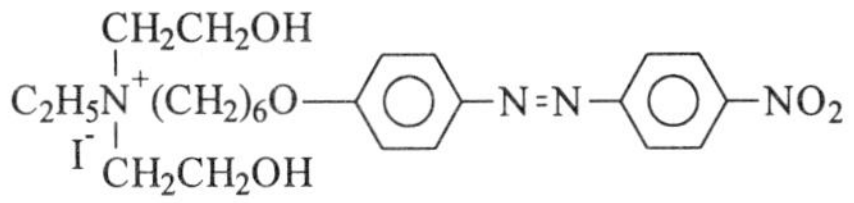

$C_2H_5N^+(CH_2CH_2OH)_2(CH_2)_6O$–Ph–N=N–Ph–$NO_2$, I^- **10a**

$C_2H_5N^+(C_2H_5)_2(CH_2)_nO$–Ph–Ph–$OCH_3$, Br^- $n = 10,12$ **10b**

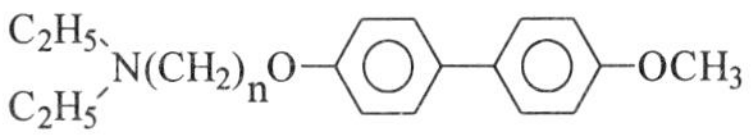

$(C_2H_5)_2N(CH_2)_nO$–Ph–Ph–OCH_3 $n = 6,8,10\text{-}12$ **11**

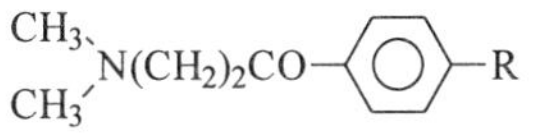

$(CH_3)_2N(CH_2)_2CO$–Ph–R $R = OC_{12}H_{25}$ = cyclohexyl–C_5H_{11} **12**

$CH_3\text{-}N^+(CH_3)_2(CH_2)_{n-1}CH_3$, Cl^- $n = 12,14,16,18$ **13**

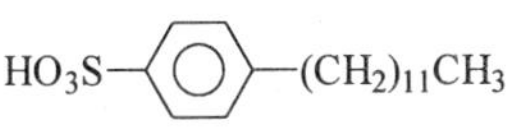

HO_3S–Ph–$(CH_2)_{11}CH_3$ **14**

Figure 3.19 Molecular components for ionic complexes of simple functional polymers and functionalized mesogens.

is 1.5 times greater than the molecular length for both the polymer complex and the ionic small molecule, an interdigitated partial bilayer structure (often referred to as smectic A_d or S_{Ad} in the liquid crystalline literature) was proposed (Figure 3.21). All of the ionic species were reported to form spontaneously a homeotropic texture, although annealing was required for the polymer complex.

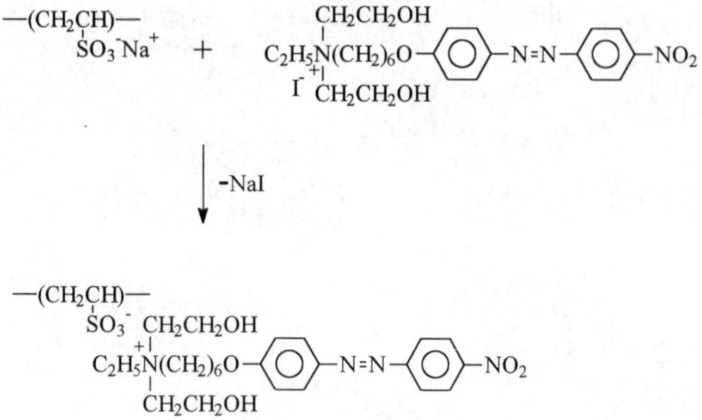

$$—(CH_2CH)— \; SO_3^- Na^+ \; + \; C_2H_5\overset{+}{N}(CH_2CH_2OH)_2(CH_2)_6O—C_6H_4—N{=}N—C_6H_4—NO_2 \; I^-$$

$$\downarrow -NaI$$

$$—(CH_2CH)— \; SO_3^- \; C_2H_5\overset{+}{N}(CH_2CH_2OH)_2(CH_2)_6O—C_6H_4—N{=}N—C_6H_4—NO_2$$

Figure 3.20 Representation of the preparation of the ionic complex PVSA + (10a) through an ion exchange reaction.

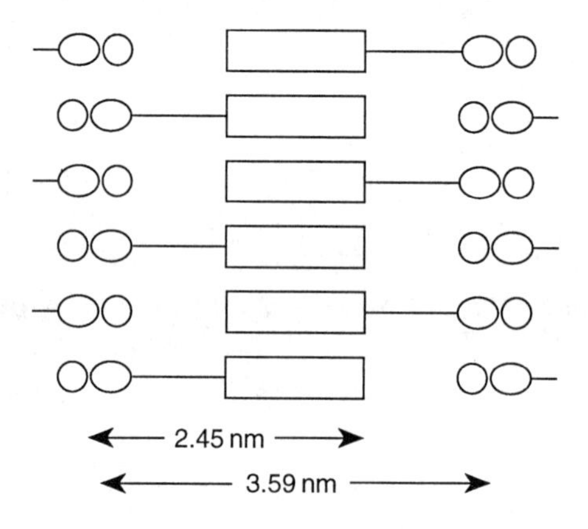

Figure 3.21 Proposed molecular packing for the smectic A phase of complex PVSA + (10a) (permission being sought from [28]).

Bazuin and Tork ([29, 30]; A. Tork and C.G. Bazuin, in preparation) have shown that it is possible to generate ionically bonded liquid crystalline PLCs from components neither of which are liquid crystalline. The most striking examples, of similar structure to that shown in Figure 3.20, are given by complexes of sulfonated polymers (PVSA and poly(styrene sulfonate) (PSSA)) with ammonium-functionalized mesogens ((10b) in Figure 3.19). Contrary to mesogen (10a), mesogens (10b) display a rather complex thermotropism with no evidence of disordered mesophases, whereas the equimolar complexes with PVSA and PSSA possess a monolayer smectic A phase over wide temperature ranges (Figure 3.22). The smectic A–isotropic transition is observed at high

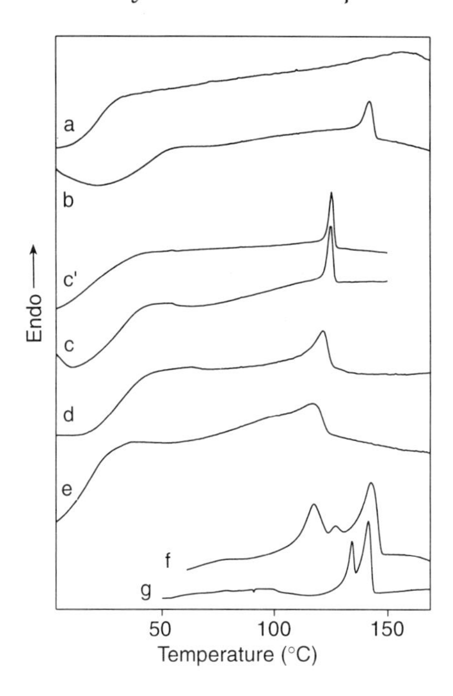

Figure 3.22 DSC thermograms [second heats; (a) $10°C\,min^{-1}$, (b)–(g) $5°C\,min^{-1}$] of rapidly cooled ($-200°C\,min^{-1}$ except (c′) $-5°C\,min^{-1}$) equimolar ionic complexes of (a) polydisperse PVSA + (10b) ($n = 12$); (b) monodisperse PSSA ($M_w = 20000$, $M_w/M_n = 1.2$) + (10b) ($n = 12$); (c) and (c′) monodisperse PSSA ($M_w = 20000$, $M_w/M_n = 1.2$ + (10b) ($n = 10$); (d) monodisperse PSSA ($M_w = 90000$, $M_w/M_n = 1.4$) + (10b) ($n = 10$); (e) polydisperse PSSA ($M_w \sim 70000$) + (310b) ($n = 10$); (f) neat (10b) ($n = 12$); and (g) neat (10b) ($n = 10$).

temperatures (typically with a rather high enthalpy, $3-5\,kJ\,mol^{-1}$, probably due to the ionic interactions); it appears to be significantly narrower when the polymers are of low polydispersity, although other factors still need to be ruled out. Slowly cooling into this phase can result in the distinct formation of bâtonnets followed by a focal conic texture as viewed through the polarizing optical microscope [29]. In some cases, significant narrowing of the broad wide-angle X-ray peak near $2\theta = 20°$, as ambient temperature is approached, may indicate a further transformation of the mesophase into a hexatic B mesophase; however, no accompanying transition can be detected by DSC [29]. A glass transition is clearly detected at lower temperatures (Figure 3.22). The disordered phase is thus frozen into the glassy state, and there is no appearance of a highly ordered or crystalline phase in the systems involving sulfonated polymers so far investigated.

When the sulfonate moiety is replaced by a carboxylate group, as in the complex shown in Figure 3.23, a disordered smectic mesophase is still obtained, but it crystallizes into or is superposed by an ordered phase before vitrification, rendering detection of a T_g much more difficult ([24, 29, 30]; A. Tork and C.G. Bazuin, in preparation). Again, the mesogens ((11) in Figure 3.19) by themselves, non-quaternized in this case and of variable spacer length, possess no disordered mesophase, exhibiting a single transition between an ordered phase (similar in nature to that of the complex, possibly smectic E) and the isotropic phase. As shown in Figure 3.24, the melt transition is relatively insensitive to the spacer length, and occurs at similar temperatures for the complexes and the neat mesogens. On the other hand, the clearing transition in the complexes generally increases with spacer length above $n = 8$, superimposed by an odd–even effect ($n = 10, 11, 12$). A minimum apparently occurs at $n = 8$. These features are frequently observed in classical MLCs. It is noteworthy that X-ray data indicate that the lamellar spacing in the disordered mesophase is between one and two molecular lengths, which again suggests an interdigitated S_{Ad} phase, although a tilted bilayer phase is not excluded.

The formation of the latter complexes can be followed by infrared spectroscopy, as it involves a proton transfer from the acid of poly(acrylic acid) to the tertiary amine of the mesogens. Complexation results in a significant reduction, or almost complete elimination, for the equimolar complexes, of the acid carbonyl band near 1700 cm^{-1} and the appearance of a new band near 1550 cm^{-1} attributed to the asymmetric carboxylate stretch. The residual carbonyl absorption that remains in the equimolar mixtures indicates the presence of a small

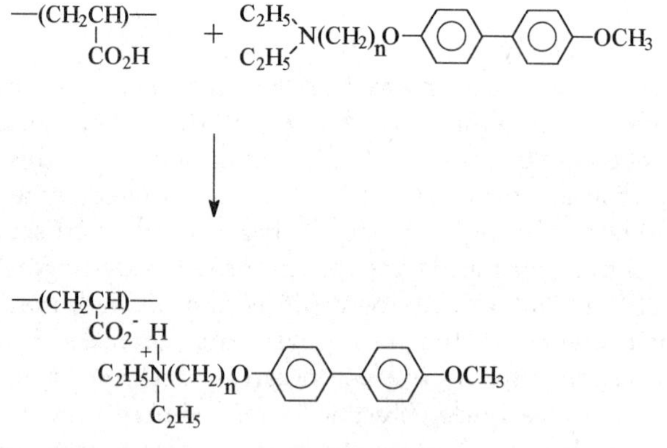

$$-(CH_2CH)- \text{ with } CO_2H \quad + \quad (C_2H_5)_2N(CH_2)_nO-C_6H_4-C_6H_4-OCH_3$$

$$\downarrow$$

$$-(CH_2CH)- \text{ with } CO_2^- \; H\overset{+}{N}(C_2H_5)_2(CH_2)_nO-C_6H_4-C_6H_4-OCH_3$$

Figure 3.23 Representation of the preparation of the ionic complex PAA + (12) through a proton transfer reaction.

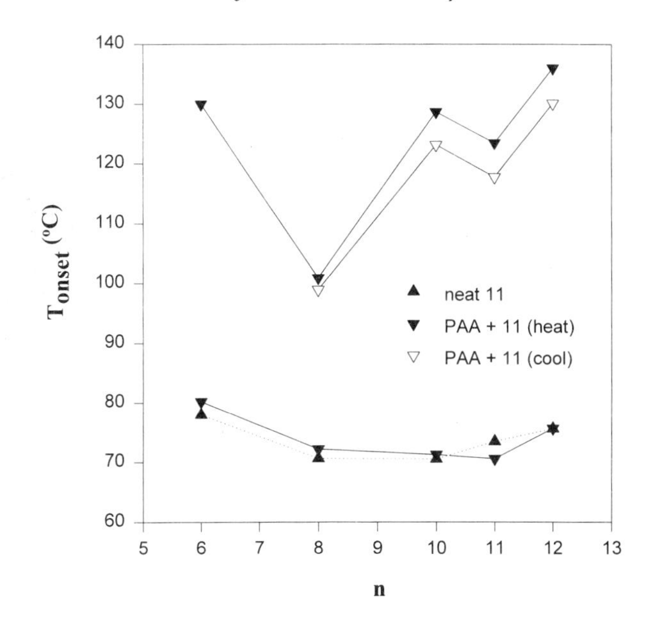

Figure 3.24 Transition temperatures as a function of spacer length for the equimolar complexes of PAA + (11).

proportion of uncomplexed mesogen. Indeed, variable amounts of residual carbonyl absorption are obtained following different preparations, but apparently with minor effects on the two transition temperatures ($\Delta T \leqslant 5°C$) and their enthalpies ([30]; A. Tork and C.G. Bazuin, in preparation). This raises the question as to the role of the uncomplexed mesogen. Does it form a separate microphase, or is it solubilized in the complex on a molecular level? The extent of complexation may also be affected by thermal history, since it may be temperature dependent and therefore can be expected to be subject to kinetic effects.

The first question is partially addressed in a communication by Tal'roze *et al.* [31] concerning complexes of tertiary amine-functionalized molecules possessing a short spacer ((12) in Figure 3.19) with PAA and with poly(methacrylic acid) (PMAA). It should be pointed out that these authors interpret the interactions as hydrogen bond complexes with strong proton polarization (partial proton transfer). They report obtaining a characteristic complex composition of two acids and one amine, following precipitation from the reaction mixture. It was concluded from X-ray data that these 2:1 complexes have crystallinities of about 2–3%, coexisting with a disordered smectic phase. The latter persists beyond the melting point of the crystalline regions (near 50°C) to about 150°C. A T_g is detected near 90°C. The lamellar spacing, again indicative of an interdigitated S_{Ad} phase, increases significantly with temperature above the T_g, but remains approximately constant below. The lamellar spacing is reduced by about 15% when the amine content is reduced to 10–20 mol% and, for a PAA complex, a second T_g is observed at temperatures corresponding to the T_g of pure PAA. These data suggest that the 'diluted' complexes are composed of microphases (undetectable by optical microscopy) of 2:1 complexes and pure PAA [28]. This renders credible, in turn, the possibility that 'concentrated' complexes are composed of microphases of characteristic complexes and excess uncomplexed mesogen. However, this remains to be determined, especially in view of the variable degrees of complexation indicated by infrared analysis of (11) + PAA mixtures, as mentioned above ([30]; A. Tork and C.G. Bazuin, in preparation).

The contrasting behavior between ionic systems involving sulfonate and carboxylate moieties has been confirmed in another comparative study [32], where ammonium-functionalized mesogens ((10a) and a variant with an ethyl carboxylate tail) are coupled to ω- and α,ω-sulfonato and carboxylato polystyrene oligomers of narrow polydispersities (also called halato(semi)telechelic polymers, or H(S)TPs). The data are summarized in Table 3.1. For the carboxylate-containing complexes, a smectic A phase is obtained for oligomer molar masses (M_n) per ionic group of less than about 8000 g mol^{-1} (above this value the complex is amorphous over the entire temperature range). The temperature of the S_A phase, where it is present, appears to be independent of molar mass. Below this range a solid phase is present. A glass transition is detectable between the melting and clearing transitions for molar masses per ionic group of above about 6000 g mol^{-1}, indicating the simultaneous

Table 3.1 Thermal behavior of ω and α,ω-sulfonato (PxSn) and -carboxylato (PxCn) halato(semi)telechelic polymers complexed with (10a) ($n = 1$ indicates ω-functionalized and $n = 2$ α,ω-functionalized polymers) [32]

Sample	M_n (g mol^{-1})	Phase transition temperatures (°C)
P1S1	700	G 50 S_A 155 I
P2S1	1000	G 74
P3S1	5500	G 85
P4S2	5500	G 85
P5C2	4700	G 50 S_A 155 I
P6C2	16 000	G + S 52 G + S_A 87 S_A 150 I
P7C1	2 000	G 50 S_A 157 I
P8C1	5 200	G 50 S_A 155 I
P9C1	6 500	G + S 50 G + S_A 85 S_A 153 I
P10C1	10 000	G 90

existence of a polymer phase and a mesogenic phase. Such microphase separation was supported by small angle neutron scattering data. In contrast, the ammonium sulfonate complexes give rise to a smectic A mesophase (in the same temperature range as for the ammonium carboxylate complexes) for the oligomer of the lowest molar mass only. At higher molar masses, amorphous materials are obtained (dispersed submicrometer size, point-like birefringent domains can, however, be observed in these materials below T_g). On the other hand, if excess mesogen is added, the amorphous materials are transformed into liquid crystalline materials, with the temperature range increasing somewhat with the amount of excess mesogen. Based on crystal close packing considerations, it was suggested that the greater difficulty of the sulfonate complexes to form ordered systems could be attributed, in part, to the greater bulkiness of the sulfonate moieties that prevents the complexed mesogens from packing as densely as in the carboxylate systems. The different interaction forces may also strongly modify the mobility of the mesogens, which may affect their ability to form mesophases [32].

Another series of ionic complexes currently under investigation, involving linear poly(ethylene imine) (PEI) and carboxylic acid-functionalized mesogens ([33]), is shown in Figure 3.25. According to infrared analysis, almost complete proton transfer occurs, resulting in protonation of the imine and carboxylation of the acid. One case (n = 6, R = $O(CH_2)_3CH_3$) parallels the hydrogen-bonded P4VP + mesogen (7a) system. Significantly, whereas the hydrogen-bonded complex is obtained only by quenching from the melt, the ionically bonded complex is obtained spontaneously. The molecular organization within smectic layers in both cases is similar, and resembles that of the neat mesogen except for the lamellar spacing, the latter being consistent with that of a single or double molecular length (depending on the thermal history and the composition) that includes the imine counterion. Preliminary DSC and X-ray data for PEI complexes with cyano-terminated mesogens, which in neat form possess a nematic phase over a narrow temperature range, indicate the existence of a smectic C phase (with suppression of the nematic phase) and, in one case, possibly a columnar phase at low temperatures.

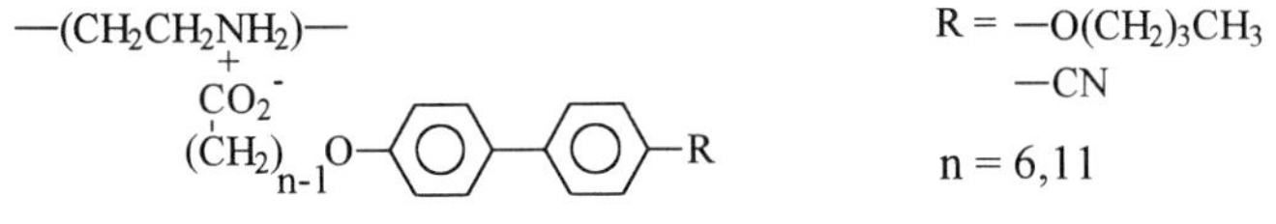

$$-(CH_2CH_2\overset{+}{N}H_2)- \quad CO_2^- \quad (CH_2)_{n-1}O-C_6H_4-C_6H_4-R$$

R = $-O(CH_2)_3CH_3$
$-CN$

n = 6,11

Figure 3.25 Representation of ionic complexes prepared from poly(ethylene imine) and carboxylic acid-functionalized mesogens.

Complexes of polyelectrolytes with oppositely charged surfactants in (aqueous) solution have received considerable attention (most often at low concentrations of surfactant, to avoid precipitation), which we will not discuss, e.g. [34]. They have also been used to form mono- and multilayer structures [35,36]. Properties of such complexes in the solid state have only recently been investigated [35–40]. They are generally prepared from aqueous solutions where cooperative electrostatic binding and precipitation occurs. Once formed, the complexes can be water insoluble, yet dissolve in other solvents. For example, complexes of poly(*N*-ethyl-4-vinylpyridinium bromide) with sodium bis(2-ethylhexyl) sulfosuccinate) [41] or with sodium dodecyl sulfate [42] are soluble in aliphatic hydrocarbon or other low-polarity solvents.

Antonietti and coworkers recently investigated equimolar polyelectrolyte + surfactant complexes of poly(*n*-alkyltrimethylammonium styrene sulfonate) [37] and poly(dodecyltrimethylammonium acrylate) [38], both involving surfactant (13) in Figure 3.19. Both types of complexes, once formed, are water insoluble, but dissolve in polar, organic solvents [43] such as 2-butanol (which is also used in the purification procedure). In the former system [37], lamellar structures, which undergo phase modification (attributed to 'rippling' or frustration) with change in alkyl chain length ($n = 12, 14, 16, 18$), are obtained. All of the systems are characterized by a relatively broad X-ray diffraction peak near $2\theta = 20°$ and a sharp diffraction peak at small angles, although only the complex with $C = 18$ is birefringent. The mesomorphic phase is thermally stable, in that neither a T_g nor any other phase transition is observed by DSC. The films themselves, which are flexible and transparent, remain mechanically stable until decomposition temperatures are reached. For the polyacrylate complex [38], which is birefringent, it was shown that, because steric effects prevent a parallel alignment of the surfactant alkyl chains in lamellar structures, a cylindrical morphology is obtained, possibly superposed by periodic fluctuations in the cylinder thickness giving rise to a face-centered cubic symmetry (Figure 3.26). Physically, this complex is a highly elastic, deformable material which undergoes necking and hardening when subjected to extension.

Another solid state polyelectrolyte + surfactant system, studied by ten Brinke and coworkers [39], involves *p*-dodecylbenzenesulfonic acid (surfactant (14) of Figure 3.19) with P4VP and P2VP. Infrared spectroscopic analyses showed that protonation of the polymers is virtually complete. The 1:1 as well as 0.5:1 (acid/VP) complexes are birefringent up to at least 200°C, whereas the 0.1:1 complex is isotropic. Again, X-ray diffraction profiles are characterized by a sharp peak at small angles and a broad peak at wide angles, indicating a lamellar structure. The layer spacing obtained suggests that the alkyl chains complexed to PVP chains in neighboring layers are interdigitated; this

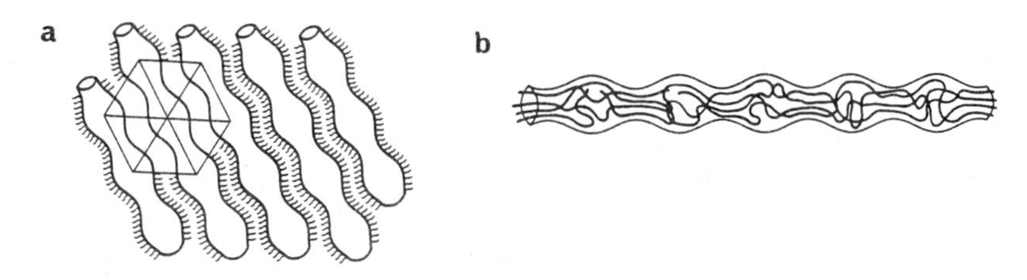

Figure 3.26 (a) Possible model for the structure of the complex PAA + (13) ($n = 12$), showing periodic fluctuations in the cylinder thickness: the alkyl chains lie perpendicular to the cylinder axis and the polymer chains are more or less aligned within the cylinders, as shown in (b) (permission being sought from [38]).

spacing decreases only a little when the surfactant content is decreased, less for the P2VP than for the P4VP complexes. It was estimated that the ordered domains encompass on the order of 5–10 layers only. Mesomorphic structures were also observed in complexes of surfactants with polypyrrole salts [44] and with synthetic polypeptides [40].

It is useful to compare the thermotropic behavior of noncovalently complexed PLCs with their analogous all-covalent PLCs. For the case of the ionically complexed comb-like PLCs, recently published series of particularly appropriate PLCs must be mentioned, one involving quaternized poly(4-vinylpyridine) [45–47], and another quaternized poly(ethylene imine) [48–50]. In both series, the quaternizing moiety is a flexible spacer (based usually on methylene, but also on ethylene oxide) ω-terminated by a mesogen. The result is a pair of ionic groups at or near the polymer backbone, just as in the ionically complexed systems described above. These systems are compared schematically in Figure 3.27.

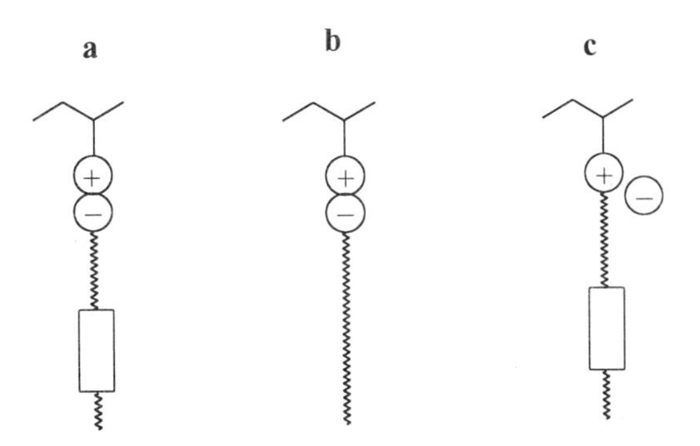

Figure 3.27 Schematic representation comparing (a) an ionic complex of a cationic polymer and an anionically functionalized mesogen, (b) an ionic complex of a cationic polymer and an anionic surfactant, and (c) a polymer to which a mesogen is covalently attached at a cationic site neutralized by a counterion.

The quaternized P4VPs ((15a) and (15b) in Figure 3.28) are characterized by monolayer smectic mesophases (A, B and E), which are stabilized, compared to their low molar mass analogs, by the connecting polymeric chains [45, 46]. It was shown that it is possible in these systems for both the ion pairs and the biphenyl groups to participate independently in the resulting structure such that ordering within the ionic sublayer is compatible with ordering within the biphenyl sublayer. Similar quaternized P4VPs, involving either cyano-terminated mesogens (which generally give rise to bilayer or partial-bilayer mesophases) or chiral mesogens (which normally produce tilted phases), are also characterized by monolayer phases, suggesting that the ionic interactions are dominant in determining the resulting molecular structure in these materials [47].

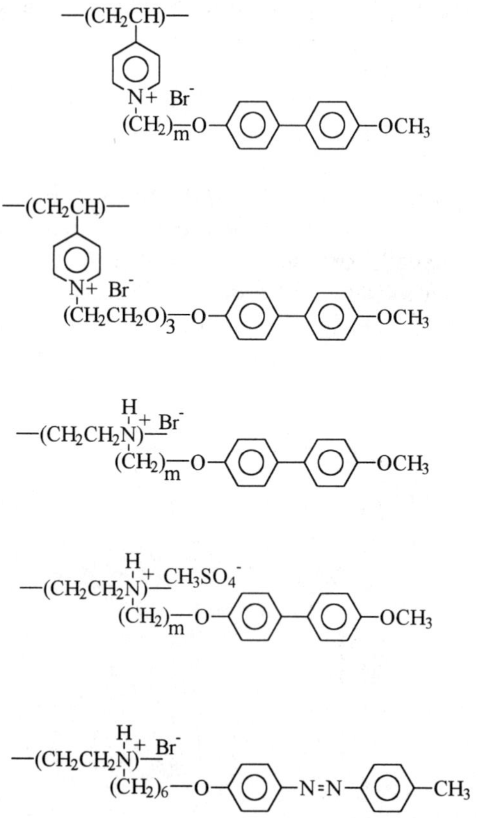

$-(CH_2CH)-$, N^+ Br^-, $(CH_2)_m-O-$, $-OCH_3$ m = 8,12 **15a**

$-(CH_2CH)-$, N^+ Br^-, $(CH_2CH_2O)_3-O-$, $-OCH_3$ **15b**

H, $-(CH_2CH_2N^+)-$ Br^-, $(CH_2)_m-O-$, $-OCH_3$ m = 6,12 **16a**

H, $-(CH_2CH_2N^+)-$ $CH_3SO_4^-$, $(CH_2)_m-O-$, $-OCH_3$ m = 6,12 **16b**

H, $-(CH_2CH_2N^+)-$ Br^-, $(CH_2)_6-O-$, $-N{=}N-$, $-CH_3$ **16c**

Figure 3.28 Molecular representations of quaternized mesogenic poly(4-vinylpyridine)s and poly(ethylene imine)s.

With a decreasing degree of quaternization (forming copolymers and corresponding to less than equimolar ratios of mesogen to polymer repeat unit in the ionic complexes), the smectic period increases somewhat. This was interpreted as a reflection of a thickening of the sublayers to accommodate free pyridine rings in addition to the quaternized parts of the backbone, possibly accompanied by tilting of the mesogenic groups [45, 46]. Mesomorphic order is lost for degrees of quaternization less than 20%, a mesogenic content similar to that observed in nonionic LC copolymers [51].

The mesomorphic behavior of the quaternized PEIs studied (both methylated and protonated), where quaternization (when 100%) occurs on every third backbone atom, generally resembles that of the quaternized P4VPs in that they possess smectic A phases [48–50]. Those with the biphenyl mesogen ((16a) and (16b) in Figure 3.28) possess monolayer smectic A phases preceded by a crystalline phase (in comparison, the corresponding alkylated PEIs, which are nonionic, display only a crystalline–isotropic transition), the crystalline phase considered to be favored by the relatively stiff backbone [49–50]. In the structural model proposed for the smectic A phase, the side chain mesogens of a given backbone segment are all located within the same smectic layer, but in a double row, ribbon-like arrangement. Partial quaternization again tends to lead to thicker smectic layers, due to exclusion of non-participating polymer segments, although competitive effects which result in thinner layers (observed in some cases) were also recognized [49,50]. In contrast to (16a) and (16b), the PEI quaternized by the azobenzene mesogen (16c) is reported to possess a bilayer smectic A phase preceded by a higher order, tilted bilayer smectic phase (with the corresponding alkylated PEI displaying a nematic phase) [48]. Liquid crystalline salts with a poly(methyl siloxane) backbone have also been reported [52].

In this section we have not considered mesomorphic ion-containing polymers such as polysoaps, although it can be argued that the counterions that are non-covalently bonded to the polymeric ion frequently play a role in the nature, range or even presence of the mesophases. As indicated in section 3.1, we prefer to restrict ourselves to PLCs composed of at least two molecular entities both of which contain significant hydrocarbon segments. The reader is referred to reference 53 for a recent review of polysoaps. Comb PLCs containing crown ether functions, whose mesomorphic behavior can be regulated by complexation with ions (sodium triflate, in particular), which is considered to be a molecular recognition process [54], should also be mentioned.

Some studies of longitudinal ion-containing PLCs involve complexed poly(ethylene oxide) backbone segments [55, 56] and ionic groups in the main chain [57–59]. The PLCs in references 57–59, in particular, will provide a good point of comparison for eventual longitudinal PLCs built from ionic bonds analogously to the hydrogen-bonded complexes in Figure 3.3.

This section has shown that the presence of ionic groups, whether in complexed systems or otherwise, tends to lead to a layered molecular organization, where an ionic sublayer alternates with a nonionic sublayer. Most frequently, a smectic A-like structure is involved. It remains to be seen to what extent other disordered mesophases can be generated in ionically complexed systems, and how various parameters affect these phases.

3.4 TRANSITION METAL OR COORDINATION PLC COMPLEXES

In principle, PLC complexes may also be prepared through the use of transition metal or coordination complexes. Like hydrogen bonds and ionic interactions, this has been a successful method, for example, for compatibilizing polymer blends [60]. Although no examples of complexes involving thermotropic mesogens have as yet been published, there are a couple of cases of complexes involving surfactants and P4VP. In the first, Belfiore *et al.* [61] showed that partial complexation occurs between zinc laurate and the pyridine of P4VP. However, (micro)phase separation takes place above about 30 mol% Zn laurate, giving rise to a system composed of an amorphous polymer-rich phase and a small molecule-rich phase, much like the hydrogen-bonded system, mesogen (7a) + P4VP, described above. In contrast, Ikkala and coworkers [62] have demonstrated that complexation of zinc dodecyl benzene sulfonate ($Zn(DBS)_2$) to P4VP (Figure 3.29) does lead to mesomorphic systems. Whereas $Zn(DBS)_2$ is birefringent up to 90°C, the $Zn(DBS)_2$/P4VP complex is birefringent to at least 200°C for Zn/VP molar ratios between 0.5 and 1.0 (at 0.25 the birefringence was reported to be faint). Infrared spectroscopic analysis was again helpful in determining that Zn coordination with the VP takes place. Disordered layer structures were detected by X-ray analysis. The

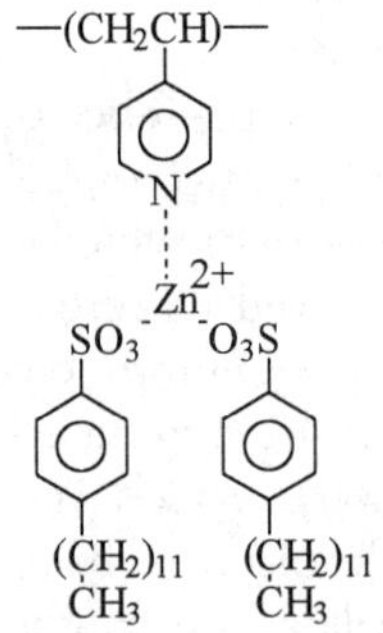

$$
\begin{array}{c}
-(CH_2CH)- \\
\text{pyridine (N)} \\
\vdots \\
Zn^{2+} \\
SO_3^- \quad {}^-O_3S \\
\text{C}_6\text{H}_4 \quad \text{C}_6\text{H}_4 \\
(CH_2)_{11} \quad (CH_2)_{11} \\
CH_3 \quad CH_3
\end{array}
$$

Figure 3.29 Schematic representation of the complexation between zinc dodecyl benzene sulfonate and poly(4-vinylpyridine).

layer thicknesses obtained are a little larger than that for neat $Zn(DBS)_2$, which is explained by the presence of the atactic polymer. This is illustrated in Figure 3.30. Obviously, coordination complexation is another promising avenue for assembling PLC materials.

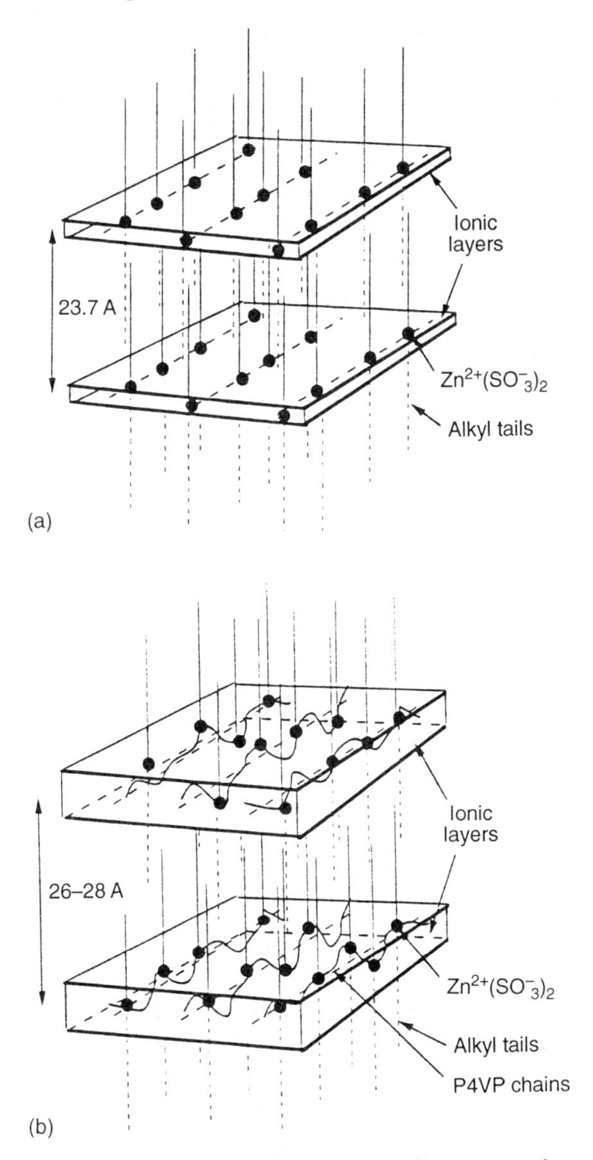

Figure 3.30 Comparison between the suggested mesomorphic structures of (a) neat $Zn(DBS)_2$ with a relatively thin ionic layer and (b) the $Zn(DBS)_2$ + P4VP complex with a relatively thick ionic layer; the black dots represent the ionic moieties $Zn^{2+}(SO_3^-)_2$ (permission being sought from [62]).

3.5 CHARGE TRANSFER AND OTHER DONOR–ACCEPTOR PLC COMPLEXES

Molecular recognition processes may also involve charge transfer or, more generally, donor–acceptor interactions. In covalently based PLCs, they have been used, for instance, to enhance mesomorphic stability and even induce mesophases, when copolymers containing separate electron donor and electron acceptor pendant chains are prepared [63–68] or when blends of the respective homopolymers are prepared [68–70].

Ringsdorf and coworkers [71] have shown that it is possible to induce liquid crystalline phases, namely discotic-columnar mesophases, by doping amorphous polymers containing disk-shaped electron donors, in either the side chain or the main chain, with a low molar mass electron acceptor, as shown in Figure 3.31. The resulting complexes can be considered as diskcomb PLCs and disk PLCs, respectively [5]. The electron-rich moiety is a triphenylene unit and the electron acceptors are fluorenone derivatives. When 20–25 mol% of 2, 4, 7-trinitrofluorenone (TNF) is added to the side chain polymethacrylate or polyacrylate

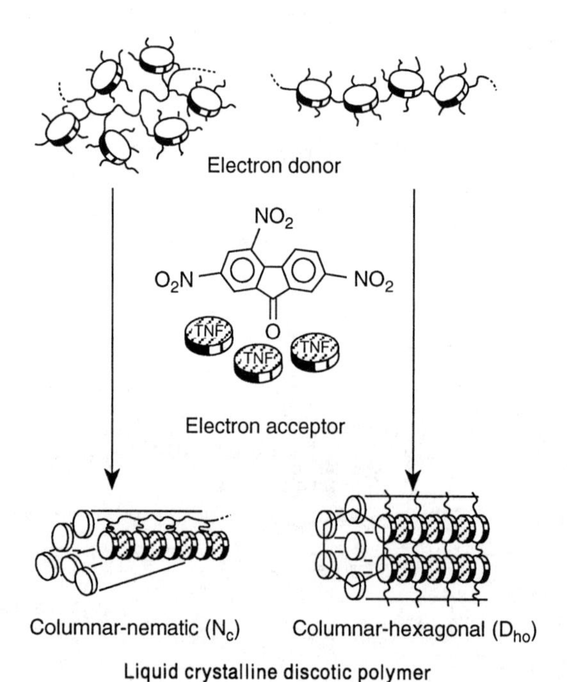

Figure 3.31 Schematic representation of the induced mesophases in amorphous discoid donor polymers via charge transfer interactions with low molar mass acceptors (permission being sought from [71]).

polymers, a nematic columnar phase is obtained (Figure 3.31). At lower concentrations of TNF, such as 10 mol%, the polymer remains amorphous; at higher concentrations, greater than about 30 mol%, phase separation occurs during preparation.

The undoped main chain polymers (polyesters) with shorter spacers (10 or 14 methylene units) possess hexagonal columnar phases. The polymer with a longer spacer (20 methylene units) is amorphous, but addition of 25 mol% or more of TNF induces an ordered hexagonal columnar mesophase (D_{ho}), as shown in Figure 3.31. A stronger electron acceptor, notably 2, 4, 7-trinitrofluoren-9-ylidene-malonodinitrile (TNF-CN), significantly increases the clearing temperature (from 83 to 165°C for a 1:2 ratio of acceptor to donor). It was shown, furthermore, that chirality can be induced (and also controlled) in the columnar structure through the use of chiral electron acceptors [72].

Complexes of polymeric acceptors and monomeric donors are also effective in inducing discotic PLCs. This was shown, in particular, for a main chain polyester into which TNF was incorporated, mixed with a triphenylene derivative [73]. In order to avoid phase separation, the donor concentration must exceed the acceptor concentration by a mole ratio of at least 3:1. It was also demonstrated that the alkyl spacer length in the main chain can be critical to the definition and stability of the discotic mesophase obtained. As illustrated in Figure 3.32, spacer lengths which are either too short or too long can inhibit the incorporation of the polymeric acceptor into the donor columns, the

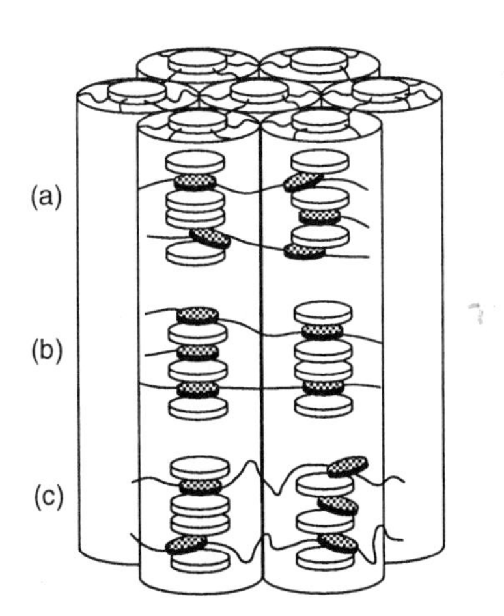

Figure 3.32 Schematic representation of the packing behavior of acceptor polymers having different spacer lengths in a discotic columnar structure with low molar mass donors: (a) spacer too short, (b) spacer of correct length, (c) spacer too long (permission being sought from [73]).

first because of insufficient play to allow such incorporation in neighboring columns, the second because of insufficient free volume between the columns (where the donor side chains are located) to accommodate the excess spacer length.

It was further demonstrated that it is possible to combine the effects of mesophase induction and compatibilization of immiscible polymers through charge transfer interactions [71]. The system for which this was illustrated is shown in Figure 3.33. The blend exhibits a nematic columnar phase which is not present in either polymer individually. This kind of blend is actually comparable to those composed of two homopolymers, one of which possesses electron acceptor groups and

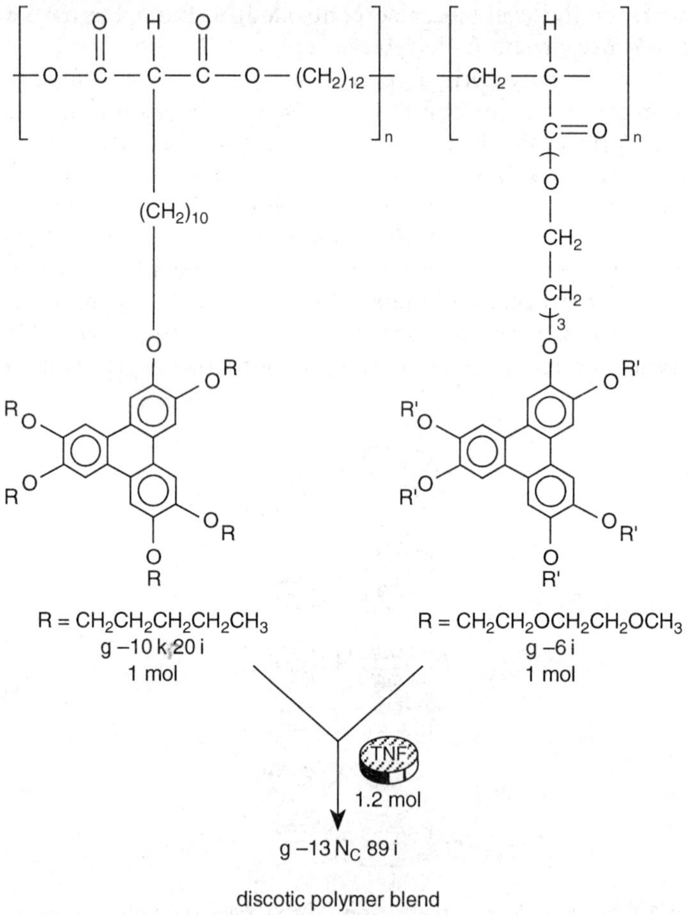

$R = CH_2CH_2CH_2CH_2CH_3$
g −10 k 20 i
1 mol

$R = CH_2CH_2OCH_2CH_2OCH_3$
g −6 i
1 mol

TNF
1.2 mol

g −13 N_C 89 i

discotic polymer blend

Figure 3.33 Combined induction of discotic mesophases and compatibilization of immiscible donor polymers through charge transfer interactions with low-molar-mass acceptors (permission being sought from [71]).

the other electron donor groups, except that no low molar mass molecule is involved in the latter. For this reason (as specified in section 3.1) we have chosen not to discuss these blends. The reader is referred to references 68–70 for details.

Another type of polymer + monomer donor–acceptor complex, involving rod-like mesogens in the side chain, has been studied by Kosaka and Uryu and coworkers [68, 70, 74]. Typical compounds are shown in Figure 3.34, where the polymers contain electron donor moieties in the side chain and the monomers contain electron acceptor groups. Mixtures of (17b) + (19) [70] and (18) + (19) [68] were investigated for various compositions and give rise to phase diagrams of the type shown in Figure 3.35 for (17b) + (19). The polymer T_g strongly decreases with the addition of the low molar mass component, and the nematic phase of the polymer component (smectic phase for (18)) transforms into the crystalline phase of the monomer component when the latter is present in high proportions. For complex (18) + (19), a nematic phase, not present in either of the pure components, is induced for a range of compositions.

Complexes of ((17a) or (17b)) + (20), where the small molecule is ionic in character as well as being an electron acceptor, were also investigated [74]. It was found that complexes with Cl^- as a counterion lead to phase separation, whereas those with Br^- and I^- usually give miscible systems. Significantly, most of the miscible complexes are

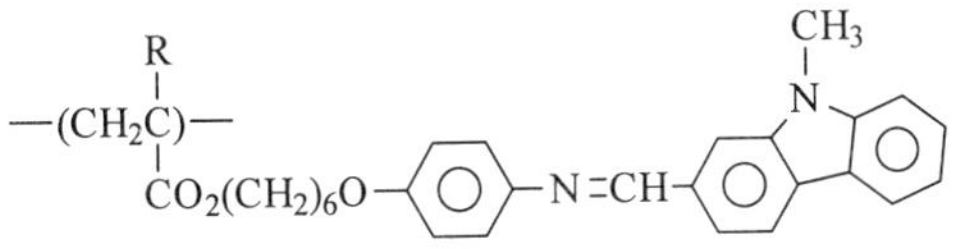

R
—(CH_2C)—
$CO_2(CH_2)_6O$–⟨◯⟩–N=CH–(N-methylcarbazolyl, CH_3 on N)

R = H **17a**
= CH_3 **17b**

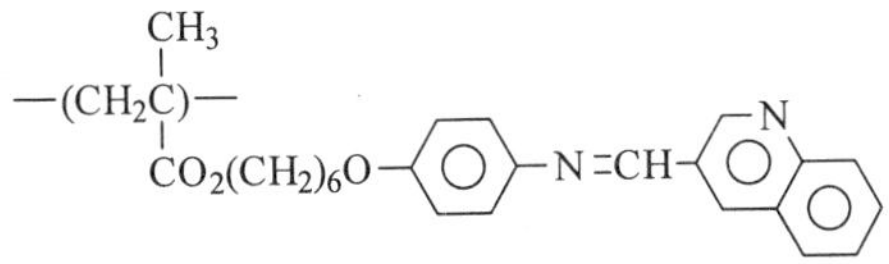

CH_3
—(CH_2C)—
$CO_2(CH_2)_6O$–⟨◯⟩–N=CH–(quinolinyl, N)

18

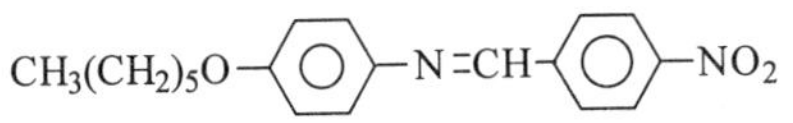

$CH_3(CH_2)_5O$–⟨◯⟩–N=CH–⟨◯⟩–NO_2 **19**

$CH_3(CH_2)_{n-1}$–$\overset{+}{N}$ (X^-)⟨◯⟩–CH=CH–⟨◯⟩–NO_2

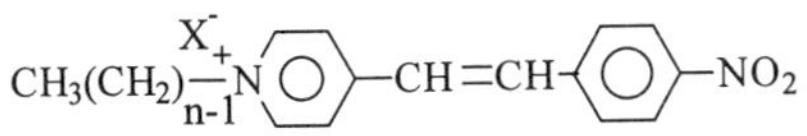

X = Cl, Br, I **20**
n = 5-10

Figure 3.34 Molecular components for comb-like PLCs assembled through donor–acceptor interactions.

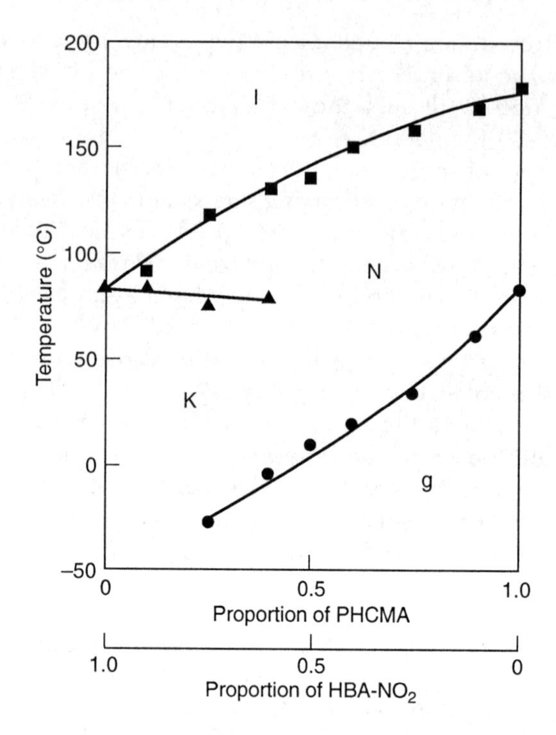

Figure 3.35 Phase diagram of donor–acceptor complexes of polymer (17b) (PHCMA) and monomer (19) (HBA–NO_2) (permission being sought from [70]).

characterized by an induced smectic A phase (in a bilayer structure) at intermediate compositions, as shown in Figure 3.36 for complex (17a) + (20) with $n = 6$ and Br^-. This is probably related to the ionic character of (20), which induces a smectic organization as noted above for ionically bonded complexes. The glass transition is only mildly affected by the small molecule component and in fact tends to increase a little, which can again be attributed to the ionic character of the complex. It was noted that the clearing temperature, studied for equimolar complexes of (17a) + (20) and (17b) + (20), increases somewhat with increasing alkyl chain length of (20), whereas the glass transition is little affected by this parameter [74].

3.6 WHY NON-COVALENT BONDS?

One of the foremost points of this review is the evident viability of using non-covalent bonds as integral elements in the assembly of PLC supramolecules. They not only modify existing mesomorphic behavior but, even more significantly, they can induce mesomorphic behavior.

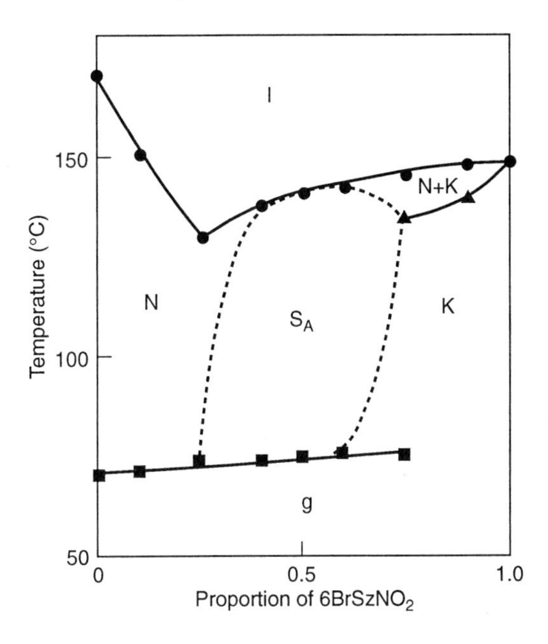

Figure 3.36 Phase diagram of ionic donor–acceptor complexes of polymer (17a) and monomer (20) ($6BrSzNO_2$) with $n = 6$ and Br^- (permission being sought from [74]).

They thus provide an entirely new tool with which to tailor PLCs for specific ends.

They are expected to possess unique properties due to the lability of the non-covalent bonds, and, from a synthetic viewpoint, many of the usual molecular parameters that control the PLC characteristics can be more easily accessible. The latter include the length and type of spacer and tail, the type of mesogen, the type of polymer backbone, and the molar mass and polydispersity of the polymer. Additional molecular parameters are introduced in conjunction with the choice of functional groups. Furthermore, a number of useful functional polymers are available either commercially or through straightforward synthetic methods. It is also a simple matter to obtain a variety of supramolecular copolymer PLCs and network PLCs, as well as, in principle, a variety of other architectures. A number of examples of these aspects have been given in this review.

The use of small molecules as at least one of the functional constituents allows a tailorable combination of polymer and monomer properties. Moreover, any temperature dependence in the lability of the non-covalent bonds will also result in a temperature-dependent polymer–monomer character. Low viscosities at higher temperatures that are imparted by the presence of the small molecules can have advantages, for instance, in the recycling or processing (including orientation) of these materials.

Systems presenting a (micro)phase-separated morphology can also be useful; for example, for polymer-dispersed liquid crystals (PDLCs) or in designing nanostructures. The potential utility of small amounts of ionically bonding components as interfacial agents for PDLCs, leading to fine, homogeneously dispersed LC droplets in a polystyrene matrix has already been illustrated [75]. An example of a liquid crystalline hydrogen-bonded polymer host–guest complex, involving a copolymer of 4-cyanophenyl-4-(6-(acryloyoxy) hexyl) benzoate and 4VP (10 or 25 mol%) blended with a carboxylic acid-functionalized azobenzene mesogen, has also been reported, and shown to be subject to reversible, isothermal phase transitions induced by sequential UV and visible light irradiation [76].

This review has presented what is a beginning in the development and understanding of supramolecular PLCs. Many questions about these materials, a number of which were posed above, must still be addressed. Their limits and their possibilities remain to be circumscribed. However, the richness of their potential is indisputable.

ACKNOWLEDGMENTS

The author is very grateful to Frank A. Brandys for invaluable help in the gathering of the references and in the preparation of the figures. Financial support to the author and coworkers from the donors of the Petroleum Research Fund administered by the American Chemical Society, from the Natural Sciences and Engineering Research Council of Canada, and from the Fonds pour la Formation de Chercheurs et Aide à la Recherche du Québec are also gratefully acknowledged.

REFERENCES

1. Lehn, J.-M. (1988) *Angew. Chem. Int. Ed. Engl.*, **27**, 89.
2. Lehn, J.-M. (1990) *Angew. Chem. Int. Ed. Engl.*, **29**, 1304.
3. Lehn, J.-M. (1993) *Makromol. Chem., Macromol. Symp.*, **69**, 1.
4. Imrie, C.T. (1995) *Trends Polymer Sci.*, **3**, 22.
5. Brostow, W. (1990) *Polymer*, **31**, 979.
6. Alexander, C., Jariwala, C.P., Lee, C.M. and Griffin, A.C. (1993) *Am. Chem. Soc. Polymer Prepr.*, **34**(1), 168.
7. Alexander, C., Jariwala, C.P., Lee, C.M. and Griffin, A.C. (1994) *Makromol. Chem. Symp.*, **77**, 283.
8. Lee, C.-M., Jariwala, C.P. and Griffin, A.C. (1994) *Polymer*, **35**, 4550.
9. Bladon, P. and Griffin, A.C. (1993) *Macromolecules*, **26**, 6604.
10. St. Pourcain, C.B. and Griffin, A.C. (1995) *Macromolecules*, **28**, 4116.
11. Wilson, L.M. (1994) *Macromolecules*, **27**, 6683.

12. Wilson, L.M. (1995) *Liq. Crystals*, **18**, 381.
13. Fouquey, C., Lehn, J.-M. and Levelut, A.-M. (1990) *Adv. Mater.*, **2**, 254.
14. Kato, T. and Fréchet, J.M.J. (1989) *Macromolecules*, **22**, 3819.
15. Kato, T., Kihara, H., Uryu, T. *et al.* (1992) *Macromolecules*, **25**, 6836.
16. Kumar, U., Kato, T. and Fréchet, J.M.J. (1992) *J. Am. Chem. Soc.*, **114**, 6630.
17. Kumar, U., Fréchet, J.M.J., Kato, T. and Ujiie, S. (1992) *Am. Chem. Soc. Polym. Mat. Sci. Eng. Prepr.*, **67**(2), 439.
18. Kumar, U., Fréchet, J.M.J., Kato, T. *et al.* (1992) *Angew. Chem. Int. Ed. Engl.*, **31**, 1531.
19. Kato, T., Kihara, H., Kumar, U. *et al.* (1994) *Angew. Chem. Int. Ed. Engl.*, **33**, 1644.
20. Kato, T., Nakano, M., Moteki, T. *et al.* (1995) *Macromolecules*, **28**, 8875.
21. Bazuin, C.G. and Brandys, F.A. (1992) *Chem. Mater.*, **4**, 970.
22. Brandys, F.A. and Bazuin, C.G. (1996) *Chem. Mater.*, **8**, 83.
23. Gavril, M. (1995) MSc Thesis, Laval University, Québec.
24. Bazuin, C.G., Brandys, F.A., Eve, T.M. and Plante, M. (1994) *Makromol. Chem. Symp.*, **84**, 183.
25. Stewart, D. and Imrie, C.T. (1995) *J. Mater. Chem.*, **5**, 223.
26. Malik, S., Dhal, P.K. and Mashelkar, R.A. (1995) *Macromolecules*, **28**, 2159.
27. Skoulios, A. and Guillon, D. (1988) *Mol. Cryst. Liq. Cryst.*, **165**, 317.
28. Ujiie, S. and Iimura, K. (1992) *Macromolecules*, **25**, 3174.
29. Bazuin, C.G. and Tork, A. (1995) *Macromolecules*, **28**, 8877.
30. Bazuin, C.G. and Tork, A. (1996) *Am. Chem. Soc. Polym. Prepr.*, **37**(1), 776.
31. Tal'roze, R.V., Kuptsov, S.A., Sycheva, T.I. *et al.* (1995) *Macromolecules*, **28**, 8689.
32. Gohy, J.F., Vanhoorne, P. and Jérôme, R. (1996) *Macromolecules*, **29**, 3376.
33. Brandys, F.A. (1996) PhD Thesis, Laval University, Quebec; Brandys, F.A., Masson, P., Guillon, D. and Bazuin, C.G. (1997) in preparation.
34. Li, Y., Xia, J. and Dubin, P.L. (1994) *Macromolecules*, **27**, 7049 and references therein.
35. Taguchi, K., Yano, S., Hiratani, K. *et al.* (1988) *Macromolecules*, **21**, 3338.
36. Taguchi, K., Yano, S., Hiratani, K. *et al.* (1991) *Macromolecules*, **24**, 5192.
37. Antonietti, M., Conrad, J. and Thünemann, A. (1994) *Macromolecules*, **27**, 6007.
38. Antonietti, M. and Conrad, J. (1994) *Angew. Chem. Int. Ed. Engl.*, **33**, 1869.
39. Ikkala, O., Ruokolainen, J., ten Brinke, G. *et al.* (1995) *Macromolecules*, **28**, 7088.
40. Ponomarenko, E.A., Tirrell, D.A. and MacKnight, W.J. (1996) *Am. Chem. Soc. Polym. Prepr.*, **37**(1), 404.
41. Kabanov, A.V., Sergeev, V.G., Foster, M.S. *et al.* (1995) *Macromolecules*, **28**, 3657.
42. Bakeev, K.N., Shu, Y.M., Zezin, A.B. *et al.* (1996) *Macromolecules*, **29**, 1320.
43. Antonietti, M., Forster, S., Zisenis, M. and Conrad, J. (1995) *Macromolecules*, **28**, 2270.
44. Wegner, G. (1986) *Makromol. Chem. Symp.*, **1**, 151.
45. Navarro-Rodriguez, D., Frère, Y. and Gramain, Ph. (1991) *Makromol. Chem.*, **192**, 2975.
46. Navarro-Rodriguez, D., Guillon, D., Skoulios, A. *et al.* (1992) *Makromol. Chem.*, **193**, 3117.
47. Chovino, C. (1994) PhD Thesis, Université Louis Pasteur de Strasbourg, France.
48. Ujiie, S. and Iimura, K. (1993) *Polymer J.*, **25**, 347.

49. Masson, P., Heinrich, B., Frère, Y. and Gramain, Ph. (1994) *Macromol. Chem. & Phys.*, **195**, 1199.
50. Masson, P., Gramain, Ph. and Guillon, D. (1995) *Macromol. Chem. & Phys.*, **196**, 3677.
51. Percec, V. and Pugh, C. (1989) in *Side Chain Liquid Crystal Polymers* (ed. C.B. McArdle), Blackie, Glasgow.
52. Ikker, A., Frère, Y., Masson, P. and Gramain, Ph. (1994) *Macromol. Chem. & Phys.*, **195**, 3799.
53. Laschewsky, A. (1995) *Adv. Polym. Sci.*, **124**, 1.
54. Percec, V., Johansson, G. and Rodenhouse, R. (1992) *Macromolecules*, **25**, 2563.
55. Dias, F.B., Voss, J.P., Batty, S.V. *et al.* (1994) *Macromol. Rapid Commun.*, **15**, 961.
56. Wright, P.V. (1995) *J. Mater. Chem.*, **5**, 1275.
57. Yousif, Y.Z., Jenkins, A.D., Walton, D.R.M. and Al-Rawi, J.M.A. (1990) *Eur. Polymer J.*, **26**, 901.
58. Blumstein, A., Cheng, P., Subramayam, S. and Clough, S.B. (1992) *Makromol. Chem. Rapid Commun.*, **13**, 67.
59. Jegal, J.G. and Blumstein, A. (1995) *J. Polymer Sci. Chem.*, **33**, 2673.
60. Bazuin, C.G. (1996) in *Polymeric Materials Encyclopedia* (ed. J.C. Salamone), CRC Press, Boca Raton, Vol. 5, 3454.
61. Belfiore, L.A., Pires, A.T.N., Wang, Y. *et al.* (1992) *Macromolecules*, **25**, 1411.
62. Ruokolainen, J., Tanner, J., ten Brinke, G. *et al.* (1995) *Macromolecules*, **28**, 7779.
63. Reck, B. and Ringsdorf, H. (1990) *Liq. Crystals*, **8**, 247.
64. Imrie, C.T., Karasz, F.E. and Attard, G.S. (1991) *Liq. Crystals*, **9**, 47.
65. Schleeh, T., Imrie, C.T., Rice, D.M. *et al.* (1993) *J. Polymer Sci. Chem.*, **31**, 1859.
66. Kosaka, Y., Kato, T. and Uryu, T. (1994) *Macromolecules*, **27**, 2658.
67. Kosaka, Y. and Uryu, T. (1995) *Macromolecules*, **28**, 870.
68. Kosaka, Y. and Uryu, T. (1995) *Macromolecules*, **28**, 8295.
69. Imrie, C.T. and Paterson, B.J.A. (1994) *Macromolecules*, **27**, 6673.
70. Kosaka, Y. and Uryu, T. (1994) *Macromolecules*, **27**, 6286.
71. Ringsdorf, H., Wüstefeld, R., Zerta, E. *et al.* (1989) *Angew. Chem. Int. Ed. Engl.*, **28**, 914.
72. Green, M.M., Ringsdorf, H., Wagner, J. and Wüstefeld, R. (1990) *Angew. Chem. Int. Ed. Engl.*, **29**, 1478.
73. Bengs, H., Renkel, R., Ringsdorf, H. *et al.* (1991) *Makromol. Chem. Rapid Commun.*, **12**, 439.
74. Kosaka, Y., Kato, T. and Uryu, T. (1995) *Macromolecules*, **28**, 7005.
75. Gohy, J.F. and Jérôme, R. (1996) *Macromol. Chem. & Phys.*, **197**, 2209.
76. Kato, T., Hirota, N., Fujishima, A. and Fréchet, J.M.J. (1996) *J. Polymer Sci. Chem.*, **34**, 57.

4

MORPHOLOGY OF THERMOTROPIC LONGITUDINAL POLYMER LIQUID CRYSTALS

Yang Zhong

4.1 INTRODUCTION

Polymer liquid crystals (PLCs), or liquid crystalline polymers (LCPs), are receiving more and more interest nowadays [1, 2] because of their unique properties compared with ordinary polymers. The most characteristic feature of PLCs is that they consist of highly anisotropic mesogens which are very sensitive to external fields, such as stress, electric or magnetic fields, and have characteristic alignment according to the external field. This feature leads to numerous valuable applications of PLCs, for example as high performance fibers and engineering plastics or functional materials.

Among the several hundreds of PLCs reported in papers and patents, thermotropic longitudinal PLCs [1] comprise a very important category. Longitudinal PLCs, formerly called main-chain PLCs, are those whose backbones consist of rod-like mesogens connected parallel to the backbone. Most thermotropic longitudinal PLCs are aromatic LC

Mechanical and Thermophysical Properties of Polymer Liquid Crystals
Edited by W. Brostow
Published in 1998 by Chapman & Hall, London.
ISBN 0 412 60900 2

polyesters, and their LC state is usually nematic. The rigid rod molecules of thermotropic longitudinal PLCs can be highly oriented along flow direction when the PLCs are processed in their LC state. The orientation can be readily maintained during solidification because of the long relaxation time of the rigid rod molecules. This leads to the products having a high strength, high modulus and a very low coefficient of linear expansion, especially in the flow direction. Furthermore, the aromatic and ester chemical groups ensure that the PLCs have excellent thermal and chemical resistance. Therefore, a large number of thermotropic longitudinal LC polyesters have been commercialized as high performance engineering plastics, high performance fibers and reinforcement of other plastics. They have great potential to be used in those situations where strength, rigidity and stability are so essential that ordinary engineering plastics are not satisfactory. In addition, they are also very suitable for precision injection molding which is generally needed in the machinery and electronics industries.

It is well known that morphology, which involves the alignment and packing of macromolecules, has a significant effect on the properties of PLCs. Morphology is closely related to molecular structures and processing histories. Morphologies based on the rigid rod molecules of thermotropic longitudinal PLCs are quite different from those of conventional flexible chain polymers. Extensive researches have been made on the morphologies of these PLCs [3]. In this paper we intend to give a review of the morphologies of thermotropic longitudinal PLCs and the relations of molecular structures and processing conditions to morphologies and properties.

The morphological differences between thermotropic longitudinal PLCs and ordinary polymers originate from the different responses of the rigid rod molecules and the flexible chain molecules to stress and temperature and from the different organization patterns of the molecules in these systems. The most frequent morphology of thermotropic longitudinal PLC products is a layered structure with the layers themselves being composed of fine fibrils of various sizes. This morphology is often referred to as hierarchical structure [4, 5]. Other featured morphologies of thermotropic longitudinal PLCs include their crystallization and multiphase behaviors. These featured morphologies of thermotropic longitudinal PLCs will be discussed in turn.

4.2 HIERARCHICAL AND FIBRILLAR STRUCTURE

The term 'hierarchical structure' was first described for aramid by Dobb *et al.* [4] and for injection molded thermotropic liquid crystalline copolyester by Weng *et al.* [5]. It has been found that no matter which LC polyester is concerned, wholly aromatic or semi-aromatic ones, similar layered and fibrillar morphologies are observed in fibers, extrudates and injection molded products. Simply, the hierarchical structure results from the flow (or stress) gradient, relaxation difficulty of the rigid rod molecules and lack of entanglement and friction among the rigid rod molecules. The formation of hierarchical structures and its relations to molecular orientation in various fabrication processes will be discussed in this section.

Practical polymer processing operations, such as extrusion and injection, are complex processes highly dependent on the conditions of shear, extensional stress and temperature. It has now been realized that shear stress and extensional stress have different effects on the orientation of PLCs. Nematic PLCs have a polydomain texture and each domain consists of mesogens with the same local orientation. The directions of these domains are randomly aligned while in a quiescent state. Only if a stress is applied are the domains oriented in one direction. As Ide and Ophir [6] and Viola *et al.* [7] have pointed out, shear stress is related to rotational motion (torque), and its application will result in a tumbling flow of PLC domains. Only once the shear stress has reached a critical value will it break down the domains and lead to a uniform molecular orientation. In contrast, extensional stress tends to orient the domains in one direction without breaking down the domains even if the stress is low. Viola *et al.* [7] also considered that shear flow induces sheet-like textures while extensional flow induces fiber-like textures. Therefore there will be differences in the hierarchical and fibrillar structures acquired in different fibrication processes.

4.2.1 Fibrillar hierarchy of PLC fibers

In fiber spinning, the dominant factor is a strong unidirectional extensional stress. Therefore, a highly fibrillar texture is generally observed in spun PLC fibers, although some differences can be noted because of variation of processing temperature and draw-down ratio. For example, spun fibers of poly(ethylene terephthalate-co-*p*-oxybenzoate) (PET/60PHB) exhibit highly fibrillar textures [8, 9]. But if the fibers are spun at lower temperature or lower drawdown ratio, there will be some sheet-like regions, and these non-fibrillar structures in the fiber are responsible for lower mechanical properties [8, 9].

A widely accepted model describing the highly fibrillar hierarchical morphology of PLC spun fibers was proposed by Sawyer *et al.* [10]. They studied highly oriented copolyester fibers and thin extruded tapes, Vectran (tradename of copolyesters composed of 2, 6-naphthyl and 1, 4-phenyl units), and aramid fibers, Kevlar. High resolution imaging methods, field emission scanning electronic microscopy (FESEM), scanning tunneling electronic microscopy (STM) and transmission electronic microscopy (TEM) were used to carry out a more detailed morphology investigation of the longitudinal PLCs. The STM images acquired by continuously scanning the same area of the specimen reveal the fibrils and the even smaller microfibrils aligned within the fibrils. The fibrils and microfibrils in fact are not perfectly oriented along the fiber axis. Local disorders, such as kink bands, local loss of microfibril contours and fibrils weaving in and out of the longitudinal plane, can be observed from the FESEM and STM micrographs. The fibrils and microfibrils are tape-like in shape and the smallest microfibrils are about 3–30 nm wide and 2–5 nm thick. In addition, the sizes of these microfibrils have been found to be similar for all thermotropic and lyotropic PLCs and for fibers with different tensile strength and modulus, such as before and after heat treatment. The TEM micrographs also show that the fibrils are very long and tend to fibrillate into even smaller units and there are no clear interfibril tie fibrils. According to these observations, Sawyer *et al.* [10] proposed a structural model for both thermotropic and lyotropic PLCs (Figure 4.1) which suggests that the PLC fibers are organized hierarchically from the elementary microfibrils 10–30 nm wide and 2–3 nm thick (with a periodic weaving texture about 50 nm across) to fibrils about 0.5 μm in size and to macrofibrils about 5.0 μm. Sawyer *et al.* [10] also related the supermolecular structure to mechanical properties. They considered the microfibrils to be the finest unit controlling mechanical properties and that it is the organization of the microfibrils that determines tensile properties since a higher degree of local order, improvement in lateral arrangement and

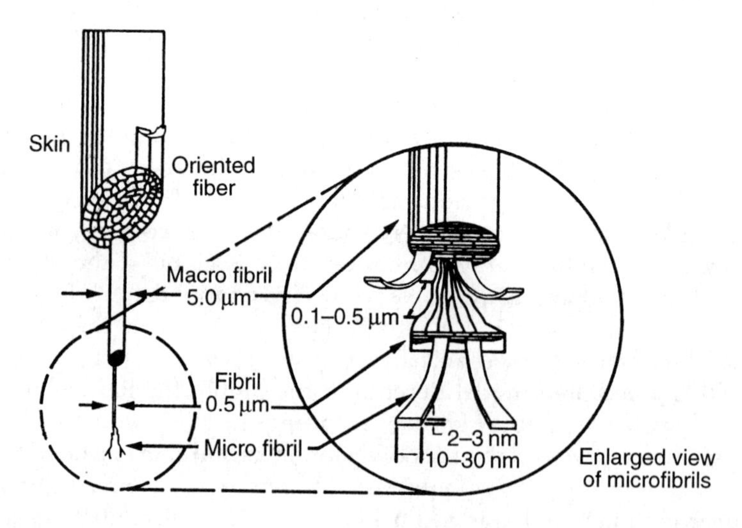

Figure 4.1 Structural model of fibers for thermotropic and lyotropic PLCs. (Reprinted with permission from [10], copyright © 1993 Chapman & Hall.)

orientation all reflect a higher tensile modulus. The lower compressive strength was attributed to the kink bands on the fiber surface.

4.2.2 Hierarchical and fibrillar structure of PLC extrudates

Extrusion of polymer involves shear flow in the die passage and some extensional flow during the drawdown stage. Extensional stress in extrusion is not as dominant as that in fiber spinning. Therefore, a less ordered fibrillar morphology is usually observed in PLC extrudates. The degree of fibrillation depends on the size, shape and aspect ratio of the die, the extrusion temperature and the drawdown ratio. The morphology of PLC extrudates, such as strands and sheets, is of a skin/core structure which is the result of the stress gradient across the die passage, and the skin/core ratio varies with temperature, drawdown ratio and thickness of the extrudates. Generally speaking, the skin/core ratio becomes lower with increasing temperature, increasing thickness of the extrudates and/or decreasing drawdown ratio. The skin/core structure can be further divided into three microscopic layers: a highly oriented outer skin, less oriented inner skin and randomly oriented core. All three sections are of ribbon-like texture except for some plate-like texture in the core region [11]. The ribbon-like texture of PLC extrudates was also described by Sawyer and Jaffe [12] as a hierarchical structure with macrofibrils about 5 μm, fibrils about 0.5 μm and microfibrils about 0.05 μm. However, the size of microfibrils formed in extrusion is supposed to be much larger than that in fiber spinning, which may be due to the decreasing extensional stress level.

Kyotani *et al.* [11] have studied the structures and properties of extruded sheets of a thermotropic polyester consisting of 73 mol% *p*-hydroxybenzoic acid and 27 mol% 2, 6-hydroxynaphthoic acid (HBA/HNA 73/27). The HBA/HNA extruded sheets have a multilayered structure. In the skin layer, dense fibrils with diameter about 0.2 μm orient almost parallel to the extrusion direction, while some plate-like regions exist in the core and the fibrils are not necessarily parallel to the extrusion direction. A schematic representation of the extruded sheet indicating the layered and fibrillar structure was proposed by Kyotani *et al.* (Figure 4.2) [11]. However, the extrudates of another LC copolyester containing a quite long aliphatic segment of seven carbon atoms, reported by Kent *et al.* [13], show much less fibrils. This indicates that the higher the rigidity of the PLC molecules, the more and thinner fibrils that will form during extrusion. Moreover, Zülle *et al.* [14] have produced extruded tubes with HBA/HNA73/27 and presented a schematic diagram to illustrate the skin/core profile of the extruded tubes (Figure 4.3).

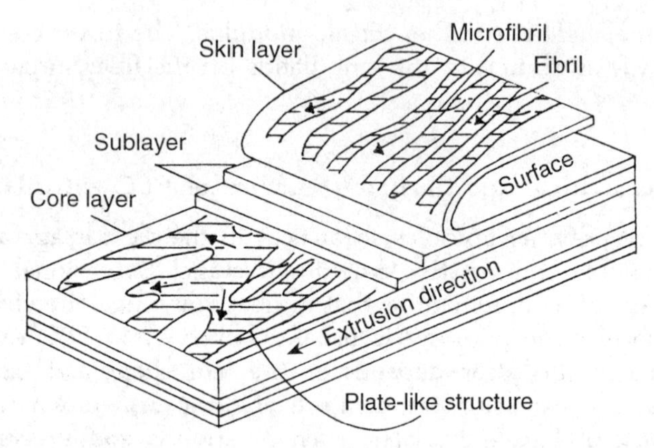

Figure 4.2 Schematic representation of the layered structure of extruded PLC sheet. The core layer contains plate-like regions, whereas the skin layer is composed of fibrous structures. The arrows in the skin and core layer surfaces indicate the direction of molecular orientation. (Reprinted with permission from [11], copyright © 1993 John Wiley & Sons, Inc.)

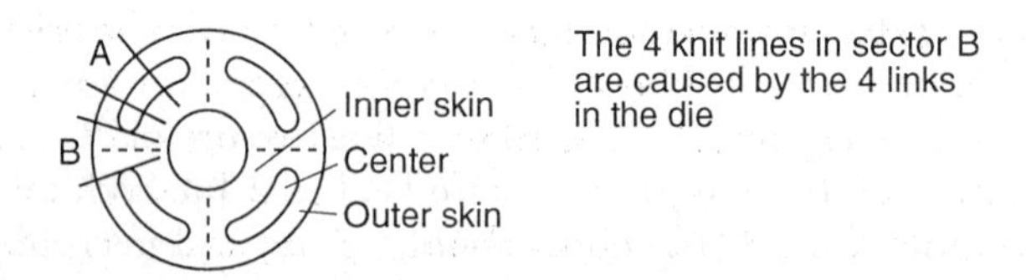

Figure 4.3 Schematic diagram of skin/core texture in the cross-section of an extruded Vectra tube. (Reprinted with permission from [14], copyright © 1993 Butterworth-Heinemann Ltd.)

Mechanical properties of extruded sheets are highly anisotropic. The strength and modulus along the extrusion direction are much higher than those perpendicular to the extrusion direction; the former increase rapidly with increasing drawdown ratio and then level off, while the latter keep almost constant with changing drawdown ratio [11]. The anisotropy results from the highly oriented skin, so that it will become less conspicuous with increasing thickness of the extruded sheets or increasing extrusion temperature and/or decreasing drawdown ratio.

4.2.3 Hierarchical structure and molecular orientation in injection molded PLCs

The morphology and properties of injection molded thermotropic longitudinal PLC products are similar to those of extrudates, i.e., macroscopic skin/core structure, microscopic hierarchically layered and fibrillar organization and anisotropic behavior. However, the injection molding process is much more complex than extrusion. The mold filling process involves complex shear and extensional stress profiles with non-isothermal condition. Therefore, the morphology and properties of injection molded PLC products are much more complex than those of extrudates. Even in the same sample, the skin/core ratio, the fibril and molecular orientation change with position. Thus, it is first necessary to characterize the morphology of injection-molded PLCs.

One aspect of morphology characterization is establishing models to describe the complex multilayered structures and molecular orientation for injection molded PLCs. Weng *et al.* [5] have studied the injection molded plaques (203 mm × 51 mm × 3 mm) of HBA/HNA 58:42 and proposed a hierarchical model (Figure 4.4) based on SEM and wide angle X-ray diffraction (WAXD) observation. They supposed in the model that the skin is composed of three subdivisions: a highly oriented top layer about 20 μm in thickness; several oriented sublayers stacked into a layer 30–50 μm beneath the top layer; and a less oriented inner zone about 500–700 μm in thickness. The core does not show hierarchical structures, and the molecular orientation in the core is parallel to the flow lines. Between the skin and the core, there is a boundary layer without known thickness.

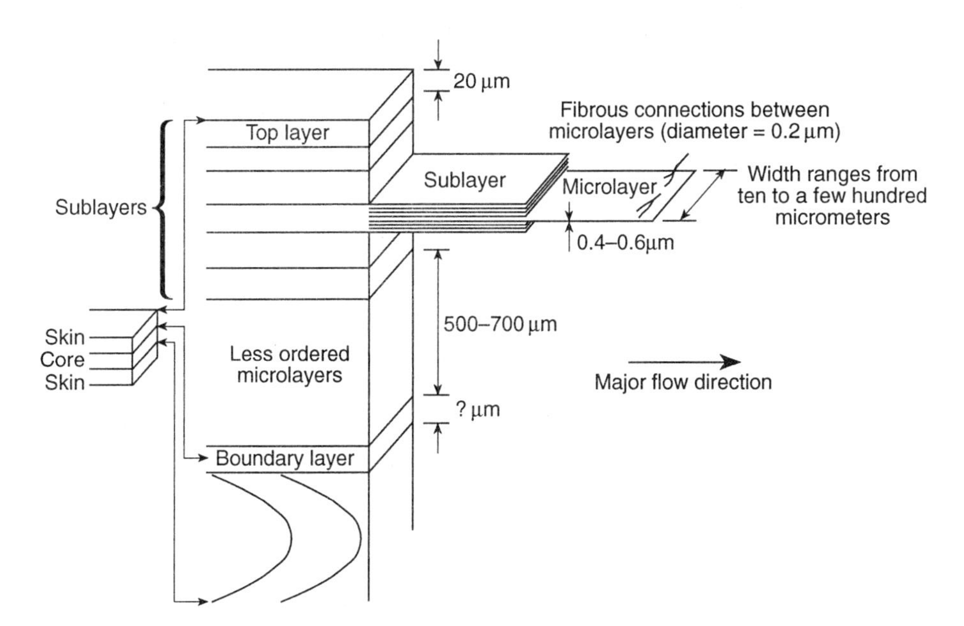

Figure 4.4 Hierarchical model of injection molded PLCs (not drawn to scale). (Reprinted with permission from [5], copyright © 1986 Chapman & Hall.)

Plummer *et al.* [15] also investigated injection molded plaques of HBA/HNA copolyester with the same thickness of 3 mm. But their HBA/HNA was 73:27 (Vectra A950), and the width of the plaque was 6 mm. Moreover, the barrel temperature and mold temperature were a little different from those used by Weng *et al.* [5]. In this case, five layers and regions were presented according to the observation of optical microscopy and SEM (Figure 4.5). The 'outer skin' layer, which extends 40–50 μm into the sample surface, is highly birefringent, indicating high orientation, and has a ridged texture when observed by SEM. The 'inner skin' layer, approximately 200 μm in total thickness, exhibits a banded texture in optical microscopy which suggests a coherent periodic variation in molecular trajectory along the flow direction, and this periodic variation of the molecular trajectory is also indicated in the longitudinal freeze fracture surfaces. The 'intermediate' region, weakly birefringent compared with the skin layers, has a tight optical texture. Both the 'outer core' and 'inner core' are of fibrillar texture under SEM. The 'inner core' shows a tight optical texture (tighter than that of the 'intermediate region') and is superimposed by concentric and parallel parabolic lines along the filling direction.

Hedmark *et al.* [16] have studied injection molded bars with different thicknesses and they believed that the detectable layers and molecular orientation inside the layers depend on the thickness of the samples.

Another aspect of morphology characterization is spectroscopic analyses of the molecular orientation. The simplest and most often used method is infrared (IR) spectroscopy. Although IR techniques cannot investigate the multilayered texture, they can readily provide vivid one- or two-dimensional profiles of molecular orientation vs. position. Bensaad *et al.* [17] used normal incidence specular reflection to characterize the orientation profile on the surface of an injection molded plaque (6 cm × 6 cm × 0.4 cm) made of a wholly aromatic thermotropic

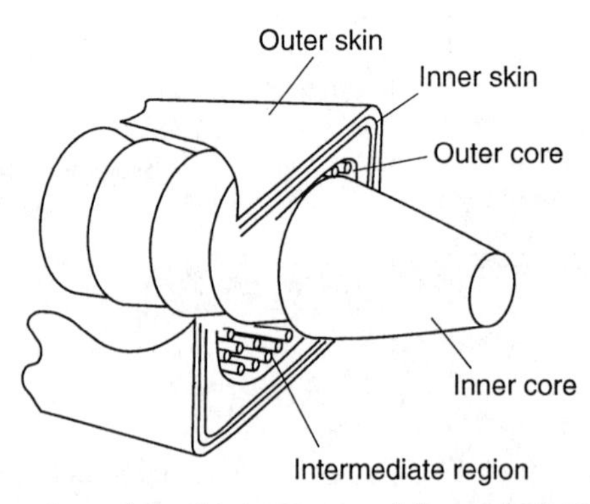

Figure 4.5 Structural model of injection molding of PLCs. (Reprinted with permission from [5], copyright © 1993 John Wiley & Sons, Inc.)

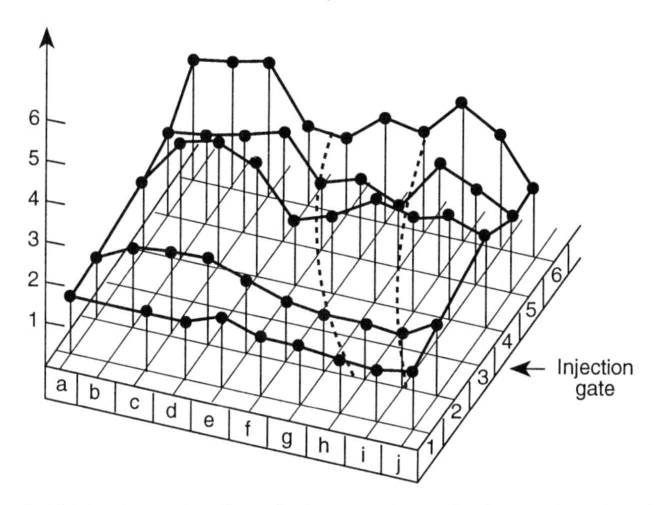

Figure 4.6 Dichroic ratio, *R*, of the $1016\,cm^{-1}$ absorption band on the opposite face with respect to the injection gate vs. position. The letters a–j and numbers 1–6 represent the positions from the gate and from the mold axis respectively. (Reprinted with permission from [17], copyright © 1993 Butterworth-Heinemann Ltd.)

copolyesteramide (Figure 4.6). The dichroic ratio, *R*, indicates the molecular orientation with respect to the filling direction by $R \gg 1$, meaning that the average molecular orientation is parallel to the filling direction, and $R \sim 1$, indicating poor orientation. Jansen *et al.* [18] used a Fourier transform IR diffuse reflectance technique to investigate the molecular orientation in injection molded flat plates (80 mm × 35 mm × 2 mm) of Vectra A950. Depth and lateral profiles of the order parameter s^* (representing the degree of orientation) have been given (Figure 4.7). From the depth profiles, it can be seen that the maximum value of order parameter is at the surface of the plate and a local maximum is at a depth between 100 µm and 200 µm. The two maxima were attributed, respectively, to the elongational flow along the mold wall and the shear flow with highest rate just beneath the solidified surface layer. It can also be shown that higher injection speed induces higher order parameters, i.e. a higher degree of orientation, but the order parameters become negative in the core region for samples with higher injection speed ($10\,cm\,s^{-1}$ in this case), indicating that the direction of molecular orientation has been reversed with respect to the flow direction. The lateral profile of s^* in the core of a plate differs from the above mentioned orientation profile in the skin of an injection molded plate (Figure 4.6) obtained by Bensaad *et al.* [17]. The former shows the highest order parameter in the center of the sample, while

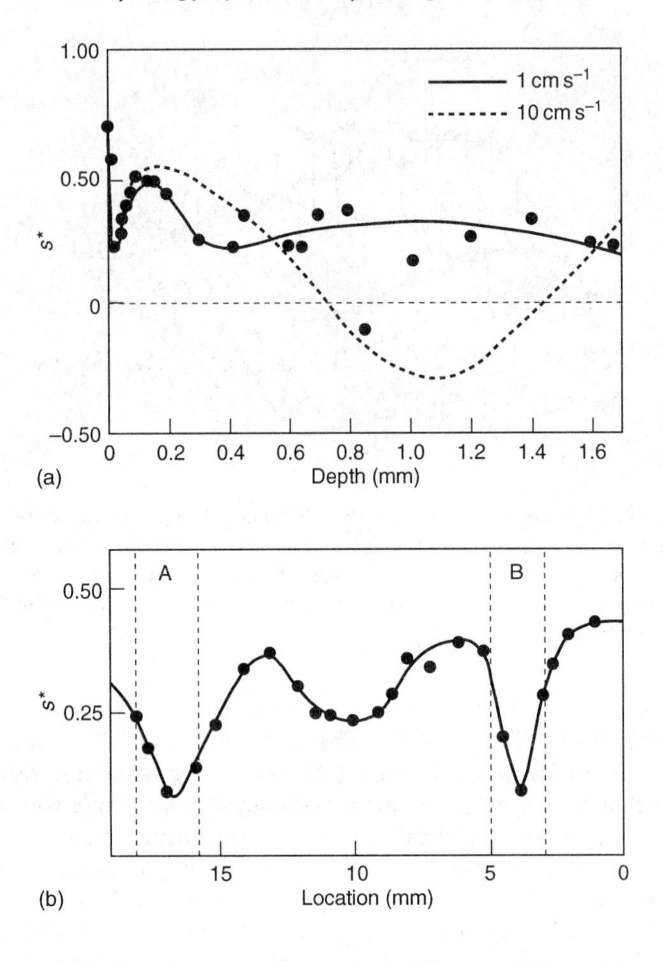

Figure 4.7 (a) Depth profiles of order parameter s^* in the center of the width and 20 mm from the injection gate of a 2 mm thick Vectra A950 plate with injection speed of 10 cm s^{-1} and 1 cm s^{-1}; (b) Lateral profile of s^* in the core of the Vectra A950 plate with injection speed of 1 cm s^{-1} (location 0 represents the sample edge). (Reprinted with permission from [18], copyright © 1994 Butterworth-Heinemann Ltd.)

the latter shows the lowest in the center. The lateral profile of s^* in the core has not been well explained by Jansen *et al.* [18].

Hierarchical structures and molecular orientation in PLC products have not yet been fully understood and the experimental results are sometimes different or even contradictory. Complete understanding of the morphology and its formation need more effective characterization means and detailed knowledge of polymer flow in molds. However, some conclusions, explanations and predictions about the morphology can be made from the already known knowledge of the mold filling process and the response of PLC molecules to the flow field.

Take an example of a flat mold with a centered end gate. Morphology is determined by the superposition of four flow situations according to Grag and Kenig [19]:

1. diverging flow in the vicinity of the gate;
2. converging flow after the melt front exceeds the half-width of the mold;
3. fountain flow at the free surface of the advancing melt front; and
4. shear flow behind the melt front.

The radially spreading flow near the entrance induces a degree of molecular orientation along the transverse direction of the principal flow in this area. The fountain flow of the melt front throws the molecules perpendicularly toward the mold wall. Molecules in which one end collides with the cool wall and is quickly frozen while the other end still being pushed forward experience an extensional flow along the filling direction in a very thin layer close to the mold wall, which results in a high degree of molecular orientation and formation of the outer skin.

Except for this thin layer, the mold cavity is primarily filled by shear flow accompanied by diverging and converging flow occurring before and after the melt front reaches the side wall, respectively. The highest shear stress and shear rate are at the inner surface of the already solidified outer skin and the highest shear stress in the mold filling process usually is enough to break down the PLC domains and to establish orientation along the filling direction. The shear-induced orientation, which is generally not so high as extension-induced one, is believed to correspond to the local orientation maximum shown in the order parameter profiles [18], to the sublayers proposed by Weng *et al.* [5] and to the inner skin reported by Plummer *et al.* [15]. In the region close to the outer skin, the melt is also rapidly cooled like quench, so that a branded texture resulting from coherent sinuous variation of molecule trajectories may occur as observed by Plummer *et al.* [15] when the stress and temperature conditions are suitable for a transient relaxation. Moreover, because of the high degree of molecular alignment and lack of entanglement among the rigid rod PLC molecules, the skin can be delaminated easily into many sublayers as reported by several groups [5, 12, 15, 16].

Farther toward the center of the mold, the shear stress gradually decreases and becomes too weak to break down the domains to induce large-scale orientation in the filling direction. The melt experiences a tumbling flow of domains. This is the formation of the intermediate region and the core. In this region, the extensional and compressive stress along the tangent of the melt front become predominant, which gives rise to molecular orientation along the flow lines of the tumbling flow. These orientation patterns can also be readily preserved as a result of the long relaxation time of PLCs, and they are responsible for the parabolic lines observed by many researchers [5, 12, 15]. The diverging flow in the vicinity of the gate may induce transverse molecular orientation in the center near the gate while the converging flow which develops once the melt front exceeds the half-width of the mold may lead to orientation along the filling direction in the region far from the gate. Therefore, transverse orientation may be expected to occur more readily in the vicinity of the gate and under conditions responsible for shorter relaxation times, such as lower mold temperature, higher injection speed or smaller gate.

Figure 4.8 is an illustration of the flow pattern and molecular orientation in an injection mold. It can be inferred from the orientation profiles of the core region that it is difficult to delaminate the core into sublayers but easy to fracture it along the flow lines. In addition, the melt front becomes progressively flatter along the filling distance, so that if an injection molded bar is fractured perpendicular to its filling direction, the core at the far end will seem larger than the parabola-shaped core at the middle. The sharpness of the boundary between the skin and the core is believed to depend on the interaction of PLC molecules. A less sharp boundary will be observed for samples made of PLCs with more bending or flexible structures. For example, we have found that no skin/core separation can be observed by the naked eye when some bisphenol-A groups are introduced to PET/0.6PHB chains.

4.2.4 Effect of processing conditions on the morphology and properties of PLCs

The effects of processing conditions and inherent nature of PLCs on the morphology and properties of injection molded PLC products have

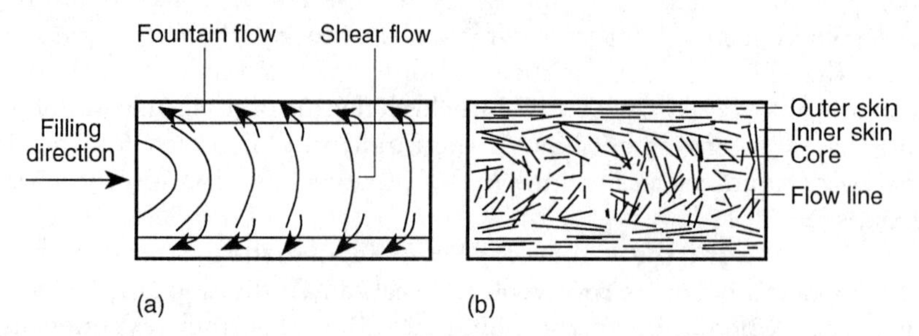

Figure 4.8 (a) Flow pattern and (b) molecular orientation in injection molded PLCs.

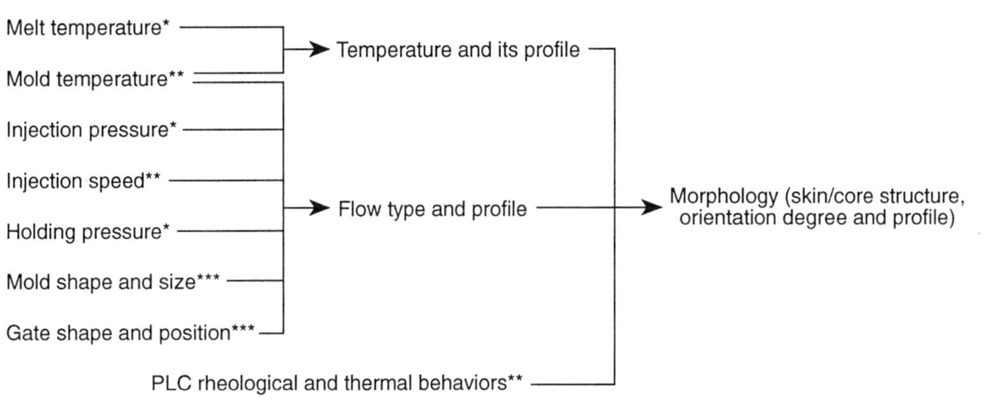

Figure 4.9 Factors directly (column 2) and indirectly (column 1) determining the morphology of injection molded PLCs. The number of asterisks at the upper right of each factor is proportional to the effectiveness of this factor.

also attracted much reasearch interest [3]. The mechanical properties of PLC products depend directly on their skin/core structure and the degree and profile of molecular orientation. Figure 4.9 demonstrates the direct and indirect factors which affect morphology formation of injection molded PLCs and the effectiveness of these factors. Although these factors are familiar in conventional plastic processing, they have much stronger effect on the morphology and properties of PLCs.

The mold and gate conditions play the most important part in morphology formation because they determine the flow profile to which the orientation of rigid rod molecules is most sensitive. For example, morphology of a dumbbell-shaped specimen is different from that of a square bar because of the strong extensional flow in the neck of the former. Molds with identical shape but different thickness also lead to morphology differences for PLC products. Zülle *et al.* [14] found five distiguishable regions in the neck of a 4 mm thick dumbbell specimen, while we found layers with similar texture of oriented fibrils running through the neck region in a 2 mm thick specimen. It has been confirmed that the proportion of skin/core will be reduced with increasing specimen thickness because of the resultant longer relaxation time. Also, it is obvious that skin layer bears most of the stress applied to a specimen because it possesses a higher tensile modulus so that the skin fracturing will determine the overall strength of a specimen [14, 15]. Thus, mechanical properties of PLC products will be increased with decreasing thickness, which can be seen from the data obtained by Duska [20] for thermotropic copolyester with 1:2:1 molar ratio of p,p'-biphenol, p-hydroxybenzoic acid and terephthalic acid (Xydar) (Tables 4.1 and 4.2). The anisotropy can be suppressed by increasing the thickness. On

Table 4.1 Skin thickness for ASTM T-bars of three sizes. (Reprinted with permission from [20], copyright © 1986 Society of Plastics Engineers, Inc.)

Bar size (in[a])	Skin thickness (in)	Percent skin	Percent core
$\frac{1}{8}$	0.010	16	84
$\frac{1}{16}$	0.012	38	62
$\frac{1}{32}$	0.009	58	42

[a] 1 in = 25.4 mm.

Table 4.2 Physical properties related to specimen thickness for injection molded unfilled and 50%-filled Xydar. (Reprinted with permission from [20], copyright © 1986 Society of Plastics Engineers, Inc.)

Properties	Specimen thickness (in[a])		
	$\frac{1}{8}$	$\frac{1}{16}$	$\frac{1}{32}$
Tensile strength (psi[b])			
Unfilled	15 840	22 140	26 500
Glass-filled	14 340	16 130	16 480
Mineral-filled	11 460	12 900	14 000
Flexural modulus (psi)			
Unfilled	1.81	1.98	3.11
Glass-filled	1.78	2.13	2.16
Mineral-filled	1.57	1.78	1.80

[a] 1 in = 25.4 mm. [b] 1 psi = 0.006895 MPa.

the other hand, even in the same specimen, anisotropy was shown to be different at various positions with the highest transverse modulus occurring close to the gate [19].

As to the gate, a rectangular gate is usually used to prevent jetting or to bring the weld lines to a non-stressed part of the specimen. Sometimes a fan-shaped gate is needed for products with large width. A multigate system is seldom used for PLCs for it will cause weld lines [21].

It has also been found that lower mold temperature, lower injection pressure and speed or lower packing pressure lead to an increase of the skin/core ratio and consequent higher mechanical properties and anisotropy [14, 20–24]. Zülle *et al.* [14] have observed three types of skin/core profile for Vectra A950 at various injection speeds (Figure 4.10) of which type 3, obtained using the lowest injection speed, has the highest skin/core ratio and tensile strength. Similar tendencies of the effects of these processing parameters on mechanical properties have been reported both by Duska [20] and Ophir *et al.* [21] (Tables

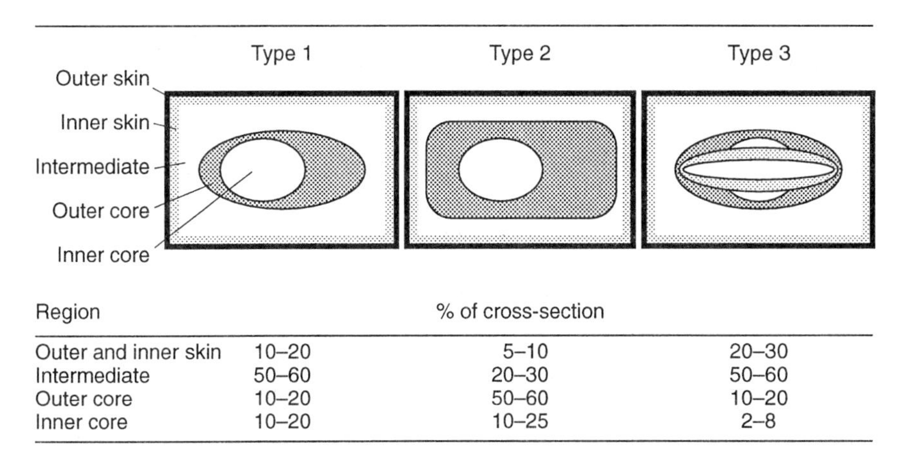

Region	% of cross-section		
Outer and inner skin	10–20	5–10	20–30
Intermediate	50–60	20–30	50–60
Outer core	10–20	50–60	10–20
Inner core	10–20	10–25	2–8

Figure 4.10 Sample morphologies in injection moldings of Vectra under different injection speed. (Reprinted with permission from [14], copyright © 1993 Butterworth-Heinemann Ltd.)

4.3 and 4.4). However, we should note from Ophir's data that the impact strength drops under conditions of slower injection speed, higher mold temperature and higher packing pressure, which cannot be interpreted satisfactorily at present. Furthermore, detailed investigations of these factors affecting the proportion of the different layers have been made by Suokas [22], Khennache and Kamal [23] and Zülle *et al.* [14]. It has been found that the effects of these factors are mainly on

Table 4.3 Processing parameters and tensile strength data for unfilled and filled Xydar. Reprinted with permission from [20], copyright © 1986 Society of Plastics Engineers, Inc.

Injection pressure (psi) [b]	Injection speed ($in^3 s^{-1}$) [a]	Mold temperature (°C)	Tensile strength [c] (psi)	Tensile strength [d] (psi)	Tensile strength [e] (psi)
13 500	2.1	150	28 390	16 740	14 300
20 300	2.1	150	26 490	15 810	13 320
13 500	4.7	150	27 670	17 470	13 100
20 300	4.7	150	24 880	16 590	13 900
13 500	2.1	260	27 740	16 800	13 970
20 300	2.1	260	22 620	15 780	13 490
13 500	4.7	260	23 400	16 640	13 610
20 300	4.7	260	23 360	15 860	13 260

[a] 1 in = 25.4 mm. [b] 1 psi = 0.006895 MPa. [c] For unfilled resin; [d] for glass-filled resin; [e] for mineral-filled resin.

Table 4.4 Processing parameters and mechanical properties for injection molded thermotropic copolyester of 60 mol% *p*-acetoxybenzoic acid, 20 mol% terephthalic acid and 20 mol% naphthalene. (Reprinted with permission from [21], copyright © 1983 Society of Plastics Engineers, Inc.)

Melt temperature (°C)	320	340	340	340	340	340
Mold temperature (°C)	40	40	40	100	100	100
Packing pressure (MPa)	28	21	39	39	28	0
Injection speed	Fast	Fast	Slow	Slow	Fast	Fast
Tensile strength (MPa)	192	173	189	208	176	153
Tensile elongation (%)	1.8	1.3	1.5	1.6	2.5	1.5
Tensile modulus (GPa)	17.2	20.0	19.3	18.6	15.2	16.5
Flex strength (MPa)	178	174	175	175	170	176
Flex modulus (GPa)	15.2	15.2	15.2	14.5	13.1	13.1
Notched impact ($J\,m^{-1}$)	395	427	230	283	283	347

the thickness of the intermediate layer and the core while there is little effect on the thickness of the skin. They all concluded that higher mold temperature results in an increase of the core thickness, decrease of the intermediate layer and almost no change in the thickness of the skin. But the effects of other factors were shown to be inconsistent [14, 22, 23].

Melt temperature has a more complicated effect. If the melt temperature is not far above the melting point of a PLC, properties of the PLC usually increase with higher melt temperature because there may be some high melting crystals not fully melted under low melt temperatures. However, with melt temperature much higher than the melting point, the properties will decrease because of greater relaxation [21, 25].

Fillers also have a great effect on the morphology and properties of PLCs. The most obvious one is that anisotropy is almost suppressed in all filled PLC systems, but the mechanism of the decrease of anisotropy varies with the shape of fillers. Ryan [26] has studied Vectra A950 filled with spherical, fibriform or flake fillers and concluded that both fabriform and flake fillers can reduce the anisotropy by enhancing the transverse properties, especially in the latter case, while spherical fillers reduce the anisotropy by weakening the long range orientation of PLC molecules along the flow direction. Another effect of filling results in a decreasing of the 'molding sensitivity' as presented by Duska [20]. The properties of neat PLC resins are very sensitive to the molding conditions, which is undesirable in practical production. With the addition of fillers, the effect of PLC molecular orientation on properties is relatively lower and then the dependence of properties on molding variables is weakened. Finally, mechanical properties, especially the impact strength, of filled PLCs are usually lower than those of neat PLCs because the interfacial adhesion between fillers and PLCs is usually quite poor [26, 27]. On the other hand, thermal properties can be improved with the addition of fillers.

Many problems concerning the morphology of PLCs and its effect on properties still remain unsolved. A more reasonable interpretation of the origin of hierarchical structures and molecular orientation in PLC products, the mechanisms of interfibrillar stress transfer and the relaxation process of the rigid rod molecules all need more investigation.

4.3 CRYSTALLIZATION BEHAVIOR

Various crystallization behaviors have been observed for different types of PLCs depending on their molecular structures. Most research interest focuses on those commercially available thermotropic longitudinal PLCs, especially the most popular HBA/HNA copolyesters. Although some inherent characteristics of HBA/HNA copolyesters are unfavorable for crystallization, such as random comonomeric sequences and molecular rigidity which make it difficult for the molecules to rearrange or fold regularly, these PLCs have a strong tendency to crystallize. Their crystallization rate is too rapid to form a completely amorphous state even in quenching. However, their crystallization behaviors are different from those of ordinary polyesters.

Blundell [28] first demonstrated that the main features of crystallization for thermotropic longitudinal PLCs are (1) small crystallites ($\ll 5$ nm for quenched samples and about 10 nm across for annealed ones); (2) limited crystallinity (about 20% in wide angle X-ray diffraction (WAXD)); (3) low surface energy of the crystallites; (4) little change of molecular conformation before and after melting and (5) low fusion entropy and enthalpy (sometimes no detectable endotherm on DSC thermograms). A model of small crystal size for extended chain polymers was proposed by Blundell [28].

Extensive work has been done more recently on further detailed investigation of the formation process and sequence arrangement of random HBA/HNA copolyesters. Various techniques such as TEM, SEM, WAXD and electron diffraction have been used to inspect the crystallite structures. There is now general recognition that the crystallites of HBA/HNA copolyester are based on lateral registration. However, different registration models have been presented, which are generally summarized into two categories [29, 30]. One is referred to as plane start registration [29], in which the common unit (the ester group for HBA/HNA) of both monomers on adjacent chains comes into register, and the register decays on moving away from the plane of registration, hence causing the formation of regions with imperfect sequence matching above and below the registration plane (Figure

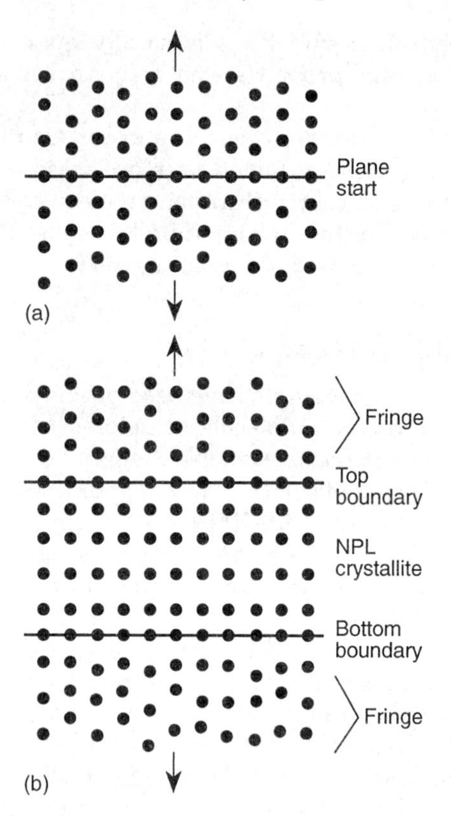

Figure 4.11 (a) Plane start model of LC copolyester crystallites; (b) model of NLP crystallites. (Reprinted with permission from [29], copyright © 1992 Butterworth-Heinemann Ltd.)

4.11a). The other registration model is based on aperiodic sequence segregation [29], in which identical monomer sequences on adjacent chains pack together not only by chance but also by axially searching along the chains. The ordered regions thus formed are known as non-periodic layer (NPL) crystallites, and there are partially ordered fringe regions above and below the crystallites (Figure 4.11b).

Hanna *et al.* [29] have investigated the crystal morphology of a low molecular weight HBA/HNA copolyester ($M_n \sim 5000$) by SEM, dark-field TEM and small-angle X-ray scattering (SAXS). The dimension and quantity of the crystallites were found to be dependent on the annealing conditions. A higher annealing temperature results in more, larger and more distinct crystallites. The typical size of crytallites in films subjected to annealing is 10–15 nm thick and up to about 300 nm long. For samples annealed at about 200°C for 1 h, the average spacing between crystallites is 38.5 nm and the crystallinity is approximately 19%, thus yielding a crystal thickness of about 7.3 nm, which is very close to the value obtained from modeling of NPL crystallites. Therefore, they concluded that the crystallization of HBA/HNA random copolyester can only be accounted for by segregation of the same monomer sequences and cannot be explained by the plane start model. Further investigation [31] was made to confirm whether the segregated and layered structures are already present in the melt before crystallization, or that there are smectic-type regions within the nematic matrix, which explains why the large-scale molecular rearrangements expected in the model of aperiodic sequence segregation can occur so rapidly during crystallization.

Hudson and Lovinger [30] also studied the crystalline morphology of HBA/HNA 73/27 copolyester via TEM and a similar etchant to that used by Hanna *et al.* [29]. Crystalline lamellae measure about 10 nm in thickness and about 100 nm laterally and are periodically spaced at approximately 34 nm. These data are quite close to those of Hanna *et al.* but a different conclusion was drawn. Since the weight average molecular length of this PLC is about 200 nm ($M_n \sim 14700$), which is six times that of the lamellar periodicity, and the chain is extended, yet having some flexibility, the chain is considered to be incorporated in several lamellae. In other words, lamellar crystals are bridged by many tie chains. Therefore, it was deduced that the plane start registration model which requires only slight axial shifts of the chain is more reasonable for this morphology. Furthermore, Hudson and Lovinger also related the lamellar structure to the disclination morphology. Since the lamellae are orthogonal to the molecular director, the disclinations can be recognized in lower magnification TEM images from the locations where the lamellar director rotates [30]. For injection molded samples, the defect density observed by this method is 10 cm^{-2} in the skin and 10^7 cm^{-2} in the core [32].

Crystallization kinetics was studied by Cheng using calorimetric analysis for random copolyesters with various HBA/HNA ratios [33]. Crystallites were expected to be formed by at least two types of molecular motion: shifting along the chain direction and rotating around the chain axis. The transition enthalpy, peak temperature of the endotherm and the exotherm, and their variation with cooling rate are different for various HBA/HNA ratios because the naphthalene ring is difficult to rotate and this leads to greater friction for the shift. Two processes were assumed to occur during cooling. One is called rapid solidification, and the other is normal crystallization at a relatively slow rate. The kinetics of the latter process was found to accord with the Avrami equation with a very low value of the Avrami index, n (see also Section 7.2).

Polymorphism is also common for thermotropic longitudinal PLCs. Two types of crystal lattice exhibit in these PLCs: orthorhombic and hexagonal ones. Their formation depends on the thermal history of PLCs. Kaito *et al.* [34] reported that the lateral packing of molecular chains in the as-extruded HBA/HNA copolyester transformed from hexagonal to orthorhombic form in the course of annealing above 230°C, while Frayer [35] observed that the core material of injection molded Xydar was mainly orthorhombic packing at room temperature, and converted into an expanded orthorhombic form, then to a pseudo-hexagonal form and finally to a regular hexagonal form with increasing temperature. The apparently contradictory results can be explained according to Kaito *et al.* [34] as being due to the aromatic planes freely distributed along the chain axis in the molten state. If the PLC is quenched, the conformation disorder will be frozen. The disordered conformation only allows the molecular chains to pack in a hexagonal lattice. Only if the temperature is high enough to permit chain rotation will the conformation transform into the energetically stable form with two-fold symmetry, which leads to the orthorhombic packing. Therefore, it is expected that a hexagonal lattice will form in quenched PLCs, while orthorhombic lattice will form in slowly cooled or annealed samples.

Recrystallization, that is the formation of high-melting crystals during partially molten state annealing, is also an important behavior for PLCs. This phenomenon has been detected both in HBA/HNA and in PET/PHB PLCs. For HBA/HNA with melting point around 280°C [36], recrystallization occurs above the melting temperature up to 320°C, and for PET/PHB ($T_m \sim 200$°C) [37], up to 265°C. In the temperature range not much higher than the apparent melting point, crystallites of long HBA segments are still unmolten, but most segments have obtained sufficient mobility to rearrange, and therefore new crystal growth can be induced by the residual HBA crystallites acting as nuclei. The crystals formed during this partially molten state have a higher fusion temperature than that of the original ones. If the PLCs are preheated above the upper temperature limit, the residual crystallites of long HBA segments will be eliminated, and no recrystallization will occur when the PLCs are annealed at the same temperature. Thus the degree and kinetics of the recrystallization depend on the thermal and orientational history of the sample as well as the HBA content and sequential randomness of the copolymer.

Besides the extended chain PLC crystals discussed above, those PLCs containing flexible aliphatic segments may form folded chain lamellar morphology [13, 38]. For a thermotropic terpolymer containing an aliphatic segment, chain folding was reported when the aliphatic segment contains more than four carbon atoms [13]. The folding conditions are dependent on the length of the aliphatic segment, ambient temperature and stress to which the PLC is subjected. For a molded bulk of this terpolymer with an aliphatic segment of seven carbon atoms, molecular chains were found to be extended in the slowly cooled interior and folded in the rapidly frozen surface region [13], which indicates that the chains are initially folded in the course of crystallization and then become extended as the temperature is high enough for the conformational transformation.

The characteristics of the crystallization behaviors of thermotropic longitudinal PLCs can be summarized as (1) small and imperfect lamellae about 10–100 nm in size; (2) limited crystallinity; (3) an increase of lamellar thickness, improvement of lateral correlation and transformation of the lattice resulting from annealing; and (4) formation of high-melting crystals in the partially molten state.

Important issues concerning the crystallization behaviors of PLCs which should be further explored include the crystallization mechanism, the relation of crystal lamellae to the microfibrils and the effect of crystallites on mechanical properties.

4.4 MULTIPHASE MORPHOLOGY

Another facet of morphology of PLCs which plays an important part in determining both the optical and mechanical properties of PLCs is the phase separation behavior. The phase separation behaviors of PLCs originate from non-uniformity of the molecular weight and chemical disorder of molecules [39]. The resultant multiphase morphologies include the phenomenon of nematic–isotropic biphase in the liquid crystalline state [39, 40] as well as the segregation of different segments in PLC products [41–43]. The latter case has been observed in block PLCs, such as PET/0.6PHB or PET/0.8PHB. Only the latter case will be discussed in this section.

The heterogeneous morphology and microphase separation process of block PLCs are closely related to the processing history and the molecular structures. For PET/PHB PLCs, PET-rich and PHB-rich phases are detected by SEM observation of etched samples [41, 42]. When the PHB content is lower than 50 mol%, the PET-rich phase is continuous, and *vice versa* when the PHB content is higher than 60 mol%. The size of the PET-rich phase was found to be 10–20 μm with 40 mol% PHB and 3–6 μm with 80 mol% PHB. In addition, the phase dimension will be influenced by the thermal and mechanical history. Joseph *et al.* have reported [41, 42] that in an injection molded plaque of PET/PHB PLCs, PHB was richer in the skin while PET was richer in the core. However, there is controversy concerning this observation [3].

Multiphase structures of two wholly aromatic blocky copolyesters with a composition similar to Xydar have been observed in our laboratory [43]. Compact spherical domains with a diameter of about

0.1 μm can be seen from SEM micrographs of the fracture surface of compression molded samples, and multitransition peaks are shown in their dynamic mechanical diagrams. Upon annealing around 200°C, which is above their glass transition temperature, these PLCs experience further phase separation, which results in an increment of the domain size and loose contact between the domains. The Young's moduli of these PLCs are found to drop slightly after annealing, although they are expected to increase because of formation of more crystallites and improvement of lateral packing during annealing. Obviously, it is the effect of phase separation that causes the modulus to fall. Therefore, it can be understood that the phase separation behaviors and the multiphase morphology are closely related to the mechanical properties of PLCs.

4.5 SUMMARY

Hierarchical structures, crystallization behaviors and multiphase morphologies of thermotropic longitudinal PLCs have been discussed in this chapter. Morphology is a bridge between molecular structures and the final properties. For PLCs, the effect of morphology is particularly important because the packing and alignment of rigid rod molecules are more sensitive to processing conditions than that of flexible chain molecules of ordinary polymers.

The products of thermotropic longitudinal PLCs have the advantages of high stability and high tensile strength and moduli, but are accompanied by the shortcomings of relatively low impact and compressive strength and large anisotropy. New molecular structures and processing technology should be developed to improve the properties of PLCs. For example, introducing short flexible branches or bending units on the rigid backbone to increase entanglement among chains. In addition, the ideas of cross (in which mesogens are cross-like instead of rod-like) or mixed (with the mesogens alternately aligned parallel and perpendicular to the backbone) PLCs [1] may be potential routes to enhance the molecular interaction as well as to reduce anisotropy. On the other hand, biaxial orientation processing techniques should also be developed for PLCs to make better use of the orientability of PLCs.

REFERENCES

1. Brostow, W. (1990) *Polymer* 31, 979.
2. Kwolek, S.L., Morgan, P.W. and Shaefgen, J.R. (1992) *High Performance Polymers & Composites*, p. 416.
3. Acierno, D. and Nobile, M.R. (1993) *Thermotropic Liq. Cryst. Polymer Blends*, 59.

4. Dobb, M.G., Hendeleh, A.M. and Saville, B.P. (1977) *J. Polymer Sci. Symp.*, **58**, 237.
5. Weng, T., Hiltner, A. and Baer, E. (1986) *J. Mater. Sci.*, **21**, 744.
6. Ide, Y. and Ophir, Z. (1983) *Polymer Eng. & Sci.*, **23**, 261.
7. Viola, G.G., Baird, D.G. and Wilkes, G.L. (1985) *Polymer Eng. & Sci.*, **25**, 888.
8. Mehta, S. and Deopura, B.L. (1993) *J. Appl. Polymer Sci.*, **47**, 857.
9. Chen, G.Y., Cuculo, J.A. and Tucker, P.A. (1988) *J. Polymer Sci. Phys.*, **26**, 1677.
10. Sawyer, L.C., Chen, R.T., Jamieson, M.G. *et al.* (1993) *J. Mater. Sci.*, **28**, 225.
11. Kyotani, M., Kaito, A. and Nakayama, K. (1993) *J. Appl. Polym. Sci.*, **47**, 2053.
12. Sawyer, L.C. and Jaffe, M. (1986) *J. Mater. Sci.*, **21**, 1897.
13. Kent, S.L., Rybnikar, F., Geil, P.H. and Carter, J.D. (1994) *Polymer*, **35**, 1869.
14. Zülle, B., Demarmels, A., Plummer, C.J.G. and Kausch,. H.-H. (1993) *Polymer*, **34**, 3628.
15. Plummer, C.J.G., Zülle, B., Demarmels, A. and Kausch, H.-H. (1993) *J. Appl. Polymer Sci.*, **48**, 751.
16. Hedmark, P.G., Lopez, J.M.R., Westdahl, M. *et al.* (1988) *Polymer Eng. & Sci.*, **28**, 1248.
17. Bensaad, S., Jasse, B. and Noël, C. (1993) *Polymer*, **34**, 1602.
18. Jansen, J.A.J., Paridaans, F.N. and Heynderickx, I.E.J. (1994) *Polymer*, **35**, 2970.
19. Grag, S.K. and Kenig, S. (1988) *High Modulus Polymers*, Marcel Dekker, New York, p. 71.
20. Duska, J.J. (1986) *Plast. Eng.*, **42**, 39.
21. Ophir, Z. and Ide, Y. (1983) *Polymer Eng. & Sci.*, **23**, 792.
22. Suokas, E. (1989) *Polymer*, **36**, 1105.
23. Khennache, G. and Kamal, M.R. (1991) *Proc. SPE ANTEK Tech. Papers*, 37th, May 1991, Montreal, p. 1063.
24. Kenig, S., Trattner, B. and Anderman, H. (1988) *Polym. Compos.*, **9**, 20.
25. Jackson, W.J. and Kuhfuss, H.F. (1976) *J. Polymer Sci. Chem.*, **14**, 2043.
26. Ryan, T.G. (1988) *Mol. Cryst. Liq. Cryst.*, **157**, 577.
27. Plummer, C.J.G., Wu, Y., Gola, M.M. and Kausch, H.-H. (1993) *Polymer Bull.*, **30**, 587.
28. Blundell, D.J. (1982) *Polymer*, **23**, 359.
29. Hanna, S., Lemmon, T.J., Spontak, R.J. and Windle, A.H. (1992) *Polymer*, **33**, 3.
30. Hudson, S.D. and Lovinger, A.J. (1993) *Polymer*, **34**, 1123.
31. Hanna, S., Romo-Uribe, A. and Windle, A.H. (1993) *Nature*, **366**, 546.
32. Hudson, S.D., Andrew, A.J., Venkataraman, S.K. *et al.* (1994) *Polymer Eng. & Sci.*, **34**, 1327.
33. Cheng, S.Z.D. (1988) *Macromolecules*, **21**, 2475.
34. Kaito, A., Kyotani, M. and Nakayama, K. (1990) *Macromolecules*, **23**, 1035.
35. Frayer, P.D. (1987) *Polymer Compos.*, **8**, 379.
36. Lin, Y.G. and Winter, H.H. (1988) *Macromolecules*, **21**, 2439.
37. Huang, K., Lin, Y.G. and Winter, H.H. (1992) *Polymer*, **33**, 4533.
38. Hudson, S.D., Thomas, E.L. and Lenz, R.W. (1987) *Mol. Cryst. Liq. Cryst.*, **153**, 63.
39. Stupp, S.I., Moore, J.S. and Martin, P.G. (1988) *Macromolecules*, **21**, 1228.
40. d'Allest, J.F., Sixou, P., Blumstein, A. and Blumstein, R.B. (1988) *Mol. Cryst. Liq. Cryst.*, **157**, 229.
41. Joseph, E., Wilkes, G.L. and Baird, D.G. (1986) *Polymer*, **26**, 689.
42. Joseph, E.G., Wilkes, G.L. and Baird, D.G. (1985) *Polymer Eng. & Sci.*, **25**, 377.
43. Zhong, Y., Xu, J. and Zeng, H. (1993) *Gaofenzi Cailiao Kexue yu Gongcheng*, **3**, 49.

5

Polymer liquid crystals in solution

Lydia Fritz, Joachim Rübner, Jürgen Springer and Dietmar Wolff

5.1 INTRODUCTION

The great interest in the application of new materials with anisotropic properties since 1980 or so has led to the stormy development of polymer liquid crystals (PLCs). With the aim of optimizing certain mechanical, optical or electrical properties, many polymers with new structures have been synthesized and tested. To understand the relationship between the properties and the molecular and supermolecular structure of this class of polymers, systematic investigations of the materials have not kept up with this dramatic development. Especially the properties of these polymers in solution have been relatively little investigated until now, even though information concerning molecular parameters such as molar mass, degree of polymerization, molecular dimensions and shape, linearity and chain branching can be accessed only through the investigation of solutions. In addition to the anisometric shape of the repeating units, the properties of mesophases are due to strong interactions between the mesogens.

Mechanical and Thermophysical Properties of Polymer Liquid Crystals
Edited by W. Brostow
Published in 1998 by Chapman & Hall, London.
ISBN 0 412 60900 2

This has been observed mainly in halogen-containing solvents, and the length of the flexible spacers also shows an influence on this effect [24, 33]. It is obvious to attribute the tendency to form aggregates to mesogen–mesogen interactions. Moreover, it is interesting to note that for the solvents where the aggregation was observed, the second osmotic virial coefficients A_2 and the radii of gyration $(\langle R_g^2 \rangle)^{0.5}$ of the SGPLCs indicate that these are thermodynamically the best solvents [15, 20]. The A_2 values of homologous series of SGPLCs become smaller with increasing molar masses, as theoretically predicted and well known for conventional polymers.

The results of the few LS measurements show unusual dependence on temperature. The A_2 values change with temperature only in thermodynamically bad solvents and remain almost constant in other cases [13, 15, 20]. A visible clouding of the solution on cooling has been observed by Ohm and others even before reaching the theta temperature. A value of $\Theta = 16 \pm 1°C$ for the theta temperature of $PMAC_6$ in toluene independent of the molar mass has been reported [13, 15, 20].

For a polymer with an almost identical constitution (the spacer consists of nine methylene units, $PMAC_9$) toluene was found to be a thermodynamically better solvent in the temperature range 16–45°C. A_2 for $PMAC_9$ ($M_w = 524\,000\,g\,mol^{-1}$, unfractionated) in toluene ($0.90 \times 10^{-4}\,cm^3\,mol\,g^{-2}$) is comparable to the A_2 value for $PMAC_6$ ($M_w = 551\,000\,g\,mol^{-1}$, unfractionated) in the thermodynamically better solvent benzene ($0.99 \times 10^{-4}\,cm^3\,mol\,g^{-2}$). In this solvent the second osmotic virial coefficient of a fractionated sample of $PMAC_6$ with a similar molar mass ($M_w = 540\,000\,g\,mol^{-1}$, $E = 1.2$ (SEC)) is $0.95 \times 10^{-4}\,cm^3\,mol\,g^{-2}$. However, the increase of the A_2 value should not be attributed to the increment of the spacer length. For instance, for a spacer homologous series of $PMAC_n$, Trapp [19] observed an odd–even effect on the A_2 values in THF. According to that the odd homologous ($n \leqslant 7$) show a stronger interaction with the solvent THF than the even ones. However, it should be mentioned, that only A_2 values for samples with relatively high but different degrees of polymerization and polydispersity indices have been compared. Nevertheless, Duran and Strazielle [24] also report a strong influence of the spacer length on the Flory–Huggins parameter χ and discuss an odd–even effect. Other authors [13] found that lengthening the tail group leads to higher values of A_2.

The same behavior as that described for the second osmotic virial coefficients has been reported for the radii of gyration determined by static light scattering. At the theta temperature of the system $PMAC_6$ dissolved in toluene, Kuhn's law ($\langle R_g^2 \rangle \sim M$) was verified [13, 15, 20]. With the improvement of the solvent quality the coils of the SGPLCs become larger, but their dimensions in solution are smaller than those of conventional polymers (for instance polystyrene) of the same molar mass. The consequence of this is that often the macromolecules behave like Rayleigh particles and then the radii of gyration cannot be determined using light scattering [23, 27, 33, 43].

In the following the radii of gyration in the unperturbed state will be compared for $PMAC_6$, for polymethylmethacrylate (PMMA) and for a polystearylmethacrylate (PSMA) with a side chain length similar to that of $PMAC_6$ (assuming an all-*trans* conformation). For samples with the same molar masses of $M_w = 1 \times 10^6\,\mathrm{g\,mol^{-1}}$ the following order was determined: $(\langle R_g^2\rangle_\Theta)^{0.5}_{PMMA} = 23.6\,\mathrm{nm} > (\langle R_g^2\rangle_\Theta)^{0.5}_{PSMA} = 21\,\mathrm{nm} > (\langle R_g^2\rangle_\Theta)^{0.5}_{PMAC_6} = 19.4\,\mathrm{nm}$. So for the same molar mass the SGPLC sample exhibits the smallest and the PMMA sample shows the largest coil dimensions. This tendency can be understood because for the same M_w the degree of polymerization is different due to different molar masses of the side groups.

Comparing samples with the same degree of polymerization P, for instance $P = 2500$, the $(\langle R_g^2\rangle_\Theta)^{0.5}$ values for PSMA and $PMAC_6$ are almost equal: $(\langle R_g^2\rangle_\Theta)^{0.5}_{PSMA} = 20.1\,\mathrm{nm} > (\langle R_g^2\rangle_\Theta)^{0.5}_{PMAC_6} = 19.7\,\mathrm{nm} > (\langle R_g^2\rangle_\Theta)^{0.5}_{PMMA} = 11.8\,\mathrm{nm}$. The origin of the small difference between PSMA and $PMAC_6$ is not quite clear. Whether it confirms the assumption of mesogen–mesogen interactions for $PMAC_6$ in solution, decreasing the volume necessary for the bulky mesogenic groups, or

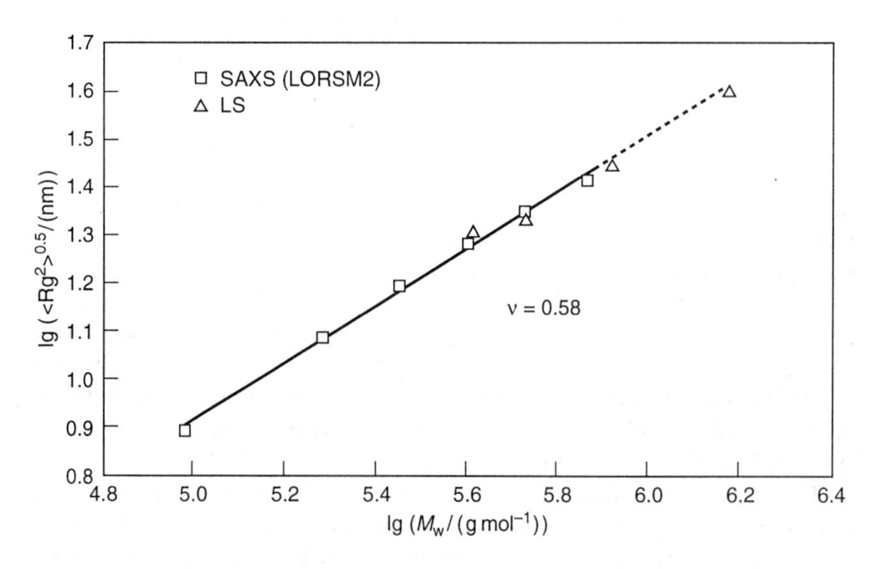

Figure 5.5 Dependence of the radius of gyration on the molar mass for $PMAC_6$ in benzene at 20°C [20].

whether the effective volume the alkyl chains of PSMA occupy is increased due to their flexibility should be investigated in the future.

Radii of gyration determined by SAXS for low molar mass PLC fractions complete the values obtained by LS measurements. In the molar mass range where it is possible to apply both methods, the results are in good agreement [20]. In Figure 5.5 the dependence of the radius of gyration on the molar mass for $PMAC_6$ at 20°C in the thermodynamically good solvent benzene is displayed. The excluded volume exponent [44] of $\nu = 0.58$ [20] is in good agreement with the value reported for linear flexible chain macromolecules in thermodynamically good solvents.

Another interesting question is if there is an influence of the mesogenic side groups on the stiffness of the backbone. This can be discussed only when the molar mass and also the coil dimensions at the theta temperature are known. Often the characteristic ratio C_∞ introduced by Flory [45], the length of the statistical or Kuhn chain segment b, the persistence length a^* [46] or the steric factor σ are discussed. All these quantities provide information on the conformational restrictions for the rotation ability of backbone segments. For all LS measurements carried out on PLCs compared to non-LC polymers with similar architecture, a drastic stiffening of the polymer backbone was observed [15, 20, 23]. The same result was obtained by viscosimetry [24, 26, 28] and also by SAXS measurements [15, 20]. The stiffening of the polymer backbone seems to be independent of the linkage of the mesogens (end-on or side-on) to the main chain, while it depends strongly on the spacer length [20, 23, 24]. For instance, for SGPLCs consisting of a polymethacrylate backbone, ethoxy spacers and methoxybiphenyl mesogens, Duran and Strazielle report values of b of 0.796 nm (spacer length $n = 1$), 1.08 nm ($n = 2$) and 1.15 nm ($n = 3$) clearly higher than the b value of 0.629 for PMMA [24]. The same tendency was also demonstrated for other polymethacrylates with alkyl spacers [20]. For example the steric factor $\sigma = 3.13$ calculated for $PMAC_6$ is higher than that of 1.87 for PMMA. In addition the persistence length of $PMAC_6$ (2.2 nm) is greater than that of PMMA (0.8 nm) [47]. Still greater values were found for the Kuhn chain segment length ($b \sim 7$ nm) and the steric factor ($\sigma = 4.2$) obtained by sedimentation measurements on an LC polymethacrylate, where the mesogenic group is side-on linked to the backbone via a long alkyl spacer ($n = 11$) [26]. However, the results discussed above do *not* allow the conclusion that the stiffening has to be ascribed to the mesogenic groups. In fact, also for non-LC polymers the chain flexibility decreases with longer and/or more bulky side groups [20, 23, 26]. Therefore, it seems that the existence of side groups is the reason for the backbone stiffening, irrespective of their mesogenic or non-mesogenic nature.

By SAXS measurements on solutions it is possible to determine also the mean square of the axial scattering mass radius $(\langle R_q^2 \rangle)^{0.5}$ that can be used to investigate the influence of the chemical structure of the side groups (for instance changes in the spacer length) on the cross-section of the chain molecule. For $PMAC_6$ in benzene at 20°C and in toluene at 17°C, the same value of $(\langle R_q^2 \rangle)^{0.5} = 1.05$ nm was determined [20]. As expected, increasing the spacer length causes an increase in the cross-section of the macromolecule. For example, for $PMAC_9$ in toluene the value of $(\langle R_q^2 \rangle)^{0.5}$ is found to be 1.17 nm. Assuming that the polymer chains are long flexible cylinders with a unique electron density, $(\langle R_q^2 \rangle)^{0.5}$ can be used to calculate the diameter d_{cyl} of these cylinders [48]. Following this idea, for $PMAC_6$ at 17°C and for $PMAC_9$ at 30°C in toluene, values of 2.98 nm and 3.30 nm are respectively calculated. Similar values (2.93 nm and 3.33 nm) are reported by Trapp [19] for the length of the side groups calculated accordingly to a Stuart–Briegleb model. This surprising agreement seems not to be in accordance with an antiparallel arrangement of side groups of adjacent repeating units but can be understood taking into account a helix-like arrangement of the side groups in solution.

Beside the results derived from static LS measurements discussed until now from dynamic LS investigations, the self-diffusion coefficient of dissolved macromolecules can be determined. It is possible to calculate the hydrodynamic radius R_h that is proportional to the radius of gyration: $\rho = (\langle R_q^2 \rangle)^{0.5}/R_h$. Since the parameter ρ depends on the density of the coil, it is possible to distinguish between compact structures, flexible statistical coils and rod-like molecules. Burchard reports ρ values for some geometries of macromolecules [49]. The few results published on PLCs are in the range from 1.6 to 2.1 independent of the kind of linkage of the mesogenic groups to the backbone [11, 27]. Such values of ρ are typical for polydispersed linear chain molecules, but it should be mentioned that the highest values were observed for halogenated solvents.

5.4 SEMIDILUTE SOLUTIONS OF SGPLCS

In semidilute solutions the macromolecules are able to interpenetrate and give rise to a network-like structure temporarily. The properties of these more concentrated solutions differ remarkably from those of dilute solutions due to the strong intermolecular interactions of the macromolecules. In LS measurements the single macromolecule is no longer involved. Structures and thermodynamic properties of the solutions due to interactions between the macromolecules can be investigated. For the analysis of static LS measurements in this concentration range, the intensity of the scattered light is extrapolated to the value $q \longrightarrow 0$ giving the inverse osmotic compressibility $(1/RT)(\partial\pi/\partial c)_T$ [44]; which is usually denoted as osmotic modulus. Following the theory, the reduced osmotic modulus $(M_w/RT)(\partial\pi/\partial c)_T$ is a function of the parameter $X = A_2M_wc$ that is proportional to c/c^*, where c^* is the concentration at which molecular coils start to overlap. Thus the parameter X reflects a thermodynamic definition of the coil overlap concentration.

In a diagram where the reduced osmotic modulus is plotted as a function of the parameter X, a universal curve is found for linear flexible chain molecules independent of the molar mass. Macromolecules of different architecture, for instance star-like, random branched or stiff macromolecules, deviate from this curve in a characteristic way. Figure 5.6 displays this for PMAC$_6$ dissolved in benzene [15].

Only for small values of X do the experimental data lie on the theoretical curve for flexible chain molecules. For values of X greater than ~ 0.1 the data show deviations. First of all, as typical for stiff macromolecules or for polydispersed samples, the measured values are always smaller than those predicted theoretically for linear flexible macromolecules. Although this negative deviation seems to be relatively

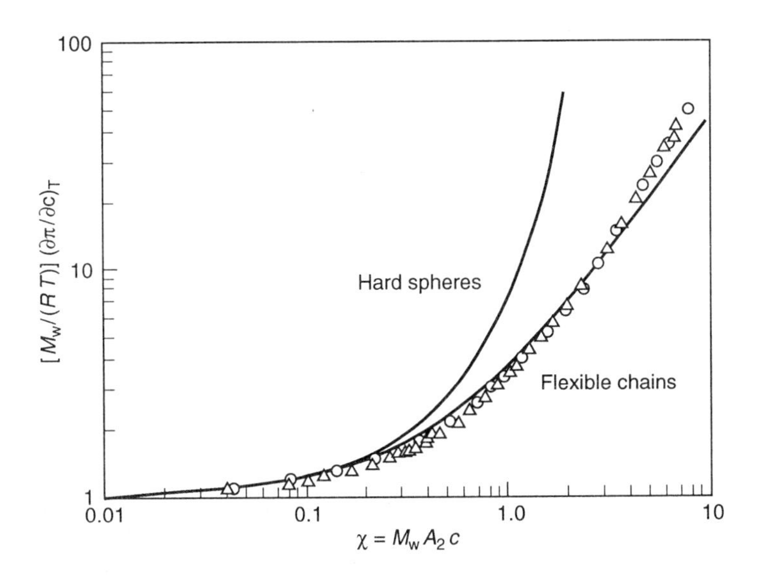

Figure 5.6 Plot of the reduced osmotic modulus $(M_w/RT)(\partial\pi/\partial c)_T$ versus the parameter $X = A_2M_wc$ for two fractions of PMAC$_6$ in benzene at 20°C. The theoretical curves for hard spheres and flexible chains are drawn as full lines [20].

small, it is significant and was observed by different groups [15, 27]. The experimental data intersect the theoretical curve for values in the range of $X \sim 2-3$ but above this range they are higher.

Richtering *et al.* [27] attributed the behavior that the measured values were smaller than the theoretical ones to the polydispersity of their radicalic synthesized LC copolymers. Taking into account the polydispersity ($E \sim 1.2-1.3$) of the samples reported in Figure 5.6, it is possible to suppose that this effect is not a consequence of the polydispersity but of the rigidity of the polymer main chain, as discussed previously for the measurements on dilute solutions. The intersection of the theoretical curve for flexible coils as well as the fact that points lay above the curve was observed for all the previous measurements on PLCs [11, 27]. However, even if not so pronounced, this effect was also found for conventional polymers like polystyrene [50].

In addition, Richtering *et al.* found an influence of the degree of polymerization on the deviation of the experimental points from the theoretical curve, ascribing it also to the larger excluded volume for PLCs due to the bulky side groups. While the kind of linking of mesogens to the backbone seems to have no or at least only a small influence [11], it is interesting to note that the osmotic moduli also show the highest values in a halogenated solvent ($CHCl_3$) [27]. The values measured in these solvents are closer to the theoretical curve for hard spheres. Therefore here the segment density should be higher.

Another property that should be pointed out, which was always found until recently in LS measurements on semidilute solutions of PLCs, is a remarkable increase of the intensity of scattered light at small angles above a certain concentration depending on the system investigated [11, 15, 20, 27]. This scattering, called excess low-angle scattering (ExLAS), was also observed for conventional polymers such as polystyrene and was attributed to an inhomogenity of the semidilute solution. Until now the exact structure of the aggregates formed by the particles was unknown. The concentration above which the ExLAS appears depends on the polymer, on the solvent and in principle also on the degree of polymerization.

Nevertheless, for three samples of a SGPLC that differ remarkably in the molar mass ($M_w = 54\,100\,\mathrm{g\,mol^{-1}}$; $168\,000\,\mathrm{g\,mol^{-1}}$; $890\,000\,\mathrm{g\,mol^{-1}}$) Richtering *et al.* observed that the degree of polymerization did not influence the occurrence of ExLAS. In $CHCl_3$ the formation of clusters is initially independent of the molar mass for concentrations higher than $c \sim 120\,\mathrm{g\,L^{-1}}$ due to a slow diffusion motion, which was investigated by dynamic LS measurements. It was supposed that the cluster formation, i.e. the increase of the size of particle aggregates, does not depend on the degree of polymerization but on the segment concentration, suggesting that specific interactions between the mesogens are responsible for it. The results actually give indication of a homogeneous and isotropic structure of the clusters, but not of a long-range LC order inside the aggregates. For the investigated LC polymethacrylate with side-on bonding of the mesogenic groups, an increase of both the size of the clusters and of their mass with increasing concentration was established. In comparison to that with static LS measurements on solutions of a polymethacrylate with end-on bonding of the mesogens to the backbone, Fritz [20] also found an increase of the mass of the clusters with increasing concentration. For the dimensions of the aggregates, however, a value of $a = 65 \pm 10$ nm independent of the concentration of the solution was reported.

5.5 CONCLUDING REMARKS

Summarizing, it was shown that to date only a few investigations on solutions of PLCs have been carried out systematically. Nevertheless, the results obtained prove that the interactions between mesogenic groups have consequences on the properties of the solutions of these molecules. PLCs in solution show a conformation that is more compact compared to that of non-LC polymers with a similar structure. The influence on the properties of the solutions resulting from that more compact conformation, however, agrees with the predictions of the theories of polymers in solution.

REFERENCES

1. Brostow, W. (1990) *Polymer*, **31**, 979.
2. Kricheldorf, H.R. and Schmidt, B. (1992) *Macromolecules*, **25**, 5471.
3. Kricheldorf, H.R. and Engelhardt, J. (1990) *J. Polymer Sci. A*, **28**, 2335.
4. Kricheldorf, H.R. and Thomsen, S.A. (1993) *Makromol. Chem. Rapid Commun.*, **14**, 395.
5. Kricheldorf, H.R., Schmidt, B. and Bürger, R. (1992) *Macromolecules*, **25**, 5465.
6. Kuhn, R., Marhold, A. and Dicke, H.R. (1990) Patentschrift DE 3712817 C2.
7. Kuhn, R., Marhold, A. and Dicke, H.R. (1990) Patentschrift US 4, **960**, 539.
8. Krömer, H., Kuhn, R., Pielartzik, H. *et al.* (1991) *Macromolecules*, **24**, 1950.
9. Kuhn, R. (1993) *J. Appl. Polymer Sci. Symp.*, **52**, 19.
10. Ueberreiter, K. (1968) in *Diffusion in Polymers* (eds J. Crank and G.S. Park), Academic Press, London.
11. Richtering, W., Gleim, W. and Burchard, W. (1992) *Macromolecules*, **25**, 3795.

12. Tsvetkov, V.N., Shtennikova, I.N. and Blumstein, A. (1985) *Polymer Liq. Cryst. Sci. Technol.*, **28**, 83.
13. Ohm, H.G., Kirste, R.G. and Oberthür, R.C. (1988) *Makromol. Chem.*, **189**, 1387.
14. Damman, S.B. and Mercx, F.P.M. (1993) *Polymer*, **34**, 2726.
15. Fritz, L. and Springer, J. (1993) *Makromol. Chem.*, **194**, 2047.
16. Springer, J. and Weigelt, F.W. (1983) *Makromol. Chem.*, **184**, 2635.
17. Cackovic, H., Springer, J. and Weigelt, F.W. (1984) *Progr. Coll. & Polymer Sci.*, **69**, 134.
18. Springer, J. and Weigelt, F.W. (1985) in *Recent Advances in LCP*, (ed. L.L. Chapoy), Elsevier Applied Science Publishers, London–New York, **14**, 233.
19. Trapp, W. (1992) Dissertation, Tech. Univ. Berlin, D83.
20. Fritz, L. (1992) Dissertation, Tech. Univ. Berlin, D83.
21. Borisova, T.I., Stepanova, T.P., Freidzon, J.S. *et al.* (1988) *Vysokomol. Soed. A*, **30**, 1754.
22. Witkowski, K., Kuten, E. and Wolinski, L. (1992) *Eur. Polymer J.*, **28**, 895.
23. Wolinski, L., Witowski, K. and Turzynski, Zb. (1990) *Eur. Polymer J.*, **26**, 521.
24. Duran, R. and Strazielle, C. (1987) *Macromolecules*, **20**, 2853.
25. Anufrieva, E.W., Pautov, W.D., Freidzon, J.S. *et al.* (1984) *Doklady Akad. Nauk USSR*, **278**, 383.
26. Lavrenko, P.N., Kolomietz, I.P. and Finkelmann, H. (1993) *Macromolecules*, **26**, 6800.
27. Richtering, W.H., Schätzle, J., Adams, J. and Burchard, W. (1989) *Coll. & Polymer Sci.*, **267**, 568.
28. Lavrenko, P.N., Finkelmann, H., Okatova, O.W. *et al.* (1993) *Vysokomol. Soed. A*, **35**, 1652.
29. Sefton, M.S. and Coles, H.J. (1985) *Polymer*, **26**, 1319.
30. Casagrande, C., Fabre, P., Veyssie, M. C. and Finkelmann, H. (1984) *Mol. Cryst. Liq. Cryst.*, **113**, 193.
31. Sigaud, G., Achard, M.F., Hardouin, F. and Gasparoux, H. (1988) *Mol. Cryst. Liq. Cryst.*, **155**, 443.
32. Handel, T.M. and Ponticello, I.S. (1987) *Macromolecules*, **20**, 264.
33. Witkowski, K. and Wolinski, L. (1991) *Eur. Polymer J.*, **27**, 687.
34. Munk, P. (1989) *Introduction to Macromolecular Science*, Wiley, New York.
35. Huglin, M.B. (1992) *Light Scattering from Polymer Solutions*, Academic Press, London.
36. Berne, B.J. and Pecora, R. (1976) *Dynamic Light Scattering*, Wiley, New York.
37. Kunz, Ch. (1979) *Synchrotron Radiation Techniques and Applications. Topics in Current Physics*, Springer, Berlin.
38. Burchard, W. (1988) *Makromol. Chem. Symp.*, **18**, 1.
39. Glatter, O. and Kratky, O. (1982) *Small Angle X-ray Scattering*, Academic Press, New York.
40. Backon, G.E. (1975) *Neutron Diffraction*, Clarendon Press, Oxford.
41. Springer, J. and Weigelt, F.W. (1983) *Makromol. Chem.*, **184**, 1489.
42. Ohm, H. (1985) Dissertation, University of Mainz.
43. Siebke, W. (1986) Dissertation, Tech. Univ. Berlin, D83.
44. de Gennes, P.-G. (1979) *Scaling Concepts in Polymer Physics*. Cornell University Press, Ithaca, New York–London.

45. Flory, P.J. (1969) *Statistical Mechanics of Chain Modules*, Interscience, New York.
46. Kratky, O. and Porod, G. (1949) *Rec. Trav. Chim. Pays-Bas*, **68**, 1106.
47. Schulz, G.V. and Kirste, R. (1961) *Z. Phys. Chem.*, **30**, 171.
48. Durchschlag, H., Puchwein, G., Kratky, O. *et al.* (1970) *J. Polymer Sci. C*, **31**, 311.
49. Burchard, W. (1987) *Macromolecules*, **11**, 455.
50. Wendt, E. (1989) Dissertation, Tech. Univ. Berlin, D83.

Part Two
Thermophysical properties

6

Memory effects in polymer liquid crystals: influence of thermal history of phase behavior

Rita B. Blumstein

6.1 INTRODUCTION

Phase behavior and morphology in conventional polymers are heavily dependent on the thermal history of the sample, as is obvious to anyone even remotely familiar with macromolecules. Polymer liquid crystals (PLCs) are clearly subject to similar constraints by virtue of their macromolecular identity. In addition, a number of thermal properties are specific to PLCs as a result of the interaction between macromolecular behavior and the molecular ordering characteristic of LC mesophases. This chapter focuses on just such features of thermal history, as revealed by the interplay of kinetic and thermodynamic factors observed in thermotropic polymers.

As a rule, the phase behavior of PLCs displays a more complex dependence on thermal history than conventional polymers because of the rich variety of phases and annealing pathways that may be encountered. Figures 6.1–6.3 may help to illustrate this phenomenon.

An isotropic (amorphous) melt may be quenched to an isotropic (pathway 2 in Figure 6.1) and/or LC (ordered) polymer glass (pathway

Mechanical and Thermophysical Properties of Polymer Liquid Crystals
Edited by W. Brostow
Published in 1998 by Chapman & Hall, London.
ISBN 0 412 60900 2

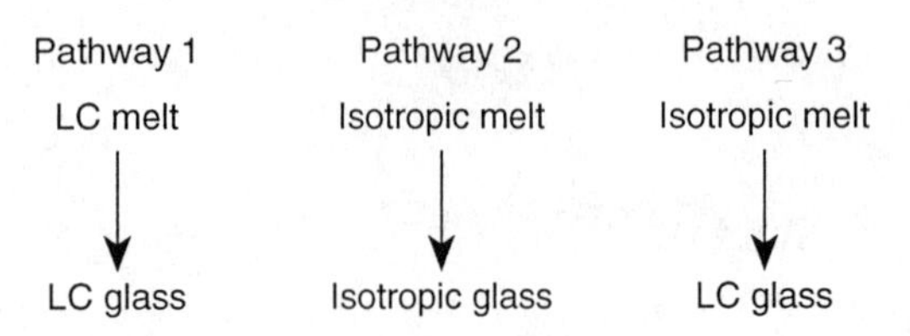

Figure 6.1 Pathways for development of LC or I glass structures.

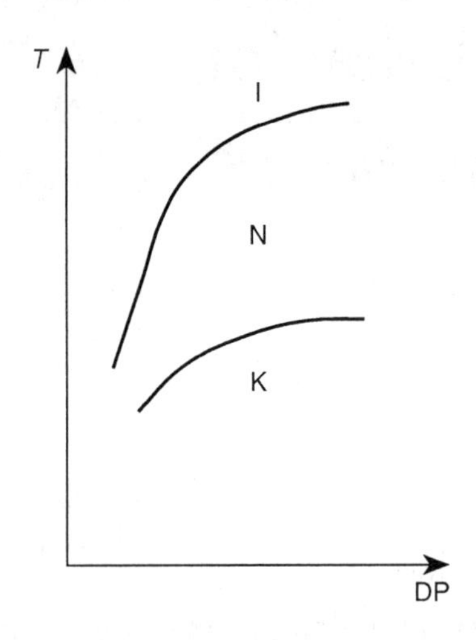

Figure 6.2 Influence of degree of polymerization on transition temperatures. Note the broad distribution of isotropization T_{ni} values that may occur in a polydisperse sample.

3). The LC glass formed directly by cooling from the mesophase via pathway 1 may be expected to have a different (perhaps more ordered and less heterogenous) structure than the LC glass formed via pathway 3; some memory of the precursor isotropic phase and isotropic + LC biphase may be quenched in the latter case. The isotropic glass 2 is metastable and may be expected to convert to the mesophase on heating, via a rearrangement process akin to cold crystallization ('cold liquid crystallization').

Liquid crystalline polymorphism is a frequent occurrence and two or more different LC phases may be exhibited following melting of the crystalline component, and prior to isotropization. It is clear that various combinations of phases may coexist and overlap, especially if phase transitions occur over a broad range of temperatures, as is often found in PLC systems. This last point is visualized in Figure 6.2, which illustrates the well known increase of phase transition temperatures

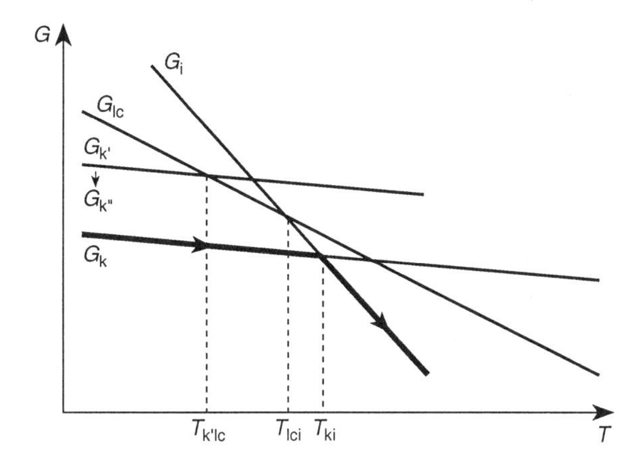

Figure 6.3 The Gibbs function diagram for a monotropic liquid crystal. G_k represents the equilibrium crystal, and $G_{k'}$ a possible metastable crystalline form. The equilibrium pathway on heating (marked by arrows) leads to direct isotropization at T_{ki}. Heating along pathway $G_{k'}$ gives a metastable enantiotropic behavior, with a mesophase at $T_{k'lc}$ and isotropization at T_{lci}. (Redrawn from data in [4])

with increasing degree of polymerization. Transition temperatures of less ordered (nematic) phases change more steeply with DP than temperatures of more ordered (smectic or crystalline) phases [1]. Very broad nematic–isotropic (N + I) biphases may ensue in some instances as a result of polydispersity in chain length [2]. In copolymers, heterogeneity in chain composition ('polyflexibility' [3]) is superimposed on polydispersity of molecular mass and this may drastically enhance the width of the N + I biphase, as will be seen later.

An interesting consequence of thermal history is revealed in the sometimes drastic difference between heating and cooling in structures where monotropic behavior is possible (i.e. a stable LC phase is formed only on cooling). Figure 6.3 [4] illustrates the origin of monotropic behavior and the possible transformation of a metastable enantiotropic phase into an equilibrium monotropic LC. Such phenomena have their origin in the progression of entropy from isotropic melt to crystal:

$$S_I > S_{LC} > S_K$$

which results in the Gibbs function diagram of Figure 6.3. On heating, the perfect crystal (characterized by free energy G_k) isotropizes directly at T_{ki}. On cooling, however, a monotropic mesophase is developed at T_{lci} and an imperfect crystal with free energy $G_{k'}$ is observed at $T_{lck'}$. The value of $G_{k'}$ tends to approach G_k upon crystal annealing, reaching $G_{k''}$ for example. A metastable enantiotropic behavior is observed as long as G_k is not recovered.

One example of such metastable enantiotropic behavior is provided by a side chain polysiloxane with the following transitions recorded on heating from the glassy state: G 5 S 54 N 112 I [5]. However, a crystal which melts above 54°C (the S–N transition) develops upon annealing the mesophase above T_g, and displaces the smectic phase which becomes monotropic.

Aging within the mesophase can occasionally result in transformation from a metastable to a different, stable LC phase, as in the change from a nematic to a smectic phase observed in copolyethers based on the diphenylethane mesogen [6, 7]. However, by far the most common consequence of thermal treatment is revealed in more subtle ways, such as changes in morphology or texture and temperatures or enthalpies of transition.

In this chapter we selectively describe the influence of thermal history within the various phases that can be present in a given sample, including biphasic regimes. Discussion is devoted primarily to main chain PLCs, although a few examples of side chain structures are briefly mentioned. Among the main chain PLCs, emphasis is placed on segmented macromolecular backbones formed by alternating mesogens and flexible spacer groups (rigid–flexible (RF) sequencing). Representative structures to be discussed are shown in Figure 6.4.

These systems are chosen to illustrate some studies of memory effects, thermal history dependence of LC transition temperatures,

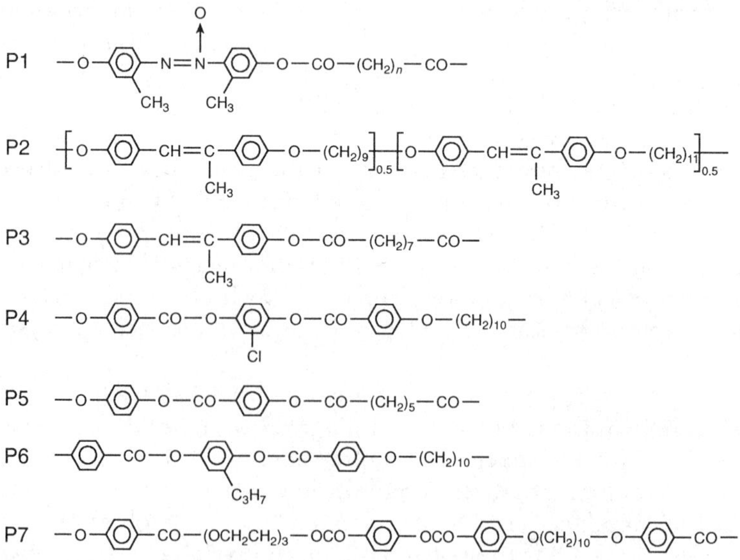

Figure 6.4 Selected RF polymers. Spacer length n and number average molecular mass M are specified in parentheses in the text, as appropriate.

enthalpy and other aspects of 'phase behavior'. Ion containing [8] or hydrogen bonded [9, 10] PLCs show remarkable thermal history dependence, but these structures are mentioned only in passing because it is difficult to segregate the contribution of mesophase order from the viscosity effects caused by exceptionally strong intermolecular interactions. Finally, fully aromatic PLCs are also mentioned rather briefly despite their obvious academic and practical interest and the many studies devoted to the effects of thermal treatment in such systems (see, for example [11] and references 3–19 cited therein). Unfortunately, thermal cycling, annealing or aging of aromatic PLCs often lead to chain degradation or reorganization (postcondensation, transesterification, transamidation, 'crystallization-induced' [12] reorganization of monomer sequence distribution, etc.). Thus it may be difficult or even impossible to single out the influence of thermal history.

Segmented RF polymers, especially polyesters, may also suffer a similar fate if transition temperatures are relatively high and one must always be cautious regarding data interpretation. Even polymers with relatively modest transition temperatures may undergo transesterification, e.g. a random chain scission and recombination at the level of the ester linkages which may result in reorganization of molecular weight or sequence distribution. A recent study of kinetics of transesterification in several (P1) polymers (Figure 6.4) provides a useful guideline in this respect [13]. Kinetics of transesterification in the nematic temperature range of this particular family of polymers has a negligible rate on the time scale of usual aging or annealing experiments. The activation energy for transesterification in the isotropic phase was found to be $157 \pm 4\,\mathrm{kJ\,mol^{-1}}$, in good agreement with values found for other polyesters such as PET. An efficient scission for each chain requires about 25 h of isothermal hold in the I phase at 160°C, but the rate constant increases drastically with temperature. Rapid transesterification reactions take place in PLCs above 200°C [14]. Even polyethers, which do not undergo transesterification, are not immune to degradation at relatively modest temperatures: polymer P2, for example, displays some degradation after 5 min of isothermal treatment in the I phase at 190°C [15].

This chapter contains the following sections: Brief overview of rigid–flexible (RF) polymers; Supercooling at the isotropic–mesophase transition; Memory of thermal history in the isotropic phase; and Memory of thermal history in the nematic–isotropic biphase and aging of the mesophase.

6.2 BRIEF OVERVIEW OF RIGID–FLEXIBLE PLCS

Our discussion in this section is limited to some characteristic features of RF PLC behavior introduced in order to provide a foundation to the subsequent sections on thermal history. The interested reader is referred to the general literature for more detailed descriptions of molecular architecture and supramolecular organization in this family of PLCs (reference 16 gives a recent review of structure–property relationships).

6.2.1 Chemical disorder in RF PLCs

Figure 6.4 shows an overly simplified chemical structure of RF PLCs which are usually composed of conformational, structural or constitutional isomers based on the repeating units shown.

(a) Conformational isomerism

The crystalline state of RF PLCs is conformationally disordered and classified as a 'condis crystal' mesophase [17]. The glassy and crystalline components of polymer P4, for example, have practically the same population of *trans* conformers within the spacer but are distinguished by the dynamics of t–g jumps [18]. The cooperative motions in condis crystals (ring flips, t–g jumps) do not change the symmetry of the three-dimensionally ordered phase. On the other hand, mesogens displaying conformational isomerism, such as diphenylethane, methylene ether or benzene ether, are in a dynamic equilibrium between an extended (anti) and a kinked (gauche) conformation and PLCs based on such moieties are in reality random copolymers with a dynamic composition of linear and bent comonomers [6, 7]. The previously mentioned nematic to smectic evolution sometimes observed upon annealing of the mesophase can be understood as a process leading to equilibration of conformer composition.

(b) Structural isomerism

Structural isomerism is a characteristic of polymers with unsymmetrical mesogen bridging groups. Structures based on moieties such as azoxybenzene, α-methyl stilbene, phenylbenzoate or azomethine are random and aregic copolymers, as a result of the random orientation of bridging groups along the chain. In the case of azoxy derivatives, the three structural isomers of twin model compounds of polymer P1 ($n = 10$; ONN, NNO and ONN-NNO) have been synthesized and phase behavior shown to be drastically dependent on isomer structure [19]. In general, it seems that the mesophase is stabilized by the presence of aregic disorder in RF PLCs [20], but systematic studies are few.

(c) Constitutional isomerism

Constitutional isomerism can be observed in polymers with an ester group in the center of the mesogen. A striking example is provided by polymer P5 which has been extensively studied by Stupp and coworkers [3, 21–23]. The repeating unit shown in Figure 6.4 contains the three moieties A (–O–Ph–O–), B (–CO–Ph–O–) and C (–CO–$(CH_2)_5$–CO–), which may be randomly placed along the chain as long as the global composition represented by formula P5 is respected (B moieties may self-condense). Transesterification at the central ester linkage, superimposed onto the random placement of the mesogen ester groups along the chain, leads to a random and aregic copolymer structure with sequences that have drastically different local bend elastic constants.

A chemically ordered and regioregular isomer of P5 has also been synthesized. Thermal data for the random and ordered P5 isomers are summarized in Table 6.1.

6.2.2 Orientational and conformational order in RF PLCs

Polymers P1 represent a model homologous series [24] in which phase behavior, molecular order and viscoelastic properties have been extensively investigated (see reference 25 and references therein for a review). The chains are inherently flexible, with Gaussian conformation in the isotropic phase, but are extended in the N phase. Mesogen and spacer order are strongly coupled. There is a strong, sustained, odd–even oscillation of nematic order, namely cybotactic nematic (akin to a fragmented smectic C) for spacer lengths n = even and conventional nematic for n = odd. Order parameters are dependent on molecular mass up to DP $\sim$ 40; for a given molecular mass, chain extension increases as the temperature decreases.

Small angle neutron scattering (SANS) experiments illustrate the drastic change in conformation occuring at the I–N transition and provide a measure of anisotropy of the radius of gyration and its increase with decreasing temperature across the nematic phase [26, 27]. The appearance of hairpin defects (abrupt changes in chain direction

Table 6.1 Thermal behavior of ordered and random constitutional isomers of polymer P5 (data from [3])

Structure	T_{kn} (°C)	T_{ni} (°C)	N + I biphase width
Ordered	203	277	Narrow
Random	146	[a]	Very broad (over 120°C)

[a] This transition is too broad to be observed by DSC.

[28, 29] in the nematic phase has been clearly demonstrated [30, 31]. Similar SANS results have recently been reported for a polyether based on a diphenylethane derivative [32].

One can expect that in the nematic glass the chains are fully extended, although to our knowledge detailed SANS studies of RF nematics in the glassy state have not been carried out. As previously mentioned, a dynamic DMR investigation of polymer P4 [18] shows that below T_g the fraction of *trans* conformers in the spacer of the glass approaches the value found in the crystal, the two phases differing mainly by the dynamics of chain segments. It was also found that the kinetics of enthalpy relaxation upon physical aging of the nematic glass is about the same for RF PLCs and for fully aromatic, rigid chains, both being orders of magnitude slower than in polystyrene [33].

6.2.3 Petransitional phenomena: the I–N transition

The isotropic phase of LCs displays strong pretransitional behavior above the mesophase clearing temperature T_c (e.g. the temperature of disappearance of the last anisotropic droplets). The angular correlations between molecular long axes that are characteristic of the mesophase order grow progressively stronger as the LC phase is approached on cooling [34, 35]. Pretransitional behavior can be revealed by monitoring the orientational order induced in the I phase upon application of an external field. It was shown via magnetic birefringence measurements that RF polymers display a much stronger pretransitional ordering than their MLC counterparts, the intensity of the effect increasing with chain length [36, 37].

The events correlated with the I–N transition on cooling occur as a two-stage process [38, 39].

1. The phase transition itself, which involves molecular ordering and segregation of anisotropic chains without large scale migration across phase boundaries. This is controlled by nucleation and growth mechanisms.
2. A second stage, which reveals strongly time dependent annealing, relaxation and phase separation (large scale demixing of isotropic and anisotropic domains) phenomena, as illustrated in subsequent sections.

As will become apparent from the examples of influence of thermal history that are provided in sections 6.3–6.5, kinetic phenomena are manifest at both stages of the I–N transition. The mesophase, as initially formed, is removed from thermodynamic and/or morphological equilibrium.

Examples of mesophase evolution during stage 2 are provided below. The phase transition (stage 1) is reflected in the transition enthalpy.

Although generally a fast process [25, 38, 39], it was monitored by isothermal differential scanning calorimetry (DSC) in a few instances. The thermal transition is completed within 1–5 min for polymers P1 [40] but requires from 100 to 1000 min for polymer P6 [41]. The value of the Avrami coefficient in P6 suggests a rod-like mechanism for the growth process. While structures P1 appear to be fairly typical, the extremely slow transition reported for polymer P6 seems to represent an exception, at least in the absence of further studies. Chain structure and molecular mass, however, play an important role in the transition kinetics, contrary to the widely held view, inherited from the literature on MLCs, that nematic ordering is always a fast process.

6.2.4 N + I biphase

The N + I biphase finds its origin in polydispersity in chain length coupled with polydispersity in chain flexibility and is observed within a temperature range delineated by $T_{n/n+i}$ and T_c. Polarizing microscopy observation of the biphase in P1 ($n = 10$, $M = 18\,700$) is illustrated in Figures 6.5a–c for three different conditions of thermal history [42]. It is apparent that textures and biphase width both depend on sample thermal treatment.

(a) Delineation of the N + I biphase

In addition to visual observation by polarizing microscopy, the biphase temperature range may be determined by methods such as DSC or broad line NMR (the fraction of nematic component present at any given temperature can be determined from the relative intensity of the narrow line associated with isotropic motion).

Small isotropic droplets, *c.* 1 μm or smaller in size, are not easily resolved by microscopy. A minor isotropic component is often found to trail at the lower temperature range within the biphase and its presence can be overlooked [25]. Yet a finely dispersed I phase can be expected to influence rheological behavior, order and orientation dynamics in the mesophase, as well as the mechanical properties of the resulting solid phase. In contrast to microscopy or DSC which detect macroscopic behavior, NMR provides a molecular or segment level view of morphology and cannot distinguish between phase and microphase separation. Thus accurate biphase delineation may be delicate to accomplish. We should further note that the customary dynamic scans do not provide the equilibrium value of biphase width but rather an apparent value as determined by thermal history (section 6.5).

(b) Molecular segregation within the biphase

Molecular mass distribution is reflected in a distribution of T_{ni} values, as illustrated on Figure 6.2. The value of T_{ni} increases with chain length up to DP $\sim$ 100, rapidly at first and then moderately. Selective partitioning of chain lengths occurs in the biphase: the longest chains are preferentially transferred into the anisotropic component. Upon heating, the shortest chains isotropize first, followed by chains of intermediate length and only the longest molecules survive within the N phase at T_c. This sequence is reversed on cooling, the nematic phase being nucleated by the longest chains.

The phase gap and selective partitioning characteristic of a given polydispersity can be predicted using a worm-like chain model where the free energy per monomer unit depends on the length of the chain (there is an additional free energy associated with each chain end) [43, 44]. A fixed (quenched [43]) polydispersity may become progressively annealed [44] if the sample is held within a temperature range where the aforementioned transesterification process is operative.

Fractionation according to chain length has been carried out for polymers P1 [2, 40, 45, 46] and P7 [47, 48] and also for a side chain polymer [49]. Based on the limited literature data available one can attempt the following generalization. Biphase width and effectiveness of fractionation are correlated and decrease in the following order: polyesters P1 (n = even), polyester-ethers P7, polyesters P1 (n = odd), polyethers = side chain polymer. This correlates roughly with the magnitude of the gradient of chemical potential ($\Delta S_{ni}/R$) between the I and N components.

Fractonation experiments require isothermal annealing in the biphase, which leads to macroscopic demixing of the I and N components (section 6.5). Large scale diffusion of chains takes place at this stage. It is driven by an interplay of factors such as the difference in specific gravity between the I and N components, interfacial tension and gradient of chemical potential across phase boundaries. Kinetic phenomena are thus expected to compete with the thermodynamic drive toward fractionation. An interesting illustration is provided by the rod-like poly(p-oxybenzoate) POB oligomers, where the compositions of the coexisting isotropic and anisotropic components of the biphase were found to be identical within experimental error [50]. This represents an unexpected departure from lattice theory predictions of selective chain length partitioning. According to Ballauff and Flory, mass transport across phase boundaries is hindered by the very small gradient of chemical potential prevalent between the I and N components in POBs and compositional equilibration in the biphase could not be observed on the time scale of the experiment.

In addition to chain length polydispersity, statistical copolymers such as the disordered polymer P5 are characterized by compositional variation among individual chains. There is a distribution in local bend elastic constants along individual chains and between chains (segments such as AC actually are not mesogens in their own right; section 6.2.1(c)). As already mentioned, this 'polyflexibility' is reflected in a very broad N + I biphase (Table 6.1). Stupp *et al.* [3] have shown that segregation within this biphase takes place according to flexibility, e.g. the more flexible chains are selectively transferred into the I component. The experimental biphase width was satisfactorily reproduced by computer simulation of a distribution of T_{ni} values calculated for a distribution of flexibilities in this system. A theoretical treatment of a polyflexible system based on the Landau–de Gennes approach makes a similar prediction [51]. Segregation by flexibility is also predicted to occur at the N–I transition in binary blends of homopolymers differing only by chain flexibility [51] and was recently observed by SANS experiments on blends of P1 (n = 10) and P7 (n = 7) (P. Sixou, personal communication).

An interesting example of fractionation according to compositional heterogeneity is provided by random copolyesters based on bromoterephthalic acid, methyl hydroquinone and hexane diol [52]. At relatively low molecular weights, where compositional heterogeneity is very marked, these copolymers show a triphase where smectic, nematic and isotropic phases are found to coexist. Fractionation reveals a selective transfer of the most rigid chains in to the smectic component. When fibers are drawn from this melt, two preferred mutually perpendicular directions of orientation, respectively corresponding to the director of the smectic and nematic components, are observed.

6.2.5 Morphology of the nematic phase

In the seemingly pure nematic phase there is an interesting difference between the broad line NMR spectra of polymers P1 and P5 (the latter both random and ordered). Below the biphase, in what appears as a homogeneosly birefringent liquid by optical microscopy (on the scale of *c.* 0.5 μm resolution), the broad line NMR spectrum of polymers P5 displays an I component indicative of fast isotropic motion, while none is observed in polymers P1. As already mentioned, NMR cannot distinguish between phase and microphase separation and the presence of the isotropic component could be explained by one or more of the following four factors:

1. an actual N + I biphase, with isotropic droplets smaller than the resolution scale of microscopy dispersed in the mesophase matrix;

2. an unstable N phase, with a spatially uniform distribution of isotropic (coiled) chains trapped in the mesophase and acting as defects;
3. disclination defects, which form at the weld lines between coalescing N droplets in the nascent mesophase;
4. local isotropic motion at the chain segment level.

If (1), (2) or (3) prevail, the morphology is metastable and will evolve upon isothermal mesophase aging. Case (1) and case (2) will be mentioned in subsequent sections. Case (3) represents the well known texture defects within PLC mesophases. The N phase in PLCs tends to be formed by coalescence of a large number of very small droplets and is initially characterized by a high density of disclination defects (around which the director orientation varies). These are progressively annihilated upon isothermal annealing [53] and visualized in the 'coarsening' of the N texture observed by microscopy or SALS.

Stupp *et al.* have proposed a model of case (4) in which the nematic phase of PLCs is viewed as tolerant of local isotropic disorder within segments of chain ('liquid fringed micelle' [3, 22]). This represents a dynamic morphology driven by 'nematic–isotropic fluctuations' with spatial and temporal periodicities that may depend on factors such as structure, temperature and molecular mass distribution. The fraction of this isotropic component in the random, aperiodic copolymer P5 increases from *c.* 0.1 to 0.3 as the molecular mass increases from 4300 to 16 700 but, for a given mass, does not appear to change appreciably with mesophase aging (isothermal annealing of the N phase) [23]. Thus it is not associated with nematic texture defects and their annihilation. Yet aging clearly leads to a mesophase morphology characterized by a more stable state with a longer range orientational order, as revealed by a drastic dependence of magnetic orientation dynamics on aging time, for example [23].

In the early days of main chain thermotropic PLCs, the concept of 'degree of liquid crystallinity' [54] was introduced to account for the seemingly biphasic behavior of many random copolyesters, especially copolyesters containing non-mesogenic co-units. Several questions were debated. Are the non mesogenic and/or the more flexible co-units rejected by the LC phase, just as non-crystallizable comonomers are in copolymer crystallization? Alternatively, are they tolerated by the LC phase because of the mild orientational constraints and rapid translational motion prevailing in a nematic medium? (See the General Discussion and references cited in reference 55.) In this chapter the phrase 'degree of liquid crystallinity' is used in its narrow sense (in analogy with the degree of crystallinity in semicrystalline polymers) in order to characterize departure from thermodynamic equilibrium as measured by the enthalpy of the LC–I transition.

The questions concerning the nature of nematic PLC morphology and of the molecular organization on the local scale still remain open. If the N phase does indeed tolerate nematic–isotropic fluctuations (the liquid fringed micelle model) is not clear why such fluctuations appear in the chemically ordered P5 and not in P1 polymers. It is clear, however, that mesophase morphology and the macroscopic properties that are affected by it are influenced by thermal history, as is illustrated below.

6.3 SUPERCOOLING AT THE ISOTROPIC–MESOPHASE TRANSITION

In this section we consider heating and cooling scans in the absence of isothermal aging within the LC + I biphase. The influence of aging in the biphase is discussed in section 6.5.

6.3.1 Cooling and heating scans

Although mesophase transitions are often characterized in the literature as 'near equilibrium' [17], supercooling at the I–LC transition is to be expected as in any nucleation and growth controlled process. It is logical to anticipate a pronounced supercooling for high molecular masses and for transitions to viscous (smectic, cholesteric) phases. One example is provided by a recent study of side chain LC polyphosphazenes containing chiral mesogens with M_w values in the range 2.06×10^5–8.04×10^5 [56]. Both temperatures and enthalpies of transition are substantially lower on cooling than on heating, as might be expected (scanning rate 10 K min^{-1}). The enthalpy change at the I–cholesteric cooling transition is *c.* 10 to 60% lower than on heating. The smaller values of ΔH recorded on cooling may signify either development of a less ordered mesophase or a residual I microphase embedded within the LC matrix (on a scale below optical microscopy resolution). Here again, as in section 6.2.5, we are dealing with unresolved questions concerning molecular organization on the local scale.

In contrast, the enthalpy values illustrated in Table 6.2 for several nematic RF PLCs are counterintuitive at first glance; e.g. supercooling is observed as expected (with $T_{in} < T_{ni}$) but ΔH_{in} is larger than ΔH_{ni}. The magnitude of supercooling and the difference between ΔH_{in} and ΔH_{ni} is relatively small for these low mass samples, but tend to increase with chain length [57].

If one assumes that a fully developed N phase (the degree of liquid crystallinity is unity) is formed on cooling, a process which is facilitated by the large drop in viscosity at the I–N transition, the smaller value of ΔH_{ni} may be attributed to an incomplete isotropization on heating. Perhaps, as a direct result of molecular segregation across the biphase,

Table 6.2 Thermal data for polymers P1 (data from [57]).

Spacer (n)	$T_{ni}(T_{in})$ (°C)	$\Delta H_{ni}(\Delta H_{in})$ (kJ mru^{-1})
5	154 (150)	1.27 (1.27)
6	224 (208)	6.22 (6.60)
7	131 (125)	2.06 (2.21)
8	166 (158)	5.79 (6.63)
9	132 (127)	2.43 (2.60)
10	154 (143)	6.10 (6.98)
11	122 (118)	3.30 (4.17)
12	133 (120)	7.09 (8.94)

Molecular masses range between 2500 and 6700. Enthalpies are given in kJ per mol of repeating unit. Values recorded on cooling are in parentheses.
Thermal history: scanning rates are 20 K min^{-1}. Samples are cooled to 20 K below T_g and heated to $T_{ni+15\,K}$. Data are from second heating and cooling runs (5 min isothermal hold at $T_{ni+15\,K}$ prior to cooling).

the I phase initially retains some 'non-equilibrium level of residual order' on the short-range scale [15]. In other words, the I phase as initially formed on heating has a non-homogeneous molecular morphology, with segregation according to chain length (or chain flexibility) on the local level. A homogeneous distribution of chains will eventually develop via a diffusion process upon isothermal aging of the I phase (the equilibrium I phase is a homogeneous mixture of the different chains that are present and is subsequently designated as a 'well isotropized I phase' [46]). The molecular morphology of the I phase is characterized by a time dependence reflected in some 'memory effects' that will be discussed below.

6.3.2 Supercooling at the I–N transition

Because formation of the N phase on cooling is nucleated by aggregates of the longest (or most rigid) chains, any investigation of supercooling should start from a well isotropized I phase. One such study involved measuring the fraction of nematic component by broad line NMR on both heating and cooling P1 polymers (see reference 46 for experimental details). A subcritical range of temperatures is observed below T_c, in the interval $T_c - T_{sc}$ (recall that T_c is the temperature of disappearance of the last nematic droplet on heating). The nematic component is still present on heating between T_{sc} and T_c but pretransitional fluctuations of chain length distribution alone cannot nucleate a nematic phase on cooling into this temperature range from a well isotropized I phase, regardless of holding time. In other words, supercooling from a well isotropized I phase appears to be of thermodynamic rather than purely kinetic origin. This seeming paradox is easily resolved if one remembers that the system is thermodynamically different on heating and cooling. On cooling from a well isotropized I phase one deals with a homogeneous mixture of chains. On heating, in contrast, a dynamic composite of metastable 'domains' is formed as a result of molecular segregation. The last surviving anisotropic droplets are the longest (and/or most rigid) and consequently have a higher clearing temperature than a homogeneous mixture diluted by shorter (or more flexible) chains. It is logical to assume that the magnitude of this subcritical temperature range would increase with increased effectiveness of molecular fractionation (section 6.2.4(b)), but definitive conclusions cannot be drawn in the absence of additional studies.

6.4 MEMORY OF THERMAL HISTORY IN THE ISOTROPIC PHASE

Although the I phase is devoid of long range order, it is characterized by a metastable morphology which is determined by the thermal history and structure of chains. Memory of this morphology is reflected in the biphase, mesophase and solid state achieved on cooling from the I precursor. The following examples illustrate such memory effects.

6.4.1 Cooling from the I phase

Figure 6.5 illustrates how isotropization (isothermal aging in the I phase) can influence transition temperatures and nematic textures, everything else being equal. Polymer P1 ($n = 10$, $M = 18\,700$) is cooled through the N + I biphase following isothermal rest in the I phase (Figure 6.5b) or immediately following a heating scan (Figure 6.5c) [42]. The differences observed for the biphase width and supercooling are specified in the figure legend. In addition, the textures observed on cooling are drastically dependent on thermal history in the I phase.

On cooling from an isotropized I state following isothermal holding at 167°C (Figure 6.5b), the anisotropic phase appears at $T_{sc} = 151$°C, with a supercooling of 11°C. The nematic domains develop instantaneously below T_{sc}, from a large number of very small anisotropic droplets uniformly distributed throughout the sample. The droplet size distribution subsequently remains strikingly uniform during their growth. This reflects an initially homogeneous distribution of the long chains which act as nuclei. On cooling from 167°C immediately following the heating scan, one observes a very different behavior. The anisotropic domains appear at 155°C as a somewhat blurred mirror image of the size, shape and spatial distribution of the last nematic droplets observed below T_c on heating. Subsequent growth radiates from this initial polydisperse domain distribution. In other words, nucleation is provided by the non-equilibrium remnants of the longest chain aggregates which are initially present in the I phase.

Crystallization from the nematic melt also appears to be facilitated by aggregates formed by the longest chains, although systematic investigations have not been carried out, to our knowledge. The morphology of the solid phase reflects a memory of the I precursor. Thus, the degree of crystallinity is higher for thermal history of Figure 6.5c than for that in Figure 6.5b and higher still if the nematic droplets surviving on heating to just below T_c are allowed to coalesce during isothermal annealing (unpublished results; section 6.5).

Temperature dependence of the fraction of isotropic component (f_i) is measured on cooling across the N + I biphase by analyzing digitized micrographs as described in reference 42. It is uneventful for the thermal history in Figure 6.5c as f_i decreases continuously between 155 and 123°C. For the history in Figure 6.5b, on the other hand, f_i rapidly falls to zero (at 1° below T_{sc} = 151°C), starts increasing to *c.* 0.2 below 148°C and falls again to zero between 143 and 133°C. A pure nematic phase, on the resolution scale of optical microscopy, is present between 150 and 148°C. This is an unstable LC phase (case (2) in section 6.2.5), with coiled isotropic chains of shorter length uniformly trapped throughout the N domains. Coalescence of coiled chains into a recognizable I phase requires large scale diffusional migration. This phase separation stage proceeds more slowly than the phase transition itself and unstable N phase is formed as a result. This texture is characterized by a high density of disclination defects, as a result of the welding of a very large number of small nematic droplets. One can speculate that isothermal aging of the unstable N phase (within the narrow temperature range 148–150°C) would lead to progressive coalescence of isotropic chains, and this would be reflected in drastic evolution of macroscopic ordering kinetics.

In the history in Figure 6.5c, on the other hand, the anisotropic and isotropic domains are 'preformed' to some extent within the precursor I phase and phase separation on cooling requires less overall chain migration. The difference in biphase width between cases in Figure 6.5b and Figure 6.5c, respectively 151–133°C and 155–123°C, is similarly explained by molecular segregation phenomena, as the lowest and highest chain lengths are allowed to undergo phase transition at temperatures somewhat closer to their equilibrium transition. This broadening of the biphasic gap becomes further amplified following isothermal annealing in the N + I biphase (see below).

6.4.2 Memory of banded textures

PLCs that have been oriented by a shear or elongational field develop a banded texture morphology when allowed to relax in the mesophase just above T_{kn} [58, 59]. Under the microscope such textures appear as an alternation of dark and light striations perpendicular to the direction of orientation. They are erased on heating, but memory of their characteristic sinusoidal supramolecular organization must temporarily remain in the I phase since they can be recovered on cooling back into the mesophase [60, 61].

Polymer P3 ($M = 56\,500$) provides a good illustration [61]. The banded textures disappear completely just above T_{kn} but reappear on cooling to room temperature. They are also recovered from the I phase even after a period of isothermal holding indicating that memory of the preceding order is stored therein, albeit on a short range scale. Memory in the I phase is facilitated by high viscosity, e.g. if the polymer has a high molecular mass as is the case for P3 ($M = 56\,500$).

6.4.3 Quenching from the isotropic state

(a) Degree of liquid crystallinity

The degree of liquid crystallinity developed upon rapid cooling of high molecular mass PLCs may be strongly affected by the extent of isothermal holding time in the I phase [15, 61]. An interesting example is provided by copolymer P2 which has a molecular mass $M_n = 31\,000$ [15]. When this polymer is held in the I phase (at 190°C, 10°C above T_{ni}) for periods ranging from 0 to 5 min, cooled rapidly to room temperature and finally reheated at $10\,K\,min^{-1}$, the values of T_{ni} and ΔH_{ni} decrease with increasing isothermal holding time, the former by some 20°C and the latter to approximately half of its initial value. Both T_{ni} and ΔH_{ni} are recovered upon subsequent isothermal aging in the N phase at 120°C. The rates of ΔH_{ni} decrease following aging in the I phase and subsequent recovery after aging in the mesophase can both be fitted to exponential kinetics, with time constants $\tau = 2.13$ min and 4.1 h, respectively. These results are observed only with fast cooling from the I phase: cooling at $20\,K\,min^{-1}$, for example, yields the highest value of T_{ni} and H_{ni} regardless of isothermal holding time (up to 5 min, since some degradation is observed for higher times of annealing). In other words, the rate of phase transition is sufficiently high to result in complete ordering of chains at cooling rates of $20\,K\,min^{-1}$, even for high chain lengths.

Similar results are observed for polymer P2 ($M = 56\,500$), where T_{kn} and ΔH_{kn} were additionally followed and found to decrease concomitantly with T_{ni} and ΔH_{ni} [61], once again illustrating the interdependence of all phases.

As previously mentioned, disclination defects disappear progressively during isothermal aging in the nematic state. Recovery of ΔH_{ni} during mesophase aging in polymer P2 is correlated by the authors with texture coarsening; e.g. disclination defects are viewed as isotropic defects affecting the degree of liquid crystallinity [15, 62]. Other studies [38–40], however, suggest a different interpretation, as will be seen below. For example, if an unstable N phase such as mentioned above were to be quenched to the nematic temperature range, annealing would result in an increase of ΔH_{ni} concomittantly with, but not necessarily correlated to the annealing out of disclination defects.

(b) Isotropic glass

There are few reported instances of actual quenching of the I melt into an isotropic (amorphous) glass (pathway 2 in Figure 6.1). LC glasses are usually developed, even when quenching is carried out from the isotropic melt (pathway 3) [63–65]. One can speculate that in high molecular mass samples a 'partial quenching' of the amorphous melt is possible, by analogy to the development of a 'partial degree of liquid crystallinity' mentioned in section 6.4.3(a). One can also speculate on the possibility of quenching into the glassy state the unstable nematic morphology discussed in section 6.4.2. Such a glass might be precursor to a poorly developed LC phase in which the 'degree of liquid crystallinity' would evolve with aging time.

Cheng *et al.* [39] report on kinetics of enthalpy relaxation of nematic poly(azomethane) glasses upon physical aging below T_g. Nematic glasses obtained by quenching the samples from the isotropic melt relax faster and display a broader distribution of relaxation times than glasses quenched from the nematic melt. This suggests that cooling from an I melt produces a more open (more heterogeneous) glass. Similar results are obtained for a side chain nematic PLC with 'side-on' attachment of mesogens (C.B. McGowan, personal communication).

6.5 MEMORY OF THERMAL HISTORY IN THE N + I BIPHASE. AGING OF THE N PHASE

We have just seen that molecular segregation on heating leads to heterogeneity of the initial I phase morphology. Segregation occurs in reverse order upon cooling to the mesophase. Thus, a nematic domain grows by adding roughly concentric layers of shorter and shorter chains onto a core containing the longest molecules. Two neighboring droplets within the N + I phase, one anisotropic and the other isotropic are characterized by a gradient of chain lengths decreasing from the core of one to the core of the other. Intermediate chain lengths are concentrated at the boundary zones. Such a gradient can occasionally be observed under the microscope or by DSC (see below) [40].

When the polymer is annealed within the biphase, one can observe the simultaneous evolution of three different processes:

1. the diffusion-controlled coalescence and demixing of anisotropic and isotropic domains;
2. homogenization of chain length distribution within their respective domain boundaries;
3. annihilation of disclination defects within the anisotropic domains.

Following equilibrium demixing (a process which typically requires from 5 to 24 h), the I and N components display sharp boundaries which are clearly preserved on cooling to the underlying nematic and solid states and upon subsequent reheating. In the semicrystalline state, for example, the formerly nematic component (which is characterized by a higher degree of crystallinity) is more brightly birefringent than the formerly isotropic component of the biphase [2]. Two cold crystallization peaks, two melting and two isotropization peaks, one for each of the former components of the biphase, can be observed by DSC on heating [2, 40, 47]. As might be expected, the isotropic phase retains the memory of demixing within the biphase and two I/N peaks are observed on cooling [40, 47].

Figure 6.6 illustrates the width of the biphase following equilibrium demixing as a function of T_s, the demixing temperature [2]. A huge broadening of the biphase, already prefigured in Figure 6.5c, is apparent. It is most pronounced when demixing is carried out at values of T_s close to $T_{n/n+i}$, where only the shortest chains are separated into the I component. No broadening or separation occurs if annealing is performed outside the biphasic range. This shows up as a discontinuity illustrated on Figure 6.6. The formerly isotropic component FI isotropizes first on heating. The boundaries of FN, the formerly nematic component, remain intact until isotropization of FI is completed [2, 47]. The same observation was made in a broad line NMR investigation of polymers P1 by combining proton and deuterium spectra and using perdeuterated *p*-azoxyanizole (PAA-d14), a close relative of the P1 mesogen, as a probe molecule [46]. The high field portion of the DMR spectrum of PAA-d14 monitors the environment as described in reference 46 and evolution of two high field triplets, one for each of the former components of the biphase, is observed.

Various stages of incomplete demixing are developed when isothermal holding time is insufficient. During homogenization of chain length distribution within the I and N domain boundaries, a gradient of chain

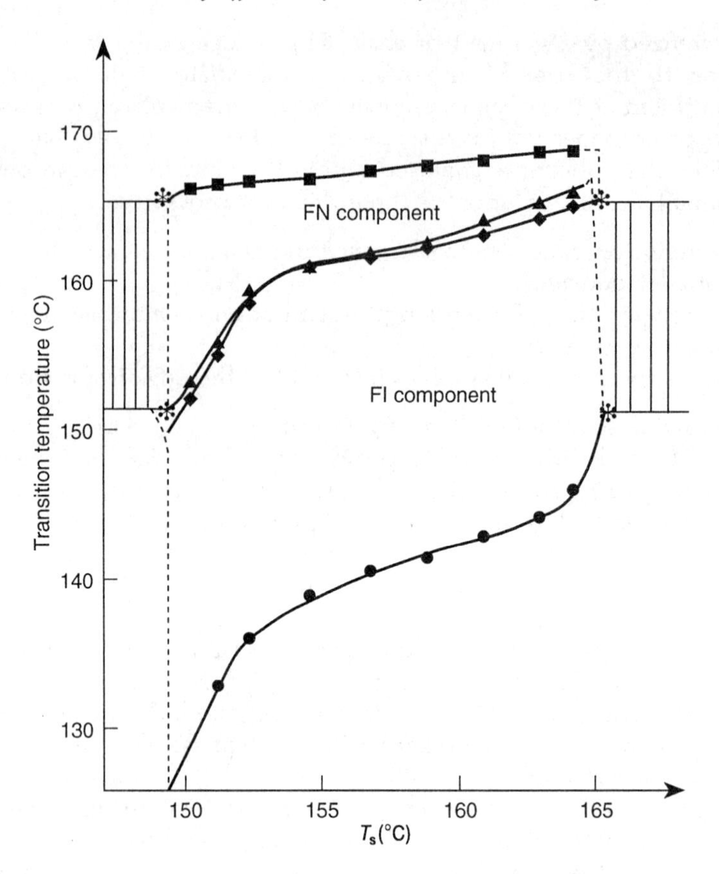

Figure 6.6 Biphasic range developed in (P1) (n = 10, M_n = 20 000) as a function of demixing temperature T_S (isothermal annealing in the N + I biphase). Heating at 5°C. Complete demixing was achieved. The beginning (●) and end (■) of the sharply distinct biphases corresponding to FI and FN (the formerly isotropic and nematic components of the initial biphase) are shown. Points represented by * illustrate the aparent discontinuity in biphase width. (Reprinted with permission from [2]. Copyright 1988 Gordon & Breach.)

lengths persists across neighboring anisotropic and isotropic domains. Three DSC peaks, corresponding to isotropization of the shortest, intermediate and longest chain lengths can sometimes be observed [40]. In general, however, incomplete demixing or even the less extensive segregation achieved during heating or cooling scans (Figure 6.5) are revealed as rather broad and irregularly shaped isotropization peaks.

Mesophase morphology becomes progressively homogenized upon isothermal aging in the nematic temperature range. The T_{ni} peak maximum moves to higher temperatures and the peak itself becomes narrower, approaching a Gaussian shape, without significant change in the value of ΔH_{ni} [40]. This evolution typically proceeds on the same time scale as the demixing experiments, as both are driven by diffusion kinetics. Memory of demixing within the I phase can be similarly erased upon annealing above T_c but the slow diffusion process is again revealed: although the two separate I/N peaks are progressively fused, the original I/N peak shape is not recovered even after several hours [40, 47].

The kinetics of the homogenization process on the molecular scale can be followed by using PAA-d14 as a probe molecule in the type of combined PMR–DMR investigation previously mentioned. A model two-component system of small molecules composed of PAA-d14 and a non-mesomorphic analog of the P1 mesogen core was studied [46]. Using the high field triplet of the DMR spectrum to monitor local concentration and order, an initial heterogeneity of the mesophase was established even after simply cooling the system across the N + I biphase. Following macroscopic demixing of the I and N components within the biphase, the nematic phase is initially demixed into two corresponding nematic components, each with its own characteristic high field triplet. As homogenization proceeds via molecular diffusion, the two triplets converge following exponential kinetics. A characteristic time of *c*. 200 min at 103°C, yielding the appropriate value of the self-diffusion coefficient of PAA in the nematic phase, was established. Since viscosity of PLCs is orders of magnitude higher, and their diffusion coefficients correspondingly lower, it is not surprising that molecular homogenization was found to be exceedingly slow, even in the absence of isothermal demixing in the N + I biphase [46]. One can generally assume that morphology remains heterogeneous on the local scale within the normally encountered experimental time frames.

As a result of molecular segregation, the underlying mesophase is always composed of metastable 'domains' (regions segregated by chain length and/or flexibility). The dimensional scale of this heterogeneity depends on chain structure, molecular weight distribution and sample history within the biphase. For polymers P1, for example, it has been shown to vary from *c*. 20 μm to macroscopic in size [46]. Morphological heterogeneity cannot be bypassed simply by avoiding the N + I biphase; similar, though much less studied, segregation phenomena are operative across the T_{kn} boundary and also upon precipitation, crystallization or film casting from solution. RF PLCs display lyotropic as well as thermotropic mesophases [66, 67] and molecular segregation undoubtedly occurs to some degree during the last stages of solvent removal. The existence of this type of segregation can sometimes be inferred from the broad and irregularly shaped DSC isotropization peak recorded on first heating if a narrower, Gaussian peak with the same value of ΔH_{ni} can be developed after aging in the mesophase (unpublished results).

We now come back to the evolution of ΔH_{ni} and the texture coarsening observed upon aging polymer P2 within the N phase as described in section 6.4.3 [15, 62]. While texture coarsening is always observed upon aging in the mesophase, ΔH_{ni} usually remains unchanged. We have just mentioned that in polymers P1 the T_{ni} peak maximum moves to higher temperatures and the peak itself sharpens without significant change in the value of ΔH_{ni}. The same observation was reported for another RF polyester [38] where the kinetics of annihilation of disclination defects was followed by microscopy and by SALS at three different temperatures after quenching from the I phase. It was noted that the kinetics of defect annihilation appears to be correlated with viscosity and is dependent on temperature whereas the rate of evolution of T_{ni} is not (at least in the fairly narrow temperature interval that was studied). Thus, texture coarsening and evolution of the isotropization peak apparently are related to two different relaxation processes. Cheng *et al.* have investigated several polyazomethines with ethyleneoxy spacers and found consistent narrowing of the isotropization peak but no change in T_{ni} [39]. ΔH_{ni} was found to be independent of mesophase aging time, except at temperatures located close to T_{ni}, where it increased with aging time. Thus, it is difficult to draw general conclusions regarding correlations between the various processes which lead to morphological and thermodynamic equilibration of mesophase textures.

We have discussed mostly the metastable morphological heterogeneity in systems where segregation of domains is driven by polydispersity in chain lengths. Morphological equilibration upon aging of such mesophases is characterized by remixing of domains and drive toward a homogeneous distribution of chain lengths. On the other hand, aging in mesophases dominated by heterogeneity of composition (polyflexibility) should lead to equilibration via a process of (micro)phase separation, that is a drive toward a stable morphological heterogeneity. The dimensional scale of this type of heterogeneity is most probably dictated by the chemical structure. In random copolyesters, for example, one might expect development of a microsegregation according to monomer sequence, as found in the 'aperiodic' crystallites which are characteristic of these PLCs [68, 69]. This type of reorganization would explain the drastic influence of aging in the mesophase on the development of crystallinity ([11] and references therein]).

If the system is characterized by a strong heterogeneity in chemical composition leading to large differences in chain flexibility, separation might ensue on a larger, perhaps macroscopic, scale. This seems to be the case of copolymers reported in reference 52 where a smectic and a nematic component were found to coexist, as mentioned in section 6.2.4. The two components were separated and found to have a very different chemical composition. In some poly(oxybenzoate-co-ethyleneterephthalate) samples aging of the nematic mesophase results in the development of an $N_1 + N_2$ phase separation. The two coexisting nematic phases have sharply defined boundaries and isotropize sequentially on heating [70, 71].

It may be, as pointed out by Stupp *et al.* [3, 23], that morphological heterogeneity similarly prevails in non-mesomorphic systems where the 'marker' provided by birefringence is lacking. Much remains to be investigated in this area, touching on fundamental questions relating to development of crystallinity and mechanical properties in PLCs, as well as their rheological and orientational behavior.

ACKNOWLEDGEMENT

The author is grateful to the National Science Foundation for support of her research on PLCs.

REFERENCES

1. Finkelmann, H. (1991) in *Liquid Crystallinity in Polymers: Principles and Fundamental Properties* (ed. A. Ciferri), VCH Publ.; Ch. 8.
2. D'Allest, J.F., Sixou, P., Blumstein, A. and Blumstein, R.B. (1988). *Mol. Cryst. Liq. Cryst.*, **157**, 229.
3. Stupp, S.I., Moore, J.S. and Martin, P.G. (1988) *Macromolecules*, **21**, 1217, 1228.
4. Keller, A., Ungar, G. and Percec, V. (1990) in *Liquid-Crystalline Polymers* (eds R.A. Weiss and C.K. Ober), in *ACS Symposium Series*, **435**, Ch. 23.
5. Stevens, H., Rehage, G. and Finkelmann, H. (1984) *Macromolecules*, **17**, 851.
6. Percec, V. (1988) *Mol. Cryst. Liq. Cryst.*, **155**, 1.
7. Percec, V. and Yourd, R. (1988) *Macromolecules*, **21**, 3379.
8. Cheng, P., Blumstein, A. and Subramanyam, S. (1995) *Mol. Cryst. Liq. Cryst.*, **269**, 1.
9. Aharoni, S.M. (1981) *J. Polymer Sci. Phys.*, **19**, 282.
10. Kato, T., Kihara, H., Uryu, T. *et al.* (1992) *Macromolecules*, **25**, 6836.
11. Leblanc, J.P., Tessier, M., Judas, D. *et al.* (1995) *Macromolecules*, **28**, 4837 (and references therein).
12. Lenz, R.W., Jin, J.I. and Feichtinger, K.A. (1983) *Polymer*, **24**, 327.
13. Li, M.H., Brûlet, A., Keller, P. *et al.* (1993) *Macromolecules*, **26**, 119.
14. Arrighi, V., Higgins, J.S., Weiss, R.A. and Cimecioglu, A.L. (1992) *Macromolecules*, **25**(20), 5297.
15. Feijoo, J.L., Ungar, G., Keller, A. and Percec, V. (1990) *Polymer*, **31**, 2019.
16. Sirigu, A. (1991) in *Liquid Crystallinity in Polymers: Principles and Fundamental Properties* (ed. A. Ciferri), VCH Publ.; Ch. 7.
17. Wunderlich, B., Möller, M., Grebowicz, J. and Baür, H. (1988) *Adv. Polymer Sci.*, Ch. 2.
18. Muller, K., Meier, P. and Kothe, G. (1985) *Progr. NMR Spectr.*, **17**, 211.
19. Blumstein, R.B., Poliks, M.D., Stickles, E.M. *et al.* (1985) *Mol. Cryst. Liq. Cryst.*, **129**, 375.
20. Ober, C., Lenz, R.W., Galli, G. and Chiellini, E. (1983) *Macromolecules*, **16**, 1034.
21. Moore, J.S. and Stupp, S.I. (1987) *Macromolecules*, **20**, 273.

22. Stupp, S.I., Wu, J.L., Moore, J.S. and Martin, P.G. (1991) *Macromolecules*, **24**, 6399.
23. Stupp, S.I., Moore, J.S. and Chen, F. (1991) *Macromolecules*, **24**, 6408.
24. Blumstein, A. and Thomas, O. (1982) *Macromolecules*, **15**, 1264.
25. Blumstein, R.B. and Blumstein, A. (1988) *Mol. Cryst. Liq. Cryst.*, 1, **165**, 361 (and references therein).
26. d'Allest, J.F., Sixou, P., Blumstein, A., Blumstein, R.B., Teixera, J. and Noirez, L. (1988) *Mol. Cryst. Liq. Cryst.*, **155**, 581.
27. d'Allest, J.F., Maissa, P., ten Bosch, A., Sixou, P., Blumstein, A., Blumstein, R.B., Teixera, J. and Noirez, L. (1988) *Phys. Rev. Lett.*, **61**, 2562.
28. Warner, M., Gunn, J.M.F., Baumgartner, A.B. and Davidson, P. (1985) *J. Phys. A.*, **18**, 3007.
29. Wang, X.J. and Warner, M. (1986) *J. Phys. A.*, **19**, 2215.
30. Li, M.H., Brûlet, A., Davidson, P. *et al.* (1993) *Phys. Rev. Lett.*, **70**(15), 2297.
31. Li, M.H., Brûlet, A., Cotton, J.P. *et al.* (1994) *J. Phys. II, France*, **4**(10), 1843.
32. Hardouin, F., Sigaud, G., Achard, M.F. *et al.* (1995) *Macromolecules*, **28**, 5427.
33. McGowan, C.B., Kim, D.Y. and Blumstein, R.B. (1992) *Macromolecules*, **25**, 4658.
34. Zadoc-Kahn, J. (1930) *Comptes Rendus*, **191**, 1002.
35. Luckhurst, G.R. (1988) *Faraday Trans.* 2, **88**(84), 961.
36. Blumstein, A., Maret, G. and Vilasagar, S. (1981) *Macromolecules*, **14**, 95.
37. Maret, G. (1983) *Polym. Prepr. Am. Chem. Soc.*, **24**(2), 249.
38. Rojstaczer, S.R. and Stein, R.S. (1990) *Macromolecules*, **23**, 4863.
39. Cheng, Z.D., Janimak, J.J., Lipinski, T.M. *et al.* (1990) *Polymer*, **31**, 1122.
40. Kim, D.Y., D'Allest, J.F., Blumstein, A. and Blumstein, R.B. (1988) *Mol. Cryst. Liq. Cryst.*, **157**, 253.
41. Battacharya, L., Misra, A., Stein, R.S. *et al.* (1986) *Polymer Bull.*, **16**, 465.
42. Nakai, A., Wang, W., Hashimoto, T., Blumstein, A. and Maeda, Y. (1994) *Macromolecules*, **27**, 6963.
43. Semenov, A.N. (1993) *Europhys. Lett.*, **21**, 37.
44. Bladon, P., Warner, M. and Crates, M.E. (1993) *Macromolecules*, **26**, 4499.
45. D'Allest, J.F., Wu, P.P., Blumstein, A. and Blumstein, R.B. (1986) *Mol. Cryst. Liq. Cryst. Lett.*, **3**, 103.
46. Esnault, P., Gauthier, M.M., Volino, F. *et al.* (1988) *Mol. Cryst. Liq. Cryst.*, **157**, 273.
47. Laus, M., Caretti, D., Angeloni, A.S. *et al.* (1991) *Macromolecules*, **24**, 1459.
48. Laus, M., Angeloni, A.S., Galli, G. and Chiellini, E. (1991) *Macromolecules*, **24**, 1459.
49. Galli, G. and Chiellini, E. (1991) *Makromol. Chem. Rapid Commun.*, **12**, 43.
50. Ballauff, M. and Flory, P.J. (1984) *Ber. Bunsenges. Phys. Chem.*, **88**, 530.
51. Fredrickson, G.H. and Leibler, L. (1990) *Macromolecules*, **23**, 531.
52. Ober, C.K., McNamee, S., Delvin, A. and Colby, R.H. (1990) in *Liquid-Crystalline Polymers* (eds R.A. Weiss and C.K. Ober), ACS Symposium Series, **435**, Ch. 17.
53. Shiwaku, T., Nakai, A., Hasegawa, H. and Hashimoto, T. (1990) *Macromolecules*, **23**, 1590.
54. Lenz, R.W. and Jin, J.I. (1981) *Macromolecules*, **14**, 1405.
55. Various authors (1985) *Faraday Disc.*, **79**, 85.
56. Allcock, H.R. and Klingenberg, E.H. (1995) *Macromolecules*, **28**, 4351.
57. Thomas, O. (1984) *PhD Dissertation*, University of Massachusetts Lowell.
58. Bedford, S.E. and Windle, A.H. (1990) *Polymer*, **31**, 616.
59. Wang, J., Battachara, S. and Labes, M.M. (1991) *Macromolecules*, **24**, 4942.

60. Liu, X., Shen, D., Shi, L. and Xu, M. (1990) *Macromolecules*, **31**, 1897.
61. Jegal, J.G. (1992) PhD Dissertation, University of Massachusetts Lowell.
62. Feijoo, J.L., Ungar, G., Owen, A.J. and Percec, V. (1988) *Mol. Cryst. Liq. Cryst.*, **155**, 487.
63. Menczel, J. and Wunderlich, B. (1981) *Polymer*, **22**, 778.
64. Sauer, B.B., Beckerbauer, R. and Wang, L. (1993) *J. Polymer Sci. Phys.*, **31**, 1861.
65. Kricheldorf, H.R., Schwarz, G., de Abajo, J. and de la Campa, J.G. (1991) *Polymer*, **32**, 942.
66. Viney, C., Yoon, D.Y., Reck, B. and Ringsdorf, H. (1989) *Macromolecules*, **22**, 4088.
67. Ratto, J.A., Volino, F. and Blumstein, R.B. (1991) *Macromolecules*, **24**, 2862.
68. Blackwell, J., Biswas, A., Gutierez, G. and Chivers, R.A. (1985) *Faraday Disc. Chem. Soc.*, **79**, 73.
69. Windle, A.H., Viney, C., Golombok, R. (1985) *Faraday Disc.*, **79**, 55.
70. Mackley, M.R., Pinaud, F. and Siekman, G. (1981) *Polymer*, **22**, 437.
71. Windle, A.H. (1985) *Faraday Disc.*, **79**, 91.

7

Longitudinal polymer liquid crystal + engineering polymer blends: miscibility and crystallization phenomena

George P. Simon

7.1 INTRODUCTION

Liquid crystalline (LC) materials, both monomeric (MLCs) and polymeric (PLCs), continue to fascinate scientists and technologists alike, with new materials being synthesized [1] at an ever increasing rate, leading to well defined structure–property relationships in main chain PLC (MCPLC) materials [2–4]. Even within a given longitudinal PLC system (usually copolyesters), properties such as modulus can now be well controlled by altering the nature and concentration of a variety of comonomers [5] with a level of chemical and analytical sophistication that allows their detailed morphological phase diagrams to be constructed as a function of comonomer feed [6–7]. As with their metallurgical predecessors, much research in commodity polymers (CPs) and engineering polymers (EPs) in recent years has concentrated on blending different plastics in an effort to widen the range of properties and processing conditions achievable, and an increasing number of these are now being commercialized [8].

Mechanical and Thermophysical Properties of Polymer Liquid Crystals
Edited by W. Brostow
Published in 1998 by Chapman & Hall, London.
ISBN 0 412 60900 2

Due to low entropies of mixing and often relatively poor (athermal) interaction between groups on the polymer chain, many EP polymer blends tend to be either partially or totally immiscible, although there are quite a significant number that are fully miscible (e.g., listings in reference 8). In either miscible or immiscible blends, synergy in physical, thermal or engineering properties is usually sought with morphology being an important variable in immiscible systems.

PLC blends have also been the subject of considerable research, due in part to the relatively high cost of most PLC production (often related to the costs in synthesizing the required monomers). Whilst PLC materials have been commercialized with varying degrees of success for the past decade, they still remain comparatively expensive compared to most EPs. Blending offers both the possibility of dilution of cost (particularly if the PLC is the majority phase) or conversely, the chance to improve the properties of another plastic if the PLC is the minority phase and thus the addition of a more expensive but useful additive may prove cost effective. As with most non-LC polymeric blends, the vast majority of PLC binary blends tend to be two-phase. Indeed, theory predicts that blends of rigid polymer chains with those that are flexible will result in low miscibility [9–10] for entropic reasons, with demixing theoretically occurring even if the blend is athermal. Blends of rigid chain polymers with each other, on the other hand, are predicted to behave in ways similar to blends of flexible EPs, with a positive Flory interaction parameter necessary to cause phase separation [10].

Immiscibility of PLC and EP phases is not necessarily an undesirable feature. One of the most often sought morphologies in PLC + EP blend research is the incorporation of a longitudinal PLC as the minor phase where, given the correct concentration, melt interaction, rheology and deformation during melt processing, it can fibrillate and act favorably as both a viscosity reducer and as a stiffening reinforcement. Brostow *et al.* [11] note that it is a fact that PLC + EP blends tend to be two-phase, in addition to their tendency to align, which leads to decreased blend viscosity. Indeed, in some systems such as an amorphous polyamide + longitudinal PLC blend [12] the viscosity of the blend can be lower than either of the two homopolymers in the melt and still result in a higher modulus material. In accordance with their generally low levels of miscibility, it is often found that longitudinal PLC + EP blends have quite high interfacial tensions and these have recently begun to be measured [13–15]. Despite this, PLC fibrils can still form in the blend if the viscosity ratio of components is right (PLC viscosity less than that of EP) and some extensional flow occurs during processing.

However, despite some improvements in low strain properties such as elastic modulus, these blends often suffer from poor interfacial adhesion and tend to be brittle (lower tensile and fracture strengths) compared with the homopolymers [16]. Indeed, poor interfacial adhesion may even lead to blend elastic moduli which are lower than the law of mixtures average of those of the two homopolymers [17, 18], and can result in a poor appearance of injection molded samples [19, 20]. Clearly a compromise between immiscibility and good adhesion is important.

There is a growing literature on the rheology and mechanical properties of longitudinal PLC + EP blends, in particular in relation to morphology in immiscible blends, and this has been well summarized recently [21–29] and it is not the purpose of this chapter to review such details. Rather, it will focus on the miscibility (full or partial) that is exhibited by some PLC + EP blend systems. Since many EPs are semicrystalline, addition of a PLC phase may lead to changes in degree of crystallization and crystallization kinetics in these blends and this will also be covered. Differing levels of crystallinity induced and the rate at which crystallization occurs can be important in processing and optimization of product properties such as stiffness, strength and skin–core morphologies. If, for example, a PLC is able to act as a nucleating agent and encourage similar levels of crystallinity on the outside and within a molded article, problems such as delamination of the skin–core region due to different crystallinities may be avoided whilst retaining the lower viscosity and higher modulus advantages of PLC blends. The interesting possibility of transcrystallinity on fibrillar morphologies also arises and may affect properties.

7.2 MISCIBILITY AND CRYSTALLIZATION PHENOMENA IN LONGITUDINAL PLC + EP BLENDS

As has already been mentioned, polymer blends generally tend to show limited miscibility, but by judicious choice of the polymer structure of the blend components, partial miscibility can often be achieved. Improved miscibility usually results from lower molecular weight components and, as will be shown, improved flexibility of the blend components can also influence this. It should also be remembered that determination of 'miscibility' is somewhat dependent on the technique that is probing the presence of domains, be it optical microscopy (OM), differential scanning calorimetry (DSC) or dynamic mechanical thermal analysis (DMTA) ([7, 30, 31]). Even in nominally miscible blends where the above widely used techniques show only one phase or a single glass transition by DSC and DMTA, techniques such as nuclear magnetic resonance (NMR) in the solid state are often still able to identify concentration fluctuations and heterogeneities on a micro-level [32] and thus miscibility is really a matter of scale of investigation.

Miscibility (or indeed lack of it) can also affect the crystallization processes of a semicrystalline EP in a variety of ways and a number of excellent reviews detail the effect of blending polymers on crystallization in a wide range of blend systems [33–36] and only some of the general aspects will be emphasized here. Since the rate of polymer crystallization occurs by a nucleation and growth mechanism, the presence of a second phase either mixed miscibly with the semicrystalline one or separate from it may affect one or both of these processes. Whilst crystallization phenomena can be measured by isothermal dilatometric experiments (changes in volume), most of the work reported and presented in this review involves DSC analysis: temperature heating scans (possibly cold crystallization exotherm and melting endotherm of crystals), cooling runs (crystallization exotherms) and isothermal crystallization in which a molten sample is quenched to a steady crystallization temperature from the melt. As a general rule, if a crystallization temperature increases with the addition of another phase, the nucleation is more heterogenous. Less undercooling is thus required to cause nucleation and the resultant crystals have higher melting points. To be strictly correct, equilibrium melting points should be determined by plots and extrapolation of crystallization temperature (T_c) vs. melting point (T_m) [37] but this is not often done. The amount of energy released during crystallization or melting is indicative of the degree of crystallinity (and should usually be normalized to the mass of the semicrystalline phase for comparison between differing blend compositions).

Isothermal DSC crystallization kinetics are usually converted to the weight fraction crystallized $X(t)$ at time t and fitted via double log plots to the *Avrami equation* [38]:

$$1 - X(t) = \exp(-Kt^n) \tag{7.1}$$

where K is the rate constant and a function of nucleation and growth rates and is sensitive to temperature, molecular weight, polymer structure, molecular weight distribution, impurities and so on. It is found [39] that the rate constant can be expressed in terms of G (crystalline growth rate), $\bar{N}$ (nucleation density) and ρ_c and ρ_a (density of the crystalline and amorphous phase), their ratio often being assumed to be approximately 1, such that

$$K = \frac{4}{3}\pi \frac{\rho_c}{\rho_a} G^3 \bar{N} \tag{7.2}$$

The Avrami exponent, n, is related to the type of crystal nucleation, growth, and the dimensionality and nature of the crystals. Often the half-time of crystallization ($t_{0.5}$) is used as an indication of the rate of overall crystal growth. OM is particularly useful in determining the kinetics of growth rate (such as radius increase with crystallization time), such as of spherulitic structures.

Microscopic examination (OM), small angle light scattering (SALS), scanning (SEM) or transmission electron microscopy (TEM) wide (WAXS) and small angle X-ray scattering (SAXS) spectroscopy are other techniques usually necessary to characterize fully the polymer crystal morphology but will not be dwelt upon in this chapter.

The crystallization of blends tends to depend on the level of mutual miscibility of the components. In miscible blends, the general result is that suppression or otherwise of crystallization with miscibility is dependent on the relative glass transition temperatures of both phases [33, 34]. For example, in a blend of an amorphous and semicrystalline polymer, if the amorphous material has the higher T_g, the miscible blend will also have a higher T_g than that of the semicrystalline homopolymer and, at a given temperature, the mobility and thus the efficacy of the semicrystalline phase molecules to crystallize is reduced. The converse is often true if the amorphous phase has a lower glass transition. Effects such as chemical interactions and other thermodynamic considerations also play a role and the depression of the melting point in a miscible blend can be used to determine the Flory interaction parameter χ [40].

Crystallization of immiscible polymer blends is also complex and can be divided into the differing effects on nucleation and growth.

- Nucleation is dependent to some extent on whether the second phase is solid or fluid at the crystallization temperatures of the crystallizing polymer (often the second phase may also possess some crystallinity, as for example do many semiflexible PLCs). If one phase is solid at the crystallization temperature, this may encourage nucleation of the other. If one is molten when the other is crystallizing, viscosity and melt miscibility considerations also are important [33, 34]. Clearly morphology induced by processing conditions is also important, especially in the case of most longitudinal PLC blends where the PLC often is the immiscible, minor phase.
- Growth of the crystalline phase in the crystallization of immiscible blends is usually retarded with the immiscible species providing a hindrance to the motion of crystallizing molecules and a coarser, spherulitic structure occurs. Also possible, and rather more complex, is co-crystallization of two crystalline species which occurs, for example, in low density polyethylene (LDPE) and ethylene/propylene copolymer (EPDM) blends [41]. This does not always occur in blends of two semicrystalline materials and in any event, the degree of crystallinity of PLCs tends to be significantly lower than that of most semicrystalline EPs which are matrices.

Most longitudinal PLCs tend to be copolyesters made of a range of chemical units such as those shown in Figure 7.1. In order to decrease the order and strong secondary interchain bonding that necessitates lyotropic PLC materials to be processed by spinning from solutions of strong acids, a range of strategies have been devised to lower the effective aspect ratio of the chains and decrease crystallinity, leading to thermotropic longitudinal PLCs. They are usually achieved by inclusion of flexible spacer groups between the mesogenic groups or copolymerization of a range of monomeric units that in some way disrupt the linearity of the chain [42].

It is thermotropics that will mainly be discussed here as they are

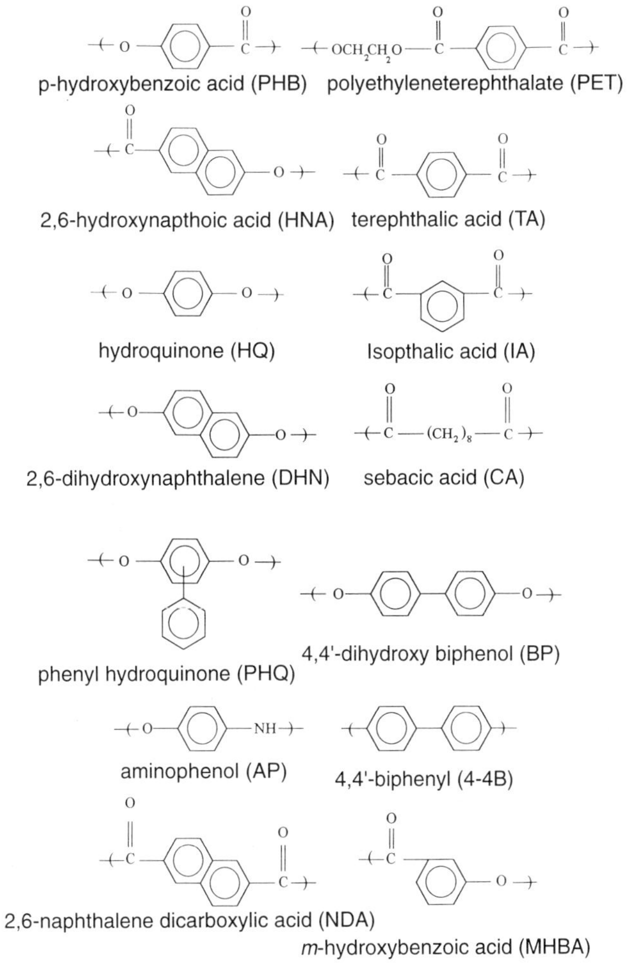

p-hydroxybenzoic acid (PHB) polyethyleneterephthalate (PET)

2,6-hydroxynapthoic acid (HNA) terephthalic acid (TA)

hydroquinone (HQ) Isopthalic acid (IA)

2,6-dihydroxynaphthalene (DHN) sebacic acid (CA)

phenyl hydroquinone (PHQ) 4,4'-dihydroxy biphenol (BP)

aminophenol (AP) 4,4'-biphenyl (4-4B)

2,6-naphthalene dicarboxylic acid (NDA) *m*-hydroxybenzoic acid (MHBA)

Figure 7.1 A range of the components from which most commercial liquid crystal polymers are synthesized.

the PLCs most commonly melt blended and processed using conventional processing equipment. Laboratory-scale solution blending can be useful if only small amounts of material (such as for optical or thermal analysis) are required or if the reaction between PLCs and another polyester – transesterification – is to be avoided. The various classes of PLC have been divided in this review into rigid longitudinal PLCs and semiflexible longitudinal PLC materials. This is a somewhat arbitrary classification of convenience although its relevance will become clearer since flexibility has implications for miscibility.

7.2.1 Blends of rigid longitudinal PLCs: HNA/*x*PHB copolyesters

The most widely used and researched rigid thermotropic PLCs have been the Vectra series by Hoechst Celanese (USA). Although section 7.2.1 will mainly discuss EP + HNA/*x*PHB blends (with Vectra A950 being the primary one), related members of the Vectra series will also be discussed, in particular the commercialized Vectra A950. Vectra A950 has the composition 27 mol% 2-hydroxy-6-naphthoic acid (HNA) and 73 mol% *p*-hydroxybenzoic acid (PHB) (HNA/0.73PHB). Vectra B950 has components 20% 4-aminophenol (AP), 20% terephthalic acid (TA) and 60% hydroxynaphthoic acid (HNA) (0.2AP/0.2TA/0.6HNA), where the acronyms are those of Figure 7.1. Earlier variants of these have been mentioned in the literature prior to Vectra commercialization and will be listed here as they are discussed in this chapter. They are Vectra RD500 (0.28HNA/0.52PHB/0.1TA/0.1HQ) [43] and Vectra LCP2000 (HNA/0.7PHB) [44].

The properties of the two commonest commercial PLC homopolymers are as follows:

- HNA/0.73PHB (Vectra A950) has a glass transition in the range of 90 to 110°C, depending on the characterization technique used [45–47], a small melting point endotherm at about 280°C with a fusion enthalpy of $1\,J\,g^{-1}$ [49] and a crystallization temperature on cooling of about 235°C [48].
- 0.2AP/0.2TA/0.6PHB (Vectra B950) has a T_g of about 132°C and a melting point of 280 to 290°C ($1-3\,J\,g^{-1}$) and a crystallization temperature of about 225 to 235°C on cooling [49, 50].

Bisphenol-A polycarbonate (PC) has been studied often with various HNA/*x*PHB longitudinal PLC materials, usually in an effort to reduce PC viscosity and aid reinforcement with the hope that compatibility between the two polyesters would be good. Rheological, morphological and mechanical property studies of this blend have been reviewed [26, 28] and most workers state that the two are largely immiscible. However, some change in T_g with composition are often observed. Kohli *et al.* [43] found in PC + 0.28HNA/0.52PHB/0.1TA/0.1HQ

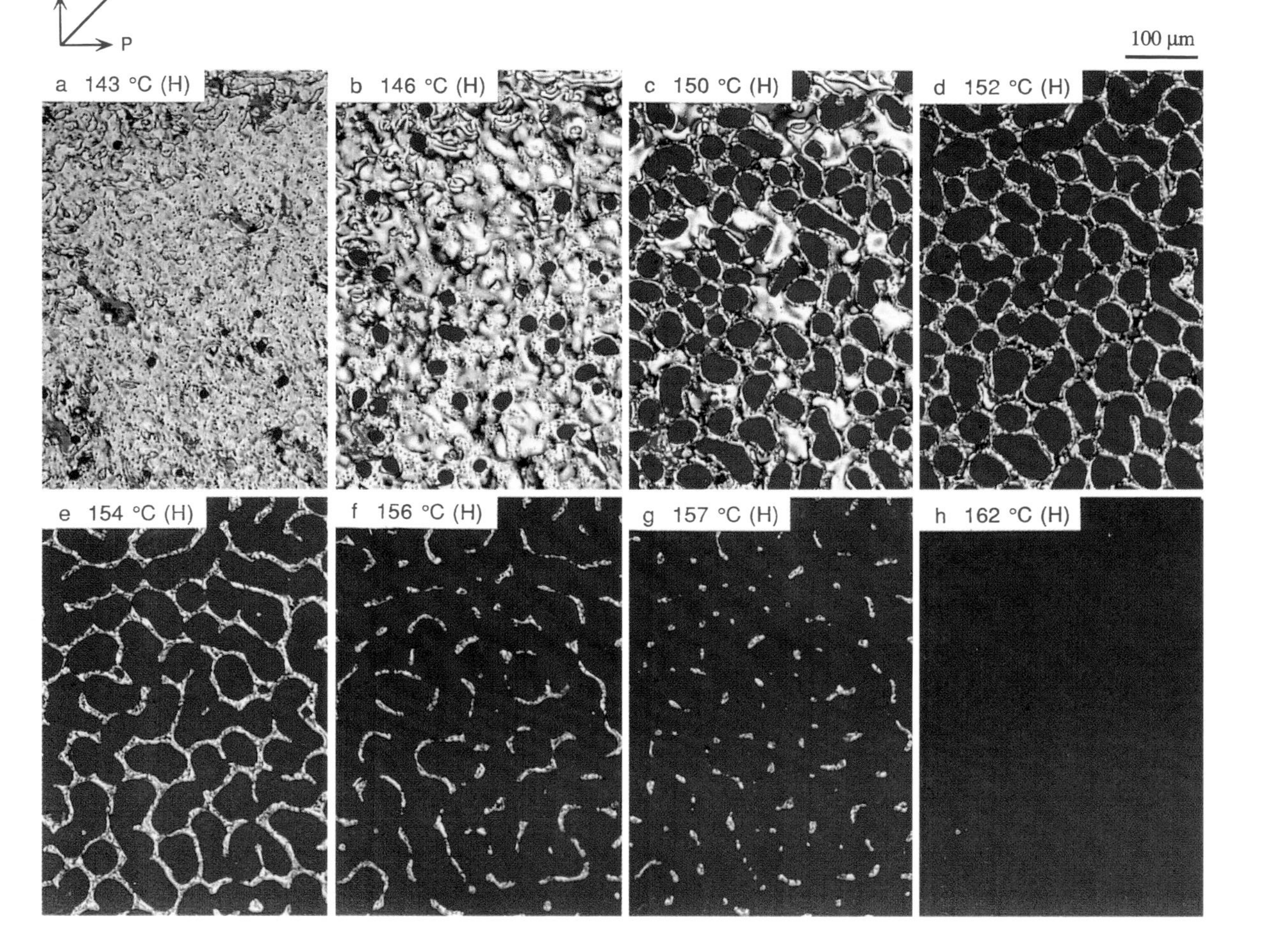

Figure 6.5(a) Textures observed for polymer (6.1) (n = 10, M_n = 18 700) on heating to the isotropic phase at 1°C min^{-1}. All micrographs focus on the same area of the same sample. Biphasic range: 143–162°C. (Reprinted with permission from [42]. Copyright 1994 American Chemical Society.)

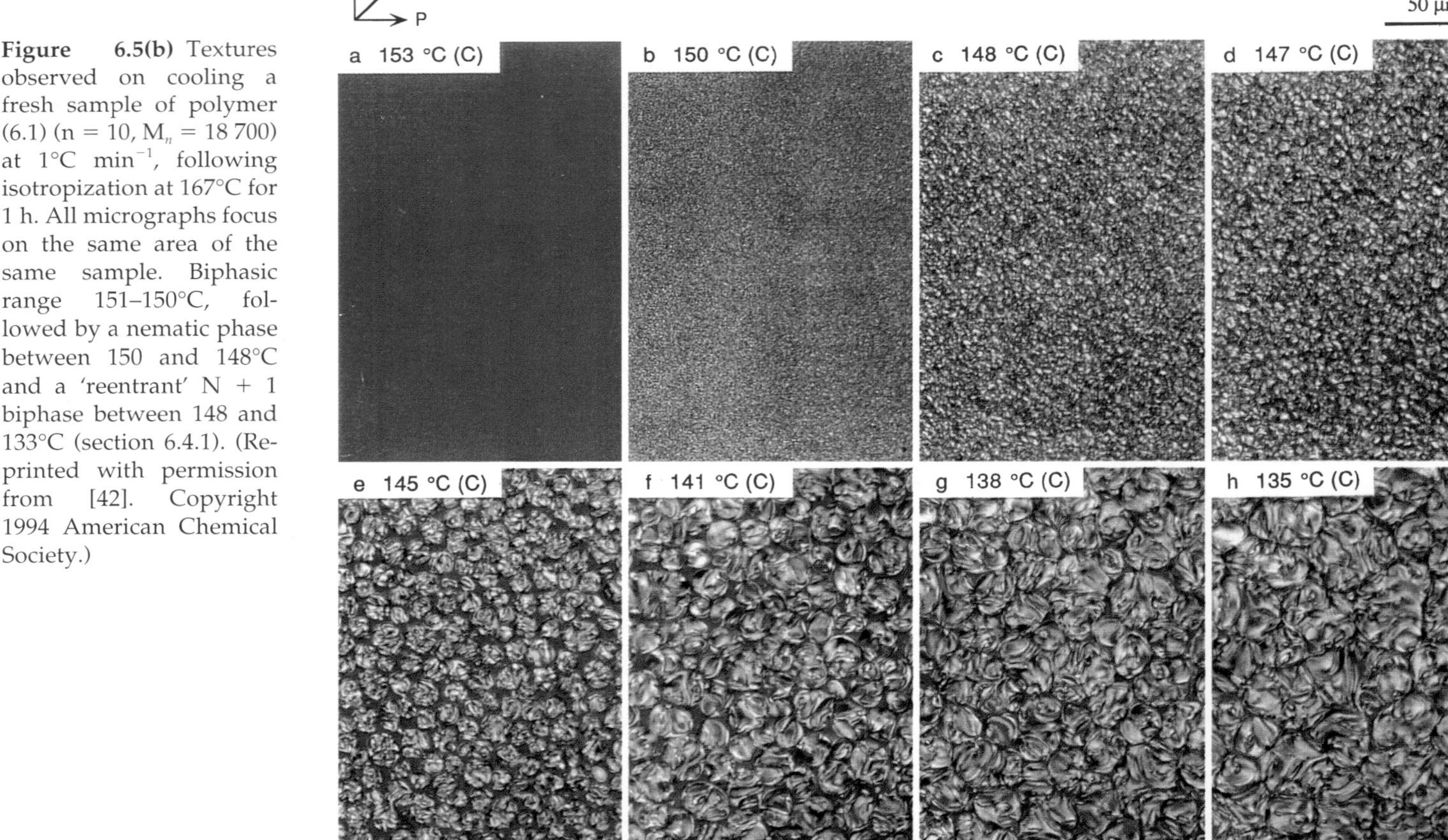

Figure 6.5(b) Textures observed on cooling a fresh sample of polymer (6.1) ($n = 10$, $M_n = 18\,700$) at 1°C min^{-1}, following isotropization at 167°C for 1 h. All micrographs focus on the same area of the same sample. Biphasic range 151–150°C, followed by a nematic phase between 150 and 148°C and a 'reentrant' N + 1 biphase between 148 and 133°C (section 6.4.1). (Reprinted with permission from [42]. Copyright 1994 American Chemical Society.)

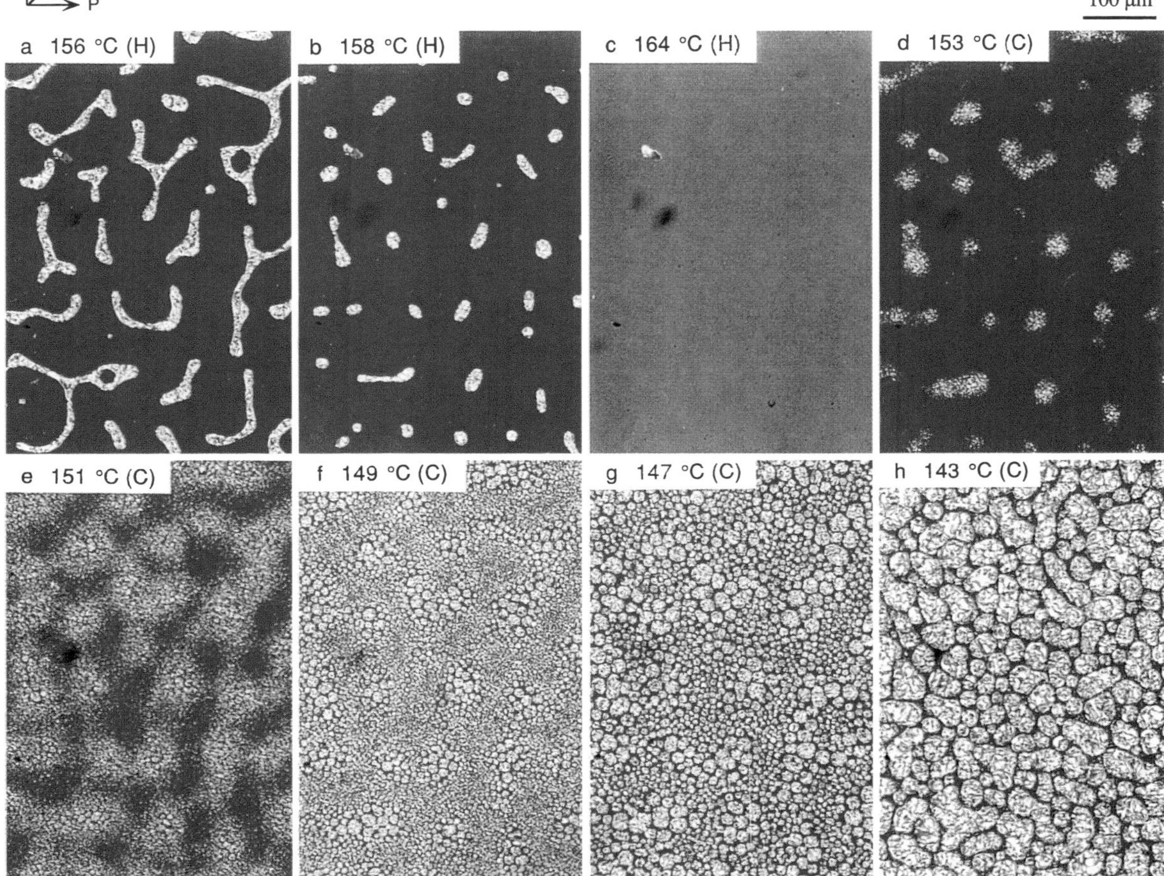

Figure 6.5(c) Textures observed on cooling a fresh sample of polymer (6.1) ($n = 10$, $M_n = 18\,700$) at 1°C min^{-1}, immediately following heating as in Figure 6.5(a) (no isotropization rest). All micrographs focus on the same area of the same sample (the small bright particle observed in the upper left corner at 164°C is a stationary impurity used as a reference point). Biphasic range: 155–123°C. (Reprinted with permission from [42]. Copyright 1994 American Chemical Society.)

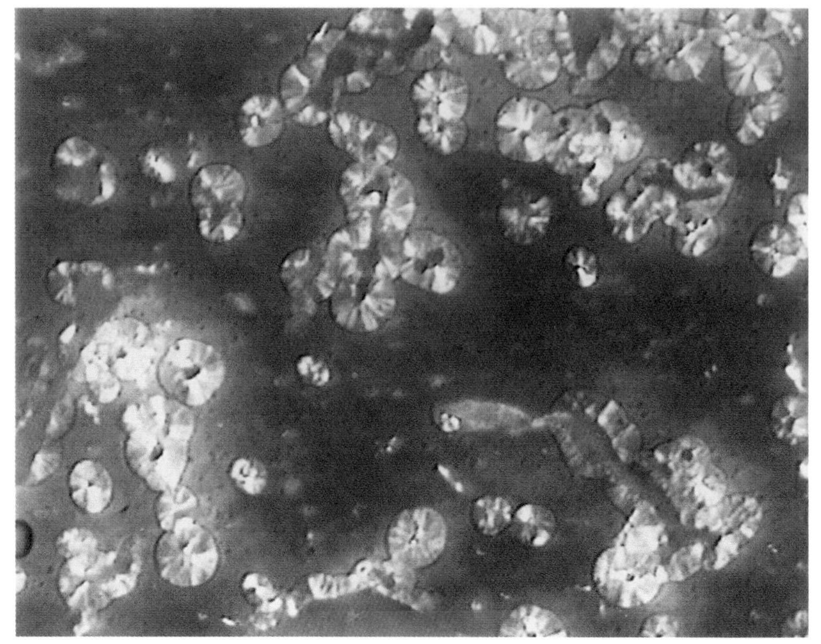

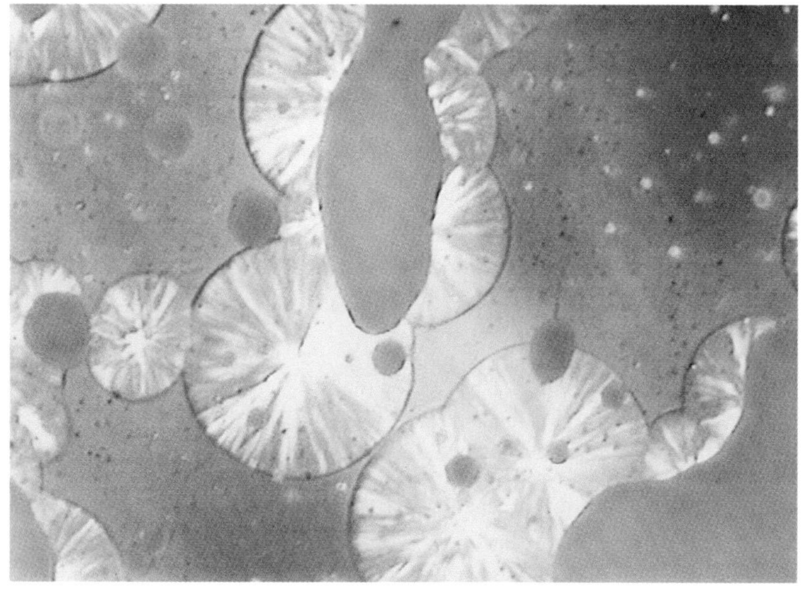

Figure 7.11 Optical microscopy of isothermally crystallized 80% PP + 20% HNA/0.73 PHB blend at 130°C. (Pictures kindly provided by Markku Heino and Tommi Vainio, Helsinki University of Technology, Finland.)

blends that the T_g of the PC (147°C), moved slightly towards that of the PLC homopolymer by about 7°C for an 80% PLC addition. The T_g of the PLC also seemed to decrease by some 3°C (this is unusual since if the blends were semicompatible, the values of T_g should move together). It is not clear if this is due to slight miscibility, some chemical reaction between the two in the melt (transesterification) or, as Kohli *et al.* [43] proposed, indicative of a small phase size of one component in the other. The melting point of the PLC component did not change with blending. If the Fox equation, which predicts T_g in miscible blend systems, is used [51], only 2% miscibility of PLC in PC is seen. Despite this slight miscibility and a favorable viscosity ratio for fibril formation (PLC viscosity much lower than PC) in systems with 0.28HNA/0.52PHB/0.1TA/0.1HQ as the minority phase, spherical droplets of 1 to 10 μm, rather than elongated fibrils, are observed. Similar behavior was found in PC + HNA/0.73PHB blends by Chapleau *et al.* [52] who observed a slight decrease of PC T_g of 2°C upon addition of 20% HNA/0.73PHB. Quite fine particles (less than 1 μm) seemed to indicate low interfacial tension and some compatability although cyrogenic fracture surfaces still show smooth debonding. For blends with less than 10% HNA/0.73PHB, the modulus was greater than that predicted by the rule of mixtures, indicating the efficacy of good interfacial adhesion in stress transfer, at least at room temperature. Smooth fracture surfaces were also seen in a PC + 0.2AP/0.2TA/0.6PHB blend [53], where interfacial tension between phases in the melt was sufficiently low to cause some fibrillation but still poor interfacial adhesion resulted.

In view of these results, it is surprising that a synergy in terms of interfacial adhesion was reported by Isayev *et al.* [44] for 90% PC + 10% HNA/0.73PHB where the blend material resulted in a greater tensile strength than that of the components (as opposed to the usual dramatic reduction in immiscible systems). Since the composition is similar to other blends, this differing behavior may be due to processing differences and, in particular, some slight chemical reaction.

It seems that despite there being some compatability in PC + HNA/xPHB and related systems and the addition of PLC to PC (particularly at low PLC levels) may be beneficial, the ultimate properties of the blends still often decrease dramatically. Amendola *et al.* [54] attempted to rectify this by using melt transesterification to produce a compatabilizer phase with the aim of better interfacial adhesion in a PC + HNA/0.73PHB blend. Although degradation and decrease in molecular weight occur during the transesterification process, it was possible to produce a copolymer in the extruder by extended melt blending of PC + HNA/0.73PHB. When the resultant copolymer is added as a third phase to PC + HNA/0.73PHB blends, increased fibrillation and adhesion result. PLC fibril pull-out was found to be reduced and that which occurred resulted in some fibrillar elongation, with fiber–matrix adhesion.

The question of whether changes in glass transition is due to phase size (as proposed by Kohli *et al.* [43]) and/or molecular miscibility, was also raised by Verghoot *et al.* [55] as the explanation for T_g changes observed in blends of HNA/0.73PHB with a Kraton ethylene/butylene (EB) and styrene (S) triblock copolymer. By DMTA the EB and S blocks can be seen separately with the EB glass transition decreasing some 17°C upon addition of 50% HNA/0.73PHB and the S block changing far less. Since it seems unlikely that the HNA/0.73PHB and the non-polar EB block are miscible or even adhere well (and this is backed up by morphological examination), it is clear that apparent miscibility (as judged by T_g values) may not be the most useful indicator of phase mixing in what are essentially immiscible systems. Even if some miscibility exists, poor interfacial properties may still result.

Given the variation in degrees of adhesion between HNA/xPHB PLC copolymers and amorphous matrices, it is of interest to see the effect of blending these PLCs on EP phase crystallinity. Poly(phenylene sulfide) (PPS) is a semicrystalline EP that has been blended with HNA/xPHB materials. PPS is a very interesting semicrystalline engineering polymer with good temperature and chemical resistance and favorable mechanical properties. As a homopolymer it has the ability to be quenched to an amorphous state, shows a cold crystallization peak if scanned by DSC and its degree of crystallinity can be well controlled [56–58]. In addition PPS is found to 'cure' under appropriate high temperature conditions rendering it crosslinked [59]. It is often used as a composite with glass or carbon fibers and crystallization phenomena in such systems have been recently studied [60–65] and this has led to the investigations of its properties blended with HNA/xPHB materials, and in particular the effect of the PLC on PPS crystallization has been examined.

PPS homopolymer has a T_g of about 88°C, a melting point of about 274°C [62] and a crystallization temperature of 229 to 235°C, all parameters depending on the grade of PPS [58]. PPS crystallization can be modeled by Avrami kinetics in terms of three-dimensional spherulites. The inclusion of a fibrous phase often reduces crystallinity to two dimensions and in addition Avrami kinetics may no longer apply [62]. Fibers are found to encourage two-dimensional transcrystallinity [63] which results in faster crystallisation but with lower ultimate levels of crystallinity. Clearly the fibers nucleate the PPS crystallinity and a slight increase in T_c is observed [58] with the type of fiber and its surface treatment affecting its nucleation efficacy. In general, non-amorphous fibers (materials other than glass) tend to be more effective in promoting crystallization [65] with carbon fibers being most effective, especially at lower degrees of graphitization. Relevant to PLC blends is the fact that sized and unsized lyotropic PLC Kevlar fibers were also found to be effective in initiating transcrystallinity and thus enhancing crystallization rates [62].

It is against this background that a number of workers have looked at blends of HNA/0.73PHB (Vectra A950) and 0.2AP/0.2TA/0.6PHB (Vectra B950) with PPS [49, 50, 66–71]. Hong *et al.* [49] found that PPS and 0.2AP/0.2TA/0.6PHB were relatively immiscible, with the T_g of PPS decreasing some 4°C with 50% addition of PLC. It was found that the crystallization rate, as measured by $t_{0.5}$ in isothermal kinetics, decreases dramatically (crystallization rate increased) with the addition of PLC up to 50% and this was ascribed to heterogeneous nucleation and an increase in the nucleation density (using equation (7.2)). The rate monotonically increased with PLC content, even up to 50% PLC as shown in Figure 7.2. There is also a decrease in supercooling (increase in T_c; Figure 7.3) and a decrease in spherulite size. The Avrami exponent n was found to remain at 3 (three-dimensional spherulites) for dispersion of PLC in PPS, in all behaving more as PPS nucleated by conventional particulate fillers [64] than by fibers.

Minkova *et al.* [50] also found an increase in the crystallization rate, decrease in supercooling and only a slight decrease in degree of

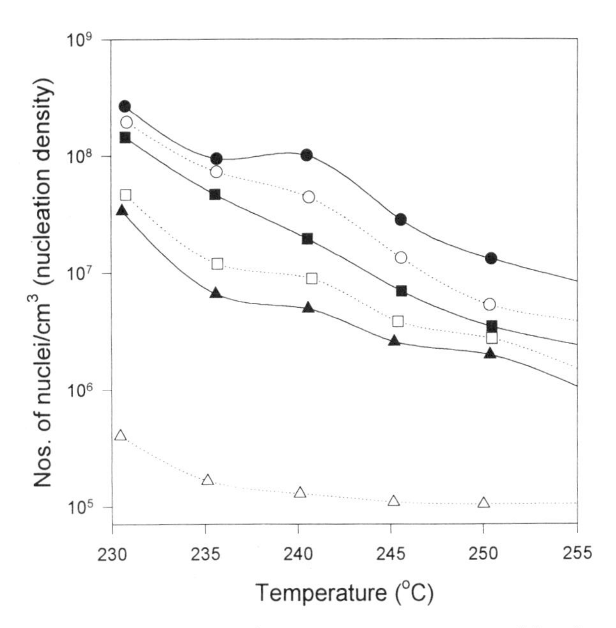

Figure 7.2 Nucleation density of PPS + HNA/0.73HBA blends as a function of temperature. Blend concentrations are (HNA/0.73HBA:PPS) ▲ (100:0), △ (97:3), ■ (95:5), □ (90:10), ○ (75:25), ● (50:50). (Redrawn from [48].)

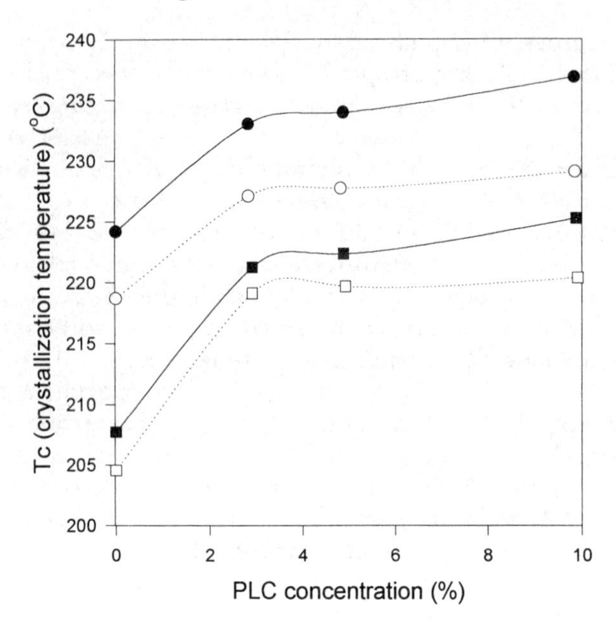

Figure 7.3 PPS crystallization temperature in PPS + HNA/0.73HBA blends as a function of PLC content for cooling rates of ● (2°C min^{-1}), ○ (4°C min^{-1}). ■ (8°C min^{-1}), □ (12°C min^{-1}). (Redrawn from [48].)

crystallinity. A slight increase in PPS T_g with PLC addition does suggest some slight interaction. A precise determination is made difficult because of the closeness in the T_g values of PLC and PPS. However, whereas Hong *et al.* [49] found a monotonic increase in overall crystallization rate (such as determined by crystallization half-time) with increasing addition of PLC up to about 50%, Minkova *et al.* [50] found a large effect on addition of 2% PLC but only a slightly increased effect at higher concentrations. Morphological studies by the same group [66, 67] on this blend demonstrate good interfacial adhesion between the two phases but no nucleation off the surface of the PLC particles is observed. This prompted the theory that the PLC is not acting simply as the nucleating phase, even though the increase in crystallization rate is due to increased nucleation and growth rate remained unaffected. Rather it has been proposed that it is dissolution into the PPS phase of impurities in the 0.2AP/0.2TA/0.6PHB that are responsible for increased nucleation and impurity concentration saturates at low levels of PLC addition. Fracture surfaces do show adherence of PLC spherical or distorted spherical particles of some 3 to 6 μm to the majority phase. However, despite indications of good adhesion in some blends, other studies (particularly with HNA/0.73PHB) have shown poor interfacial adhesion and problems of porosity and decreased thermal stability of the blend compared to homopolymers. This has been attributed to some unspecified chemical interaction or to PPS thermal degradation [70] and thus these blends may be somewhat problematic. Similar results were found in non-isothermal situations (more relevant to most processing conditions) where addition of 2 to 50% 0.2AP/0.2TA/0.6PHB led to a threefold increase in crystallization rate, although the nature of the crystal growth process and the final crystal geometry remained unchanged [71].

Other semicrystalline EP + HNA/xPHB blends have been examined. A detailed study was performed by Sharma *et al.* [72] where 5, 10 and 15 wt% of HNA/0.73PHB was melt blended with poly(ethylene terephthalate) (PET). The values of PET homopolymer T_g and T_m range from 65 to 85°C and 235 to 258°C, respectively, with a T_c of *c.* 145°C–179°C [7, 72–77]. Addition of the PLC resulted in a slight increase in PET melting point (about 6°C) whilst the crystallization temperature is constant at 18°C for all three HNA/0.73PHB concentrations, with its maximum efficacy at and below 5% addition (as with PPS blend studies mentioned earlier). The heat of fusion increases on addition of 5% (due to increased nucleation) but decreases for 10–15%, due to retardation by the HNA/0.73PHB on crystal growth. Melot and MacKnight [78] also showed a monotonic increase in crystallization rate with the addition of HNA/xPHB copolymer to PET with optical microscopy confirming the nucleating effect of the particles, although overall levels of crystallinity are little changed. The blends seem quite immiscible and yet good adhesion between the phases occurs [79].

Contrasting somewhat to these previous studies is that of Kim and Denn [45] in which the melting point of PET decreases with HNA/0.73PHB addition, and quite dramatically above about 60% HNA/0.73PHB. This, along with very good adhesion of spherical HNA/0.73PHB particles to the PET matrix, suggested the possibility of miscibility. Transesterification or impurities/gas evolution could also be involved and lead to such a variation of results.

Polybutylene terephthalate (PBT) is more flexible than PET and has lower transition temperatures (T_g = 31°C and T_m = 225°C) [80] and has also been blended with HNA/xPHB polymers. Increasing flexibility of the EP phase would not be expected to improve miscibility with a rigid PLC, and indeed this has been found to be the case. Heino and Seppälä [81] blended HNA/0.73PHB with PBT and despite finding good reinforcement in the flow direction of blend fibres, SEM studies inidicate only moderate adhesion. Engberg *et al.* [82] studied the crystallization phenomena of PBT + HNA/0.73PHB and demonstrated complete immiscibility and, perhaps a little surprisingly, no interaction of the two phases at all. In fact, increasing the HNA/0.73PHB content did not change melting or crystallization temperatures, and the Avrami exponent of PBT crystallization or spherulite size and OM showed no nucleation on PBT phase boundaries. This demonstrates that some other factor, such as perhaps the precise chemical nature or microstructure of the minor phase, may be necessary to encourage nucleation.

A blend of HNA/0.73PHB with a more rigid semicrystalline polymer, poly(ether ether ketone) (PEEK) with a T_g of 151°C and T_m of 338°C, showed slightly greater miscibility (about a 10°C decrease in PEEK T_g for high contents) than with more flexible EPs and resulted in improved crystallization processes [48]. Small additions (up to 2.5% of HNA/0.73PHB) result in a slight decrease in PEEK melting and cold crystallization temperature with a slight (but significant) increase in fusion and cold crystallization energies with a maxima at 2.5% PLC. The glass transition of the HNA/0.73PHB does not change, as is often the case in these blends. If there is some miscibility in these rigid PLC + EP blends, it involves dissolution of the PLC into the EP phase.

7.2.2 Blends of rigid longitudinal PLCs: non-HNA/*x*PHB copolyesters

Quite early solution blend work by Takayanagi *et al.* [83] of rigid aromatic polyamides (poly(*p*-phenylene terephthalate)) (PPTA) and flexible aliphatic polyamides, nylon 66 and nylon 6 (the latter will be discussed here) has been instrumental in developing the concept of 'molecular composites' and although this lies more in the realm of lyotropic blends, it is often cited in the literature with regard to the effect of the PLC on the crystallization of other components in blends. PPTA has the structure

$$\left[NH-C_6H_4-NH-CO-C_6H_4-CO-NH-C_6H_4-NH \right]$$

In nylon 6 blends with PPTA, some miscibility is observed (the glass transition of nylon 6 increasing from 60 to about 90°C) and the main effects of blending on nylon 6 crystallinity can be summarized as

- cooling curves for nylon 6 demonstrate primary crystallization temperatures at 210°C and secondary crystallization at 195°C (compared to 180°C for nylon 6 homopolymer), due to increased nucleation in the blends which require less undercooling to stimulate crystal growth;
- X-ray diffraction patterns are less broad for blends compared to homopolymers, which indicates higher levels of better defined crystals;
- the Avrami exponent changes from $n = 4$ (pure nylon 6) indicating

homogeneous, three-dimensional spherulitic crystallization) to $n = 2$ (blend) heterogeneous nucleus followed by two-dimensional growth.

It is proposed that the nylon 6 molecules nucleate and grow as lamellar crystals on the microfibrils in a chain-folded manner shown in Figure 7.4. Overall the effect is highly desirable as increased modulus, heat distortion temperature and yield stress result due to the resultant morphology and increased crystallinity.

PLCs have been shown to nucleate systems which are hard to crystallize by any other means. Polyethersulfone (PES) is a high temperature thermoplastic which, even when cast from solution, is unable to crystallize [84], as opposed to, for example, PC. However, when solution blended with 30 wt% wholly aromatic longitudinal PLC (structure not reported), spherulites of the PES can be formed and it is suggested that the combined effect of solvent molecules to 'lubricate' the motion of the crystallizing PES molecules and a nucleant (PLC) is required, either condition on its own not being sufficient. Similar attempts to crystallize the blend from the melt failed.

As with HNA/xPHB blends, other rigid rod PLCs should also be able to affect the crystallization process in blends of semicrystallizing components and, given that different effects have been reported for

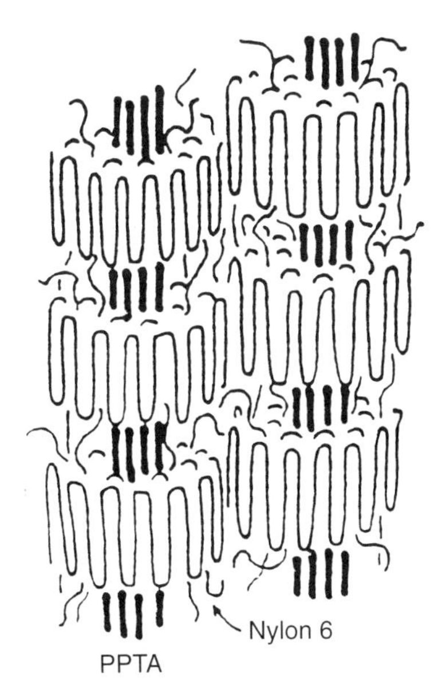

Figure 7.4 Proposed microfibrillar structure of chain-folded nylon 6 initiated and crystallized on PPTA molecular chains. (Reprinted from reference 83, p. 608 by courtesy of Marcel Dekker Inc.)

HNA/xPHB blends (see above), subtle changes in copolymer composition may be sufficient to effect crystallization to differing degrees and can be compared with HNA/xPHB effects. One such rigid longitudinal PLC is 0.4TA/0.1BP/0.1IA/0.4CHQ, where CHQ is chlorohydroquinone and has been blended with PET [85]. It is an interesting blend component because, as well as a T_g of about 110°C and a melting point into the nematic phase at 315°C, it also seems to show some resemblance to a smectic phase (rare in rigid PLCs) and clears into isotropic at about 421°C. As with the work of Sharma *et al.* [72] and Melot and MacKnight [78], addition of the 0.4TA/0.1BP/0.1IA/0.4CHQ PLC is found to cause a slight increase in the PET crystallization temperature [85], acting as a nucleation promoter. Its efficacy is reduced at above 10 wt% due to interference by the PLC in crystallization. The materials do seem to be slightly miscible, but once again this may be due to some chemical reaction.

Particularly of interest in PLC blends is their combination with new, high performance plastics (which tend to be more expensive than other engineering polymers) since, if properties can be improved, the addition of PLC and its cost may be justified. Two such reported blends are of an Amoco (USA) PLC, Xydar, which is a random copolymer of PHB, TA and BP, and NEW-TPI, a semicrystalline polyimide [86, 87] of structure

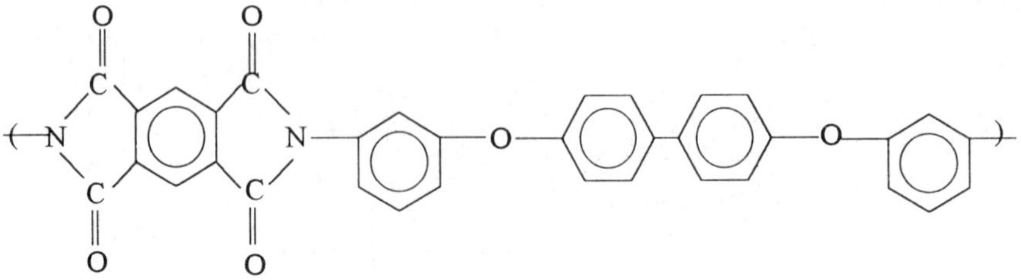

Xydar has the structural formula of 0.68PHB/0.16BP/0.16TA and a melting point of about 420°C [88]. In the case of the blend with NEW-TPI which has a T_g of 240 to 250°C dependent on crystallinity, a T_m of 381–389°C with a maximum crystallization rate at 330°C [87, 89], although the blends are immiscible, addition of some 10–30% 0.68PHB/0.16BP/0.16TA promotes early initiation of cold crystallization at temperatures some 33°C lower than for neat polyimide [87], substantially increasing the modulus [86]. Cooling scans of NEW-TPI at 10°C min^{-1} in DSC show no crystallization peak unless blended with the PLC. Clearly, addition of PLC has a role to play in further improving properties of such new high temperature materials.

Of particular interest in this class of materials is work by Baird and coworkers [90–94] who found good miscibility between a commercial polyetherimide (PEI) and the early commercial PLC HX-series (DuPont, USA). The two PLC materials were an amorphous PLC (HX1000) with composition of the copolymer PHQ/HQ/PEHQ (PEHQ being phenyletherhydroquinone whose precise composition is not known) with a T_g of 185°C [91, 93] and a semicrystalline material (HX4000) mentioned above with composition of TA/HQ/PHQ (where PHQ is phenylhydroquinone) and $T_g = 225$°C and $T_m = 311$°C. It was found that TA/HQ/PHQ blended with PEI (T_g of 228°C) resulted in two phases, a TA/HQ/PHQ phase, whose glass transition as measured by DMTA remained unchanged, and a PEI loss peak that moved to lower temperatures due to partial miscibility with TA/HQ/PHQ [91]. A similar effect is seen with the PHQ/HQ/PEHQ + PEI blend [91] and both show excellent signs of good intermixing and adhesion. Both the minor phase and the matrix phase in which it resides appear deformed when the failure surfaces are examined by SEM. Annealing above the glass transitions of the phases demonstrated that the phase miscibility was stable and not simply kinetically entrapped due to quenching.

The importance of such results is seen in that tensile modulus of both blends, as a function of composition [93], lies above the rule of mixtures between the homopolymers (as in the case of TA/HQ/PHQ + PEI; Figure 7.5) and indeed, in the PHQ/HQ/PEHQ + PEI blends, some compositions are above that of the PLC itself (Figure 7.6). The advantage of such results has been well demonstrated by Bafna *et al.* [90] where it is shown that some

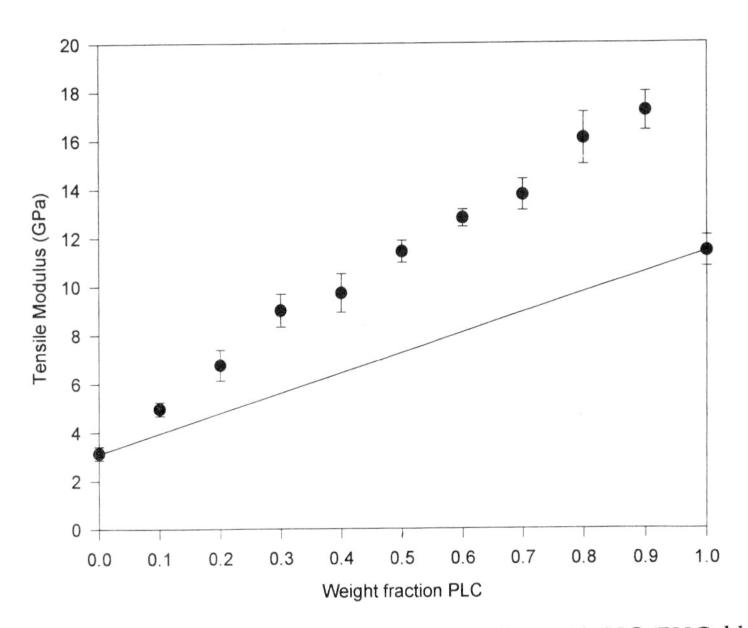

Figure 7.5 Tensile modulus of PLC content in PEI + TA/HQ/PHQ blends (see text for details of structure) at room temperature. (Redrawn from [92].)

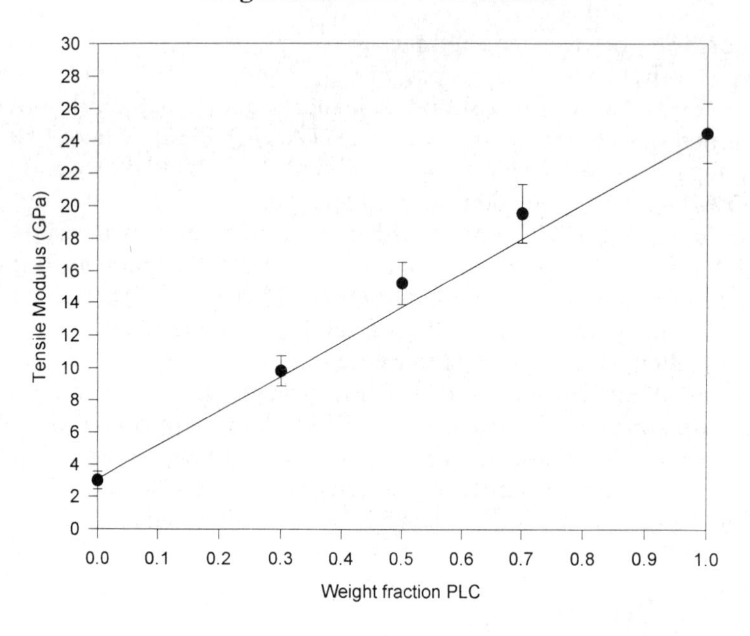

Figure 7.6 Tensile modulus of PLC content in PEI + PHQ/HQ/PEHQ blends (see text for details of structure) at room temperature. (Redrawn from [92].)

compositions are comparable in modulus and strength to glass-filled PEI yet with easier processability and appearance when injection molded. Further investigation of these blends has been carried out by positron annihilation spectroscopy (PALS) [95] which is a method of free volume determination in polymers and involves injection of subatomic particles and the measurement of their decay times. It was found to be sensitive to the degree of miscibility of the PLC materials in which it was found that the free volume properties of blends were dominated by the PLC material in the case where the PLC + EP blend was more miscible. Such a technique has only recently been applied systematically to polymer blends [96, 97] and it appears as though the size of free volume cavities is less in blends than an arithmetic mean of that of the homopolymers. This fits in well with the view of tighter packing and higher densities in miscible polymer blends and it has been shown in isotropic blends that impact strength in blends with lower than average free volume cavity size have dramatically lower fracture toughness [98]. PALS has also recently been applied to PLC + PLC polymer blends where it was able to give insights into the packing of rigid chains [99].

The blend of HX4000 (TA/HQ/PHQ) with PEEK is an interesting one because it is one of the few blends where the effect of blending on crystallization of the PLC phase can be examined due to a significant crystallization peak at approximately 275°C which can be observed even in 50% blends [100], PLC phases generally showing very low levels of crystallinity. As with PEEK + HNA/0.73PHB blends mentioned above [48], PEEK + HX4000 seem partially miscible and this tends to retard PEEK crystallization (more undercooling required). Conversely, the crystallization peak of the PLC phase increases due to promotion of crystallization. This, in addition to a resultant increase in melt temperature of the PLC crystals, indicates that PEEK aids crystallization of the HX4000 PLC phase [100].

As if such structure–property relationships were not complex enough in the binary blend HX-series blends, the number of possible combinations can be further increased by the introduction of another polymeric phase which has some miscibility with each of the other two. Bretas and Baird [94] have achieved this with the ternary TA/HQ/PHQ + PEI + PEEK system. As mentioned already, TA/HQ/PHQ and PEI show miscibility, PEEK + PEI is a well known miscible blend [101, 102] and it is found that PEEK and TA/HQ/PHQ show full miscibility up to 50% TA/HQ/PHQ and partial at concentrations thereafter [94]. By judicious manipulation of relative concentrations it is possible to obtain substantial synergy in ternary systems (particularly those with high TA/HQ/PHQ concentration) in terms of both tensile strength and modulus. Such synergy seems to depend not just on miscibility but also on increased PEEK crystallinity (probably due to a lower, miscible blend glass transition and thus greater mobility of the crystallizing species).

Clearly from these examples, miscibility has a role to play both in adhesion and promotion of crystallinity, affecting both blend moduli and strength. Although more rigid EPs such as PEEK and, to a lesser extent, PC show some miscibility (whilst not necessarily guaranteeing good interfacial adhesion), it would be desirable to improve miscibility of the PLC with the EP. Even though increased PLC flexibility may help achieve this, it would be advantageous if this could be achieved whilst maintaining the rigid rod nature of the PLC to be blended. To avoid such contrasting requirements, Ballauff [10] suggested that, given the very poor entropic drive for mixing where one of the components is liquid crystalline, improved miscibility or adhesion could be achieved by blending EPs with 'hairy' rod-like PLC macromolecules. These are mesogenic macromolecules with many attached, flexible side chain units protruding. From a fundamental point of view, it would seem likely that if the mesogenic macromolecule is surrounded by a 'solvent' of covalently connected flexible units, intermingling, entanglement and some miscibility with a surrounding EP phase may be encouraged, with PLC properties being maintained.

This was attempted by Heitz *et al.* [103] who synthesized a rigid chain copolymer of terephthalic acid and *t*-butylhydroquinone with attached polystyrene macromer side chains of a variety of molecular weights (1000 to 20 000 g mol^{-1}). Such polymer molecules, when blended with polystyrene (PS), showed a dramatic increase in viscosity, even at PLC concentrations as low as 10%. This, along with a three-dimensional network (rather than the dispersed morphology expected at such low concentrations), implies quite high compatability and phase interaction. This was reflected in improved moduli of the blends (compared to similar PLC + PS blends where no such macromers were attached to the rigid chain) although the effect was moderate. The fact that the addition of PLCs resulted in an increase in viscosity, rather than the usual favorable decrease, illustrates the problem of phases adhering too well in the molten state.

As well as long flexible 'hairs', shorter rigid pendant groups may also be effective. Schleeh *et al.* [104] prepared a series of copolymers of rigid units and laterally substituted alkyloxy PLCs (terphthaloyl groups – $O(CH_2)_4CH_3$). Such short units pendant to the rigid, aromatic units can also be effective in promoting miscibility with the EP phase (in this case PET), by reducing PLC crystallinity and directly interacting with the EP phase. OM and WAX techniques show that the contribution of the side chains is indeed to cause the molecules to become entangled with the PET and 'force' miscibility or inhibit phase separation, unless higher shearing stresses or greater temperatures are used. Claßen *et al.* [105] synthesized a rigid PLC with bulky substituents (*t*-butyl moiety) and a T_g of 150°C. It was blended with a range of amorphous EPs (PC, PMMA, PS and PES). Blends with PC and PMMA showed optical transparency for concentrations below 8 and 4%, respectively, and very fine droplets of about 0.2 to 0.3 μm at higher concentrations. Alternately, PS and PES were immiscible at all concentrations with particles of 1 μm diameter. It appeared that the efficacy of the materials ability to reinforce and raise blend modulus was related to this compatability (which was quantified by the use of solubility parameters [105]). Whilst it is usually reasoned that low strain measurements such as modulus would be relatively independent of the amount of interphase and interfacial adhesion in polymer blends, it is proposed that in this situation where there is a lower concentration of entanglements (such as in a PLC blend) compatability may play a greater role in such properties, as well as influencing failure [105].

Other workers have tried to force miscibility by use of much more rigid pendant groups [106, 107]. Land *et al.* [106] synthesized disubstituted *para*-linked aromatic PLC polyesters with either pendant phenyl or biphenyl groups. Homopolymers with the shorter side chains (phenyl groups) had lower glass transition and higher clearing point temperatures than the biphenyl-pendant homopolymers. Solution blends of PC and the biphenyl-substituted material resulted in a homogeneous mixture which showed an intermediate T_g. Closer TEM examination showed a very fine dispersion of PLC domains rather than molecular miscibility. This high level of mixing was ascribed to the reduction in crystallinity caused by the substituted chain since, without these groups, the rigid main chain material would have had a high level of crystallinity (being regular and very linear). This cannot be the only consideration because, as mentioned above, even in blends with rigid PLCs and low crystallinity such as HNA/xPHB, blends with fine dispersions and high miscibility are rarely observed. Therefore the phenyl groups must also be actively involved in the improved adhesion, perhaps by greater intra- and intermolecular interaction (by 'hooking' the EP molecules to themselves), which would explain increased blend yield stress compared to that of the PC homopolymer. Increases in yield stress are often seen in miscible systems and are ascribed to decreased chain mobility due to favorable interactions and entanglements [108, 109] but this may also lead to reduced impact strength.

In a similar manner, Kricheldorf *et al.* [107, 110] synthesized a wholly aromatic PLC with rigid aromatic side groups connected to the central mesogenic core by short alkyl spacers and solution blended this with poly(ε-caprolactone) (PCL), with the resultant blend showing unusual phase behavior. Two phases could be observed, a neat PCL phase and a slightly mixed PLC + PCL phase. This second mixed phase was termed 'lyotropic' because of the ease of its alignment by shear deformation – something not possible in the PLC homopolymer. There was a resultant large increase in modulus (doubling on addition of 1% PLC) which indicated that the very fine dispersion achievable was able to influence the mechanical properties dramatically.

7.2.3 Blends of semiflexible longitudinal PLCs: PET/xPHB copolyesters

Copolymers of PET and PHB (PET/xPHB) are another of the most widely researched and blended longitudinal PLC materials. First reported by Jackson and Kuhfuss [111] for a range of compositions, 29 mol% of PHB or more was found to be required for liquid crystallinity whilst the PET/0.6PHB material demonstrated the best mechanical properties. Therefore, it has been this latter material which has been commercialized, originally in the form of XG7 (Eastman Kodak, USA) and more recently as Rodrun LC3000 (Unitika, Japan). There have been many thermal, structural and relaxational studies of this material [77, 112–114] for a range of copolymer compositions and much of this work has recently been brought together in a phase diagram by Brostow *et al.* [7] (see also section 9.11 by Hess and López). The copolymerization of the two units is clearly non-random and this has led to models of a phase separated flexible 'PET-rich' matrix with regions of rigid 'PHB-rich' material, the latter demonstrating liquid crystallinity. The terms 'rich' indicate that some degree of mixing between various phases exists. Attempts to characterize definitively this material is made difficult not just by its chemical complexity but by differences in materials due to the nature of processing, rates of cooling, analysis techniques and so on [7].

As a result, a range of phase temperatures have been reported, not all transitions or relaxations being observed by all groups. Most workers find a glass transition of the PET-rich phase between approximately 50 and 70°C [7, 77, 79, 113–123]. Less often observed is the glass transition of the PHB-rich phase at approximately 160 to 187°C [116, 118, 119]. Some researchers [119] have observed, using DMTA, a relaxation at 90°C ascribed to the motion of PET moieties in the PHB-rich phase and one at 125°C that may be the T_g of PHB units residing in the PET-rich phase, although if there was true miscibility of those units in the other phase, individual values of T_g would not be seen. The melting point of the PET-rich phase is usually between about 190 and 215°C (with an energy of approximately 2 to 6 J g^{-1}) and T_c of between 114°C and 152°C [7, 77, 80, 120], whilst that of the PHB phase is usually not seen with thermal degradation taking over at high temperatures. However, some workers have seen a small isotropization peak at about 250 to 260°C [124 and references therein] whilst others report one at 454°C [119]. These transitions can be compared with those of the PET homopolymer mentioned earlier: T_g of approximately 70°C, T_m of about 240°C and T_c about 145°C [75–80].

Studies of PET/xPHB copolymers as a function of x have shown that the T_g of the PET-rich phase does not change much, increasing slightly for high PHB content [7, 73, 77, 111–113]. Likewise, the melting point is reduced with increased PHB content, perhaps due to the presence and influence of PHB in the PET-rich phase [77] with a quite high melting point of 311°C observed in the PET/0.8PHB sample [7]. The results of this compositional work also indicate that the solubility of PHB in the PET-rich phase is lower than PET in the PHB-rich phase [7].

Because of the PET-rich phase in this longitudinal PLC, it is useful, where possible, to mention the compatability of PET homopolymer with the various EPs when discussing the EP + PET/xPHB blends. In this way useful comparisons between miscibility properties of PET homopolymer and the PET-rich phase of PET/xPHB material can be made.

Kimura and Porter [116] have investigated blends of PET/0.6PHB with PBT homopolymer. This may be expected to be a blend with good properties because miscibility between PET and PBT homopolymers has already been reported [125], and a single composition glass transition observed in the amorphous phase of the blend of the homopolymers. As already mentioned, PBT has a lower T_g and T_m than PET homopolymer due to its longer, flexible spacer unit and when blended with PET results in an increase in PET crystallization and reduced PBT crystallization, due mainly to the change in glass transition temperature [34]. In PBT + PET/0.6PHB the PBT + PET-rich phase seems miscible although the PHB-rich phase glass transition remains unchanged at 117°C [116]. Crystallization phenomena of the PBT are not altered with the melting point and degree of PBT crystallinity remaining constant. Mehta and Deopura [80] examined the same blend in spun fibers and found that, despite little change in the melting peak temperature of PBT, there was a substantial decrease in the crystallization endotherm position, and thus PLC inhibits crystallization. Avrami analysis indicated that the addition of PET/0.6PHB dramatically slowed down the PBT crystallization rate, possibly for reasons of reduced chain mobility, as mentioned for PBT + PET blends, although Mehta and Depoura [80] also suggest interference in crystallization due to the rigid PHB units in the blend. Ajji *et al.* [115] found in drawing experiments of the PBT + PET/0.6PHB blend that, despite some miscibility, spherical particles of the PLC in PBT fibers did not readily extend upon drawing. It may be concluded that miscibility with one phase is not necessarily sufficient to ensure good adhesion on a larger scale. In addition, the loss of PBT crystallinity with the addition of PLC also helped to negate the increase in modulus hoped for from the presence of aligned PLC fibrils.

Given the presence of a PET-rich phase in the PET/0.6PHB copolymer, blending with PET homopolymer is an obvious experiment and has been performed by a number of workers [72–74, 76, 117, 118, 126–131]. Joseph *et al.* [128, 129] and Brostow *et al.* [117] in studying this blend found that the phases were essentially immiscible with the values of T_g little changed. Zhuang *et al.* [73, 74] noted that the blends did seem partially miscible, especially at high concentrations of PET/0.6PHB, with a slight depression in PLC T_g observed. This indicated some small miscibility with the PET-rich phase, as expected, but conclusive results are difficult due to the similarities in T_g of the PET homopolymer and PLC copolymer phase. Zhuang *et al.* [73, 74], Brostow *et al.* [117] and Joseph *et al.* [128, 129] all observed that the addition of the PET/0.6PHB acted to lower the PET melting point, increased the rate of crystallization, the final degree of crystallinity and decreased the width of the crystallization exotherm, ascribing this to a combined nucleation/diluent phenomena of the LCP phase. The diluent aspect caused a depression in crystallization at high PLC concentrations and similar results have been reported recently for a PET blend with a mixture of PET/0.6PHB and PET/0.8PHB [131].

Bhattacharya *et al.* [75] carried out detailed work on low concentrations of PET + PET/0.6PHB blends to probe further the nucleating effect of the PLC on PET crystallization (less than 50% PLC) and found dramatic enhancement in crystallization (as seen from the rapid decreases in values of $t_{0.5}$ from Avrami analysis after 10% addition; Figure 7.7) and little change for further additions. SALS indicated the presence of anisotropic LC domains in the PET melt, which probably results in enhanced nucleation. Unusually, the size of spherulites was greater on PLC addition (the opposite would be expected if they were nucleating directly from PLC fragments) and suggests incorporation of PLC into the growing PET spherulites. Changes in the lamellar layer spacing of PET due to PLC addition would also seem to indicate this. The partial miscibility of the PET + PET/0.6PHB blend has been attributed to the difficulty in obtaining good PLC fibrils. Whilst clearly rheology and processing conditions also have a role to play, Sukhadia *et al.* [131] believe that too much miscibility, as in this system, may result in difficulty in obtaining distinct phases which can be readily flow aligned. The converse is also problematic, values of interfacial tensions which are too high can also make formation of fibrils difficult. Whilst Zhuang *et al.* [74] was able to produce a very fine spherical PLC morphology, adhesion was generally poor and a smooth fracture surface resulted [76].

Brostow *et al.* [117] made use of the range of PET/xPHB copolymeric and PET homopolymers available to perform an intrinsically interesting experiment that is relevant to many copolymer and related polymer blend systems. It addresses the issue of the comparative effect of polymer connectivity (primary bonding) compared to interchain bonding, entanglement and coupling (secondary bonding) on engineering polymer mechanical properties. In this work [117], compositions of differing PHB

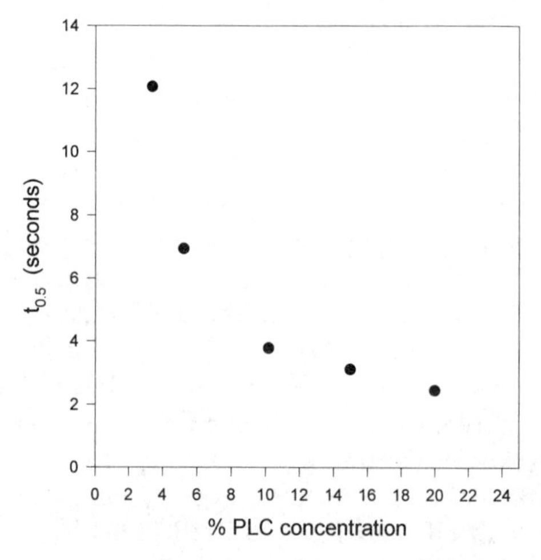

Figure 7.7 Halftime of crystallization for blend of PET + PET/0.6PHB as a function of PLC concentration. Crystallization temperature is 100°C. (Redrawn from [74].)

copolymer composition were blended with PET to the same PHB composition of another PET/xPHB copolymer in an effort to see how their mechanical properties differ. It was found, as expected, that copolymers of a given composition tended to perform better when applied to external loads (such as tensile strength) compared to the blends of the same PHB content with interfacial regions acting as sites of weakness. It should be noted that other morphological changes such as crystallinity would also result and influence properties. Interfaces and their potential weaknesses were less important with blends of high PLC content where the PLC was the dominant, continuous phase.

As with all polyester blends, in particular those with PET, differences in sample preparation (melt blend vs. solution casting), thermal history and possibly transesterification, add to the complexities of these materials [131]. Ou and Lin [130] demonstrated that improved compatability due to continued reaction in the melt blender further improved the propensity of the PLC to act as a nucleating agent, no doubt copolymerization in the melt resulting in a refinement of blend morphology. Blending for 50 min at 300°C results in decreases in T_g and T_c of some 8°C and 19°C, respectively, whilst leaving T_m unchanged.

As with other PLCs, PC has also been quite extensively blended with PET/0.6PHB [73, 74, 118–120, 127, 132–134] and once again offers the possibility of good interfacial adhesion. With PET homopolymer, PC has been shown to form a partially miscible blend in which a single glass transition is observed between amorphous PET and PC for part of the blend composition range, in particular at high PET concentrations. Murff *et al.* [135] found that PC + PET blends of between 60 and 90 wt% PET were miscible and Nassar *et al.* [136] found similarly between 70 and 90% PET. Overall, crystallinity was inhibited in the blends – particularly in the miscible regions, and higher temperatures for crystallinity were required due to the greater T_g resulting from molecular mixing of the PC phase.

In blends of PC with PET/0.6PHB, the general result is that there is some miscibility of the PET-rich phase in the PC but little dissolution of the PC into the PET-rich phase. Nobile *et al.* [119] found a slight decrease of a few degrees in the T_g of the PC in a melt blend and demonstrated that it was not due to chemical transesterification. Jung and Kim [120] noted a similar decrease in PC T_g from 149 to 139°C (80% PET/0.6PHB) whilst the PET-rich phase T_g remained at approximately 56°C. Brostow *et al.* [127] have shown in a phase diagram of the PC + PET/0.6PHB blend that three glass transition temperatures can be seen, a PC-rich blend phase, a PC phase miscible with another constituent and the unchanged PET-rich phase of the PLC and they propose that the interaction is between the PC and PHB-rich phase [127]. Brostow *et al.* [137] note that this system is somewhat more complex than others due to the two-phase nature of the PLC itself and that miscibilities of all the various phases must be taken into account when assessing properties.

Information relevant to miscibility can be obtained from transesterification behavior of these materials [118] since in immiscible and semimiscible blends, transesterification occurs largely at the interface of the potentially reactive species. When PC + PET/0.6PHB samples were transesterified at elevated temperatures, the resultant T_g was greater than both that of the PC and PET-rich phase of the PET/0.6PHB, leading to the conclusion that PHB-rich phase was involved in the transesterification process. This most likely occurred at the PHB–island interface and is possibly indicative of some slight miscibility between PC and the PHB-rich phase [127]. However, similar work with blends of PET/0.35PHB (less PHB phase) showed a T_g intermediate between PC and that of the PET-rich phase, indicating that with PLC copolymers of lower PHB concentration the reaction was between the PC and PET-rich phases [118].

Whilst it is difficult to see the small endotherm exhibited by the melting of the PET-rich phase, both Zhuang *et al.* [73, 74] and Jung *et al.* [120] were able to observe with DSC that its position changed little with addition of the PC, even up to 90% PC addition. Interphase adhesion in the solid state seemed good between PC and PLC, with fracture surfaces of spun fibers with the PLC as the minority phase showing good extension with little fiber pull-out.

Apart from DSC and DMTA tests, dimensional stability studies of drawn fibers and sheets of homopolymers and blends have also been used to good effect to characterize the interaction between the phases [138]. These studies are performed on drawn fibers and quantify frozen-in stresses and orientation of the molecules by annealing them at temperatures above T_g (and T_m, if relevant). Liquid crystalline materials by themselves tend to have low levels of shrinkage due to lower entropic driving forces than random coil thermoplastics, and other mechanisms of shrinkage such as domain reorientation do not involve large deformations. Conceivably then, inclusion of a fibrillar PLC phase in another isotropic polymer fiber (such as PC) should reduce shrinkage. However, the addition of PET/0.6PHB in PC results in a greater shrinkage and this is thought to be indicative of the PET-rich phase 'plasticizing' the PC phase due to partial miscibility.

It should be noted that whilst the previous PET/0.6PHB + PC studies are mostly melt blended and have shown only partial immiscibility at best, solution blending of PC and PET/0.6PHB can lead to a miscible blend film [134]. Heating of the sample yields a cloud-point temperature curve, as generally seen in miscible blends and occurs at the point when some melting of the PET-rich phase has occurred and molecular mobility is sufficient for phase separation. In these solution-cast blends, as in cases discussed below, the apparent miscibility is not reversible and the single phase can be thought to be kinetically trapped. This is due in part to the low entropies of mixing of flexible coils and rigid macromolecules and the propensity of the PLCs, once phase separated, to form liquid crystalline domains of local order which would encourage exclusion of other molecules.

Similar variation in apparent miscibility between melt vs. solution blending of EP + PET/0.6PHB has been demonstrated in PLC blend systems with EPs other than PC. Whilst Carfagna *et al.* [132] found melt blending of PEI with PET/0.6PHB to be an immiscible blend (no change in values of T_g), Zheng and Kyu [139, 140] found a single, composition-dependent glass transition temperature in optically clear films cast from a phenol + tetrachloroethane mixed solvent. In this system (as with the miscible, solution-blended PC + PET/0.6PHB mentioned above [134]) phase separation remains retarded upon elevation of temperature. In a 60% PEI + 40% PET/0.6PHB blend with a T_g of about 145°C, phase segregation seems to occur rapidly to a limited degree above T_g but even temperatures up to 240°C do not result in further growth of the dispersed phase. Only above 260°C does growth of the particles in the fine dispersion finally become enhanced. This inhibited growth is ascribed to the strength of self-association of the anisotropic LC phase, preventing further phase separation until temperatures above 260°C. It appears necessary to cause melting of the LC phase (PHB) segments to cause demixing, 260°C being of the order of the PHB melting point mentioned earlier.

Li *et al.* [122] found some partial compatability between PES and PET/0.6PHB on melt blending and a little more on solution casting, although miscibility of the components in the mixed solvent used was not great. In both cases it seemed as though the PLC was dissolved into the PES phase, with rather less PES in the PLC phase. Fracture morphology examinations showed smooth cavities of about 1 μm from where PLC particles had been readily detached. Zheng *et al.* [124] were able to codissolve both materials and produce a miscible, kinetically trapped blend which, as in the PEI + PET/0.6PHB solution blends, readily separates and grows in size at temperatures greater than the 260°C PHB melting level.

Individual studies exist for a range of other blends with engineering thermoplastics, with a variety of behaviors observed. A particularly striking example of partial miscibility is that of PET/0.6PHB + PAR where PAR signifies polyarylate [141]. The polyarylate phase was miscible with (and its glass transition reduced by) dissolution of the PET-rich phase into it. Simple Flory–Fox calculations demonstrated that 30% of the PET-rich phase was dissolved in the PAR, whilst conversely there seemed to be no solubility of the PAR in the PET-rich material and two glass transitions were observed, as demonstrated in Figure 7.8. This miscibility is perhaps not surprising given the similarity of the arylate mer unit to that of bisphenol-A polycarbonate (the arylate being somewhat more rigid and having a higher T_g). Nonetheless, the actual magnitude of decrease in T_g with PLC composition is more dramatic than in the case of PC blends. This may be due in part to the fact that this was solution blended (although it should be noted that unlike the solution blends mentioned in the previous paragraphs, a single trapped miscible phase was not observed).

The differences in miscibility between the PET/xPHB material and various materials – particularly polar ones – are clearly complex, with both chain polarity and flexibility playing a role. This can be illustrated by considering miscibility differences between PC, PAR (both previously discussed) and phenoxy resin (the polyhydroxyether of bisphenol-A) (PR) with the homopolymer structures and glass transitions being PR (T_g = 100°C):

$$-\!\!\left(\!\!-C_6H_4-C(CH_3)_2-C_6H_4-O-CH_2-CH(OH)-CH_2-O-\right)_n$$

PC (T_g = 150°C):

$$-\!\!\left(\!\!-O-C_6H_4-C(CH_3)_2-C_6H_4-O-C(=O)-\right)_n$$

and PAR (T_g = 190°C):

$$-\!\!\left(\!\!-C(=O)-C_6H_4-C(=O)-O-C_6H_4-C(CH_3)_2-C_6H_4-O-\right)_n$$

PAR is more rigid than PC and has a high glass transition whilst PR is more flexible than either and the carbonyl group has been replaced by a hydroxyl. Kodama [121] studied the PR + PET/0.6PHB blend and found total immiscibility and poor adhesion between the two phases. As mentioned above, the most rigid molecule – the PAR – seems to form semimiscible blends. Judging by these few examples, it appears as though effective polarity and perhaps the chain stiffness may help

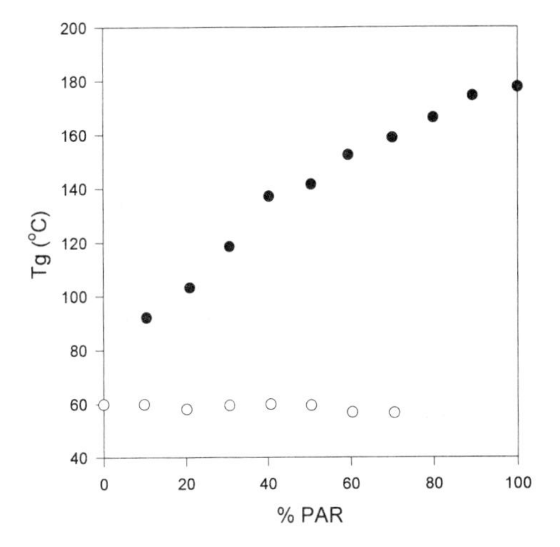

Figure 7.8 Glass transition temperatures of two phases observed in blends of PAR + PET/0.6PHB PLC as a function of PLC content. ● is the PAR-rich phase and ○ is the PET-rich phase of the PLC. (Redrawn from [136].)

determine which material will most favorably interact with PET/xPHB. Given these general conclusions, it would be thought that relatively flexible, less polar polymers such as polystyrene (PS) would show complete immiscibility, and this is observed, with very coarse, discrete phase morphologies being apparent in PS + PET/0.6PHB blends [73, 74].

7.2.4 Blends of semiflexible PLCs: non-PET/PHB copolyesters

Whilst blends of various engineering thermoplastics with PET/0.6PHB are perhaps the most commercially relevant of semiflexible, longitudinal PLC blends, it is of interest to review work done where the flexibility and chemical structure of the semiflexible material is altered and structure–property relations with regards to miscibility, interfacial adhesion and crystallization phenomena determined. This section will discuss blends in terms of the structure of the PLC, rather than the EP with which they are blended.

Paci *et al.* [142] blended a semiflexible, longitudinal PLC copolymer of poly(biphenyl-4, 4′-ylene sebacate) (PB8) with chemical structure

$$-\!\!\left(\text{O}-\text{C}_6\text{H}_4-\text{C}_6\text{H}_4-\text{O}-\underset{\substack{\|\\ \text{O}}}{\text{C}}-(\text{CH}_2)_8-\underset{\substack{\|\\ \text{O}}}{\text{C}}\right)\!\!-$$

with PBT. PB8 is clearly more flexible than PET/0.6PHB and displays an indistinct melting point between 170 and 230°C (indicative of only very weak crystallinity), after which a smectic phase persists until clearing at 280°C. A glass transition is not observed. It could be expected that its flexibility would promote miscibility with PBT homopolymer which has a similar T_m of 227.5° and indeed, a dramatic effect is found both on the crystallization properties of the PBT (a reduction in the temperature and energy of melting and in the temperature and rate of crystallization) and a change in the position of the PB8 clearing temperature. The latter information, in concert with other observations, leads to the conclusion that the two materials are miscible in the isotropic melt and that this retards the crystallization of the PET phase. Thus some miscibility in the melt results in a variation in the phase transitions of the PB8 and the crystallization of PBT.

This same PLC material was studied by Jo *et al.* [143] in a solution blend with PC and demonstrates complex phase behavior, not least because this is one of only a few PLCs which show measurable crystallinity. This blend was found to be miscible up to 40 wt% addition of PB8 where a single glass transition was observed. At these concentrations the miscibility of PB8 in PC seems greater than the corresponding miscibility of PC with the commercial PET/0.6PHB discussed in the previous section and cloud points are readily observable. At higher PLC concentrations (50–80 wt% PB8) it becomes difficult to see the glass transition temperature but the material appears to be heterogeneous, becoming single phase (miscible) at temperatures approximately 80°C above the PB8 clearing point in the isotropic state. This is indicative of the presence of an upper critical solution temperature (UCST), above which a single phase is observed. At these higher PLC concentrations the melting and crystallization temperature of the PLC changes little, although the crystallization rate of the PLC is found to decrease upon blending with the higher glass transition PC. At very high PLC content (10 wt% PC) very much lower rates of PB8 crystallization occur compared to other blends. In addition, at this concentration the isotropization temperature is above the UCST seen in the 50–80 wt% PB8 materials and the decrease in the crystallization rate is due to this enhanced miscibility.

PLC polymeric molecules with even longer flexible chains have been synthesized with a rigid mesogenic unit of greater aspect ratio [144, 145], e.g. HTH10

$$-\!\left[\underset{\|}{\overset{}{C}}(=O)-C_6H_4-O-C(=O)-C_6H_4-C(=O)-O-C_6H_4-C(=O)-O(CH_2)_{10}O \right]\!-$$

and blended with PBT. The HTH10 homopolymer has properties similar to that of the PB8 with a T_g of some 4°C, a crystal-to-smectic temperature of 192°C and a clearing point of 260°C. A high level of miscibility is found in the solid state with PBT for quenched samples, with a single composition-dependent T_g which is indicative of miscibility. With low contents of HTH10 (less than 30%), the PLC is excluded between PBT spherulites and found to depress the melting point, decreasing both the crystal growth and overall crystallization rate. By performing isothermal DSC and OM crystallization studies, the growth rate G and the nucleation rate N can be determined according to equation (7.2). Whereas the radial growth rate decreased by a factor of about two with 10% addition of HTH10, the nucleation rate was reduced two orders of magnitude for the same addition. Qualitatively, the nature of the spherulite growth is unaltered and a value of the Avrami exponent $n = 2.8$ indicates athermal nucleation. This clearly shows that the miscibility of the two phases in the melt and its effect on nucleation density is the primary factor in decreasing the overall crystallization rate. This drop in nucleation and growth occurs despite the reduction in T_g of the material by the addition of the HTH10, which could have been expected to encourage crystallization.

Whilst the above examples seem to indicate that miscibility is related to PLC flexibility, it would be useful to be able to blend an EP with an homologous series of semiflexible PLCs with sequentially longer spacer chains and monitor how this affects miscibility. This was reported recently by Yan and He [146] who synthesized a PLC homopolymer HQ(n)

$$\left[\overset{\displaystyle O}{\overset{\|}{C}} - C_6H_4 - \overset{\displaystyle O}{\overset{\|}{C}} - O - C_6H_4 - O - \overset{\displaystyle O}{\overset{\|}{C}} - C_6H_4 - \overset{\displaystyle O}{\overset{\|}{C}} \left(OCH_2CH_2O - \overset{\displaystyle O}{\overset{\|}{C}} - C_6H_4 - \overset{\displaystyle O}{\overset{\|}{C}} \right)_n OCH_2CH_2O \right]$$

with varying lengths of flexible segment ($n = 1, 2, 4, 8$ and 14) and produced blends with PET. It was found that increasing the value of n and hence the flexibility did indeed result in increased miscibility. If $n \geqslant 4$ full miscibility occurs up to 10% HQ(n) with a very fine dispersion of particles of approximately 0.5 μm for greater concentrations. Small amounts (< 10%) of flexible PLC HQ(4) seem to promote crystallization and increase crystallization temperature and this effect increases with PLCs with higher values of n, greater amounts seem to discourage crystallization of the PET phase. Even though it is suggested that the small (miscible) amounts of PLC encourage crystallization due to nucleation, it could be the low glass transition of the HQ(n) series that encourages crystallization. Although the T_g is not reported (probably not observed) the melting point of the HQ(n) series is lower than that of PET homopolymer and it is likely that the T_g is also lower. Similar results were found by Lin *et al.* [147] whose blend system of semiflexible PLC + PET was miscible for PLC content below 35% and immiscible for concentrations above. Blends with a low (and thus miscible) concentration of 10% PLC dramatically increased the crystallization temperature and enhanced crystallinity, opposite to previous examples because in this case the T_g of the PLC is lower than that of the crystallizing species and molecular mobility is enhanced. As well as Yan and He [146], Lin *et al.* [147] also suggest the possibility of some nucleation due to the presence of PLC particles which are of a very small size, just above the level of a molecular dispersion.

A sizeable body of work has examined the blending of a flexible PLC (H10), with a structure related to that of HTH10:

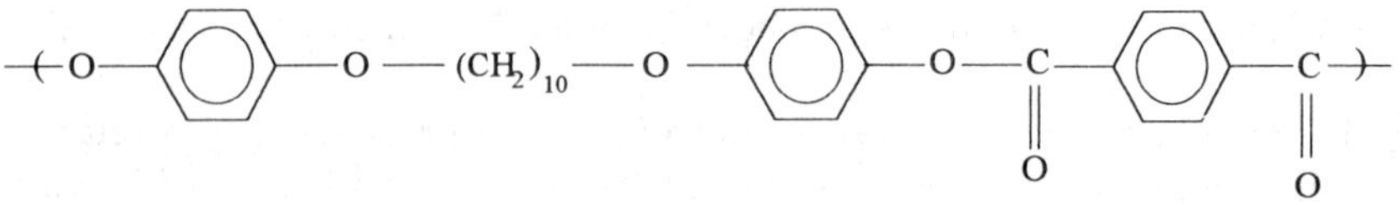

$$-\!\!\left(\!O-C_6H_4-O-(CH_2)_{10}-O-C_6H_4-O-\underset{\underset{O}{\|}}{C}-C_6H_4-\underset{\underset{O}{\|}}{C}\!\right)\!\!-$$

with a range of EPs, PET [148], nylon 66 [149] and PC [150]. H10 proved immiscible with PET, with an invariance of blend glass transitions with composition. Greatly nucleated crystallization was observed with a much higher crystallization temperature of PET in the blend and a more complex, two-stage cooling exotherm, indicative of enhanced heterogenous nucleation on the surface of the PLC. Clearly good contact between the phases is necessary and this is seen in SEM photographs of the blend failure surfaces. This polymer also showed good adhesion with nylon 66 [149] and blends of the two allowed production of fibers which were still able to deform in a ductile fashion, without brittle fracture at the interface, leading to an increase in breaking strength. Similarly, blends with polycarbonate [150] show slight miscibility of the PLC in the PC phase and the PLC particles consequently deform with fiber spinning. Other than improved (greater) fibrillar aspect ratio, Shin and Chung [150] note that it is possible that fibrils can be obtained with quite low PLC concentration (in this work 0.1–1 μm fibrils are seen with 10% PLC), indicating the efficacy of long spacers on inducing miscibility and lowering interfacial tension.

There are other studies of different semiflexible PLC + semicrystalline

EP blends which illustrate a number of the features already discussed and will not be outlined here [151–153].

7.3 GENERAL COMMENTS ON MISCIBILITY AND CRYSTALLINITY OF LONGITUDINAL + EP BLENDS

Based on the results of the rigid and semiflexible PLC materials that have been blended with EPs, improving properties of blends by mixing with PLCs is clearly a fine balance between a number of different considerations. Whilst the precise interplay of various PLC structural elements with miscibility and other mechanical properties is not totally clear, what is apparent is that some miscibility is desirable in the PLC + EP blend to encourage good interfacial mechanical properties and that more flexibility in the spacer between mesogens encourages such mixing. Too much flexibility in the PLC, however, reduces favorable PLC properties (such as low viscosity and high modulus and strength) and in the extreme case, liquid crystallinity may be lost. Observation of changes in T_g are not in themselves good indicators of interfacial adhesion in the solid state in these systems. In a partially miscible A + B blend, if insufficient B is soluble in the A-rich phase, even though a decrease in the T_g of the A-rich phase may occur, that phase may still not be sufficiently 'B-like' to adhere well to the B-rich phase in the solid state. An additional complexity is that, as discussed, small phase sizes may lead to changes in T_g due to surface forces without the need for miscibility on a molecular level. If miscibility is observed, care should be taken to ensure that it is not a chemical process such as transesterification that has occurred, and solution blending assists in verifying this aspect. However, if transesterification is well controlled, it may present one way of improving interphase adhesion.

Although increased flexibility of the PLC tends to lead to greater miscibility, the alternative approach (increasing the aromatic group content and rigidity of the EP phase) such as in PLC blends with PEEK [48] may also show some success. As mentioned in section 7.1, theoretical work by Ballauf [10] indicates that the most extreme example of this, blends of two longitudinal PLCs, should result in greater possibilities of molecular mixing and indeed this has been verified experimentally [154–156] and represents an interesting area of PLC research. In situations where poor miscibility occurs, it could be argued that the synthesis and adhesion of a compatibilizer phase is a cheaper and more straightforward solution than developing new, novel PLC homopolymers, scientifically interesting though this latter approach is. Thus compatibilization has become an important area of research in PLC + EP blends [157–160] in recent times. However, in terms of approaching the miscibility problem by new polymer synthesis, the design and synthesis of 'hairy' PLC molecules is an intellectually pleasing route that does seem to show promise. Rather than diluting the rigid nature of the mesogenic units and the good properties that result, chain linearity can be maintained whilst improving interfacial properties. In a sense, 'hairy' PLCs supply their own covalently bonded compatabilizer on a molecular level and should lead to a finer dispersion with stronger interfacial bonding, not requiring a third compatabilizer component to be dispersed into the blend.

The manner in which the blending is done (by solution or in the melt) can also affect the final phase outcome. In PLC blends, as with EP blends like PMMA/PC, it is possible by solution blending to entrap kinetically the different molecules, preventing them phase separating due to rapid solvent removal, and single T_g and LCST temperatures (as in miscible blends) may result. Whilst this is clearly of academic interest, applications of such blends may even be technologically feasible on a small scale if the blends are to be used in thin film applications such as for their barrier or permeability properties. The blends would need to have glass transitions well above operating temperatures so that the stability of the entrapped miscible system is assured. Another method of forcing minimal phase separation and a very fine dispersion has been reported [161–163] and involves synthesizing the second PLC phase within the EP matrix. This approach is known as '*in situ* blending' and excellent properties also often result.

With regard to crystallization of a semicrystalline phase in the presence of a PLC phase, many of the considerations raised by Nadkarni and Jog [33, 34] for isotropic EP blends and reviewed in section 7.2 above must also be considered in PLC systems. If a semiflexible PLC material is synthesized to promote miscibility and adhesion it may turn out that the miscibility between PLC and crystallizing polymer results in a decrease in crystallization rate due, for example, to higher blend glass transitions compared to that of the crystallizable homopolymer, depending on the values of T_g of the components. Systems which are immiscible and show high degrees of phase separation are often necessary for promotion of nucleation of crystallization. Studies in which the same semicrystalline EP is used with different PLCs are very instructive in this regard. Sharma *et al.* [72] blended the same PET homopolymer with both rigid HNA/0.73PHB copolymer and a semiflexible PLC, PET/0.8PHB material. The slightly more miscible, semiflexible PET/0.8PHB causes much lower crystallization temperatures (i.e. depression, indicative of miscibility) whilst the addition of rigid chain HNA/0.73PHB increases the clearing point (i.e. a nucleating agent) and it this latter blend which results in greater crystallinity. In another study by Mélot and MacKnight [78] similar results are observed. Both a rigid chain

HNA/0.73PHB and a semiflexible PLC of structure similar to HTH10 but with four flexible units in the spacer, TR-4,

$$-\!\!\left(\!\underset{\underset{O}{\|}}{C}-\langle\bigcirc\rangle-O-\underset{\underset{O}{\|}}{C}-\langle\bigcirc\rangle-\underset{\underset{O}{\|}}{C}-O-\langle\bigcirc\rangle-\underset{\underset{O}{\|}}{C}-O(CH_2)_4O\right)\!\!-$$

are mixed separately with PET. It is seen that although both are essentially immiscible, the flexible PLC is slightly more miscible, forming a much finer dispersion. Both blend systems have slightly increased nucleation due to the dispersed phase. In the case of the HNA/0.73PHB, excessive nucleation occurs at the interface and the spherulitic structure is coarser than for the semiflexible additive. At higher concentrations of the semiflexible PLC (about 20%), crystallization is impeded due to miscibility whereas that with the HNA/0.73PHB additive is still increasing, even at 20% addition (Figures 7.9 and 7.10).

Part of the reason for the difficulty in making hard and fast rules in these systems is that both nucleation and growth contribute to overall crystallization rate and blending has the potential to be affective either in subtle and possibly independent ways. This is illustrated by the work of Van Ende *et al.* [164] who found that a semicrystalline polymer formed a miscible blend with a flexible EP and this resulted in increased nucleation density and growth rate whilst the addition of a more rigid PLC of similar chemical structure but only partial miscibility was

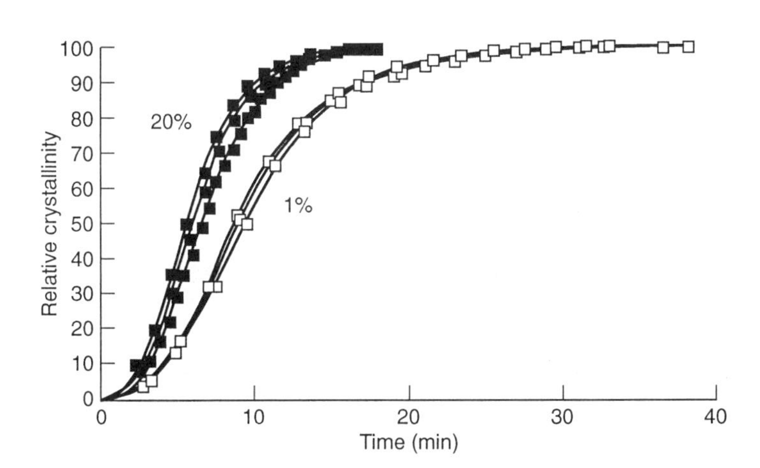

Figure 7.9 Crystallization kinetics (crystallinity vs. time) for HNA/0.73PHB at 232°C, wt% PLC is shown. (Permission granted for reproduction of D. Mélot and W.J. MacKnight (1992) *Polymer Adv. Technol.*, **3**, 382. Copyright John Wiley and Sons Ltd.)

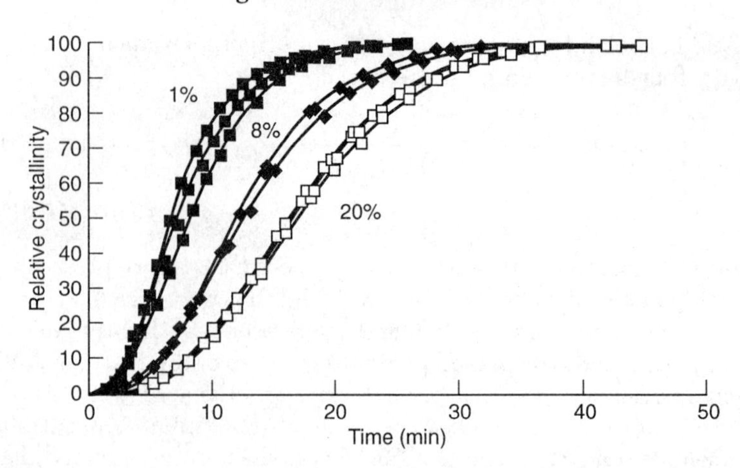

Figure 7.10 Crystallization kinetics (crystallinity vs. time) for TR-4 at 232°C, wt% PLC is shown. Structure of TR-4 is shown in text. (Permission granted for reproduction of D. Mélot and W.J. MacKnight (1992) *Polymer Adv. Technol.*, **3**, 383. Copyright John Wiley and Sons Ltd.)

responsible for a decrease in crystallization due to a decrease in nucleation density, the opposite to the ideas expressed above. Such results may be explained by the fact that what appears miscible by some techniques may simply be phase dispersion on a fine level, although it may be sufficiently coarse to encourage nucleation of crystallization.

When considering overall crystallization rate in immiscible blends, there is also often a trade-off between increased nucleation due to PLC addition (leading to an increase in crystallization rate) and increasing dilution of the crystallizing species and its growth rate (resulting in a decrease in crystallization rate). Although blending often results in little change or only a slight decrease in the final degree of crystallinity, in some cases it has been observed to lead to an overall crystallinity increase, as seen in a solution blend of PEEK and a copolymer of PHB/TA/RE (where RE is resorcinol) [165, 166]. A maximum overall crystallization rate at about 20 wt% PLC due to nucleation and dilution effects with PLC addition was observed. It also may hasten the attainment of high levels of crystallinity and could thus be useful in rapid cooling situations (injection molding) and for this reason immiscible PLC blends are worth pursuing.

Another important factor in immiscible PLC + EP blends is whether the crystallizing EP phase is crystallizing in the presence of a solid (glassy, crystalline) or flexible (liquid crystalline) PLC phase and this can be assessed by looking at the various crystallization temperatures of PLCs and the EPs (preferably at cooling rates relevant to the processing technique) with which they are blended. This can be illustrated using examples already reviewed in this chapter, those of EP + longitudinal rigid chain (HNA/0.73PHB) blends, chosen because they tend to be almost totally immiscible. Increased crystallization rate is often found if the included LC phase is already a semicrystalline solid at the point of thermoplastic crystallization [72, 78, 85] as with PET. This is also the case if both components have similar crystallization temperatures (as is the case of HNA/0.73PHB + PPS blends) [66, 67]. Conversely, if the PLC is in the LC (fluid) state at the thermoplastic crystallization temperature (as with a HNA/0.73PHB + PEEK blend where the PEEK T_c is some 55°C above that of the PLC phase [48]), the change to crystallization rate behavior was only modest. (To confuse the issue, a PBT + HNA/0.73PHB blend where the PLC phase is also solid during crystallization of the PBT [82] showed no effect either way on crystallization.)

It is clear that a whole range of factors – miscibility, interfacial tension, dilution effects, component viscosities and morphology – are important in miscibility and crystallization phenomena in PLC + EP blends with patterns and other issues starting to emerge. For example, it may be highly desirable to encourage transcrystallinity in these blends as has been found in carbon fiber/PPS composites [167], since transcrystallinity can improve mechanical properties and could be likely, particularly in fiber samples where the high level of elongational flow may encourage extended fibrillar phases. Recent unpublished optical microscopy crystallization studies by Heino and Vainio with polypropylene PP + HNA/0.73PHB blends (Figure 7.11) demonstrates that the PLC phase can indeed stimulate a range of crystallite superstructures from spherulites to transcrystalline regions, dependent on size and shape of the PP included phase. The influence of such miscibility and crystallinity on mechanical properties is not, however, always easy to discern because altering chemical structure to the influence miscibility usually means that other properties also change. One way to gage the influence of miscibility on mechanical properties is to include a compatibilizer phase which locates at the PLC/EP interface. Heino and coworkers recently attempted the compatabilization of a blend of HNA/0.73PHB with polypropylene [160]. Whilst standard maleic anhydride-grafted polypropylenes were found to be of limited use, it was found that small additions of a reactive ethylene-based terpolymer were found to improve the fineness of PLC dispersion due to decreased interfacial tension [15] and improved impact strength [160]. The change in morphology can be clearly seen in Figures 7.12a and 7.12b which are 30 wt% PLC (HNA/0.73PHB) + PP without and with compatabilizer, respectively. Whilst roughly spherical dispersions of some 50 μm are seen without compatabilizer, its addition dramatically reduces the included phase size. Clearly some reaction between the phases is in evidence. Whilst impact strength increased by a factor of three, the

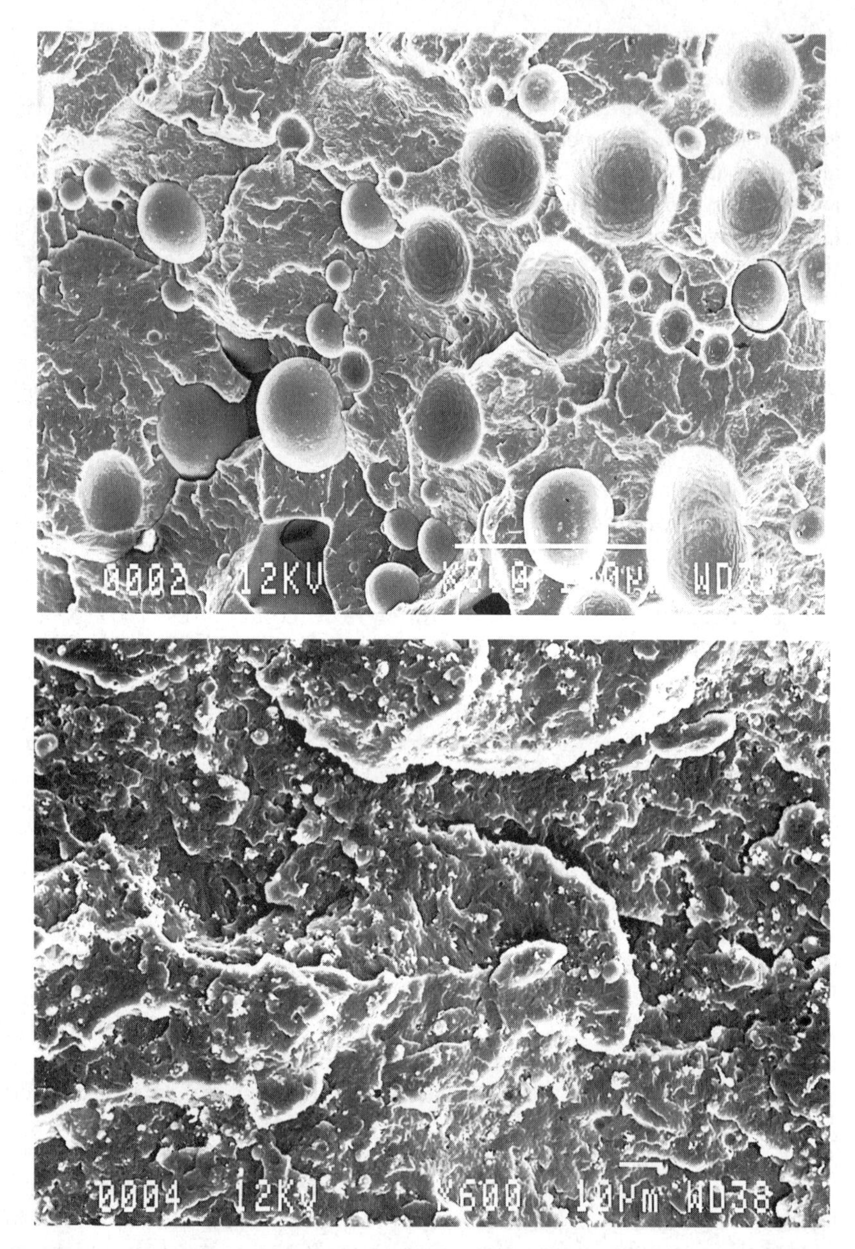

Figure 7.12 SEM micrograph of fractured surface of a 70% PP + 30% HNA/0.73PHB blend with (a) no compatibilizer and (b) with 7% reactive ethylene-based terpolymer compatibilizer. (Reprinted from [160], M. Heino and J. Seppälä, *J. Appl. Polymer Sci.*, **48**, 1677 (1993). © 1993, reprinted by permission of John Wiley and Sons, Inc. Original negative kindly provided by Markku Heino, Helsinki University of Technology, Finland.)

tensile modulus was found to fall. However, this decrease in flexibility could be limited if the concentration of compatabilizer was not too great [160]. This was likewise found for a PP-graft–epoxy material which, when blended with HNA/0.73PHB and PP, dramatically improved impact strength [159], making use of the reactive functionalities that exist at the ends of polyester materials, the family to which the majority of main chain PLCs belong. Such manipulation of the interface by addition of a small amount of a third phase is clearly a major direction of future research, as opposed to chemically altering the matrix or PLC phase. Only small amounts of material are required if a compatabilizer is used and both melt properties (morphology development and dispersion) and solid state properties (failure stress and fracture toughness) can be improved. The nature of the flow fields and processing that the blends experience is also always going to be instrumental in determining the final blend properties.

REFERENCES

1. Vill, V. (1994) *Adv. Mater.*, **6**, 527.
2. Collyer, A.A. (1989) *Mater. Sci. Technol.*, **5**, 309.
3. MacDonald, W.A. (1991) in *High Value Polymers, Royal Soc. Chem.*, London, *Spec. Publ.*, **87**, 428.
4. MacDonald, W.A. (1992) in *Liquid Crystal Polymers: From Structures to Applications* (ed. A.A. Collyer), Elsevier Applied Science, London, Ch. 8.
5. Calundann, G.W. and Jaffe, M. (1982) in *Proceedings of the Robert A. Welch Conference on Chemical Research XXVI: Synthetic Polymers*, p. 47.
6. Hess, M. and Lopez, B. (1997) Phase diagrams of polymer liquid crystals and polymer liquid crystal blends: relation to mechanical properties, in this volume.
7. Brostow, W., Hess, M. and Lopez, B.L. (1994) *Macromolecules*, **27**, 2262.
8. Utracki, L.A. (1990) *Polymer Alloys and Blends*, Hanser, Munich.
9. Flory, P.J. (1978) *Macromolecules*, **11**, 1138.
10. Ballauff, M. (1990) *Polymers Adv. Technol.*, **1**, 109.
11. Brostow, W., Sterzynski, T. and Triouleyre, S. (1996) *Polymer*, **37**, 1561.
12. Siegmann, A., Dagan, A. and Kenig, S. (1985) *Polymer*, **26**, 1325.
13. Kenig, S. (1991) *Polymers Adv. Technol.*, **2**, 201.
14. Kenig, S. (1994) in *Abstracts of ANTEC 94*, Vol. II, p. 1590.
15. Kirjava, J., Rundqvist, T., Holsti-Miettinen, R. (1995) *J. Appl. Polymer Sci.*, **55**, 1069.
16. Kiss, G. (1987) *Polymer Eng. & Sci.*, **27**, 410.
17. Bassett, B.R. and Yee, A.F. (1990) *Polymer Compos.*, **11**, 10.
18. Crevecoeur, G. and Groeninckx, G. (1990) *Polymer Eng. & Sci.*, **30**, 532.
19. Done, D., Sukhadia, A.M., Datta, A. and Baird, D.G. (1990) *SPE Tech. Pap.*, **49**, 933.
20. Datta, A., Sukhiada, A.M., Desouza, J.P. and Baird, D.G. (1991) *SPE Tech. Pap.*, **49**, 913.
21. Brown, C.S. and Alder, P.T. (1993) in *Polymer Blends and Alloys*, (eds P.S. Hope and M.J. Folkes), Blackie, London, Ch. 8.
22. Dutta, D., Fruitwala, H., Kohli, A. and Weiss, R.A. (1990) *Polymer Eng. & Sci.*, **30**, 1005.

23. Hawksworth, M., Hull, J.B. and Collyer, A.A. (1993) in *Processing and Properties of Liquid Crystalline Polymers and LCP Based Blends* (eds D. Acierno and F.P. La Mantia), ChemTec, Basel, p. 65.
24. Jaffe, M. (1990) in *Proceedings of COMPALLOY '90*, March 7–9, New Orleans, Louisiana, USA, 1990, p. 243.
25. Kulichikhin, V.G. and Plate, N.A. (1991) *Polymer Sci. USSR*, **33**, 1.
26. La Mantia, F.P. (ed.) (1993) *Thermotropic Liquid Crystal Polymer Blends*, Technomic, Basel.
27. Pawlikowski, G.T., Dutta, D. and Weiss, R.A. (1991) *Ann. Rev. Mater. Sci.*, **21**, 159.
28. Roetting, O. and Hinrichsen, G. (1994) *Polymers Adv. Technol.*, **13**, 57.
29. Weiss, R.A., Huh, W. and Nicolais, L. (1988) in *High Modulus Polymers. Approaches to Design and Development*, Marcel Dekker, New York, Ch. 5.
30. Rostami, S. (1992) *Multicomponent Polymer Systems* (eds. I.S. Miles and S. Rostami), Longman Scientific and Technical, New York, Ch. 3.
31. Stoelting, J., Karasz, F.E. and MacKnight, W.J. (1970) *Polymer Eng. & Sci.*, **10**, 133.
32. Miller, J.B., McGrath, K.J., Roland, C.M. *et al.* (1990) *Macromolecules*, **23**, 4543.
33. Nadkarni, V.M. and Jog, J.P. (1989) in *Handbook of Polymer Science and Technology. Volume 4: Composites and Specialty Applications*, (ed. N.P. Cheremisinoff), Marcel Dekker, New York, Ch. 3.
34. Nadkarni, V.M. and Jog, J.P. (1992) in *Two Phase Polymer Systems*, (ed L.A. Utracki, Hanser, Munich, Ch. 8.
35. Stein, R.S. (1994) *Mater. Res. Soc. Proc.*, **321**, 531.
36. Utracki, L.A. (1990) *Polymer Alloys and Blends*, Hanser, Munich, p. 52.
37. Hoffman, J.D. and Weeks, J.J. (1962) *J. Res. Nat. Bur. Stand. US*, **66**, 13.
38. Avrami, M. (1941) *J. Chem. Phys.*, **9**, 177.
39. Mandelkern, L. (1964) *Crystallization of Polymers*, McGraw-Hill, New York.
40. Nishi, T. and Wang, T.T. (1975) *Macromolecules*, **8**, 909.
41. Starkweather, H.W. (1980) *J. Appl. Polymer Sci.*, **25**, 139.
42. Donald, A.M. and Windle, A.H. (1992) *Liquid Crystalline Polymers*, Cambridge University Press, Cambridge.
43. Kohli, A., Chung, N. and Weiss, R.A. (1989) *Polymer Eng. & Sci.*, **29**, 573.
44. Isayev, A.I. and Modic, M. (1987) *Polymer Compos.*, **8**, 158.
45. Kim, W.N. and Denn, M.M. (1992) *J. Rheol.*, **36**, 1477.
46. Wissbrun, K.F. and Yoon, H.N. (1989) *Polymer*, **30**, 2194.
47. Troughton, M.J., Davies, G.R. and Ward, I.M. (1989) *Polymer*, **30**, 58.
48. Mehta, A. and Isayev, A.I. (1991) *Polymer Eng. Sci.*, **31**, 963.
49. Hong, S.M., Kim, B.C., Kim, K.U. and Chung, I.J. (1992) *Polymer J.*, **24**, 727.
50. Minkova, L.I., De Petris, S., Paci, M. *et al.* (1993) *Polymer Eng. & Sci.*, **32**, 57.
51. Fox, T.G. (1956) *Am. Phys. Soc.*, **1**, 123.
52. Chapleau, N., Carreau, P.J., Peleteiro, C. *et al.* (1992) *Polymer Eng. & Sci.*, **32**, 1876.
53. Valenza, A., Citta, V., Pedretti, U. *et al.* (1993) in *Processing and Properties of Liquid Crystalline Polymers and LCP Based Blends* (eds D. Acierno and F.P. La Mantia), ChemTec, Basel, p. 175.
54. Amendola, E., Carfagna, C., Netti, P. *et al.* (1993) *J. Appl. Polymer Sci.*, **50**, 83.
55. Verghoot, H., Langelaan, H.C., Van Dam, J. and Posthuma De Boer, A. (1993) *Polymer Eng. & Sci.*, **33**, 754.
56. Chung, J.S. and Cebe, P. (1992) *J. Polymer Sci. Phys.*, **30**, 162.
57. Brady, D.G. (1976) *J. Appl. Polymer Sci.*, **20**, 2541.

58. Jog, J.P. and Nadkarni, V.M. (1985) *J. Appl. Polymer Sci.*, **30**, 997.
59. Hawkins, R.T. (1976) *Macromolecules*, **9**, 189.
60. Cole, K.C. and Noel, D. and Hechler, J.-J. (1990) *J. Appl. Polymer Sci.*, **39**, 1887.
61. De Porter, J. and Baird, D.G. (1993) *Polymer Compos.*, **14**, 201.
62. Desio, G.P. and Rebendfeld, L. (1992) *J. Appl. Polymer Sci.*, **44**, 1989.
63. Desio, G.P. and Rebendfeld, L. (1992) *J. Appl. Polymer Sci.*, **45**, 2005.
64. Song, S.S., White, J.L. and Cakmak, M. (1990) *Polymer Eng. & Sci.*, **30**, 944.
65. Zeng, H. and Ho, G. (1984) *Angew. Chem.*, **127**, 103.
66. Minkova, L.I., De Petris, S., Paci, M. *et al.* (1993) in *Processing and Properties of Liquid Crystalline Polymers and LCP Based Blends*, ChemTec, Basel, p. 153.
67. Pracella, M., Manganini, P.L. and Minkova, L.I. (1992) *Polymer Networks & Blends*, **2**, 225.
68. Seppala, J., Heino, M. and Kapanen, C. (1992) *J. Appl. Polymer Sci.*, **44**, 1051.
69. Shonaike, G.O., Yamaguchi, S.Y., Ohta, M. *et al.* (1994) *Eur. Polymer J.*, **30**, 413.
70. Subramanian, P.R. and Isayev, A.I. (1991) *Polymer*, **32**, 1961.
71. Valenza, A., La Mantia, F.P., Minkova, L.I., De Petris, S. *et al.* (1994) *J. Appl. Polymer Sci.*, **52**, 1653.
72. Sharma, S.K., Tendolkar, A. and Misra, A. (1988) *Mol. Cryst. Liq. Cryst. Inc. Nonlin. Opt.*, **157**, 597.
73. Zhuang, P., Kyu, T. and White, J. (1988) *Abstracts of ANTEC 88*, p. 1237.
74. Zhuang, P., Kyu, T. and White, J.L. (1988) *Polymer Eng. & Sci.*, **28**, 1095.
75. Battacharya, S.K., Tendolkar, A. and Misra, A. (1987) *Mol. Cryst. Liq. Cryst.*, **153**, 501.
76. Amano, M. and Nakagawa, K. (1987) *Polymer*, **28**, 263.
77. Joseph, E., Wilkes, G.L. and Baird, D.G. (1985) *Polymer*, **26**, 689.
78. Mélot, D. and MacKnight, W.J. (1992) *Polymers Adv. Technol.*, **3**, 383.
79. La Mantia, F.P., Cangialosi, F., Pedretti, U. and Roggero, A. (1993) *Eur. Polymer J.*, **29**, 671.
80. Mehta, S. and Depoura, B.L. (1993) *J. Thermal Anal.*, **40**, 597.
81. Heino, M.T. and Seppala, J.V. (1993) *Polymer Bull.*, **30**, 353.
82. Engberg, K., Ehblad, M., Werner, P.-E. and Gedde, U.W. (1994) *Polymer Eng. & Sci.*, **34**, 1346.
83. Takayanagi, M., Ogata, T., Morikawa, M. and Kai, T. (1980) *J. Macromol. Sci. Phys.*, **B17**(4), 591.
84. Xu, J., Zhou, S. and Zeng, H. (1991) *Polymer Commun.*, **32**, 337.
85. Mehta, S. and Depoura, B.L. (1991) *Polymer Bull.*, **26**, 571.
86. Kalika, D.S., Yoon, D.Y., Iannelli, P. and Parrish, W. (1991) *Macromolecules*, **24**, 3413.
87. Aihara, Y. and Cebe, P. (1994) *Polymer Eng. & Sci.*, **34**, 1275.
88. Blizard, K., Haghighat, R., Lusignea, R. and Connell, J. (1992) *Proc. of ANTEC '92*, p. 2263.
89. Huo, P.P., Friler, J.B. and Cebe, P. (1993) *Polymer*, **34**, 4387.
90. Bafna, S.S., De Souza, J.P. and Baird, D.G. (1993) *Polymer News*, **17**, 345.
91. Bafna, S.S., Sun, T. and Baird, D.G. (1993) *Polymer*, **34**, 708.
92. Baird, D.G. (1994) *Polymer News*, **19**, 309.
93. Baird, D.G., Bafna, S.S., De Souza, J.P. and Sun, T. (1993) *Polymer Compos.*, **14**, 214.
94. Bretas, R.E.S. and Baird, D.G. (1992) *Polymer*, **33**, 5233.
95. Naslund, R.A. and Jones, P.L. (1992) *Proc. Mater. Res. Soc.*, **274**, 53.
96. Simon, G.P., Zipper, M.D. and Hill, A.J. (1994) *J. Appl. Polymer Sci.*, **52**, 1191.

97. Zipper, M.D., Simon, G.P., Cherry, P. and Hill, A.J. (1994) *J. Polymer Science, Phys.*, **52**, 1237.
98. Zipper, M.D., Simon, G.P., Stack, G.M. *et al.* (1995) *Polymer Int.*, **36**, 127.
99. McCullagh, C.M., Yu, Z., Jamieson, A.M. *et al.* (1995) *Macromolecules*, **28**, 6100.
100. de Carvalho, B. and Bretas, R.E.S. (1995) *J. Appl. Polymer Sci.*, **55**, 233.
101. Goodwin, A.A. and Hay, J.N. (1989) *Polymer Commun.*, **30**, 288.
102. Grevecoeur, G. and Groenickx, G. (1991) *Macromolecules*, **24**, 1190.
103. Heitz, T., Rohrbach, P. and Höcker, H. (1989) *Makromol. Chem.*, **190**, 3295.
104. Schleeh, T., Salamon, L., Hinrichsen, G. and Koßmehl, G. (1993) *Makromol. Chem.*, **194**, 2771.
105. Claßen, M.S., Schmidt, H.-W. and Wendorff, J.H. (1990) *Polymers Adv. Technol.*, **1**, 143.
106. Land, H.-T., Heitz, W., Karbach, A. and Pielartzik, H. (1992) *Makromol. Chem.*, **193**, 2571.
107. Kricheldorf, H.R., Engelhardt, J. and Weegen-Schultz, B. (1991) *Makromol. Chem.*, **192**, 645.
108. Yee, A.F. and Maxwell, M.A. (1979) *J. Macromol. Sci. Phys.*, **B17**, 543.
109. Zipper, M.D., Simon, G.P., Stack, G.M. *et al.* (in press) *Polymer Int.*
110. Taesler, C., Kricheldorf, H.R. and Petermann, J. (1994) *J. Mater. Sci.*, **29**, 3017.
111. Jackson, Jr, W.J. and Kuhfuss, H.F. (1976) *J. Polymer Sci. Phys.*, **14**, 2043.
112. Menczel, J. and Wunderlich, B. (1980) *J. Polymer Sci. Phys.*, **18**, 1433.
113. Meesiri, W., Menczel, J., Grrault, U. and Wunderlich, B. (1982) *J. Polymer Sci. Phys.*, **20**, 719.
114. Buhner, S., Chen, D., Gehrke, R. and Zachmann, H.H. (1988) *Mol. Cryst. Liq. Cryst.*, **155**, 357.
115. Ajji, A., Brisson, J. and Qu, Y. (1992) *J. Polymer Sci. Phys. Ed.*, **30**, 505.
116. Kimura, M. and Porter, R.S. (1984) *J. Polymer Sci. Phys. Ed.*, **22**, 1697.
117. Brostow, W., Dziemianowicz, T.S., Romanski, J. and Werber, W. (1988) *Polymer Eng. & Sci.*, **28**, 785.
118. Friedrich, K., Hess, M. and Kosfeld, R. (1988) *Makromol. Chem., Symp.*, **16**, 251.
119. Nobile, M.R., Amendola, E., Nicolais, L. *et al.* (1989) *Polymer Eng. & Sci.*, **29**, 244.
120. Jung, S.H. and Kim, S.C. (1988) *Polymer J.*, **1**, 73.
121. Kodama, M. (1992) *Polymer Eng. & Sci.*, **32**, 267.
122. Li, W., Jin, X., Guang, Li. and Jiang, (1994) *Eur. Polymer J.*, **30**, 325.
123. Laivins, G.V. (1989) *Macromolecules*, **22**, 3974.
124. Zheng, J.Q. and Kyu, T. (1992) *Polymer Eng. & Sci.*, **32**, 1004.
125. Escala, A. and Stein, R.S. (1979) in *Multiphase Polymers* (eds S.L. Cooper and G.M. Ester), *Adv. Chem. Ser.*, **176**, American Chemical Society, Washington, DC, p. 455.
126. Akemi, N., Shiwaku, T., Hasegawa, H. and Hashimoto, T. (1986) *Macromolecules*, **19**, 3008.
127. Brostow, W., Dziemianowicz, T.S., Hess, M. and Kosfeld, R. (1990) in *Liquid-Crystalline Polymers* (eds R.A. Weiss and C.K. Ober), American Chemical Society, Washington, DC, Ch. 28.
128. Joseph, E.G., Wilkes, G.L. and Baird, D.G. (1983) *Polymer Prepr. Am. Chem. Soc.*, **24**, 304.
129. Joseph, E.G., Wilkes, G.L. and Baird, D.G. (1985) in *Polymeric Liquid Crystals* (ed. F. Blumstein), Plenum, New York, p. 197.
130. Ou, C.-F. and Lin, C.-C. (1994) *J. Appl. Polymer Sci.*, **54**, 1223.

131. Sukhadia, A.M., Done, D. and Baird, D.G. (1990) *Polymer Eng. & Sci.*, **30**, 519.
132. Carfagna, C., Amendola, E. and Nicolais, L. (1992) *Int. J. Mater. Prod. Tech.*, **7**, 205.
133. Kosfeld, R., Hess, M. and Friedrich, K. (1987) *Mater. Chem. & Phys.*, **18**, 93.
134. Kyu, T. and Zhuang, P. (1988) *Polymer Commun.*, **29**, 99.
135. Murff, S., Barlow, J.W. and Paul, D.R. (1984) *Macromol. Chem.*, **185**, 1041.
136. Nassar, T.R., Paul, D.R. and Barlow, J.W. (1989) *J. Appl. Polymer Sci.*, **23**, 85.
137. Brostow, W., Hess, M., López, B.L. and Sterzynski, T. (1996) *Polymer*, **37**, 1551.
138. Apicella, A., Nicodemo, L., Nicolais, L. and Weiss, R.A. (1989) in *Handbook of Polymer Science and Technology. Volume 2: Performance Properties of Plastics and Elastomers*, Marcel Dekker, New York, Ch. 20.
139. Zheng, J.Q. and Kyu, T. (1989) *Polymer Prepr. Am. Chem. Soc.*, **30**, 550.
140. Zheng, J.Q. and Kyu, T. (1990) in *Liquid Crystal Polymers* (eds R.A. Weiss and C.K. Ober), American Chemical Society, Washington, DC, Ch. 31.
141. Wang, L.-H. and Porter, R.S. (1993) *J. Polymer Sci. Phys.*, **31**, 1067.
142. Paci, M., Barone, C. and Magagnini, P.L. (1987) *J. Polymer Sci. Phys.*, **25**, 1595.
143. Jo, W.H., Yim, H.Y., Kwon, I.H. and Son, T.W. (1992) *Polymer J.*, **24**, 519.
144. Pracella, M., Chiellini, E. and Dainelli, D. (1986) *Makromol. Chem.*, **187**, 2387.
145. Pracella, M., Chiellini, E. and Dainelli, D. (1989) *Makromol. Chem.*, **190**, 175.
146. Yan, Q. and He, J. (1994) *Polymer J.*, **26**, 1309.
147. Lin, Y.G., Lee, H.W., Winter, H.H. *et al.* (1993) *Polymer*, **34**, 4703.
148. Shin, B.Y. and Chung, I.J. (1990) *Polymer Eng. & Sci.*, **30**, 13.
149. Shin, B.Y. and Chung, I.N. (1990) *Polymer Eng. & Sci.*, **30**, 22.
150. Shin, B.Y. and Chung, I.J. (1989) *Polymer J.*, **21**, 851.
151. Pedretti, U., Roggero, A., La Mantia, F.P. and Maganini, P.L. (1993) *Proceedings of ANTEC 93*, Vol. II, p. 1706.
152. Shin, B.Y., Jang, S.H., Chung, I.J. and Kim, B.S. (1992) *Polymer Eng. & Sci.*, **32**, 73.
153. Kim, Y.S. and Chung, I.J. (1991) *Polymer J.*, **23**, 1339.
154. DeMeuse, M.T. and Jaffe, M. (1988) *Mol. Cryst. Liq. Cryst. Inc. Nonlin. Opt.*, **157**, 535.
155. DeMeuse, M.T. and Jaffe, M. (1990) in *Liquid Crystal Polymers* (eds R.A. Weiss and C.K. Ober), American Chemical Society, Washington, DC, Ch. 30.
156. De Meuse, M.T. (1990) *Polymer Adv. Technol.*, **1**, 81.
157. Datta, A., Chen, H.H. and Baird, D.G. (1993) *Polymer*, **34**, 759.
158. Kobayashi, T., Sato, M., Takeno, N. and Mukaida, K. (1993) *Eur. Polymer J.*, **29**, 1625.
159. Holsti-Miettinen, R.M., Heino, M.T. and Seppälä, J.V. (1995) *J. Appl. Polymer Sci.*, **57**, 573.
160. Heino, M. and Seppälä, J. (1993) *J. Appl. Polymer Sci.*, **48**, 1677.
161. Gupta, B., Calundann, G., Charbonneau, L.F. *et al.* (1994) *J. Appl. Polymer Sci.*, **53**, 575.
162. Ogata, N. (1993) in *Liquid Crystalline Polymers, Proceedings of the International Workshop on Liquid Crystalline Polymers*, WLCP 93, Capri, Italy, June 1–4, 1993 (ed. C. Carfagna), Pergamon, Oxford, 1993, Ch. 14.
163. Ogata, N., Sanui, K. and Itaya, H. (1990) *Polymer J.*, **22**, 85.
164. Van Ende, P., Groeninckx, G., Reynaers, H. and Samyn, C. (1992) *Polymer*, **33**, 3598.
165. Zhong, Y., Xu, J. and Zeng, H. (1992) *Polymer J.*, **24**, 999.
166. Zhong, Y., Xu, J. and Zeng, H. (1992) *Polymer*, **33**, 3893.
167. Huang, Y. and Petermann, ,J. (1995) *J. Appl. Polymer Sci.*, **55**, 981.

8

Thermal expansivity

Ram Prakash Singh

8.1 INTRODUCTION

Interest in the field of polymer liquid crystals (PLCs) has attained the status of forefront of research activity with the first report of ultrahigh modulus, high strength fibers spun from nematic dopes in the patent literature in 1972 [1]. Since then we have seen tremendous growth of technology of high performance PLCs [2–10]. The liquid crystalline phase exists for certain polymers (lyotropic PLCs) over a specified concentration range in solution. On the other hand it exists for thermotropic PLCs (TPLCs) only within certain temperature intervals. The latter can be processed in the molten state. Their melt viscosity is low and they are capable of forming highly oriented crystalline-like structures when subjected to shear and elongation above their melting point. Due to their rod-like rigid molecular structure and high stiffness, PLCs form fiber chains in the final product and thus provide the neat resin its self-reinforcing properties, in most cases exceeding those of conventional fiber-reinforced thermoplastics. They are also endowed with high resistance to chemicals, UV light and ionizing radiation along with extremely low combustibility and smoke generation [10]. Their applications are proliferating from aerospace and military fields to electronics and photonics. However, even now their cost is high. This has been taken care of by blending TPLCs with low-cost engineering

Mechanical and Thermophysical Properties of Polymer Liquid Crystals
Edited by W. Brostow
Published in 1998 by Chapman & Hall, London.
ISBN 0 412 60900 2

polymers [11–14]. The latter serve as matrix materials for these composites designated as *in situ* composites. The advantage of these *in situ* composites is that the problems occurring during the melt processing of conventional short fiber-reinforced composites, such as fiber breakage, wear of equipment and rise in viscosity, can be avoided [15]. The melt viscosity of the blend is lowered significantly by addition of TPLC which can be elongated and fibrillized during processing [16, 17]. The fibrils, having a greater length and smaller diameter than for instance glass fibers, can effectively reinforce the matrix.

The evolution from highly oriented polymers to *in situ* composites through PLCs has a common feature in that they all have negative thermal expansivity (as defined in section 8.1.1) in the orientation direction of the order of $-10^{-5}\,K^{-1}$ in a wide range below and around room temperature. As in oriented flexible polymers, these negative values are attributed to a decrease of the fully extended conformation of the chains induced by thermal vibration [19–21]. Various models and theories have been proposed semi-empirically as well as based on fundamental principles. In the following, a detailed discussion of the subject is presented.

8.1.1 Definitions

The isobaric expansivity (or expansivity in short: formerly called the 'coefficient of thermal expansion') is defined by [22]

$$\alpha = v^{-1}\left(\frac{\partial v}{\partial T}\right)_P = \left(\frac{\partial \ln v}{\partial T}\right)_P \tag{8.1}$$

where v is volume, T is temperature and P is pressure. Similarly linear isobaric expansivity, can be defined by [23];

$$\alpha_L = l^{-1}\left(\frac{\partial l}{\partial T}\right)_P \tag{8.2}$$

Here l is the length of the sample. For cubic and isotropic solids, α and α_L are related by the following equation [23]:

$$\alpha = 3\alpha_L \tag{8.3}$$

The microscopic theory of expansivity was initially developed by Mie [24] and Grüneisen [25–27] for atomic solids. The isobaric expansivity for solids is extensively reviewed by Barron *et al.* [23].

8.2 EXPERIMENTAL DETERMINATION

α_L and α are measured by determining the extension of the rod of polymeric or composite material or increase in the volume of a sample in a dilatometer over a desired range of temperature [28]. Two kinds of dilatometers have been used [29]. Some dilatomers are designed for absolute measurement of linear isobaric expansivity in which linear dimensions of the samples are directly measured at various temperatures. Other dilatomers have been designed for relative measurements in which the expansivities are determined by comparison with a reference standard with a known expansivity. The minute change in length and volume can be measured by optical interferometers or electrically by making them change effective area or spacing in a parallel plate capacitor [22] Choy *et al.* [30] have measured the change in length of the samples at the heating rate of $10\,\mathrm{K\,min^{-1}}$ with a Perkin-Elmer TMA-7 thermomechanical analyzer. For plastics and composites, ASTM methods D696-44 (linear) and D864-52 (cubic) describe the thermal expansivity measurements. Amatuni [31] and Barron *et al.* [23] have described various instruments and methods for thermal expansivity measurements.

8.3 POLYMERS IN GENERAL

The α_L of solids mainly depends on the strength of interaction between its constituent atoms. For solids in which atoms are held together by weak van der Waals forces, a thermal expansivity of the order of $10^{-4}\,\mathrm{K^{-1}}$ is found. On the other hand for covalently bonded solids like diamond, thermal expansivity of the order of $10^{-6}\,\mathrm{K^{-1}}$ is reported. In case of polymers, constituent atoms are covalently bonded along the chain direction and perpendicular to the chain, the weak van der Waals interaction exists in between the chains. Hence, a large anisotropy in thermal expansivity is expected in crystalline and drawn polymers [32].

Even metals with a one-dimensional chain-like structure along the *c*-axis, such as Te and Se, were found to possess negative values of α_L along the *c*-axis (α_c). For Te and Se the reported α_c values are $-1.6 \times 10^{-6}\,\mathrm{K^{-1}}$ [33, 34] and $-17.89 \times 10^{-6}\,\mathrm{K^{-1}}$ [33], respectively. Earlier Cole and Holmes [35] and Swan [36] reported negative α_L along the *c*-axis of a single orthorhombic polyethylene crystal. Later on, in the 1970s, Kobayashi and Keller [37] and Davis *et al.* [38] carried out precise X-ray diffraction measurements of the thermal expansion along the *c*-axis of a polyethylene crystal and confirmed the negative sign for α_c. The α_L of a polyethylene single crystal along the *a*, *b* and *c* axes are found to be [38] $\alpha_a = 20 \times 10^{-5}\,\mathrm{K^{-1}}$, $\alpha_b = 6.4 \times 10^{-5}\,\mathrm{K^{-1}}$ and $\alpha_c = -1.3 \times 10^{-5}\,\mathrm{K^{-1}}$. The negative α_L along the chain direction has been reported for a number of polymer crystals such as cotton cellulose [39], nylon 6 [39, 40], isotropic polypropylene [41] poly(oxy methylene), [42] polychloroprene [39] and poly(ethylene terephthalate) [39]. The universal feature of the negative expansivity along the chain axis ($\alpha_{\parallel}^{C}$) is that its values range from -1×10^{-5} to $-5 \times 10^{-5}\,K^{-1}$. The mechanism of such thermal contraction has been considered by some workers [37, 38, 43] who ascribe the negative $\alpha_{\parallel}^{C}$ to internal rotational motion around C—C bonds caused by thermal agitation. Chen *et al.* [44] carried out a lattice dynamical calculation for a planar zigzag polyethylene molecule under zero tension taking into account lateral fluctuation of the chain. According to them [44], the negative expansion phenomenon arises from the effective shortening along the chain direction as a result of the bending and torsional motions of the chains for both flexible and rigid chain polymers. Lacks and Rutledge [45] carried out simulation of crystalline polyethylene at a finite temperature using a molecular mechanics force field for the interatomic potential and quasi-harmonic lattice dynamics for the vibrational Helmholtz function. The α_L determined by direct minimization of A is in excellent agreement with experimental results for temperatures up to 250 K and remains in excellent agreement throughout the range of temperatures for which experimental results exist. For polyethylene (PE), axial contraction increases the entropy thereby leading to negative axial thermal expansion. In contrast for isotactic polypropylene (IPP) [46], axial contraction increases the entropy which alone would lead to a positive axial thermal expansion. The negative axial thermal expansion in IPP occurs as an elastic response to the positive transverse thermal expansion and is based on potential energy effects.

For both amorphous and crystalline polymers, as the polymers are oriented, the chains become increasingly aligned along the draw axis, which leads to a large drop in the expansivity along the draw axis ($\alpha_{\parallel}$) and a slight increase in the perpendicular direction ($\alpha_{\perp}$) [32]. The decrease in $\alpha_{\parallel}$ with increasing draw ratio λ is much faster for crystalline regions because of the easy rotation of crystalline regions towards the draw axis. While $\alpha_{\perp}$ increases with temperature, $\alpha_{\parallel}$ tends to decrease above the glass transition temperature. $\alpha_{\parallel}$ decreases to values typical of polymer crystals ($-1 \times 10^{-5}\,K^{-1}$) for highly crystalline polymers [47, 48]. This has been attributed to the constraining effect of the crystalline bridges connecting the crystalline blocks. For polymers of lower crystallinity, $\alpha_{\parallel}$ may become an order of magnitude more negative and this remarkable phenomenon is attributed to the rubber-elastic contraction of taut molecules [32, 48].

The values of α and α_L of polymer single crystals [32–46], amorphous and crystalline drawn plastics [32, 47, 48] and drawn and undrawn speciality polymers [49–52] have been extensively studied in the last two decades. A brief description as a prelude to the expansivities of polymer liquid crystals, their blends and composites is discussed in the following.

8.3.1 Polymer crystals

The negative expansivity in the direction of the chain axis in crystalline lattices has been found for many polymers. There is only a small variation in values for various polymer crystals which points towards a universal mechanism independent of their chemical structure. The α_L values of various polymer crystals have been tabulated in Table 8.1. Polyethylene crystals were first and extensively investigated by various workers over an extended range of temperature. Lifshitz [53] predicted the negative thermal expansion of the chain-like structures in the 1950s. He showed that the excitement of bending waves in such structures with the dispersion relationship $W \sim K^2$ must lead to negative expansivities along the chains and in the plane of the layers due to the specific membrane effect. Usually, during thermal expansion the distances between particles are increased and correspondingly the frequencies are decreased. In contrast, during thermal expansion along the chain axis or in the plane of the layers, the frequencies of the long bending waves are increased, i.e. the vibrations become more rigid. Due to this membrane effect, the excitement of the bending waves is accompanied by a contraction of dimensions along the chains and in the plane of the layers. According to Lifshitz [53], the thermal contraction of the chain-like structures is not connected with the molecular features of these structures but is a result only of strong anisotropy in inter- and intra-chain interactions which lead to the generation of bending waves and an unusual dispersion relationship in the long-wave part of the vibration spectra. Earlier work suggested that the negativity is

Table 8.1 Thermal expansivity of polymer crystals

Polymer	$\alpha_{\parallel}^{C}$ (10^{-5} K^{-1})	$\alpha_{\perp}^{C}$ (10^{-5} K^{-1})	Temperature range (°C)	Reference
Polyethylene	-1.2	$\alpha_a^C = 14.7$	20 to 60	37
	-1.8	$\alpha_b^C = 6.0$	20 to 120	37
	-1.1 to -1.3	–	-130 to 60	38
Polyethylene (terephthalate (PET)	-2.2	$\alpha_a^C = 17.1$ $\alpha_b^C = 12.7$	-196 to 20	39
Nylon 6 (PA-6)	-4.5	$\alpha_a^C = 6.0$ $\alpha_b^C = 23.4$	-196 to 20	39
Cotton cellulose	-4.5	$\alpha_a^C = 6.0$ $\alpha_b^C = 2.8$	-196 to 20	39
Polychloroprene (PCP)	-4.1	$\alpha_a^C = 13.1$	-196 to 20	39

mainly due to the torsional modes of a planar zigzag, which can be characterized by the rotational angle between consecutive planes defined by pairs of C—C bonds. Kobayashi and Keller [37] postulated that the rotational angle is equal to 3°, which was considered to be 13° to explain the shortening of PE polymers found experimentally in the temperature range 20–100°C by Baugham [43]. Later on, dynamical models were investigated based on the anharmonicity of the interatomic potential incorporating both torsional and bending modes. The $\alpha_{\parallel}^{C}$ was evaluated in terms of the known force constants of crystals. Two models were considered: (1) a single linear chain embedded in a passive medium and (2) a lattice of linear chains. The linear models [54], without torsional modes but with twice as many bending modes, are able to give a reasonable estimate of the expansivity ($\alpha_{\parallel}^{C} \sim -1 \times 10^{-5}$ K). In actual polymer crystals, the chains are arranged in planar zigzag or helical conformations, and therefore torsional modes should also be considered. Such analysis was undertaken by Kan [55], Chen *et al.* [44, 54] and Barron [56]. Chen *et al.* [44, 54] considered a planar zigzag model taking into account both the torsional and bending waves. They showed that torsional modes are responsible for slightly more than half of the negative expansivity with the rest being attributed to bending modes. Using values of force constants for polyethylene, it was found that $\alpha_{\parallel}^{C} = -1.3 \times 10^{-5}\,K^{-1}$ in good agreement with experimental results. The expansivity is found to be independent of the mass of the atomic groups on the chain and only weakly dependent on the interchain van der Waals interaction.

We have already mentioned the molecular mechanics by Lacks and Rutledge [45]. As is evident from Figure 8.1, their calculated $\alpha_{\parallel}^{C}$ are in perfect agreement with experimental results of White and Choy [57] and in good agreement with the results of Davis *et al.* [38]. It is further shown by Lacks and Rutledge [45] that α_{bulk} are in excellent agreement with results of Engeln *et al.* [58] up to temperatures above 250 K. The experiments of Engeln *et al.* [58] are expected to be the most accurate available as their sample was 98% crystalline and had a lamella thickness of 4400 Å. The sample of Davis *et al.* [38] on the other hand had a lamella thickness of only 385 Å. The decreasing agreement of the calculated values of α_L above 250 K (*c.* $\frac{2}{3}$ of the melting temperature of PE) is most probably due to worsening of quasi-harmonic approximation. Further, it could be also due to the decrease in accuracy of the Karasawa, Dasgupta and Goddard (KDG) force field as the system moves farther from the potential energy minimum or due to inaccuracies in the experimental results with temperature or a combination of these reasons.

Lacks and Rutledge [46] carried out simulation for isotactic polypropylene (IPP) using the Sorensen, Lian, Kesner and Boyd (SLKB*) force field for the temperature range 0–350 K. They observed that although the

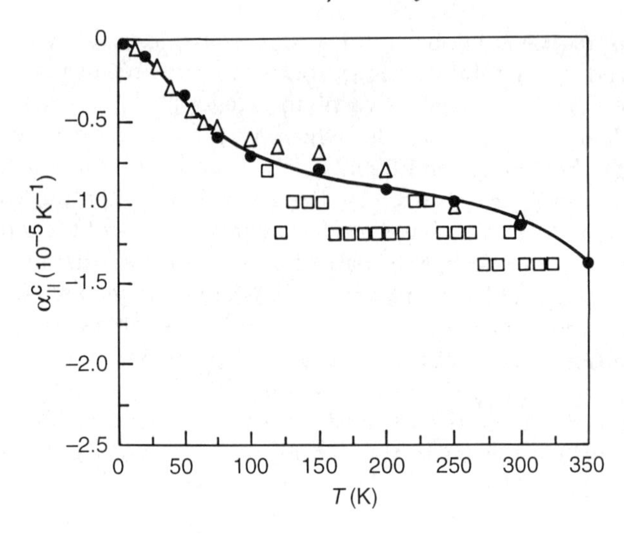

Figure 8.1 Thermal expansivity of polymer crystals in the direction of the chain axis ($\alpha^{C}_{//}$) as a function of temperature. Filled circles are a result of simulations with a KDG force field [45]; open squares [38]; open triangles [57]. (Adapted from [46]).

axial thermal behavior is very similar for both PE and IPP, the causes of their negative axial thermal expansion are completely different. For PE the axial contraction increases the entropy, thereby leading to negative axial thermal expansion. They also pointed out there is an additional contribution to the negative thermal expansion from elastic coupling causing approximately one-third of the negative axial thermal expansion. In contrast for IPP, the axial Grüneisen parameter is positive, which could lead to a positive axial thermal expansion. Hence the negative thermal expansion observed by Choy *et al.* [47] must be due to the coupling involving the off-diagonal compliance moduli. The static lattice off-diagonal compliance moduli are found to be negative and relatively large for IPP. S_{13}, S_{23} and S_{53} total about 90% of the magnitude of the diagonal modulus S_{33}, and are of opposite sign. Since the static lattice moduli are determined only by the potential energy surface, the negative axial thermal expansion of IPP is due to potential energy effects rather than entropic effects as in PE. Napolitano *et al.* [59] measured α_L of IPP along the *a* and *b* axes at room temperature for unoriented samples of α_2 structure of IPP to be 0.62×10^{-4} and $1.5 \times 10^{-4}\,K^{-1}$. They also found for the α_1 structure of IPP that α_L values differ by up to a factor of two for oriented samples. Hence α_L values reported by them can only be considered approximate. Choy *et al.* [47] measured the α_L of IPP along the *c*-axis at 280 K to be $-1.0 \times 10^{-4} K^{-1}$, and although they have not specified the crystal structure, their values are in reasonable agreement with the calculations of Lacks and Rutledge [46].

The negative expansivity along the chain direction ($\alpha_{\parallel}^c$) seems to be a universal phenomenon with planar zigzag chains as it is exhibited by a variety of polymers of different chemical compositions, i.e. PE [37, 38, 60], ethylene tetrafluoroethylene alternating copolymer [61], PET [36], nylon 6 [39, 40], poly(vinyl alcohol) [62] and polydiacetylene [63]. The transverse expansivity $\alpha_{\perp}^c$ is positive and larger by an order of magnitude in these crystals.

A number of studies have been carried out on the helical chain structure. Two crystals i.e. polyoxymethylene (POM) and poly(4-methylpentene 1) (P4MP1) have been extensively studied by Choy and Nakafuku [64] and White *et al.* [42], respectively. Choy and Nakafuku [64] studied the thermal expansion of the crystal lattice of POM using X-ray diffraction. They reported that $\alpha_{\parallel}^c$ is negative and the values increase from 0.22 to $0.85 \times 10^{-5} K^{-1}$ as the temperature rises from 160 to 400 K. In the case of P4MP1, the $\alpha_{\perp}^C$ value is close to that of POM; the $\alpha_{\parallel}^c$ value is not only positive in the temperature range -160 to $+160°C$ but its value is larger by an order of magnitude. $\alpha_{\parallel}^C$ depends strongly on the magnitude of the elastic anisotropy according to Choy and Nakafuku [64]. For POM which has anisotropy factor ($E_{\parallel}^c/E^c$) of 7, $\alpha_{\parallel}^c$ is negative and its value is one-tenth that of $\alpha_{\perp}^c$. For P4MP1 with an $E_{\parallel}^c/E^c$ value of 2.3, $\alpha_{\parallel}^c$ has a large positive value that is only 20% lower than $\alpha_{\perp}^c$. Due to the presence of bulky side groups, P4MP1 has a loose spiral chain structure. As a result of the open structure in the crystal, molecular motions other than vibration may be excited as the temperature increase and these may be responsible for the stepwise increase in expansivity below room temperature.

Among high performance polymers, single crystals of poly(ether-ether-ketone) (PEEK) and stretched and unstretched films of polyimides have been extensively studied for thermal expansivity. Thermal expansivities of an orthorhombic unit cell of PEEK along the *a*-, *b*- and *c*-axes have been determined from 160 to 470 K by Choy and Leung [52] using wide angle X-ray diffraction. The α_L values along the *a*- and *b*-axes are positive while the expansivity along the *c* chain direction is negative. The values at 300 K are as follows; $\alpha_a^c = 11.3 \times 10^{-5} K^{-1}$, $\alpha_b^c = 4.6 \times 10^{-5} K^{-1}$ and $\alpha_c^c = -1.4 \times 10^{-5} K^{-1}$. The estimated anisotropy is about 10. Thus it is greater than the elastic anisotropy of 7 which is sufficient to produce a negative axial expansivity [54]. Their values are in reasonable agreement with those obtained by other workers [65, 66].

In general, polyimides have relatively high glass transition temperatures, high planarization, high processability and low dielectric and thermal expansion parameters. All these attributes are crucial to device fabrication, performance and reliability. However several problems may arise from the stress on the polyimide film when it is coated on other materials such as silicon, due to the mismatch in thermal expansion values.

The latter problem may lead to warping, bending, delamination or cracking in a given layered structure or device composed of different materials. Hence the thermal expansivity of polyimides has been extensively studied [49–51, 67–69]. It has been found that some of the polyimides have α and α_L as large as that of copper or aluminum and some as small as that of silicon or even quartz [49–51]. The α values of various polyimides (PIs) depend strongly on their molecular structure. PI films were cured either on or off the substrate. The α_L was temperature dependent, so an average value from 50 to 250°C was reported. Bent structures generally have higher $\alpha_{\perp}$. For PIs containing benzophenone tetra-carboxylic dianhydride (BTDA), values of α and α_L were in the range $3-5 \times 10^{-5}\,K^{-1}$ while PIs containing PMDA, which is more rigid group, had lower α_L in the range $1-2 \times 10^{-5}\,K^{-1}$. The polyimides obtained from pyromellitic anhydride and aromatic diamines which were constituted only of benzene rings fused at *para* positions have the lowest α_L values [49–51]. The α_L values of crystalline polyimides have not been extensively studied. Brillhart and Cebe [69] have reported α_L of two semicrystalline thermoplastic polyimides i.e. NEW-TPI (Mitsui Toatsu Chemical Co.) and LARC-CPI (NASA Langley Research Centre) by wide angle X-ray scattering (WAXS) over an extended temperature range (25–325°C). Films were treated in 1-methyl-2-pyrrolidinon (NMP) at the reflex temperature, washed and then dried under constraint. The films treated in this manner were highly oriented with the *c*-axis preferentially aligned normal to the films. The thermal expansion along the *c*-axis is observed to be smaller than 8×10^{-6}°C for LARC-CPI and for NEW-TPI may be weakly negative. From WAXS of unoriented films, the variation of *a* and *b* lattice parameters were deduced as a function of temperature, as given in Table 8.2. LARC-CPI has larger crystal isobaric expansivity than

Table 8.2 Variation of lattice parameters *a* and *b* as a function of temperature *T* from wide angle X-ray scattering (WAXS) of unoriented films

Polymer	*a*	*b*
NEW-TPI	$7.82\,(1 + 93 \times 10^{-6}T)$	$7.82\,(1 + 93 \times 10^{-6}T)$
LARC-CPI	$8.06\,(1 + 117 \times 10^{-6}T)$	$7.82\,(1 + 93 \times 10^{-6}T)$

NEW-TPI indicating higher contraction of LARC-CPI crystals after melt processing [69].

8.4 ORIENTED POLYMERS

In any undrawn crystalline polymer, α will be generally positive because the two $\alpha_\perp^c$ values are positive and considerably larger than $\alpha_\parallel^C$. Moreover, there is a statistical arrangement of crystallites and accordingly of the axes of chains in bulk crystalline polymers, hence thermal expansivity along any direction happens to be positive. The crystalline regions are generally accompanied by amorphous regions, accordingly the additive principle provides the following expression for the expansivity of the crystalline polymers [70]:

$$\alpha = \bar{\alpha}_a(1 - W) + \bar{\alpha}_c W$$

where $\bar{\alpha}_a$ and $\bar{\alpha}_c$ are the expansivities of amorphous and crystalline polymers, respectively. At $T < T_g$, $\bar{\alpha}_a \sim \bar{\alpha}_c$ and at $T > T_g$, $\bar{\alpha}_a = (3-4)\,\bar{\alpha}_c > 0$, hence amorphous regions make significant constribution to the total expansivity. On the basis of the above it may be concluded that the negative thermal expansion along the chain axes in a crystalline lattice will have little influence on macroscopic thermal properties.

However, in uniaxially oriented semicrystalline polymers the orientation produces significant anisotropy in the linear isobaric expansivity of polymers. This anisotropy is caused by two major factors [47]: the alignment of covalently bonded chains in the crystalline blocks along the draw direction and the increase in the volume fraction V_f of the intercrystalline bridges formed during the drawing process. Thus, due to the presence of the drawn amorphous regions, the contraction is not only the result of the negative contribution of the crystallites alone but also the intercrystalline bridges constrain the expansion of the amorphous region between the crystallites. Consequently drawn crystalline polymers exhibit considerable contraction upon heating. The expansivity of drawn polymers has been extensively reviewed by Choy [32]. In the following, the essential features are described together with various models developed to explain the expansivity of oriented polymers.

8.4.1 Aggregate model

It has been observed that the expansivity of amorphous polymers is drastically affected by orientation. $\alpha_\perp$ increases by 10–30% with increasing draw ratio while $\alpha_\parallel$ exhibits a larger decrease [32]. The thermal expansion behavior of amorphous polymers is reasonably described by the aggregate model [71–73] using the orientation function determined by nuclear magnetic resonance measurements [74, 75]. According to the aggregate model [71–73], an isotropic polymer is regarded as a random aggregate of axially symmetric anisotropic intrinsic units whose properties (such as expansivity) are those of fully oriented materials. As the polymer is drawn, the intrinsic units orient towards the draw direction. The degree of molecular orientation can be characterized by orientation functions. The thermal expansion of a partially oriented sample can be calculated by using the series model which gives the upper bound or by the parallel model which involves summation of the inverse of the expansivities giving the lower bound. For a series model the expansivities are given by [71]

$$\alpha_{\parallel} = \tfrac{1}{3}[(1 + 2s)\,\alpha_{\parallel}^{U} + 2(1 - s)\,\alpha_{\perp}^{U}] \tag{8.4}$$

$$\alpha_{\perp} = \tfrac{1}{3}[(1 - s)\,\alpha_{\parallel}^{U} + (2 + s)\,\alpha_{\perp}^{U}] \tag{8.5}$$

where $\alpha_{\parallel}^{U}$ and $\alpha_{\perp}^{U}$ are the thermal expansivities of intrinsic units and s is the orientation function (order parameter) given by following relation:

$$s = \tfrac{1}{2}\overline{(3\cos^2\theta - 1)} \tag{8.6}$$

where θ is the angle between the draw direction and the symmetry axis of the intrinsic unit and the bar denotes the average over the aggregate. For isotropic material $s = 0$, hence

$$\alpha_{\text{iso}} = \tfrac{1}{3}(\alpha_{\parallel} + 2\,\alpha_{\perp}) = \tfrac{1}{3}(\alpha_{\parallel}^{U} + 2\,\alpha_{\perp}^{U}) \tag{8.7}$$

The expressions for the parallel model can be obtained by replacing all α's in equation (8.4)–(8.6) by α^{-1}.

In most of the reported cases [74, 75], a gentle rise in $\alpha_{\perp}$ with increasing λ closely follows the prediction of the series model (upper bound). $\alpha_{\parallel}$ decreases much faster than predicted by the series model; its value at $\lambda = 3$–4, is about half-way between the upper and lower bounds. For slightly crystalline polymers (such as PVC and PC), $\alpha_{\parallel}$ has a stronger dependence and this is probably because of the presence of intercrystalline tie molecules which can effectively constrain the expansion along the draw direction.

8.4.2 Dispersed crystallite and intercrystalline bridge models

In the case of crystalline polymers such as high density polyethylene (HDPE), the effect of orientation on the morphology has been extensively studied [32, 48]. The isotropic sample consists of crystalline lamellae (thickness 100–400 Å) embedded in an amorphous matrix. Each lamella is composed of a mosaic of crystalline blocks of lateral dimension 100–200 Å with boundaries defined by dislocations. The lamellae are randomly oriented and generally arrange themselves end to end in ribbon-like structures which grow out from nucleating centers to form

Also for polymer solutions, the properties depend on interactions between solvent and dissolved components and vary strongly with concentration and temperature. Therefore, beside the molecular analysis, there is the important question for the understanding of the relationship between structure and properties, whether and how PLCs in a solution can be distinguished from those containing non-liquid crystalline polymers with a similar molecular architecture. Another interesting question is, to what extent the conformation of the macromolecules in solution is influenced by interactions between mesogenic groups. As a consequence of that, the hydro- and thermodynamic properties of the solution should also be affected.

In the following, attention is focused only on some results obtained for solutions of thermotropic PLCs, and particularly on those of side group polymer liquid crystals (SGPLCs) since main chain polymer liquid crystals (MCPLCs) are normally difficult to dissolve and at least are insoluble in non-protonated solvents. The inclusion of lyotropic systems would be beyond the scope of this chapter and also would give no contribution to solving the question of how the properties of solutions of PLCs differ from those of non-LC polymers with similar molecular design.

5.2 CONSTITUTION AND SOLUBILITY

The LC state supposes rigid and anisometric shaped molecules. In MCPLCs the mesogens form the polymer backbone, while in SGPLCs the mesogenic groups are components of the side groups. In addition to this fundamental difference there are a great number of structural variations [1].

From theoretical considerations it follows that rod-like molecules must form mesophases in solution starting from a particular length-to-width axial ratio and above a critical concentration. Therefore MCPLCs, because of their shape, should be predestined to build those lyotropic mesophases, but the linear links of the mesogenic units lead to a low solubility or insolubility even for low degrees of polymerization (>10). The entropic gain derived from the mixture of an MCPLC with a low molar mass component is very small. So only a few exotic solvents or mixtures of solvents are known in which MCPLCs can be dissolved. For this reason the properties of MCPLCs in solution until now were rarely investigated, except in works concerning the lyotropic phases, i.e. phases with a high polymer concentration. In addition to the limited solubility due to the chain rigidity, the MCPLCs are generally distinguished by high melting temperatures. To improve the solubility of these polymers and/or to decrease their melting temperatures, flexible structures (spacers) were introduced into the backbone enhancing the chain flexibility or lowering the linearity by employing zigzag units of the chain. Greater solubility and decreased melting temperatures were also obtained by the attachment of side groups to the backbone, increasing the chain to chain distances and decreasing the interaction between the chains. However, to achieve this goal, the disadvantageous effects on the properties of these polymers as high performance materials must be taken into account. Some examples of the concept explained above and which could be important for future investigations of solution properties are shown in Table 5.1.

In 1990, 3,5-bis(trifluoromethyl)phenol (BTFMP) was found to be a solvent for aromatic polyesters [6–9]. This discovery made it possible to perform for the first time the fractionation and molecular characterization of a thermotropic MCLC copolyester. Table 5.2 reports results for a typical random copolyester consisting of 70 mol% of *p*-hydroxybenzoic acid and 30 mol% of 2-hydroxy-6-naphtho acid moieties.

In contrast to MCPLCs, SGPLCs are soluble in many common solvents for polymers because of the coiled structure of the backbone. Consequently, more experimental results have been published concerning the properties of SGPLCs in solution. Therefore in the following we focus on the presentation of the properties of these systems. It should be mentioned that to our knowledge almost no investigations have been reported on the dissolution process of SGPLCs, such as swelling of polymers and the kinetics [10] of dissolution. In the following an overview of some investigated systems of SGPLCs and solvents is given.

The most systematically studied SGPLCs in solution are polymethacrylates and polyacrylates with n methylene units making up an alkyl spacer. Both can have different forms with terminal (end-on; Figure 5.1, Table 5.3) and lateral (side-on) connection of the side groups to the backbone. The mesogens mostly consist of two phenyl rings often connected via an ester linkage. To increase the length to width ratio of the rod-like side groups a third phenyl ring can be introduced. In general the tail groups have an alkoxy structure with a different length m. The majority of the results, however, has been reported for polymethacrylates $PMAC_n$ and polyacrylates PAC_n with methoxy tail groups and ester linkages but with variable spacer length n. Furthermore, the behavior of polymethacrylates with an oxethyl spacer in tetrahydrofurane (THF) and $CHCl_3$ [24] was studied. Also, results for solutions of polymethacrylates with cholesteryl units in the mesogen and *t*-butyl tail groups in $CHCl_3$, decane and toluene [21, 25] were reported. In a few papers the properties of polymethyacrylates with

Table 5.1 Constitution and solubility of MCPLCs, $c_{poly} \sim 10$ wt%

Polymer type	Example	Solvent								Ref.
		H_2SO_4 (conc.)	NMP + 5 wt% LiCl	NMP	DMSO	DMF	*m*-Cresol	CH_2Cl_2 + 20 vol% TFA	CH_2Cl_2	
Polyaramide	A	–	–	–	–	–	–	–	–	2
Copolyester	B 1	+	–	–	–	–	–	–	–	3
Copolyester	B 2	+	–	–	–	+	–	–	–	3
Copolyester	B 3	+	–	–	–	–	+	–	+	3
Polyaramide imide	C	+	+	+	–	–	–	–	–	4
Polyaramide	D 1	+	+	–	–	–	–	–	–	2
Polyaramide	D 2	+	+	+	–	–	–	–	–	2
Polyaramide	D 3	+	+	+	+	–	–	–	–	2
Polyaramide	D 4	+	+	+	+	–	–	+	–	5
Copolyaramide	E	+	+	+	+	+	+	+	–	5

NMP: *N*-methyl-2-pyrrolidone; DMSO: dimethyl sulfoxyde; DMF: *N,N*-dimethyl formamide; TFA: trifluoroacetic acid.

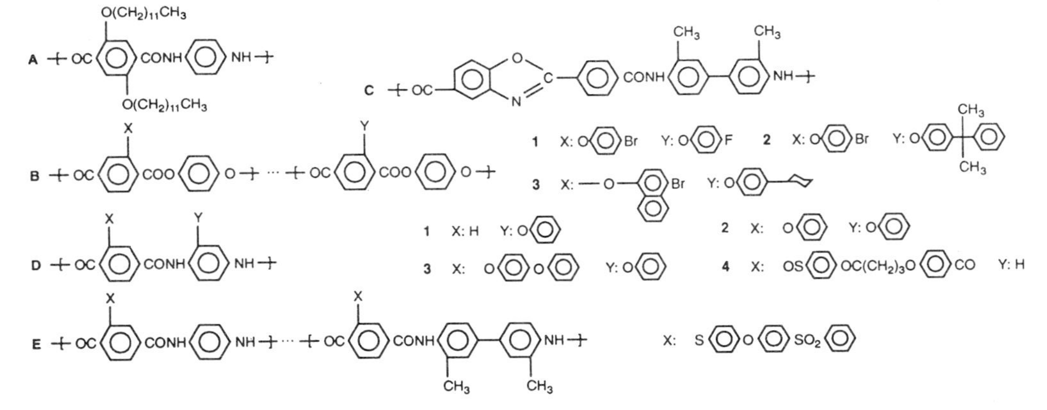

Table 5.2 Light scattering, photon correlation spectroscopy, viscosity and size-exclusion chromatography data of a fractionated random copolyester (for structure and composition see text) in BTFMP at 60°C [8]

M_w ($g\,mol^{-1}$)	A_2 ($10^{-3}\,mol\,cm^3\,g^{-2}$)	R_g (nm)	R_h (nm)	$\rho = R_g/R_h$	$[\eta]$ ($10^2\,m^3\,g^{-1}$)	$M_{w,calc}$ [a] ($g\,mol^{-1}$)	$M_{n,calc}$ [a] ($g\,mol^{-1}$)	$(M_w/M_n)_{calc}$ [a]
43 000	8.6	38	13	2.9	8.0	40 900	19 300	2.11
36 000	7.8	33	13	2.5	6.4	46 000	20 200	2.28
31 000	7.8	29	11	2.6	6.1	30 900	16 300	1.90
25 000	6.9	22	10	2.2	5.2	26 000	18 800	1.39
13 000	7.0	13	6	2.2	2.9	14 000	11 100	1.26
7 500	6.4	—	4.5	—	1.78	8 000	6 100	1.30

[a] Data from size exclusion chromatography ('universal' calibration curve: $[\eta] \cdot M_w$ vs. elution volume, $[\eta] = 9.93 \times 10^{-4} M_w^{0.842}$ (BTFMP, 60°C)). $(\partial n/\partial c)_T(\lambda = 514.5\,nm) = 0.236 \pm 0.002\,cm^3\,g^{-1}$.

M_w: weight average molar mass; M_n: number average molar mass; A_2: second osmotic virial coefficient; R_g: radius of gyration; R_h: hydrodynamic radius; ρ: ratio of R_g and R_h; $[\eta]$: intrinsic viscosity.

Table 5.3 Investigated systems of SGPLCs and solvents

n	Z	F	m	Solvent	Ref.
			Polymethacrylates		
–	COO	OC_mH_{2m+1}	1, 9, 12, 16	THF	11, 12
–	COO	CN		CCl_4, DMA	12
–	COO	OC_mH_{2m+1}	3, 6, 9	CCl_4, DMA	12
2, 6, 10	COO	OC_mH_{2m+1}	1, 4, 6	Benzene, chloro-, dichlorobenzene, tert-butylbenzene, $CHCl_3$, CH_2Cl_2, trifluoroethanol, toluene, THF, 1-Cl-pentane, -heptane	12–20
11	–	CN		toluene	21
			Polyacrylates		
–	COO	OC_mH_{2m+1}	1, 2, 3, 4	CCl_4, DMA	12
2, 6	COO	OCH_3		$CHCl_3$, CH_2Cl_2	22, 23

side-on connected mesogens dissolved in $CHCl_3$, CCl_4, benzene and toluene [11, 26–28] were discussed.

Polymers containing siloxane backbones (for general structure see Figure 5.1) are another class of SGPLCs whose behavior in solution was investigated, inducing LC properties in mixtures using low molar mass LCs, mostly nematics [29–31], as solvents (for example see Figure 5.2). Therefore the results of this work do *not* help to clarify the relationship between the structure of SGPLCs and the hydro- and thermodynamic properties of their solutions. The same holds also for the studies carried out on poly(1,1-dimethyl-3-imidazolylpropyl) acrylamides [32], polyolefinesulfones [33] and poly(ω-alkoxy-benzoyl-phenylamino) ethylenes [12].

In addition to the thermodynamic quality of the solvent [34], the properties of a polymeric solution depend strongly on the concentration of the dissolved macromolecules. In a dilute solution the macromolecules are isolated from each other and consequently the interactions between the diluted polymer molecules play a minor role. Increasing the polymer content, a critical concentration is reached which gives rise to contact between different polymeric coils leading to a penetration at higher

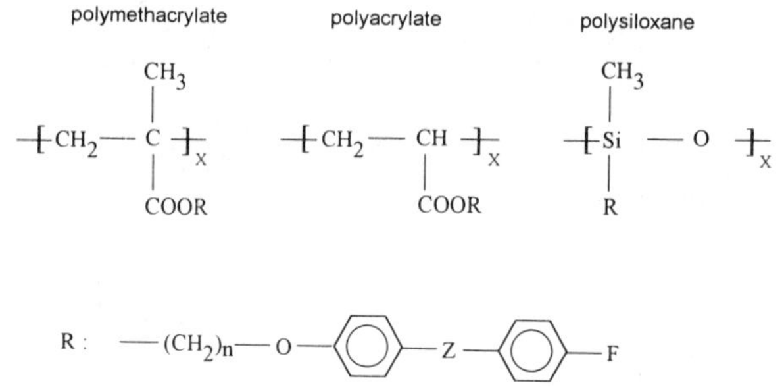

polymethacrylate polyacrylate polysiloxane

$$-[CH_2-C(CH_3)(COOR)]_x- \qquad -[CH_2-CH(COOR)]_x- \qquad -[Si(CH_3)(R)-O]_x-$$

$$R: \; -(CH_2)_n-O-C_6H_4-Z-C_6H_4-F$$

Figure 5.1 Structural formulae for SGPLCs with end-on attached side groups.

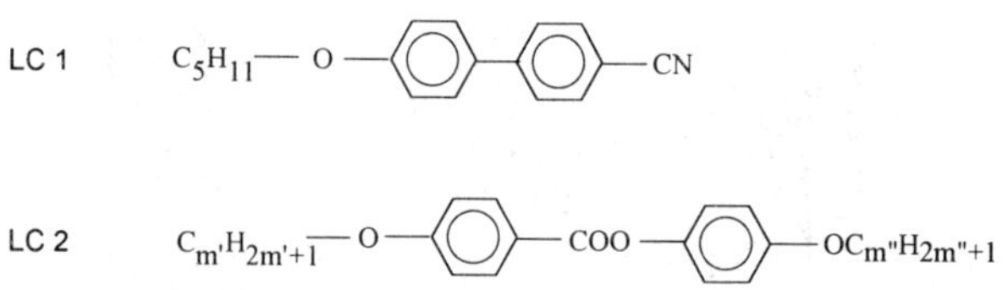

LC 1 $C_5H_{11}-O-C_6H_4-C_6H_4-CN$

LC 2 $C_{m'}H_{2m'+1}-O-C_6H_4-COO-C_6H_4-OC_{m''}H_{2m''+1}$

Figure 5.2 Examples of MLCs that were used as solvents for PLCs (LC1: 4-cyano-4′-pentyloxy-biphenyl [29], LC2: 4-alkyloxy-4′-alkoxyphenyl-benzoate [30, 31]).

concentrations. The properties of such a semidilute solution are quite different from those of dilute solutions.

5.3. DILUTE SOLUTIONS OF SGPLCS

The properties of the scattering due to the interaction of electromagnetic radiation and matter depend on the size, shape and, in the case of time-resolved detection of the scattering intensity (dynamic light scattering), the mobility of the scattering particles. Scattering methods are thus particularly suitable for determining the molar mass, the radius of gyration and the diffusion coefficient of macromolecules in solution. From the concentration dependence of the scattering intensity some information on the thermodynamic properties of the solution can be obtained. Depending on the wavelength of the primary beam a distinction is made between light scattering (LS) [35–38], small angle X-ray scattering (SAXS) [39] and small angle neutron scattering (SANS) [40]. Static LS measurements for dilute solutions are normally used according to the extrapolation procedure given by Zimm, that is based on the following equation:

$$\frac{Kc}{R(q)} = \frac{1}{M_w \cdot P(q)} + 2A_2c \tag{5.1}$$

where K is an optical constant, c is the concentration, $R(q)$ is the Rayleigh ratio, q is the value of the scattering vector $q = (4\pi/\lambda)\sin\theta$ (λ is the wavelength of incident light in the solution; 2θ is the angle between the scattered and incident light), M_w is the weight-average molar mass, $P(q)$ is the structure factor and A_2 is the second osmotic virial coefficient. For the structure factor $P(q)$ the following equation holds relating it to the mean square radius of gyration $\langle R_g^2\rangle$

$$P(q) = \left(1 + \frac{\langle R_g^2\rangle}{3} q^2\right)^{-1} \tag{5.2}$$

A typical Zimm plot is given in Figure 5.3 for a fraction ($M_w = 540\,000\ \mathrm{g\,mol^{-1}}$) of $PMAC_6$ in benzene.

The optical constant depends on the radiation source employed. To calculate this constant for LS measurements the refractive index increment $(\partial n/\partial c)_T$ has to be determined, for the analysis of the SAXS and SANS measurements the partial specific volume v_2 of the polymer in solution must be known. With the exception of [41] the few results published on SGPLCs in solution concerning these molecular constants are in relatively good agreement, as shown in Table 5.4.

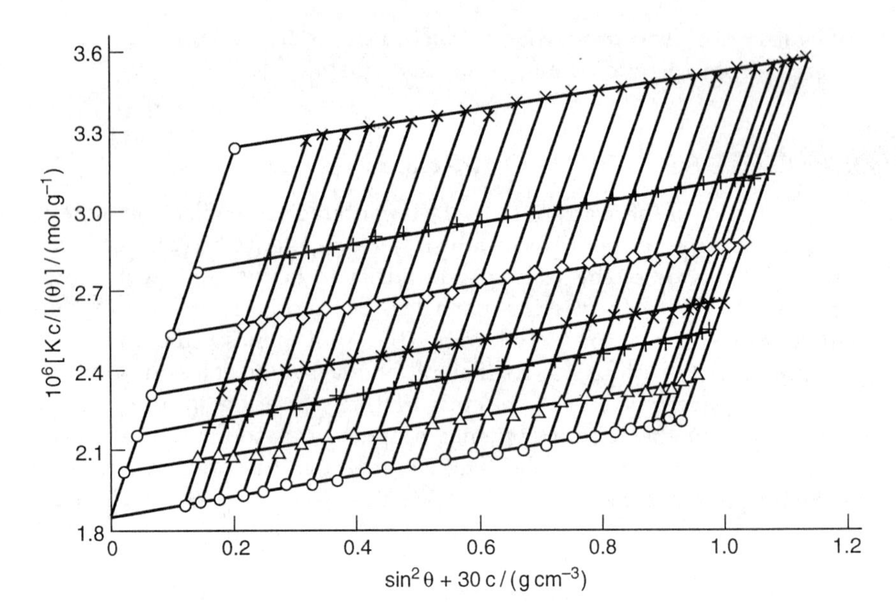

Figure 5.3 Zimm plot of a fraction of $PMAC_6$ in benzene at 20°C [15] (c: 0.85 $g\,L^{-1}$; 1.53 $g\,L^{-1}$; 2.42 $g\,L^{-1}$; 3.56 $g\,L^{-1}$; 5.14 $g\,L^{-1}$; 7.05 $g\,L^{-1}$).

For the spacer homologous series presented, the refractive index increments always become smaller with increasing spacer length. The result obtained by Duran and Strazielle [24] is very striking. They found that the refractive index increments of random copolymers consisting of comonomers with different spacer lengths and a composition of 1:1 are significantly smaller than those of the respective homopolymers (Figure 5.4).

Two results concerning the partial specific volumes of SGPLCs must be emphasized. In comparison with the specific volumes of an homologous series of polyalkylmethacrylates (for instance at $T = 20°C$, $v_2 \sim 0.920\ cm^3\,g^{-1}$ (polybutylmethacrylate/toluene), $v_2 \sim 1.045\ cm^3\,g^{-1}$ (polydodecylmethacrylate/toluene) [20]) the values for SGPLCs are significantly smaller. Therefore it can be argued that strong interactions exist between the mesogenic groups. The observation that random copolymers, where mesogens can scarcely interact because of their different spacer lengths, occupy larger specific volumes than the homopolymers, supports this assumption (Figure 5.4).

The second result that should be mentioned is the agreement with the known properties of conventional polymers, that the partial specific volume of a polymer increases with increasing side group length (the

Table 5.4 Results of the investigation of samples PMAC$_n$ in different solvents

Polymer PMAC$_n$	Solvent	T (°C)	λ_0 (nm)	$(\partial n/\partial c)_T$ (cm^3 g^{-1})	v_2 (cm^3 g^{-1})	Ref.
1	THF	25	633	0.1655		19
2				0.1671		
3				0.1643		
4				0.1627		
5				0.1678		
6		20	546	0.1690	0.788	15, 20
		25	633	0.1622		19
			546	0.076		41
			436	0.170		13
		40	546	0.1761		15, 20
	Toluene	25	633	0.0864	0.813 ([42])	19
			436	0.0815		7
		15	546	0.0770	0.807 (17°C)	15, 20
		22		0.0794		
		30		0.0820		
		40		0.0861		
		50	436	0.0899		13
	Benzene	20	546	0.0741	0.815	15, 20
		25	436	0.0816		13
			546	0.079		41
		35		0.082		
		45		0.087		
		55		0.090		
		40		0.0826		15, 20
	1,1,2-Trichloroethane	20	546	0.0966		
		40		0.1023		
		60		0.1110		
	Trichloromethane	25	633	0.1257		19
			546	0.081		41
		35		0.083		
		45		0.085		
		55		0.088		
	Chlorobenzene	25	436	0.0581		13
		50		0.0649		
	Trifluoroethanol	25	436	0.261		

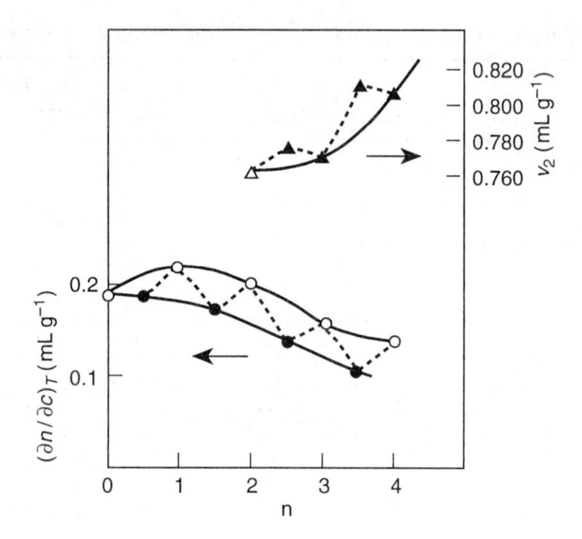

Figure 5.4 Refractive index increments $(\partial n/\partial c)_T$ at $\lambda = 546$ nm in THF and partial specific volumes v_2 at 25°C in THF [24]. Unfilled and filled symbols represent homopolymers and copolymers, respectively. The dashed line is a guide to show the shift in the curves and has no physical significance.

spacer in the present case). However, molar masses determined by scattering methods are always found to be higher than those observed by size exclusion chromatography (SEC). From this observation it can be concluded that SGPLCs in solution, independently of the kind of bonding of the mesogenic groups to the polymer backbone (for example mesogens side-on or end-on linked), occupy a smaller hydrodynamic volume than non-liquid crystalline polymers with a very similar molecular architecture. SEC provides correct values of molar masses only if previously calibrated with corresponding PLC standards or when a 'universal' calibration [34] has been established with the aid of structurally similar polymers.

In spite of the errors in absolute values resulting from the SEC method using an incorrect calibration, the analysis of elution diagrams shows, for example for the polydispersity index $E = M_w/M_n$, good agreement with the values determined by absolute methods [20, 24, 26].

Different results from LS measurements carried out for the same SGPLC in several solutions have not been understood until now. While some authors [11, 13, 27] always obtain linear Zimm diagrams and measured molar masses independent of the solvent, others [15, 20, 24, 33] report on increasing scattering intensities for small scattering angles, suggesting a certain tendency of the SGPLCs in solution to aggregate.

spherulites. As the polymer is drawn, the crystalline orientation function s_c increases rapidly and reaches about 0.9 at $\lambda = 5$. The spherulite structure is deformed and gradually broken up. The crystalline blocks are pulled out of the lamellae incorporated in microfibrils, which therefore consist of alternating crystalline and amorphous regions. A large number of taut tie molecules originating from partial chain unfolding connect the crystalline blocks. The crystalline phase, therefore, is essentially continuous along the draw direction with the degree of continuity increasing with increasing λ. An intercrystalline material may be regarded as consisting of the following three components [48]: A, amorphous material which includes floating chains, cilia and loops; TM, tie molecules which increase both in number and tautness with increasing λ; and B, intercrystalline bridges which are essentially crystalline in nature and may be formed by the coalescence of adjacent taut tie molecules (Figure 8.2).

At low draw ratios where the intercrystalline bridges might be negligible, the drawn polymer can be regarded as a two-phase composite consisting of crystalline blocks embedded in an isotropic amorphous matrix. The drawing process mainly increases the alignment of the crystalline chain axes of these blocks along the draw direction. Levin [76] has derived a general formula for a two-phase composite with fully oriented inclusions allowing for the most general anisotropy in both thermal and mechanical properties of each phase. The formula requires a knowledge of the complete compliance tensors of both phases and of the composite itself which in general are not available experimentally.

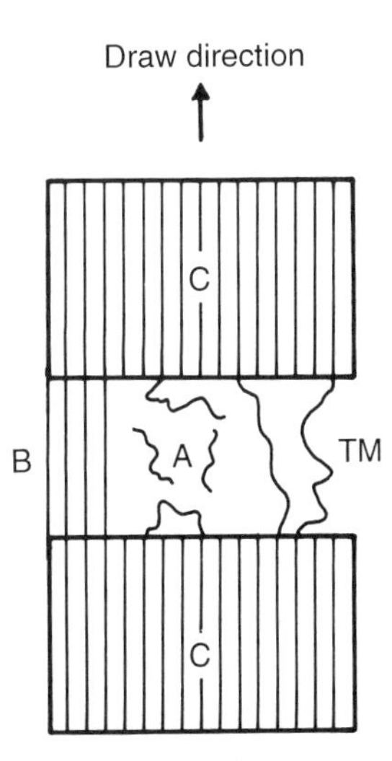

Figure 8.2 Schematic diagram indicating the structure of a highly oriented semicrystalline polymer. A, amorphous region; B, intercrystalline bridges; C, chain-folded crystal block; TM, tie molecules.

Hence, for a rough estimate, it is assumed that the expansivity tensor of a composite $\boldsymbol{\alpha}'_{ij}$ with fully aligned crystallites can be obtained by the linear interpolation between those of its component phases, i.e.

$$\boldsymbol{\alpha}'_{ij} = V\boldsymbol{\alpha}^{\mathrm{C}}_{ij} + (1 - V)\,\boldsymbol{\delta}_{ij}\boldsymbol{\alpha}^{\mathrm{a}} \tag{8.8}$$

where V is the volume fraction crystallinity and α^{C}_{ij} (which is diagonal with elements ($\boldsymbol{\alpha}^{C}_{\perp}$, $\boldsymbol{\alpha}^{C}_{\perp}$ and $\boldsymbol{\alpha}^{C}_{\parallel}$) and $\boldsymbol{\delta}_{ij}\boldsymbol{\alpha}^{\mathrm{a}}$ are the expansivity tensors of crystalline and the (isotropic) amorphous phase, respectively. Equation (8.8) implies that the two phases expand independently, which is strictly valid only above the major amorphous relaxation where the shear modulus of the amorphous region is very low. Taking into account the distribution in the chain orientation of the crystalline blocks, the expansivities of the oriented polymers are given by following equations [47]:

$$\alpha_{\parallel} = V\alpha^{\mathrm{C}} + (1 - V)\,\alpha^{\mathrm{a}} - \tfrac{2}{3}Vs_{\mathrm{c}}(\alpha^{\mathrm{C}}_{\perp} - \alpha^{\mathrm{C}}_{\parallel}) \tag{8.9}$$

$$\alpha_{\perp} = V\alpha^{\mathrm{C}} + (1 - V)\alpha^{\mathrm{a}} - \tfrac{1}{3}Vs_{\mathrm{c}}(\alpha^{\mathrm{C}}_{\perp} - \alpha^{\mathrm{C}}_{\parallel}) \tag{8.10}$$

where $\alpha^{\mathrm{C}} = \tfrac{1}{3}(\alpha^{\mathrm{C}}_{\parallel} + 2\alpha^{\mathrm{C}}_{\perp})$ is the average expansivity of the crystallites and s_{c} is the crystalline orientation function.

For an isotropic sample, $s_{\mathrm{c}} = 0$ and equations (8.9) and (8.10) both reduce to

$$\alpha_{\mathrm{iso}} = V\alpha^{\mathrm{C}} + (1 - V)\,\alpha^{\mathrm{a}} \tag{8.11}$$

The equations (8.9)–(8.11) can also be combined to yield

$$\alpha_{\mathrm{iso}} = -\tfrac{1}{3}(\alpha_{\parallel} + 2\alpha_{\perp}) \tag{8.12}$$

i.e. α_{iso} is equal to the average linear expansivity $\bar{\alpha}_{\lambda}$ of the oriented sample. The same result was obtained in series coupling in the aggregate model. It is apparent from equations (8.9) and (8.10) that the dependence on λ of the expansivities arises solely from the partial orientation of the chain axis as described by $s_{\mathrm{c}}(\lambda)$. Above $\lambda = 5$, the effect becomes saturated when the chains are largely aligned along the draw direction. Although the above dispersed crystallite model explains the observed s dependence on λ of α_{λ}, its prediction for $\alpha_{\parallel}$ is much higher than experimental results which underlines the importance of intercrystalline bridges, particularly above $\lambda > 5$ [32].

At $\lambda > 5$, nearly all crystalline blocks become oriented ($s_{\mathrm{c}} \sim 1$). Then α_{λ} and $\alpha_{\parallel}$ become independent of λ and the appearance of an intercrystalline bridge becomes a reality. Then the oriented structure can be represented schematically by Figure 8.3. Gibson *et al.* [77] have suggested on the basis of statistical consideration that the volume fraction of intercrystalline bridge material is

$$V_{\mathrm{f}} = V_{\mathrm{p}}(2 - p) \tag{8.13}$$

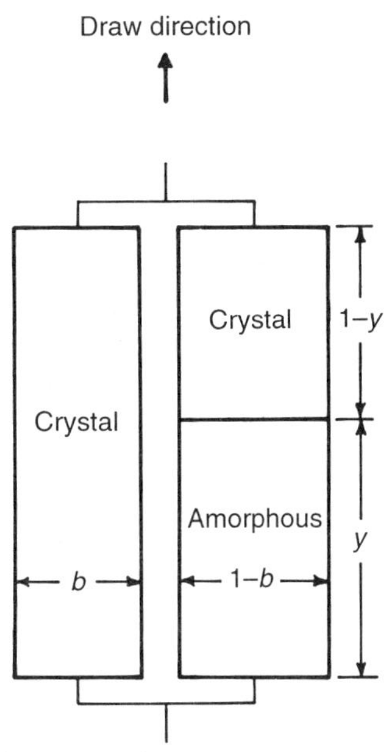

Figure 8.3 Schematic diagram indicating a version of the Takayanagi model.

where p is the probability of a chain within a crystallite traversing the intervening disordered region and entering the next crystallite on the stack. p can be evaluated from the observed average crystal length D_{002} and the long period L through the relation $D_{002} = L(1 + p)/(1 - p)$. Gibson *et al.* [77] have tabulated values of p for drawn and extruded HDPE. Now it is simple to calculate the expansivity. In this version of the Takayanagi model [78], the chain folded crystalline block thermally expands in series with the amorphous region along the draw direction but the entire stack is constrained by the intercrystalline bridges in parallel with it. This simple calculation leads to the following relations [47, 79]:

$$\alpha^{\parallel} = \alpha_{\parallel}^{C} + q(\alpha^{a} - \alpha_{\parallel}^{C}) \tag{8.14}$$

$$\alpha_{\perp} = \alpha_{\perp}^{C} + \frac{\alpha^{a} - \alpha^{C}}{\eta - 1} \tag{8.15}$$

where

$$\eta = -\frac{V - V_{f}}{1 - V}\frac{E_{\perp}^{C}}{E^{a}} \tag{8.16}$$

$$q = \left[(1 - V)^{-1} + V_{f}(1 - V_{f})^{-1}\left(\frac{E_{\parallel}^{C}}{E^{a}} - 1\right)\right]^{-1} \tag{8.17}$$

where E^a, $E_\perp^C$ and $E_\parallel^C$ are the Young modulus of the amorphous region and of the crystalline region perpendicular to the *c*-axis, respectively. Since $E_\parallel^C \gg E^a$, $q \ll 1$ and the constraining effect of the intercrystalline bridges then becomes predominant and $\alpha_\parallel$ approaches $\alpha_\parallel^C$.

For crystalline polymers, a slight increase in $\alpha_\perp$ and a sharp drop in $\alpha_\parallel$ with increasing λ can be attributed to two factors, the alignment of chains in the crystalline blocks and production of intercrystalline bridges [47, 48]. As far as $\alpha_\perp$ is concerned, the oriented polymers at all draw ratios may be treated adequately as a composite with partially aligned crystalline blocks embedded in an isotropic amorphous phase. The behavior of $\alpha_\parallel$ is very sensitive to the presence of intercrystalline bridges. For ultra-oriented HDPE ($\lambda = 18$) the bridge fraction ($b = 0.36$) is sufficiently large for $\alpha_\parallel$ at room temperature to be only 10% above $\alpha_\parallel^c$. For highly crystalline polymers such as HDPE, $\alpha_\parallel$ becomes negative throughout the range 120–300 K when $\lambda > 4$. The intercrystalline bridges having a high tensile modulus constrain the expansion of the amorphous region between crystallites. With increasing λ, the fraction of bridges increases and this gives rise to more severe constraints on the amorphous region, hence the expansivity along the draw direction will approach $\alpha_\parallel^C = -1.3 \times 10^{-5}\,K^{-1}$, the expansivity of the crystallites along the chain axis. The behavior of branched LDPE was found to be entirely different [80]. At low temperature, $\alpha_\parallel$ for a sample with $\lambda = 4$ is positive but $\alpha_\parallel$ decreases so sharply above 250 K that its value 320 K is about 30 times more negative than $\alpha_\parallel^C$. The strong temperature dependence and large negative value provide strong evidence that for LDPE the rubber-elastic effect is fully operative at 130 K above the amorphous transition at T_s. $\alpha_\perp$ also exhibits a sharp rise in the same temperature range so that the average expansivity $\bar{\alpha}$ is not much different from the expansivity of an isotropic sample. Similar behavior has been observed for a number of polymers (nylon 6, PVF_2 and PCTFE) of low crystallinity ($V = 0.34$–0.46) [48]. It was realized by various workers [81, 82] that the large negative thermal expansion along the draw direction arises from the rubber-elastic contraction of intercrystalline tie molecules.

8.4.3 Rubber-elastic effect

A quantitative treatment of the rubber-elastic effect has recently been proposed by Choy *et al.* [48], which has led to an understanding of the drastic difference in the expansivities of LDPE and HDPE. According to them, an oriented polymer may be considered as a composite made up of four phases, i.e. the crystallites, the amorphous region, bridges and tie molecules. The internal strain due to rubber-like tie molecules is responsible for the high temperature behavior of $\alpha_{\parallel}$ in semicrystalline polymers.

The polymer may be considered as a parallel combination of bridges, amorphous phase and tie molecules which are connected in series with the crystalline block as depicted in Figure 8.4. Since the emphasis is on the effect of tie molecules, the amorphous and bridge phases will be competely regarded as a single phase denoted by S; the effective tensile modulus $E_{\parallel}^{s}$ and expansivity $\alpha_{\parallel}^{s}$ may be calculated as [48]

$$E_{\parallel}^{s} = \frac{(bE_{\parallel}^{c} + aE^{a})}{b + a} \sim bE_{\parallel}^{c} \tag{8.18}$$

$$\alpha_{\parallel}^{s} = \frac{(bE_{\parallel}^{c}\alpha_{\parallel}^{c} + aE^{a}\alpha^{a})}{bE_{\parallel}^{c} + aE_{a}} \tag{8.19}$$

where E^{a} is the tensile modulus of the amorphous region, and a and b are the volume fractions of the amorphous and intercrystalline bridge portions of the polymer, respectively. The tensile modulus of the bridge is taken to be the same as that of the crystalline block $E_{\parallel}^{c}$. The two phases S and TM must be mutually constrained to have the same length l; the phases would assume their respective lengths l_{s} and l_{t} if constraints are removed. By equating the constraining forces on the both phases, it can be easily shown that the expansivity of the intercrystalline material is

$$\alpha_{\parallel}^{m} = \alpha_{\parallel}^{s} - \frac{d\eta}{dT} \tag{8.20}$$

where $\eta = tE_{\parallel}^{t}/bE_{\parallel}^{c}$, $E_{\parallel}^{t}$ being the modulus of the tie molecules. In the derivation of equation (8.20), it has been assumed that b, $t \ll 1$ and

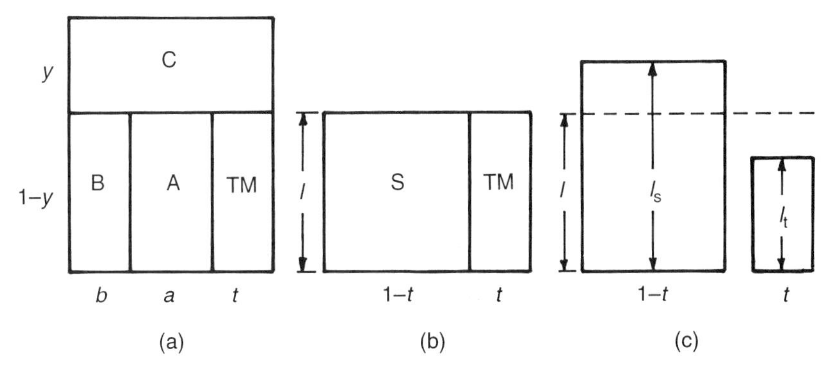

Figure 8.4 (a) Schematic diagram of an oriented polymer consisting of an amorphous region (A), intercrystalline bridges (B), crystalline blocks (C) and tie molecules (TM) with volume fractions $a + b + t = 1$. (b) Parallel combination of two phases. (c) Phases TM and S after removal of constraint.

$aE^a, tE^t \ll bE^c_{\parallel}$. The expansivity of the entire composite can be obtained by adding crystalline phase in series, i.e.

$$\alpha_{\parallel} = y\alpha^c_{\parallel} + (1-y)\,\alpha^s_{\parallel} - (1-y)\frac{d\eta}{dT} \tag{8.21}$$

Above the amorphous transition, the modulus of the tie molecules increases with temperature ($E^t_{\parallel}$ is proportional to T) according to entropic theory of rubber elasticity. $E^c_{\parallel}$ is nearly constant so

$$\eta = \frac{tE^t}{bE^c} \propto T$$

and

$$\frac{d\eta}{dT} \sim \frac{\eta}{T} = \frac{1}{T}\frac{tE^t}{tbE^c_{\parallel}} > 0 \tag{8.22}$$

The tie molecules therefore make a negative contribution to $\alpha_{\parallel}$ through the last term in equation (8.21).

Though the necessary input parameters of equation (8.21) are not known accurately enough to allow detailed comparison with experimental data, equation (8.21) can be used to determine factors controlling the magnitude of the rubber-elastic contraction [32, 48]. For example Choy *et al.* [48] consider the case of an LDPE sample with $\lambda = 4.2$ for which $\alpha_{\parallel} = -40 \times 10^{-5}\,K^{-1}$ at 320 K. The rubber-elastic effect is presumably fully appropriate at this temperature and equation (8.21) may be expected to apply. The first two terms of equation (8.21) are of the order of $10^{-5}\,K^{-1}$ and may be neglected. Using equation (8.21) for the last term and $y = 0.4$ yields

$$tE^t/bE^c_{\parallel} = 0.2 \tag{8.23}$$

The relation for the modulus E of the composite in the above series–parallel model is given by:

$$\frac{1}{E} = \frac{y}{E^c_{\parallel}} + \frac{1-y}{bE^c_{\parallel} + aE^a + tE^t} \tag{8.24}$$

$$\sim \frac{y}{E^c_{\parallel}} + \frac{1-y}{bE^c_{\parallel} + tE^t} \tag{8.25}$$

The observed value of $E \sim 0.6\,GN\,m^{-2}$ gives

$$bE^c_{\parallel} + tE^t \sim 0.36\,GN\,m^{-2} \tag{8.26}$$

Using both equation (8.21) and (8.23) leads to

$$bE^c_{\parallel} \sim 0.3\,GN\,m^{-2} \tag{8.27}$$

$$tE^t \sim 0.07\,GN\,m^{-2} \tag{8.28}$$

which gives a bridge fraction $b = 10^{-3}$, $t = 0.1$ for $E^{c}_{\parallel} \sim 255\,\mathrm{GN\,m^{-2}}$ and $E^{t}_{\parallel} \sim 0.75\,\mathrm{GN\,m^{-2}}$.

It is evident from this result that the tendency of the tie molecules to contract is resisted mainly by the bridges and it is the very small bridge/tie molecular ratio in this case ($b/t = 0.04$) which leads to the large negative expansivity [32, 48]. In contrast consider HDPE ($\lambda = 4.2$) for which the thermal expansivity at 320 K is $-1.0 \times 10^{-5}\,\mathrm{K^{-1}}$ [32, 48]. Taking into account the results of wide angle X-ray studies and statistical considerations, the estimate $b \sim 0.04$ can be made which is consistent with the observed modulus, $E \sim 10\,\mathrm{GN\,m^{-2}}$. Similar analysis using equations (8.21)–(8.23) then provides $t \sim 0.02$, i.e. b/t is of the order of unity. It can be concluded that the rubber-elastic effect is stronger for polymers with lower crystallinity primarily because of the small bridge/tie molecule ratio in such materials. Further, this effect is more pronounced at about 130 K above T_s (or T_g) so that the phenomenon of a large contraction is readily observable for polymers having low T_s (or T_g) and high melting point [32, 48]. The rubber-elastic contraction is exhibited in stretched crystallizable elastomers at $\lambda < 1$ [83]. In semicrystalline polymers, if the conformational mobility is strongly restricted by intermolecular interactions, $\alpha_{\parallel} < \alpha^{C}_{\parallel}$. Thus in two-phase polymers, the behavior of their amorphous regions above T_g is governed by conformational changes strongly restricted by interchain interaction [70, 84].

8.5 POLYMER LIQUID CRYSTALS

Among high performance materials, thermotropic polymer liquid crystals, having high stiffness and strength combined with light weight and dimensional stability, have attracted recent scientific and technological activity and attention. Their superior mechanical and thermal properties are due to the molecular orientations induced by shear and elongational stresses applied when they are extruded or molded. When TPLC is drawn from the melt, the elongational stress causes the orientation of molecular chains leading to the formation of a microfibrillar structure with microfibrils aligning along the draw direction. Thus the drawn materials have very high axial stiffness and strength. Though structural and elastic properties of thermotropic PLCs (TPLCs) have been successfully explained in terms of molecular architecture [85], the thermal dependence of these properties has not been extensively studied and only recently has much progress been made [32]. In particular, the effect of orientation on expansivity of TPLC has not been studied extensively, and only a few reports exist [19–21, 86–97].

Polymeric solids in general exhibit α_L of the order of $5-10 \times 10^{-5}\,\mathrm{K^{-1}}$. It is well known that polymeric films or fibers can have a negative α_L in the orientation direction. For highly oriented polymer liquid crystals and aromatic polyamides (all without side chains), α_L values of the order of $-10^{-5}\,K^{-1}$ are found at a wide range of temperatures below and above room temperature [19–21]. Similar effects have been reported in oriented films of rigid rod polymer liquid crystals with flexible side chains [86]. Like flexible chain polymers, PLCs also exhibit a negative expansion phenomenon. It was earlier contemplated that the thermal contraction allows increased vibrational motion transverse to the main chain axis while maintaining approximately constant bond lengths, i.e. the entropy increases with the axial contraction because of a decrease in the frequencies of the transverse vibrational motion [32]. As mentioned above Lacks and Rutledge [87] have calculated thermal expansion based on first principles using molecular mechanics force fields for the interatomic interactions and a quasi-harmonic approximation for the vibrational Helmholtz function. The negative axial thermal expansion for poly(*p*-phenylene terephthalamide) (PPTA) and poly(*p*-benzamide) (PBA) is attributed primarily to elastic coupling to the thermal stresses transverse to the chain axis [87]. According to Lacks and Rutledge, the increase in entropy associated with chain contraction makes a minor contribution (*c.* 25%) to the negative axial thermal expansion. In the following, all the major investigations carried out on the thermal expansion of polymer liquid crystals, particularly TPLCs, are given in detail. The dependence of α_L on temperature and draw ratio is also discussed.

The first report on the α_L of TPLCs was made by Jackson and Kuhfuss. [88]. They observed that in injection-molded copolyesters containing 40–90 mol% of *p*-hydroxybenzoic acid (PHB), the linear thermal expansions were highly anisotropic. They also observed that the linear thermal expansion is zero along the flow but not across the flow. The anisotropy in linear thermal expansion is due to the orientation of polymer chains during molding.

Takeuchi *et al.* [20] measured α_L in the extruded rods of thermotropic nematic copolyester consisting of 40 mol% poly(ethylene terephthalate) (PET) and 60 mol% PHB. PET/PHB copolyester rods exhibiting the Young moduli from 3.3 to 34.5 GPa were made by shear and elongational induced molecular orientation. α_L decreased drastically from $2.2 \times 10^{-5}\,K^{-1}$ and levelled off at $-6 \times 10^{-6}\,K^{-1}$. α_L changed linearly with the reciprocal of the Young modulus. This result is explained with an aggregate model by assuming both a series connection and parallel connection of the aggregate. In a nematic PLC there exist domains with aligned molecules similar to the molecular bundles in amorphous polymers, and hence the aggregate model was adopted.

Rojstaczer *et al.* [89] measured the α_L of Kevlar 49 fiber from 20 to 150°C. It shows two regimes with a slope change around 90°C, depicting

$\alpha_L = -5.7 \times 10^{-6}\,°C^{-1}$ for the temperature range of 20 to 80°C and $\alpha_L = -6.3 \times 10^{-6}\,°C^{-1}$ for the temperature range of 100 to 150°C. The α_L of Kevlar 49 (obtained by them) gave excellent agreement with measured values for unidirectional Kevlar/epoxy composites in terms of the Schapery model, which is an improvement over the reported axial α_L of Kevlar fiber and those of Kevlar epoxy composites reported by Strife and Prewo [90].

Ii *et al.* [21, 91–93] measured α_L of highly oriented fibers of poly(*p*-benzamide) (PBA) and poly(*p*-phenyleneterephthalamide) (PPTA). Negative thermal expansion $(-7$ to $-11) \times 10^{-6}\,K^{-1}$ for PBA and $(-6$ to $-9) \times 10^{-6}\,K^{-1}$ for PPTA have been found in the temperature range 170–630 K in the annealed samples. Similar thermal contraction for the *c*-axis of PBA and PPTA crystals has been observed based on X-ray diffraction [91]. For the annealed PBA sample, the lattice contraction from 300 to 600 K is 0.23% giving an average α_L, i.e. $\alpha_c = 7.7 \times 10^{-6}\,K^{-1}$. For the annealed sample of PPTA, the lattice contraction from 300 to 600 K is 0.087%, thus $\alpha_c = -2.9 \times 10^{6}\,K^{-1}$. The α values of the annealed sample are estimated as $\alpha = -8.5 \times 10^{-6}\,K^{-1}$ for PBA and $\alpha = -7.6 \times 10^{-6}\,K^{-1}$ for PPTA in the range 300–600 K. In contrast to the *c*-axis, the lateral spacing, *a*, *b* and d_{110} increase with a rise of temperature, giving a normal positive α. For PBA, $\alpha_a = 7.0 \times 10^{-5}\,K^{-1}$ between 300 and 500 K and $\alpha_a = 9.8 \times 10^{-5}\,K^{-1}$ between 500 and 700 K, while $\alpha_b = 4.1$ and 8.7 respectively in the respective temperature ranges. Similarly for PPTA $\alpha_a = 8.3$ and 12.0 and $\alpha_b = 4.7$ and 6.9 in temperature ranges of 300–500 K and 500–700 K, respectively. Ii *et al.* [92] measured the α_L values of PBA and PPTA fibers under a stress of 0.05 GPa. They found the average α_L along the chain axis to be $-6.7 \times 10^{-6}\,K^{-1}$ for the lattice and $-8.9 \times 10^{-6}\,K^{-1}$ for the bulk in the temperature range 200–600 K.

Ii *et al.* [93] proposed a theoretical treatment based on a string-in-medium model, in which a flexible string is embedded in a medium and fluctuates thermally in an external field of extensional stress. An expression for the strain ε along the chain is derived as a function of stress σ and temperature T:

$$\varepsilon(\sigma, T) = \frac{\sigma}{E_0} - \frac{1}{2}\Sigma \frac{U_k}{\rho V^2 + \sigma} \tag{8.29}$$

where E_0 is the Young modulus under no lateral thermal agitation, U_k is the total energy of each transverse acoustic mode, V is its phase velocity along the chain, ρ is the density and the summation runs over all transverse modes. From this expression, the expression of α_L, $\alpha_L(=\partial\varepsilon/\partial T)$ is derived. The equation for α_L is also interpreted in terms of Grüneisen theory and corrected slightly by a positive contribution due to the thermally excited longitudinal modes. The equations are found to reproduce quantitatively the experimental results.

Green *et al.* [19] reported α_L values of a series of highly oriented thermotropic liquid crystal copolyesters with random phenylene–naphthalene main chain sequences over extended temperature range. In spite of differences in composition and production conditions, all LC copolyesters have similar expansion–temperature curves. All the curves have a pronounced transition at about 100°C, while in several cases a subsidiary transition occurs at about room temperature. Below the subsidiary transition, all samples have a similar low temperature $\alpha_{\parallel}^{c}$ of $-7 \times 10^{-6}\,K^{-1}$. In the intermediate range between the transitions, there are only small differences in the slopes of the expansion–temperature curves. The main transition occurs at about 100°C and the axial expansions become increasingly negative, decreasing in an approximately linear fashion to the maximum measured temperature. It seems that the increasingly negative axial expansions of the materials above this temperature are generated by segmental motions of the highly oriented main chain. The close agreement between the X-ray and macroscopic measurements shows that the observed macroscopic expansion results from a shortening of the axial projection of the monomer units.

Buijs and Damman [18] measured the α_L of oriented films of poly(*p*-phenylene-2, 5-didodecyloxy-terephthalate) (PTA12HQ) in three layered structures B, A and L_f. A combination of flexible aliphatic side chains and an aromatic polyester rigid backbone gives rise to layered structures in which main layers are separated by aliphatic side chains. The above mentioned polymer exhibits three different room temperature structures based on the amount of ordering of side and main chains [94]. Modification B is a highly crystalline structure in three dimensions having both main chain and side chain ordering. In modification A the microcrystals of the side chains are located between main chain layers somewhat remote from the polymer backbone. In phase L_f the structure with the lowest ordering is formed by quenching from the layered mesophase L_m. Upon heating, all the three room temperature structures transform to the intermediate phase A in which side chains have lost most of their ordering. At higher temperatures, the main chains lose their three-dimensional ordering and biaxial layered mesophase L_m is formed. At still higher temperatures, the parallel ordering of main chains is lost and an isotropic melt is formed. In unidirectionally oriented film or tape, the oriented rigid main chain layers provide a negative contribution to α while the side chains make a positive contribution. The resulting α depends on the details of the main and side chain packing and the low temperature values between $\alpha = +0.3 \times 10^{-5}\,K^{-1}$ (A and L_f) and $\alpha = -1.2 \times 10^{-5}\,K^{-1}$ (B) are found in highly oriented films. The above results are interpreted in terms of a molecular laminate model in which rigid main chain layers are separated by the aliphatic side chains. The model also explains the thermal expansion behavior of undrawn film.

It is well known now that the longitudinal PLCs exhibit a high modulus and a negative thermal expansion in the direction of the macroscopic orientation. In the direction transverse to the orientation of the chains, the properties are poor. These differences can be reduced by the introduction of the crosslinks between the chains which improve the dimensional stability of these ordered systems. A number of crosslinked oriented polymer networks have been produced and their thermal expansion has been measured [95, 96]. In the first report, densely crosslinked oriented polymer networks are produced by *in situ* photopolymerization of oriented LC diacrylates [95]. In the monomeric state the mesomorphic nematic diacrylate is oriented and polymerized by exposure to ultraviolet light. The orientation of the rigid central aromatic part of the monomer persists during the free radical chain crosslinking of the acrylate double bonds at both ends of the rod-like molecule. The acrylate groups are spacered from the mesogenic moiety in order to provide sufficient conformational mobility for polymerization without distortion of the orientation of mesogenic units. The resulting coatings or free standing films exhibit anisotropic properties which are maintained up to high temperatures. The α_L values of uniaxially oriented networks in the direction of the molecular orientation are slightly positive at temperatures below the glass transition of the networks and become negative as soon as this transition sets in. By introducing a molecular helix in the network perpendicular to the film surface, a low thermal expansion is established in two directions in the plane of the film and coatings. In this case an effectively zero expansion results over a wide temperature range. Jahromi *et al.* [96] reported the thermal expansion in highly crosslinked epoxy networks with a high macroscopic order produced by copolymerization with aromatic amines. LC diepoxide was the best monomer for producing ordered networks by copolymerization with aromatic diamines because of its broad nematic range. Three types of aromatic diamines were used for copolymerization. The orientational order increased during the chain extension process and it became irreversibly fixed as a result of the crosslinking reaction. The α_L values are found to be highly anisotropic and independent of gel point (T_{cure}). Thus the ordered, highly crosslinked epoxy–amine network provided a combination of good mechanical and thermal properties of conventional epoxy networks with anisotropic behavior connected with the high macroscopic order.

The thermal expansivity and thermal conductivity of thermotropic liquid crystalline copolyesteramides (Vectra B950 and Vectra A950), with a draw ratio λ from 1.3 to 15, have been measured parallel and perpendicular to the draw direction from 120 to 430 K in detail [97]. Further comparative study of these properties has been undertaken with that of a number of flexible chain polymers such as high density polyethylene (PE), polyoxymethylene (POM) and polypropylene (PP). The salient aspects of this detailed investigation [97] follow below.

The expansivities of the rod and core samples of Vectra B950, parallel ($\alpha_{\parallel}$) and perpendicular ($\alpha_{\perp}$) to the draw ratio exhibit opposite behavior. While the axial expansivity $\alpha_{\parallel}$ drops sharply with increasing draw ratio whereas the transverse expansivity $\alpha_{\perp}$ shows a slight increase. The saturation effect for the thermal expansivity occurs at a lower λ than the thermal conductivity. The saturation values of $\alpha_{\parallel}$ and $\alpha_{\perp}$ above $\lambda = 2.6$ are -0.85×10^{-5} and $8.0 \times 10^{-5}\,K^{-1}$, respectively. It is also observed that at $\lambda < 2$, $\alpha_{\parallel}$ is much lower than that for the core because the skin layer not only has a negative $\alpha_{\parallel}$ but also has high axial stiffness, and thus constrains the expansion of the core region in the rod. On comparison of the draw ratio dependence of $\alpha_{\parallel}$ for Vectra B950 with PE, POM and PP, it is found that in all cases the thermal expansivity decreases with increasing λ. However, the effect of drawing is strongest for Vectra B950. Here $\alpha_{\parallel}$ becomes negative at a low draw ratio of 1.3. At high λ, $\alpha_{\parallel}$ for all polymers is negative and of similar magnitude. The drastic drop in axial expansivity at low λ and its saturation at $\lambda > 4$ arise from the corresponding increase in the degree of chain orientation evidenced by wide angle X-ray diffraction. The temperature dependence of $\alpha_{\parallel}$ and $\alpha_{\perp}$ of extruded rods of Vectra B950 and A950 at various draw ratios has been studied in detail. It is observed that $\alpha_{\parallel}$ decreases as the temperature increases from 120 to 270 K. Near this temperature there is a transition and the slope of the curves change from negative to positive. A more pronounced transition occurs at about 300 K above which $\alpha_{\parallel}$ decreases sharply. At about 300 K, large scale segmental motions of the polymer chains start to take place. The transition is similar to the glass transition of conventional polymers. The transition at 270 K has been attributed to the local motion of naphthalene units. The thermal expansivity of highly oriented samples ($\lambda \geqslant 3$) of both Vectra B950 and Vectra A950 have a similar magnitude and temperature dependence. The high temperature transition in Vectra B950 occurs at 25 K higher, indicating the higher glass transition of the material. The $\alpha_{\perp}$ value of B950 is lower by 10–20%, which arises from the stronger interchain interaction due to the presence of hydrogen bonds. The oriented samples of both flexible and rigid chain polymers exhibit negative thermal expansion along the draw direction. Below the glass transition, both LCP and flexible polymers have similar $\alpha_{\parallel}$, indicating the same microscopic mechanism. Above the glass transition an additional mechanism may come into play. Most of the tie molecules in POM and PP have an amorphous nature and their tendency to assume a more crumpled conformation leads to a sharp drop in $\alpha_{\parallel}$. This effect is also observed in Vectra which indicates that the tie molecules between the crystalline lamellae in this class of TPLCs possess considerable semiflexibility (Figure 8.5).

The aggregate model [71–73] has been applied to explain the thermal expansion behavior of TPLC. In this model an isotropic polymer is regarded as a random aggregate of axially symmetric units whose properties are those of the fully oriented materials. When polymer is drawn, the intrinsic units rotate towards the draw direction though the units themselves remain unchanged. Then the thermal conductivity of a partially oriented sample can be calculated by using either the parallel or the series model. In the case of thermal conductivity, the parallel and series models provide the upper and lower theoretical bounds. There is agreement between theory and experiment in the general trends of the draw ratio dependence, i.e. a sharp rise in $K_{\parallel}$ and a moderate fall in $K_{\perp}$. However the $K_{\parallel}$ data follow closely the upper bound whereas the $K_{\perp}$ data lie between the two bounds. Although the draw ratio dependence of the expansivity is reproduced by the series model, the observed $\alpha_{\parallel}$ drops more rapidly than the prediction.

At high orientation, the axial thermal conductivity of Vectra B950

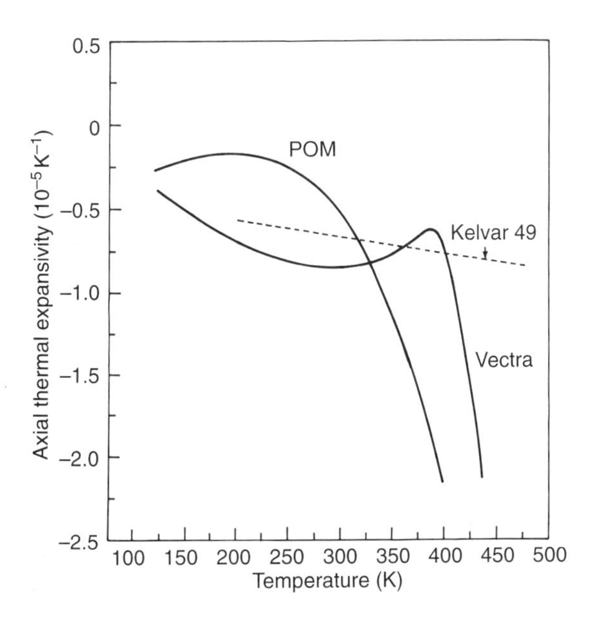

Figure 8.5 Temperature dependence of the axial thermal expansivity of Vectra B950 ($\lambda = 4.6$), Kevlar 49, polyoxymethylene (POM; $\lambda = 20$). (Adapted from [15].)

is slightly higher than that of polypropylene but is an order of magnitude lower than that of polyethylene.

8.6 POLYMER COMPOSITES AND POLYMER LIQUID CRYSTAL + THERMOPLASTIC BLENDS

Polymeric materials have relatively large thermal expansion. However, by incorporating fillers of low α in typical plastics, it is possible to produce a composite having a value of α only one-fifth of the unfilled plastics. Recently the thermal expansivity of a number of *in situ* composites of polymer liquid crystals and engineering plastics has been studied [14, 16, 98, 99]. Choy *et al.* [99] have attempted to correlate the thermal expansivity of a blend with those of its constituents using the Schapery equation for continuous fiber reinforced composites [100] as the PLC fibrils in blends studied are essentially continuous at the draw ratio of $\lambda = 15$. Other authors [14, 99] observed that the Takayanagi model [101] explains the thermal expansion.

The thermal expansion of multiphase materials has been reviewed by Holliday and Robinson [28] and by Sideridis [102] and Vaidya and Chawla [103]. The thermal expansivity of a composite is actually a thermochemical parameter and depends not only on the expansivities but also on the elastic constants of constituent phases, as well as on shape and other aspects of the particle geometry of reinforcing materials. In general, a composite is prepared from a thermoplastic or thermoset at a temperature above ambient. In the process of cooling, each phase will shrink but the shrinkage of the matrix will be restrained by the filler, thus setting up compressive stresses across the interface which may be relieved or reduced by non-elastic deformations in the matrix. When the composite is heated, the matrix tends to expand more than the filler, and if the interface is capable of transmitting the stresses which are set up, the expansion of the matrix will be reduced. Though a general theory which takes into account all the factors is still absent, it is possible to estimate quite correctly the thermal expansion of anisotropic unidirectional or bidirectional composites with polymer matrices.

If there is no adhesion between the two phases, if α of the matrix (α_m) is greater than α of the filler (α_p) and if there is no residual compressive strains across the interface, then on heating the composite the matrix will expand away from the particles. In this case, α of the composite (α'_c) is equal to α_m i.e. $\alpha'_c = \alpha_m$ and is independent of composition. When the polymer ceases to be continuous phase because of the increase in the volume fraction of other phases

$$\alpha'_c = \alpha_p \tag{8.30}$$

When there is no interphase interaction and each phase expands unhampered by the other, i.e. the matrix is behaving like a liquid, α of a composite follows the simple law of mixtures [28] given by

$$\alpha_c' = V_m \alpha_m + \alpha_p V_p$$

or

$$\alpha_c' = \alpha_m + V_p(\alpha_p - \alpha_m) \tag{8.31}$$

where V_p and V_m are the volume fractions of the filler and the matrix. Natural rubber filled with sodium chloride follows the above equation.

Kerner [104] made the first sophisticated analysis of thermoelastic properties of composite media using a model which had been considered earlier by van der Poel for calculation of the mechanical properties of composite materials. Here the dispersed phase has been assumed for spherical particles. Kerner's model accounts for both the shear and isotactic stresses developed in the component phases and gives α_c' for the composite:

$$\alpha_c' = \alpha_m V_m + \alpha_p V_m - (\alpha_m - \alpha_p) \cdot \frac{1/K_m' - 1/K_p'}{\dfrac{V_m}{K_p'} + \dfrac{V_m}{K_m'} + \dfrac{3}{4} G_m} \tag{8.32}$$

where K', and G are the bulk and shear moduli, respectively, and the subscripts c, m and p refer to composite, matrix and particle, respectively.

Other proposed models for particulate composites are variants of Kerner's model or have specific temperature ranges or volume fractions of reinforcements over which they can be used.

When the restriction on the shape and size was ignored, several approaches ranging from the wholly empirical to sophisticated analyses based on applied mechanics have been developed. Turner [105] has used an equal strain approach for calculating α_L of mixtures involving the density, modulus of elasticity, α_L and volume fractions of the constituents. According to Turner [11], the α_L value of a particulate composite is given by

$$\alpha_c' = -\frac{\alpha_m V_m K_m' + \alpha_p V_p K_p'}{V_m K_m' + V_p K_p'} \tag{8.33}$$

It is apparent by the inspection that equation (8.33) reduces to a percentage by volume calculation if the phases have the same bulk modulus. If assumptions are valid, the equation is also applicable to anisotropic composites.

Thomas [106] proposed a totally empirical equation based on the concept of the mixture rule. According to Thomas

$$\alpha_c'^a = \alpha_p^a V_p + \alpha_m^a V_m \tag{8.34}$$

where a may vary from -1 to $+1$ depending on the particulate system. At these limits, the equation takes the following forms:

$$a = -1 \quad \frac{1}{\alpha_c'} = \frac{V_m}{\alpha_m} + \frac{V_p}{\alpha_p} \tag{8.35}$$

$$a = +1 \quad \alpha_c' = V_m \alpha_m + V_p \alpha_p \tag{8.36}$$

These equations are analogous to the Reuss and Voigt bounds for bulk moduli and represent such wide extremes that most data can be accommodated within them.

Cribb [107] has adopted an approach in which no limitations are made on the shape and size of the fillers. The phases are supposed to be homogeneous, isotropic and linearly elastic. The simplicity of this approach is attractive but it converts the problem of calculating α_c' to the related question of calculating the bulk modulus of a composite. The Cribb equation is given as

$$\alpha_c' = \theta_1 \alpha_m + \theta_2 \alpha_p \tag{8.37}$$

where

$$\theta_1 = \frac{K_m'(K_c' - K_p')}{K_c(K_m' - K_p')} \tag{8.38}$$

and

$$\theta_2 = \frac{K_p'(K_m' - K_c')}{K_c'(K_m' - K_p')} \tag{8.39}$$

This equation can be used if K_c' is known or if the ability to calculate it from the properties and volume fractions of the individual components is acquired.

Schapery [100] has obtained bonds for the effective α_L for both isotropic and anisotropic composites consisting of isotropic phases by employing extremum principles of thermoplasticity. The exact solutions were derived in particular cases. For a unidirectional composite, the following expressions are obtained. The α ($\alpha_{\|}$) for a fiber-reinforced composite in the longitudinal direction is given by the following equation

$$\alpha_{\|} = \frac{\alpha_p V_p E_p + \alpha_m V_m E_m}{V_p E_p + V_m E_m} \tag{8.40}$$

where E is the Young modulus and subscript $||$, p and m represent the composite in the longitudinal direction, the fiber and matrix, respectively. Schapery also gives the α_L of a fiber-reinforced composite in the transverse direction as

$$\alpha_{\perp} = (1 + \nu_m)\alpha_m V_m + (1 + \nu_p)\alpha_p V_p - \alpha_{||}\nu_c \tag{8.41}$$

and

$$\nu_c = \nu_p V_p + \nu_m V_m \tag{8.42}$$

where ν is the Poisson ratio.

Other models have also been proposed in addition to Schapery's model. Chamis [108] has given an expression for $\alpha_{||}$ in the longitudinal direction identical to that of Schapery. The $\alpha_{\perp}$ in the transverse direction due to Chamis [108] is

$$\alpha_{\perp} = \alpha_p (V_p)^{0.5} + [1 - (V_p)]^{0.5}\left[1 + V_p \nu_m \frac{E_p}{E_c}\right]\alpha_m \tag{8.43}$$

Rosen and Hashin [109] extended the work of Levin [110] to derive expressions for the α_L of unidirectional fiber reinforced composites as

$$\alpha_{||} = \bar{\alpha} + \left[(\alpha_p - \alpha_m)/\left(-\frac{1}{K'_p} - \frac{1}{K'_m}\right)\right]\left\{[3(1 - 2\nu_c)/E_c] - \left(\frac{1}{K'_c}\right)]\right\} \tag{8.44}$$

$$\alpha_{\perp} = \bar{\alpha} + \left[(\alpha_p - \alpha_m)/\left(\frac{1}{K'_p} - \frac{1}{K'_m}\right)\right]\left\{[3/2K'_c] - (1 - 2\nu_c) - \frac{\nu_c}{E_c}\right] - [1/K'_c]\right\} \tag{8.45}$$

where $\bar{\alpha}$ is the α_L calculated from the rule of mixtures and K' is the bulk modulus.

The predictions given by Van Fo Fy [111] are expressed as follows:

$$\alpha_{||} = \alpha_m - (\alpha_m - \alpha_p) \cdot \frac{(1 + \nu_m)E_p \nu_p - (1 + \nu_p)(E_L - E_m \nu_m)}{(\nu_m - \nu_p)E_L} \tag{8.46}$$

and

$$\alpha_{\perp} = \alpha_m + (\alpha_m - \alpha_L)\alpha_{LT} - (\alpha_m - \alpha_p)\cdot(1 + \nu_p)\frac{\nu_m - \nu_{LT}}{\nu_m - \nu_p} \tag{8.47}$$

where E_L and ν_{LT} are the elastic modulus and Poisson ratio for the direction of fibers. However the values predicted by these expressions are very sensitive to variations in E_L and deviations well within the experimental error may cause considerable discrepancies. Several investigators [112, 113] have used finite element analyses to obtain $\alpha_\perp$ in unidirectional composites.

Chamberlain [114] derived an alternate model for transverse expansion of unidirectional composites using thick-walled cylinder equations for the case of a fiber embedded in a cylindrical matrix section. The radial displacements on the outside of the cylinder were related to α_L in the radial direction. The expression for $\alpha_\perp$ is given by the following equation:

$$\alpha_\perp = \frac{2(\alpha_p - \alpha_m) V_p}{\nu_m(F - 1 + \nu_m) + (F + \nu_p)} + E_m(1 - \nu_{LT})(F - 1 + \nu_m)/E_p \tag{8.48}$$

where F is a packing factor equal to 0.9096 for hexagonal packaging and 0.7854 for square packing. Schneider [115] considered a hexagonal arrangement of cylindrical fiber–matrix elements consisting of a fiber surrounded by a cylindrical matrix jacket. The express for $\alpha_\parallel$ is equivalent to Schapery's relation. For the transverse direction, he derived the following equation:

$$\alpha_\perp = \alpha_m - (\alpha_m - \alpha_p) \cdot \left[\frac{2(1 + \nu_m)(\nu_m^2 - 1)C}{(1 + 1.1\nu_p)/(1.1\,V_p - 1) - \nu_m + 2\nu_m^2 C} - \frac{\nu_m(E_p/E_m)}{1/C + E_p/E_m} \right] \tag{8.49}$$

$$\text{where } C = \frac{1.1\nu_p}{1 - 1.1\nu_p}$$

The thermal expansion behavior of unidirectional composites has been studied also by using the concept of the boundary interphase [102]. Interphase material was assumed to be inhomogeneous with properties varying from the fiber surface to the matrix. Further, assuming perfect bonding at all interfaces, fiber and matrix capable of carrying tensile stresses, the interface only carrying shear stresses, and homogeneity and isotropy of matrix and fiber phases, Sideridis [102] used linear, hyperbolic and parabolic variations to derive expressions for α_L. A thermal analysis method due to Lipatov [116] for particulate composites was adopted for cylindrical inclusions in order to find the extent of the interphase layer. Results for E-glass fiber–epoxy resin composites derived from three different laws differ from the respective values derived by Schapery and Van Fo Fy when the fiber volume fraction increases, which leads to an increment in the volume fraction of the interphase.

Holliday and Robinson [28] compiled the α values of a large number of composites and observed the following. It is possible to reduce markedly the thermal expansion of plastics by use of the correct filler.

The greater effect is caused by glass fiber and fabrics while powdered fillers are less effective. The behavior of PTFE is highly anomalous. In several cases there is reasonable agreement between the theoretical predictions and experimental data. However, general agreement can be reached when due consideration of particle size and shape, interfacial area and the polarity of polymers and fillers are taken into account.

A large number of studies have been made on α of polymer matrix composites of epoxy, polyimide and other polymers reinforced with carbon and Kevlar fibers [117–123]. The carbon and graphite fibers are characterized by slightly negative α_L in their axial directions and very large positive $\alpha_\perp$ in their radial directions [122–124]. The unidirectional and bidirectional composites based on them are found to demonstrate highly anisotropic thermal expansion behavior. In unidirectional composites, α_L increased from a value of $-2.2 \times 10^{-6}\,K^{-1}$ along the fiber axis to a value of $79.3 \times 10^{-6}\,K^{-1}$ perpendicular to the fibers. The thermal expansion behavior of Kevlar–epoxy composites could be predicted from the thermoelastic properties of fiber and resin. In the case of carbon–epoxy unidirectional composites, the α_θ in a direction making an angle with the fiber direction has the following equation [102, 117]:

$$\alpha_\theta = \alpha_\parallel \cos^2\theta + \alpha_\perp \sin^2\theta$$

In recent studies of metal matrix composites by Vaidya and Chawla [103], Kerner's model appears to predict the thermal expansion of the particulate composites quite reasonably. Turner's model does not do so well. Kerner's model is close to a rule of mixtures when the constraint term is small. In the case of fiber reinforced metal matrix composites, Schapery's model comes close to the experimental results. The observations of Sideridis [102] are similar in the case of fiber reinforced epoxy matrix reinforced with long E-glass fibers. However, his model gives good agreement between theoretical values ($\alpha_\parallel = 6.6 \times 10^{-6}\,K^{-1}$) and experimental values ($\alpha_\parallel = 6.5 \times 10^{-6}\,K^{-1}$) for $V_f = 0.647$. The respective values in the transverse direction were 25.33×10^{-6} and $25.8 \times 10^{-6}\,K^{-1}$. In the case of α_θ, there is good agreement between theory and experiment. His theoretical findings also agree with the experimental results of Clements and Moore [125].

The thermal expansion in PLC and thermoplastics blends has not been extensively investigated, and has been reported only in few PLC blend systems.

Dutta *et al.* [16] reported the thermal expansion in prepreg film and crossply laminates of prepreg films. The films were prepared from a blend containing 10 wt% of thermotropic LC copolyester (Vectra A950) and 90 wt% polycarbonate (General Electric Corporation bisphenol-A polycarbonate, PC2). The highest order parameter of the PLC phase L57 ($s = 0.45$) was obtained at a draw ratio of the film $(\lambda)_{film} = 7.6$.

The angular dependence of α for the prepreg film $(\lambda)_{film} = 7.6$ and a 0/90° crossply laminate were reported. For prepreg, the angle 0° was the draw direction and 90° was the transverse direction. The results are shown in Figure 8.6. The curve is the lamination theory prediction using the method of Halpin and Pagano [126].

$$\alpha_{laminate} = \frac{A_{22}R_1 + A_{12}R_2}{A_{11}A_{22} + A_{12}^2} \tag{8.50}$$

where

$$A_{ij} = \sum_{h=1}^{n} \overline{Q}_{ij}(\theta)(h_k - h_{k-1}) \tag{8.51}$$

$$R_1 = J_1 H_2 + J_2 H_1 \tag{8.52}$$

$$R_2 = J_1 H_2 - J_2 H_1 \tag{8.53}$$

where n is the number of layers and h_k the thickness of the kth layer. The components of the lamina stiffness matrix $\overline{Q}_{ij}$ are functions of the angular orientation and the mechanical properties of the prepreg (Appendix A of reference 16). J_1 and J_2 are invariants dependent on material functions, and H_1 and H_2 are summation terms involving individual lamina thickness and fiber orientation angles with respect to

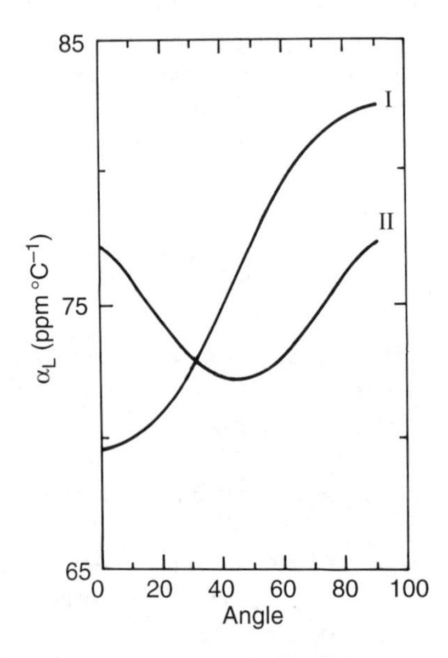

Figure 8.6 Thermal expansivity vs. angle for film prepreg, $(DR)_{film} = 7.6$ (I) and crossply laminates of 10% A950/90% PC2 (II) according to prediction of laminate theory. (Adapted from [16].)

the laminate axis (Appendix B of reference 16). There is good agreement between experimental data and theoretical predictions except for crossply laminates. The discrepancy in the latter case may be due to the errors in the alignment of these piles during the lamination process as pointed out by the investigators [16]. There is anisotropy of *c.* 20% in the expansivity of the drawn film.

Engberg and coworkers [14, 98] investigated the blends of thermotropic PLCs with fully amorphous polymers: polyethersulfone (PES), polycarbonate (PC) and aromatic poly(ester carbonate) (APEC) [98] and semicrystalline polymers like polybutylene terephthalate (PBT) [14]. α_L was calculated according to the Takayanaga model [101] as follows:

$$\alpha_{\parallel} = \alpha_{1,\parallel} + q(\alpha_2 - \alpha_{1,\parallel}) \tag{8.54}$$

$$\alpha_{\perp} = \alpha_{1,\perp} + \frac{1}{p+1}(\alpha_2 - \alpha_{1,\perp}) \tag{8.55}$$

where

$$q = \frac{1}{\left[\dfrac{1}{1-V_1} + \dfrac{bV_1}{1-bV_1}\left(\dfrac{E_{1,\parallel}}{E_2} - 1\right)\right]} \tag{8.56}$$

$$p = \frac{V_1(1-l)E_{1,\perp}}{(1-V_1)E_2} \tag{8.57}$$

where E_2 and α_2 are the modulus and the α_L of the thermoplastics, $E_{1,\parallel}$ and $E_{\perp}$, are the moduli of the PLC parallel to and perpendicular to the melt flow direction, and $\alpha_{1,\parallel}$ and $\alpha_{1,\perp}$ are the α_L of the PLC parallel to and perpendicular to the melt flow direction. For l please refer to the Takayanaga model [101].

PLC and PES were found to be incompatible and formed a self-reinforced system. The chain orientation of the PLC component was constant in the melt flow direction (Hermans orientation function, defined by equation 8.6, $s = 0.39 \pm 0.03$) in these blends. The α_L parallel to the melt flow direction was always lower than that in the lateral direction. Values of $\alpha_{\parallel}$ of PES + PLC blends decreased markedly in an almost linear manner with increasing PLC content. There is moderate increase in $\alpha_{\perp}$. A comparison with the Takayanaga model [101] gives the best fit value for $b = 0.50$. A slight deviation between predicted and experimental values was obtained for the blends with 60 to 80% PLC, showing a higher b value due to the greater reinforcement.

The transesterification reactions were observed for all the other blends [14, 98]. The transesterification reactions led to randomization of the constituents of the PC + PLC and APEC + PLC blends. The effect was more pronounced in PC + PLC blends and $\alpha_{\parallel}$ and $\alpha_{\perp}$ parallel and perpendicular to the melt flow direction were essentially similar to those of pure PC. There was only a slight decrease in $\alpha_{\parallel}$ and increase in $\alpha_{\perp}$ with increase in concentration of PLC. In APEC + PLC blends, the α_{L} was only moderately lower along the flow direction ($\alpha_{\parallel}$) in comparison with $\alpha_{\perp}$ in the blends having PLC content between 0 and 27%. The thermal expansion behavior was anisotropic in blends containing more than 30% PLC with a large thermal expansion in the transverse direction and a much lower value parallel to the major flow direction. $\alpha_{\parallel}$ was very low for the APEC + PLC blend with 67% PLC corresponding to large values of b. Since the composite equation requires the properties of the constituents to be invariant and not to depend on the composition of blends, the thermal expansion for APEC + PLC blends in the whole composition range could not be adopted to a single composite equation.

The differential scanning calorimetry (DSC) and optimal microscopic studies showed that the crystallization of PBT was unaffected by presence of PLC [14]. The PBT component was unoriented in the injection molded samples. The blends with less than 28 vol% PLC exhibited the same α_{L} as PBT. In this composition range, the α_{L} parallel to the melt flow direction ($\alpha_{\parallel}$) was actually greater than $\alpha_{\perp}$, which is due to mechanical anisotropy in the samples. In higher PLC containing blends, $\alpha_{\parallel}$ decreased significantly while $\alpha_{\perp}$ was fairly constant with an increase in PLC content. The blends containing more than 48% PLC exhibited approximately the same $\alpha_{\parallel}$ as the pure PLC yielding the best fit value of $b = 1$ on application of the Takayanaga model. There is a similar thermal expansion behavior in APEC + PLC and PBT + PLC blends. The APEC + PLC blends in which the PLC formed the continuous phase also exhibited much smaller $\alpha_{\parallel}$ values. Thus thermal expansion data of PBT + PLC blends indicate limited transesterification reactions occurring at phase boundaries leading to improved interfacial bonding.

Thermal expansion was reported for the PLC copolyester Vectra B950 and polycarbonate blends [99]. The composite rods were prepared by feeding a mixture of PLC and PC pellets through an extruder and drawing the resulting extrudate. $\alpha_{\parallel}$ of PC does not exhibit any change on drawing. For PC/B950 blends, $\alpha_{\parallel}$ drops sharply with increasing λ. In the case of blends with high PLC content ($V_{f} = 0.60$ for PLC), $\alpha_{\parallel}$ is negative even at a low draw ratio of 1.3; Vectra B950 itself has negative expansivity at a draw ratio of < 1.3. It has been found [97] that at $\lambda < 2$ the highly oriented thin skin layer of B950 has a negative expansivity and high stiffness, thus constraining the expansion of the core region of the extruded rod. According to Choy *et al.* [99] in the blends having 60 and 80 vol% PLC, the PLC phase is continuous and negative expansivity of the blends is due to skin–core morphology.

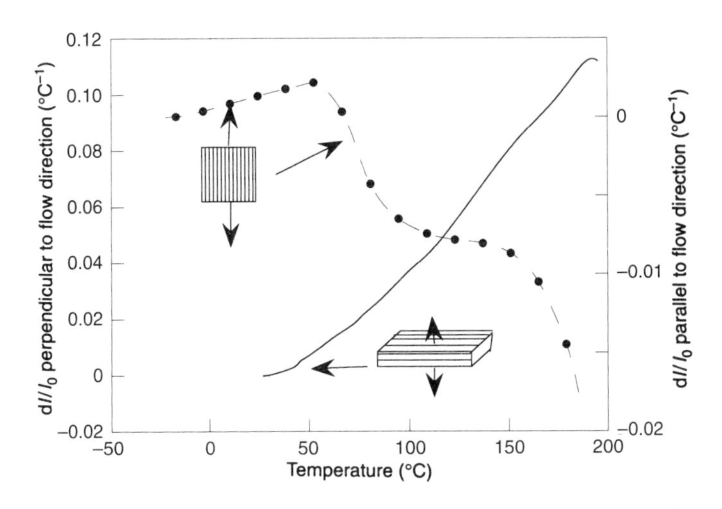

Figure 8.7 Variation of linear expansivity in directions parallel and perpendicular to the flow vs. temperature for PET/0.6PHB; after [127].

For extruder film of PC + Vectra A950 with 10% Vectra A950 Dutta *et al.* [16] obtained $\alpha_{\parallel} = 7.0 \times 10^{-5}\,K^{-1}$ and $\alpha_{\perp} = 8.2 \times 10^{-5}\,K^{-1}$. Values of $\alpha_{\parallel}$ and $\alpha_{\perp}$ reported by Engberg *et al.* [14, 98] are of the same order in the case of PC + A950 blends of similar composition. However, the $\alpha_{\parallel}$ and $\alpha_{\perp}$ values reported by Choy *et al.* [99] are lower than those obtained by Dutta *et al.* [16] and Engberg *et al.* [14, 98] and may be due to molecular orientation induced more easily in Vectra B950 than in Vectra A950.

Brostow *et al.* (127) reported the linear expansivity along the flow and perpendicular to the flow direction in PET/0.6PHB over a temperature range of −44 to 180°C under varying stress levels up to 56 KPa in 8 KPa step increments. There was no perceptible change in α_L; however, the stress levels affect $\alpha_{\parallel}$. As is evident from Figure 8.7, $\alpha_{\perp}$ goes on increasing with temperature. There is a drastic fall in $\alpha_{\parallel}$ at the T_g values of PET and PHB as observed in other cases due to the rubber-elastic effect (48).

8.7 CONCLUSIONS AND PREDICTIONS FOR THE FUTURE

The negative expansivity is an universal phenomenon in polymer crystals, semicrystalline polymers, oriented or unoriented polymers, PLC and PLC + thermoplastic blends. The rigid rod PLC molecule is highly anisotropic. The axial expansivity of a PLC is −5 to $-10^{-6}\,K^{-1}$ while the radial expansivity is highly positive ($\sim 70 \times 10^{-6}\,K^{-1}$). PLCs have low dielectric constants (<3) and low water absorption ($<0.1\%$). The axial expansivities of PLCs are lower than conventional reinforced or unreinforced polymers and are comparable to those for metals (Figure 8.20 of reference [128]). They have good chemical resistance and high mechanical strength and modulus. They can also be processed by conventional processing techniques such as extrusion, injection molding, etc. Biaxial films can be drawn from PLC + engineering thermoplastic blends having α_L less than $1 \times 10^{-6} K^{-1}$ with superior physical properties for electronic packaging [129]. The PLCs are suitable for thin walled electronic connectors subjected to surface mount technology (SMT) soldering. The circuit board fabricators can improve the dimensional stability of components [130].

PLCs can be crosslinked, blended with interacting or non-interacting engineering plastics and can also be reinforced to tailor-make materials with zero or any matching low linear expansivity. The necessary phenomenological and fundamental theoretical framework is emerging for the above. The relevance of the existing framework for oriented or unoriented polymers and composites will buttress such development, as is evident from the preceding discussion in this chapter. More and more experimental data and relevant theory will be forthcoming in the coming years. The applications of PLCs and their blends with engineering and speciality thermoplastics will unfold in the near future in various sectors such as the aerospace, aircraft, automobile, chemical, electrical and electronics industries [129–133].

REFERENCES

1. Kwolek, S.L. (1971) US Patent 3600350; (1972) US Patent 3671542.
2. Priestley, E.B., Wojtowicz, P.J. and Sheng, P. (1972) *Introduction to Liquid Crystals,* Plenum Press, New York.
3. Blumstein, A. (ed.) (1978) *Liquid Crystalline Order in Polymer,* Academic Press, New York.
4. Meyer, R.B. (1982) *Polymer Liquid Crystals,* Academic Press, New York.
5. Brostow, W. (1988) *Kunststoffe,* **78**(5), 411.
6. Davies, G.R. and Ward, I.M. (1988) in *High Modulus Polymers* (eds Zachariades, A. and Porter, R.S.), Marcel Dekker, New York.
7. Zachariades, A. and Porter, R.S. (eds) (1983) *The Strength and Stiffness of Polymers,* Marcel Dekker, New York.
8. Zacchariades, A. and Porter, R.S. (eds) (1988) *High Modulus Polymers,* Marcel Dekker, New York.
9. Brostow, W. (1990) *Polymer,* **31**, 979.
10. Isayev, A.I. and Limtasari, T. (1992) in *International Encyclopedia of Composites* (ed. S.M. Lee), Vol. 3, VCH Publishers, New York, p. 55.
11. Kiss, G. (1987) *Polymer Eng. & Sci.,* **27**, 410.
12. Crevecoeur, G. and Groeninckx, G. (1990) *Polymer Eng. & Sci.,* **30**, 532.
13. Isayev, A.I. and Subramanian, P.R. (1992) *Polymer Eng. & Sci.,* **32**, 85.

14. Engberg, K., Ekbald, M., Werner, P.E. and Gedde, U.W. (1994) *Polymer Eng. & Sci.*, **34**, 1346.
15. La Mantia, F.P. (ed.) (1992) *Thermotropic Liquid Crystal Polymer Blends*, Technomic, Lancaster, PA.
16. Dutta, D., Weiss, R.A. and Kristal, K. (1992) *Polymer Compos.*, **13**, 394.
17. Minkova, L.T., Petris, S.De., Paci, M. *et al.* (1993) in *Processing and Properties of Liquid Crystalline Polymers and LCP Based Blends* (eds D. Acierno and F.P. La Mantia), Chem. Tech. Publ., Toronto; Brostow, W., Sterzynski, T. and Triouleyre, S. (1996) *Polymer*, **37**, 1561.
18. Buijs, J.A.H.M. and Damman, S.B. (1994) *J. Polymer Sci. Phys.*, **32**, 851.
19. Green, D.I., Orchard, G.A.J., Davies, G.R. and Ward, I.M. (1990) *J. Polymer Sci. Phys.*, **28**, 2225.
20. Takeuchi, Y., Yamamoto, F. and Shuto, Y. (1986) *Macromolecules*, **19**, 2059.
21. Ii, T., Tashiro, K., Kabayashi, M. and Tadokoro, H. (1986) *Macromolecules*, **19**, 1809.
22. McGlashan, M.L. (1979) *Chemical Thermodynamics*, Academic Press, London–New York, p. 89.
23. Barron, T.H.K., Collins, J.G. and White, G.K. (1980) *Adv. Phys.*, **29**, 609.
24. Mie, G. (1903) *Ann. Phys.*, **11**, 473.
25. Grüneisen, E. (1912) *Ann. Phys.*, **39**, 257.
26. Grüneisen, E. and Groens, E. (1924) *Z. Phys.*, **29**, 541.
27. Grüneisen, E. (1926) *Handb. Phys.*, **10**, 1.
28. Holliday, L. and Robinson, J.D. (1977) in *Polymer Engineering Composites* (ed. M.O.W. Richardson), Applied Science Publ., London, p. 263.
29. Perepechko, I. (1980) *Low Temperature Properties of Polymers*, Mir Publ., Moscow, p. 82.
30. Choy, C.L., Chem, F.C. and Ong, E.L. (1979) *Polymer*, **20**, 1191.
31. Amatuni, A.M. (1972) *Methods and Instruments for Determination of Thermal Coefficient of Linear Expansion of Materials*, Izd. Standartov, Moscow (in Russian).
32. Choy, C.L. (1982) in *Development in Oriented Polymer–1* (ed. I.M. Ward) Applied Science Publ., London, Ch. 4, p. 121.
33. Bridgman, P.W. (1925) *Proc. Am. Acad. Arts Sci.*, **60**, 305.
34. Straumanis, M.Z. (1940) *Kristallogr.*, **102**, 432.
35. Cole, E.A. and Holmes, D.R. (1960) *J. Polymer Sci.*, **46**, 245.
36. Swan, P.J. (1962) *J. Polym. Sci.*, **56**, 403.
37. Kobayashi, Y. and Keller, A. (1970) *Polymer*, **11**, 114.
38. Davis, G.T., Ebby, R.K. and Colson, J.P. (1970) *J. Appl. Phys.*, **37**, 4316.
39. Wakelin, J.H., Sutherland, A. and Beck, L.R. (1960) *J. Polymer Sci.*, **139**, 278.
40. Miyasaka, K., Isomoto, T., Koganewa, H. *et al.* (1980) *J. Polymer Sci. Phys.*, **18**, 1047.
41. Nakamae, K., Nishino, T., Hata, K. and Matsumoto, T. (1983) *Polymer Prepr. Japan.*, **32**, 2421; (1984) *Meeting of the Society of Polym. Sci. Japan*, Kobe, p. 39.
42. White, G.K., Smith, T.F. and Birch, J.A. (1976) *J. Chem. Phys.*, **65**, 554.
43. Baughman, R.H. (1973) *J. Chem. Phys.*, **65**, 2973.
44. Chen, F.C., Choy, C.L., Wong, S.P. and Young, K. (1981) *J. Polymer Sci.: Phys.*, **19**, 971.
45. Lacks, D.J. and Rutledge, G.C. (1994) *J. Phys. Chem.*, **98**, 1222.
46. Lacks. D.J. and Rutledge, G.C. (1994) *Chem. Eng. & Sci.*, **49**, 2881.
47. Choy, C.L., Chen, F.C. and Ong, E.L. (1979) *Polymer*, **20**, 1191.
48. Choy, C.L., Chen, F.C. and Young, K. (1981) *J. Polymer Sci. Phys.*, **19**, 335.

49. Numata, S., Oohara, S., Fujisaki, K. *et al.* (1986) *J. Appl. Polymer Sci.*, **31**, 101.
50. Numata, S., Fujisaki, K. and Kinjo, N. (1987) *Polymer*, **28**, 2282.
51. Numata, S., Kinjo, N. and Makino, D. (1988) *Polymer Eng. & Sci.*, **28**, 906.
52. Choy, C.L. and Leung, W.P. (1990) *J. Polymer Sci. Phys.*, **28**, 1965.
53. Lifshitz, I.M. (1952) *J. Exp. Theor. Phys.*, **22**, 475.
54. Chen, F.C., Roy, C.L. and Young, K. (1980) *J. Polymer Sci. Phys.*, **18**, 2313.
55. Kan, K.N. (1975) *Theoretical Questions of Thermal Expansion of Polymers*, Izd. LGU Leningrad (in Russian).
56. Barron, T.H.K. (1955) *Phil. Mag.*, **46**, 720.
57. White, G.K. and Choy, C.L. (1984) *J. Polymer Sci. Phys.*, **22**, 835.
58. Engeln, I., Meissner, M. and Pape, H.E. (1985) *Polymer*, **26**, 364.
59. Napolitano, R., Pirozzi, B. and Varriale, V. (1990) *J. Polymer Sci. Phys.*, **28**, 139.
60. Dadobaev, G. and Slutsker, A.I. (1981) *Sov. Phys. Solid State*, **23**, 1131.
61. Tanigami, T., Yamaura, K., Matzuzawa, S. *et al.* (1986) *Polymer*, **27**, 1521.
62. Shirakashi, K., Ishikawa, K. and Miyasaka, K. (1964) *Kobunshi Kagaku*, **21**, 588.
63. Baughman, R.H. and Turi, E.A. (1981) *J. Polymer Sci. Phys.*, **19**, 971.
64. Choy, C.L. and Nakafuku, C. (1988) *J. Polymer Sci. Phys.*, **26**, 921.
65. Zoller, P., Kehl, T.A., Starkwether, Jr, H.W. and Jones, G.A. (1989) *J. Polymer Sci. Phys.*, **27**, 993.
66. Langford, J.I. and Lloyd, J.K. (1989) *Polymer*, **30**, 489.
67. Jou, J.H., Huang, P.-H., Chen, H.C. and Liao, C.-N. (1992) *Polymer*, **33**, 967.
68. Pottiger, M.T., Coburn, J.C. and Edman, J.R. (1994) *J. Polymer Sci. Phys.*, **32**, 825.
69. Brillhart, M.V. and Cebe, P. (1995) *J. Polymer Sci. Phys.*, **33**, 927.
70. Godovsky, Y.K. (1992) *Thermophysical Properties of Polymers*, Springer-Verlag, Berlin–Heidelberg–New York.
71. Hennig, J. (1967) *J. Polymer Sci C.*, **16**, 2751.
72. Kausch-Blecken von Schmeling, H.H. (1970) *Kolloid Z.*, **237**, 251.
73. Ward, I.M. (1979) *Mechanical Properties of Solid Polymers*, Wiley Interscience, New York.
74. Kashiwagi, M., Folkes, M. and Ward, I.M. (1971) *Polymer*, **12**, 697.
75. Kashiwagi, M. and Ward, I.M. (1972) *Polymer*, **13**, 145.
76. Levin, V.M. (1967) *Inzh. Zh. Mekh. Tverd. Tela.*, **2**, 88.
77. Gibson, A.G., Davies, G.R. and Ward, I.M. (1978) *Polymer*, **19**, 683.
78. Takayanagi, M., Imada, K. and Kajiyama, K. (1966) *J. Polymer Sci. C.*, **15**, 263.
79. Capiati, N.J. and Porter, R.S. (1977) *J. Polymer Sci. Phys.*, **15**, 1427.
80. Kim, B.H. and Batist, R.De. (1973) *J. Polymer Sci. Lett.*, **11**, 121.
81. Kozlov, P.V., Kaimin, N.F. and Kargin, V.A. (1966) *Dokl. Akad. Nauk SSSR*, **167**, 1321.
82. Malinskii, Y.M., Guzeev, V.V., Zubov, Y.A. and Kargin, V.A. (1964) *Vysokomol. Soed.*, **6**, 1116.
83. Godovsky, Y.K. (1987) *Progr. Colloid & Polymer Sci.*, **75**, 70.
84. Choy, C.L., Ito, M. and Porter, R.S. (1983) *J. Polymer Sci. Phys.*, **19**, 395.
85. Tashiro, K. (1993) *Prog. Polymer Sci.*, **18**, 377.
86. Buijs, J.A.H.M. and Damman, S.B. (1994) *J. Polymer Sci. Phys.*, **32**, 851.
87. Lacks, D.J. and Rutledge, G.C. (1994) *Macromolecules*, **27**, 7197.
88. Jackson, W.J. and Kuhfuss, H.F. (1976) *J. Polymer Sci. Chem.*, **14**, 2043.
89. Rojstaczer, S., Cohn, D. and Marom, G. (1984) *J. Mater. Sci. Lett.*, **3**, 1028.
90. Strife, J.R. and Prewo, K.M. (1979) *J. Compos. Mater.*, **13**, 264.
91. Ii, T., Tashiro, K., Kobayashi, M. and Tadokoro, H. (1986) *Macromolecules*, **19**, 1772.

92. Ii, T., Tashiro, K., Kobayashi, M. and Tadokoro, H. (1987) *Macromolecules*, **20**, 347.
93. Ii, T., Tashiro, K., Kobayashi, M. and Tadokoro, H. (1987) *Macromolecules*, **20**, 552.
94. Damman, S.B. and Vroege, G.J. (1993) *Polymer*, **34**, 2732.
95. Broer, D.J. and Mol, G.N. (1991) *Polymer Eng. & Sci.*, **31**, 625.
96. Jahromi, S., Kuipers, W.A.G., Norder, B. and Mijs, W.J. (1995) *Macromolecules*, **58**, 2201.
97. Choy, C.L., Wong, Y.U., Lau, K.W.E. *et al.* (1995) *J. Polymer Sci. Phys.*, **33**, 2055.
98. Engberg, K., Stromberg, O., Martinson, J. and Gedde, U.W. (1994) *Polymer Eng. & Sci.*, **34**, 1336.
99. Choy, C.L., Lau, K.W.E., Wong, Y.W. *et al.* (1996) *Polymer Eng. & Sci.*, **36**, 827.
100. Schapery, R.A. (1968) *J. Compos. Mater.*, **2**, 380.
101. Takayanaga, M., Imada, K. and Kajiyama, T. (1966) *J. Polymer Sci., Symp.*, **15**, 263.
102. Sideridis, E. (1964) *Compos. Sci. Technol.*, **51**, 301.
103. Vaidya, R.U. and Chawla, K.K. (1994) *Compos. Sci. Technol.*, **50**, 13.
104. Kerner, E.H. (1956) *Proc. Phys. Soc.*, **69B**, 808.
105. Turner, P.S. (1946) *J. Res. NBS.*, **37**, 239.
106. Thomas, J.P. (1960) AD 287-826, General Dynamics, Fort Worth, TX.
107. Cribb, L. (1968) *Nature*, **220**, 576.
108. Chemis, C.C. (1984) *SAMPE Quart.*, **15**, 14.
109. Rosen, B.W. and Hasin, Z. (1970) *Int. J. Eng. Sci.*, **8**, 157.
110. Levin, V.M. (1968) *Mech. Solids*, **2**, 58.
111. Van Fo Fy, G.A. (1965) *Thermal Strain and Stresses in Glass-Fibre-Reinforced Media, Prikl. Mekh. Teor, Fiz.*, **4**, 118.
112. Adams, D.F. and Craneo, D.A. (1984) *Composites*, **15**, 181.
113. Foye, R.L. (1973) *J. Compos. Mater.*, **7**, 178.
114. Chamberlain, N.J. (1968) BAC Report SON (P) 33.
115. Schneider, W. (1971) *Kunststoffe*, **61**, 33.
116. Lipatov, Y.S. (1977) *Physical Chemistry of Filled Polymers*, Khimiya, Moscow (in Russian). Translated to English by R.J. Mosely, *Int. Polym. Sci. Technol. Monogr.*, **2**.
117. Fahmy, A.H. and Ragai, A.N. (1970) *J. Appl. Phys.*, **41**, 5112.
118. Rogers, K.F., Phillips, L.N., Kingston-Lee, D.N. *et al.* (1977) *J. Mater. Sci.*, **12**, 718.
119. Yate, B., Overy, M.J., Sergent, J.P. *et al.* (1978) *J. Mater. Sci.*, **13**, 433.
120. Ishikawa, T., Koyama, K. and Kohayaski, S. (1978) *J. Compos. Mater.*, **12**, 153.
121. Stufe, J.F. and Pravo, K.M. (1979) *J. Compos. Mater.*, **13**, 264.
122. Tompkins, S.S. (1987) *Int. J. Thermophys.*, **8**(1), 119.
123. Bowles, D.E. and Tompkins, S.S. (1989) *J. Compos. Mater.*, **23**, 270.
124. Vyshvajuk, V.I., Alypov, V.T. and Vishnevskii, Z.N. (1982) *Mekh. Compos. Mater..*, **N6**, 1102.
125. Clements, L.L. and Moore, R.L. (1978) *Composites*, **1**, 73.
126. Halpin, J.C. (1984) *Primer on Composite Materials*, Technomic Publ., Lancaster, PA.
127. Brostow, W., D'Souza, N.A. and Maswood, S. (1996) International Conference on Polymer Characterization, University of North Texas, Denton, Paper P-14a.

128. MacDonald, W.A. (1992) in *Liquid Crystal Polymers: From Structures to Applications* (ed. A.A. Collyer), *Polymer Liquid Crystals Series*, Vol. 1, Elsevier Applied Science, London–New York, Ch. 8.
129. Noll, T.E., Blizard, K., Jayaraj, K. and Rubin, L.S. (1994) NASA Contract Report NASA-CR-198338.
130. Hoechst Celanese Corporation (1996) *Modern Plastics*, May, 104.
131. Chung, T.-S., Calundann, G.W. and East, A. (1989) in *Handbook of Polymer Science and Technology*, Vol. 2 (ed. N.S. Cheremisinoff), Marcel Dekker, New York, p. 625.
132. Jansson, J.-F. (1992) in *Liquid Crystal Polymers: From Structure to Applications* (ed. A.A. Collyer), *Polymer Liquid Crystals Series*, Vol. 1, Elsevier Applied Science, London–New York, Ch. 9.
133. Wong, C.P. (1993) *Polymers for Electronics*, Academic Press, Harcourt Brace Jovanovich Publ., New York.

Part Three
Mechanical properties

9

Phase diagrams of polymer liquid crystals and polymer liquid crystal blends: relation to mechanical properties

Michael Hess and Betty L. López

9.1 INTRODUCTION

The application of unmodified polymers is limited by the glass transition temperature (T_g), the melting temperature (T_m), the thermal stability, the yield stress, its elongation at break or other properties. The processability may be limited by thermal degradation, chemical reactions, crystallization or the viscosity of the material. Consequently, it is common practice to modify engineering plastics (EP) to obtain a material with the desired properties. This can not only be achieved by adding low molar mass modifiers and processing aids but also by blending with other polymers and by providing the system with a filler. Following these lines it is possible, for example, to enhance the modulus of poly(propylene) by filling it with kaolin and to amplify its low temperature impact resistance by adding a rubber to the system. The physicochemical properties of these types of complex heterogeneous systems depend on the composition as well as on the morphology.

Mechanical and Thermophysical Properties of Polymer Liquid Crystals
Edited by W. Brostow
Published in 1998 by Chapman & Hall, London.
ISBN 0 412 60900 2

Therefore, knowledge of the temperature and concentration dependencies of the interactions of the constituents provides important information which can be used to manipulate the properties of the polymer system. The following sections of this chapter will discuss the potentialities of reinforcing engineering plastics with special emphasis on the capabilities provided by blending with polymer liquid crystals (PLCs). A short review will be given on the present state of the art and the analytical tools available; finally, a chosen example will be discussed.

9.2 HETEROGENEOUS VS. MOLECULAR COMPOSITES

An anisotropic reinforcement of a polymer can be achieved by incorporating anisotropic fillers, e.g. platelets, short or long fibers, etc., in the matrix polymer. If the fibers have higher mechanical values for the yield strength or the modulus or other physical quantities, this will be reflected in a certain way by the properties of the resulting reinforced composite. If the filling process does not cause voids, crazes or other failures in the system, certain properties of the composite will be better compared with the reinforced system. Other properties, however, may be negatively affected by the presence of a filler, e.g. the impact resistance if the fibers are brittle. The modulus of a reinforced polymer will be improved anisotropically in the fiber direction. The extent to which this happens strongly depends on the interfacial fiber–matrix interaction, on the volume fraction of the reinforcing species and its mechanical properties, on the degree of fiber orientation and on the quality of dispersion of the filler within the polymer matrix. Agglomerations of filler particles are mechanical weak points in a composite.

The well known Tsai–Halpin equation [1] describes the dependence of the modulus of a composite on that of the basis materials, the volume fraction of the fibers and their aspect ratio. In these composites the fibers carry the load and the matrix distributes it. Fiber reinforced thermosets have a wide range of technical applications and play an important role in self-supporting, low-weight constructions.

$$E = E_m \frac{1 + \zeta \eta \phi_f}{1 + \eta \phi_f} \tag{9.1}$$

with

$$\eta = \frac{\dfrac{E_f}{E_m} - 1}{\dfrac{E_f}{E_m} - \zeta} \tag{9.2}$$

and

$$\zeta = 2\frac{L}{D} \tag{9.3}$$

where E is the elastic modulus of the composite in the fiber direction, E_m is the (isotropic) modulus of the matrix, E_f is the modulus of the reinforcing fibers, ϕ_f is the volume fraction of the fibers, L is the fiber length and D its diameter. When the aspect ratio $\zeta/2$ approaches infinity, equation (9.1) reduces to

$$E = (1 - \phi_f)E_m + \phi_f E_f \tag{9.4}$$

Equation (9.4) is often used as a crude mixing rule; the closer L/D approaches infinity, the better results it provides.

The main drawback of these composites lies in their processability: high quality laminates often have to be formed by hand, short-fiber filled EPs can be sprayed or extruded but there is high abrasion of the processing equipment. Another disadvantage of these composites might occur when the composite has to be recycled again and the filler has to be removed from the polymer.

A large step forward concerning high-modulus polymer materials was made after the first polymers with thermotropic liquid crystalline properties that carried the mesogenic groups exclusively in the main chain has been reported, for example by Roviello and Sirigu [2]. A number of high modulus fibers based on lyotropic aromatic poly(amides) [3–7] were already known. Reasonable progress in developing thermotropic PLC of technological importance started, based on copolymers from *p*-hydroxybenzoic acid (PHB) and poly(ethyleneterephthalate) (PET) [8–10]. These linear polymer liquid crystals belong to the type α (longitudinal) as classified by Brostow [11]. Reviews on these types of PLCs are given by Jin and coworkers [12, 13] and others [14, 15]. The thermal behavior of these types of polymers has been reviewed by Wunderlich and Grebowicz [16]. In the following text the nomenclature PET/0.6PHB (or COP) means a copolyester with 60 mol% PHB, almost randomly distributed in the polymer. Many comparable PLCs with the mesogenic group in the main chain but with an arrangement other than the type α are also known: the pure main chain types (β, γ, ζS, ζR), the mixed types which also contain nonlinear mesogens in the main chain (λ1, λ2, λ3) and double types (ψ1, ψ2) which also have side chain mesogens [11].

This great variety of rigid or semirigid polymers opens a wide field for the creation of new polymers by combining them with EPs. Combining here means producing a blend. Such a blend may be a phase separated system or a miscible system in the thermodynamic sense, such that

$$\Delta G_{mix} < 0 \tag{9.5}$$

$$\left(\frac{\partial^2 \Delta G_{mix}}{\partial \phi_2^2}\right)_{T,p} > 0 \tag{9.6}$$

ΔG_{mix} is the Gibbs free energy of mixing and ϕ is the volume fraction of the component under consideration.

The term 'miscibility' should be preferred rather than the term 'compatibility'. 'Compatibility' is a notion widely used in materials technology and has a more general meaning. It is not unambiguously used and has a diversity of meanings, often simply saying that certain desired properties appear upon combining different materials. The following is restricted to studies of blends that exclusively contain longitudinal PLCs, Brostow's type α.

The studies of Wierschke [17] claim to predict high tensile properties for rod-like molecules from quantum mechanical studies. However, his computations deal with single molecules, while a molecule is always stronger than a microscopic material. Further advantages are to be expected from PLCs since their melt viscosities decrease in the mesophase. In a polymer blend this creates a processing aid, somewhat similar to drag reducing additives [18]. Moreover, the disadvantage of the high prices of PLCs might be compensated thus since a small amount of PLC material is sufficient to achieve the desired properties of the resulting blend. After this quite optimistic view the question arises as to which effects govern the properties of the blend, along with the properties of the pure components, and how they can be manipulated.

Different types of systems have been investigated by a number of scientists during the past years [19–49], and the results are sometimes contradictory. Different mechanisms act in a complex manner [30, 41, 47], and the processing is very important [37, 50]. In general it turns out that the formation of fibrillar structures is important for a reinforcing effect of the PLC in a blend with an EP [25], but this is not a necessary precondition to obtain a reinforcing effect [21]. This leads to the fact that a hierarchy of structures and not a simple architecture is formed [51, 52]. Good adhesion between the fibrils and the EP matrix is necessary [23, 28, 29]. The miscibility of the PLC and the EP phase must not decrease significantly the fiber stability. The fibrils should be well dispersed in the matrix and have a high degree of orientation [36, 37, 39]. The reinforcing effect is sometimes describable [49] by equation (9.1) and is comparable to a reinforcement by short fibers. Extensional flow results in a better orientation of the fibers than shear flow [46, 48, 49].

Interfacial properties are crucial for the material properties because the interphases either couple or decouple neighboring phases from one another. Manipulation of interfacial properties therefore can be a tool for materials engineering. It is well known that compatibilizers such as short-chained copolymers can increase the interactions between neighboring phases [33, 36]. However, reactions or processing at elevated temperatures can also lead to better coupling between neighboring phases [47, 53]. A general review of miscibility effects in blends containing copolymers is given by Roe and Rigby [54]. The correlation between structure and adhesion between polymers is reviewed by Wu [55].

9.3 MICROSCOPY

Scanning electron microscopy (SEM) shows the formation of fibrillar structures formed by the PLC phases in injection molded specimens, particularly in regions not far from the wall of the mold. The effect becomes more pronounced if the concentration of PLC in the blend increases. Along the gradient of shear towards the center of the sample the shear stress changes gradually. Kuhfuss and Jackson [10] reported a shear-dependent viscosity of PLC phases: with increasing shear, a decrease of the melt viscosity by several orders of magnitude was observed. Wissbrun [56] has formulated a theory of PLC flow that predicts the shear dependence of the viscosity, as indeed is observed. The result of this behavior is a decrease in the fibrillation tendency of the PLC-rich phase so that the fibrils degenerate to ellipsoidal structures with decreasing aspect ratio far away from the walls. Depending on the deforming forces, spherical phases (islands) have also been observed with SEM [54, 39].

The formation of the islands of any shape in the melt can be explained by differences of viscosity and surface tension of the LC-poor and LC-rich phases under the influence of a mechanical field gradient. The LC-poor matrix in most cases has a higher melt viscosity compared with the PLC [50]. A shear flow applied to the highly viscous matrix will then cause the deformation of the islands which can finally result in the formation of the desired fibrillar structures. At this point it is important for the material's properties to determine the strength of the interfacial contact between matrix and fibrils (see below), because this governs the paths of an applied force through the specimen. The fibrils were already predicted by the qualitative model of Sawyer and Jaffe [51]. By contrast, in the elongational flow near the center of the conduit no such deformation can occur so that the islands can relax and can obtain their equilibrium shape, which is only dictated by the difference of the surface tensions. Moreover, the lack of shear stress in the center results in a relatively homogeneous distribution of the low-viscosity islands in the highly viscous matrix. The structure of the islands has been the subject of X-ray experiments (SAXS and WAXS at the HASYLAB of Deutsches Elektronen Synchrotron, DESY, at Hamburg; section 9.4), and has been discussed in connection with the formation of hierarchical structures [20]. The facts outlined above are an example of the application of rule 5 in reference 52 (see also section 9.5) which deals with the formation of such structures and states that by assembling entities in a specific way one can achieve properties which a system of unassembled entities does not have.

In contrast to specimens that were not prepared by extrusion or injection molding but were heated and pressed without appreciable shear stress, there is no significant contrast of the LC-poor and LC-rich phases in SEM of broken samples. This is true, independent of the composition. The existence of different phases, however, can be shown with polarized light microscopy (see below).

Kuhfuss and Jackson's polyester containing 60 mol% PHB (COP) [8–10] can serve as a PLC model. Pure COP kept under pressure in a mold shows small fibrillar structures at break (Figure 9.1). This diminishes quickly with an increasing amount of PC and is completely lost in that

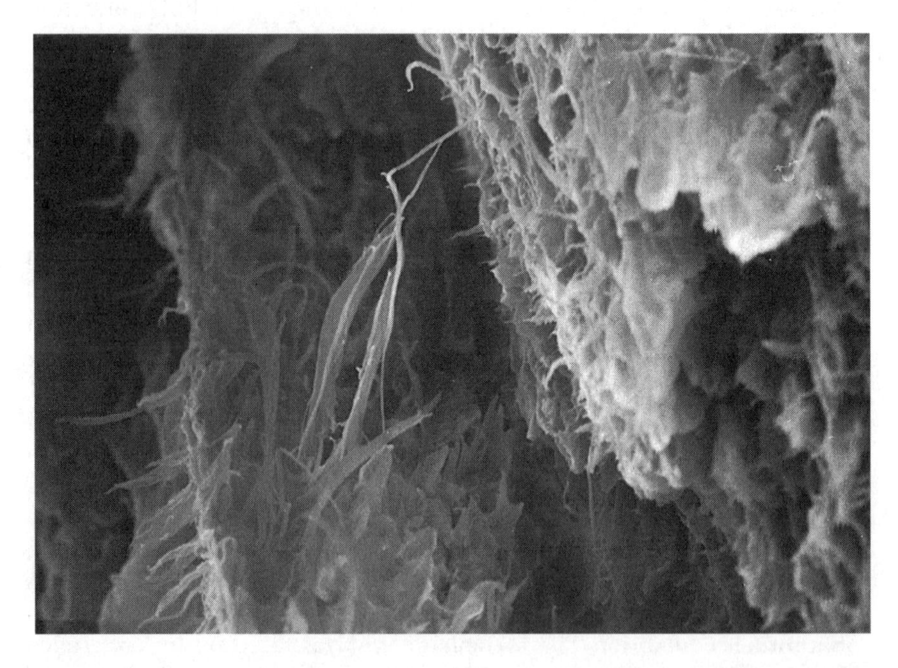

Figure 9.1 Scanning electron micrograph (SEM) of pure PET/0.6 PHB (COP) sputtered with Au. Fracture surface of a sample pressed for 16 h at 210–220°C.

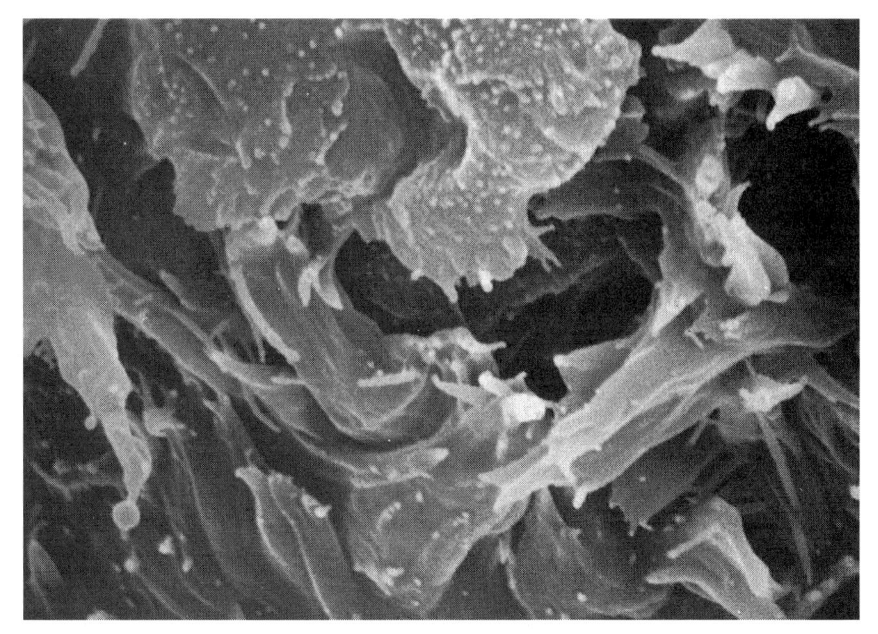

Figure 9.2 SEM of COP + 20% PC sputtered with Au. Fracture surface of a sample pressed for 16 h at 210–220°C.

manner if there is more than 20% PC present in the blend (e.g. Figure 9.2).

Semicrystalline blends – for example COP + 50% PC, 210–220°C/ 4 MPa/10 h – show the typical view of a smoothly cut brittle fracture in the crystalline phases and the deformational fracture of the amorphous parts (Figure 9.2).

Extrusion of the samples changes their morphology as already mentioned above. The COP phase forms fibers which are split up into a large number of thin fibers; these are split again to smaller fibrils. The fibers are embedded in less oriented material (Figures 9.3 and 9.4).

In contrast to the samples that were simply pressed without application of a shear stress, these fibers can be found in blends with up to 90% PC by mass. In blends containing *c.* 20 mass% PC the fibers are covered with a thin layer of PC which mantles the fibrils and glues them to each other (see Figures 9.5 and 9.6). The result is that in a fracture experiment these fibers do not split up into smaller fibers and fibrils but they persist as a complete bundle with smooth fiber ends indicating an elastic deformation at fracture. The PC covered fibers are still visible in a blend containing 33 mass% COP (Figure 9.7). Now the fibers show a brittle fracture while the PC still is ductile. A partial fiber pull-out is observed, indicating poorer adhesion of the phases.

Figure 9.3 SEM of pure (COP) sputtered with Au. Fracture surface of a sample extruded at 245°C.

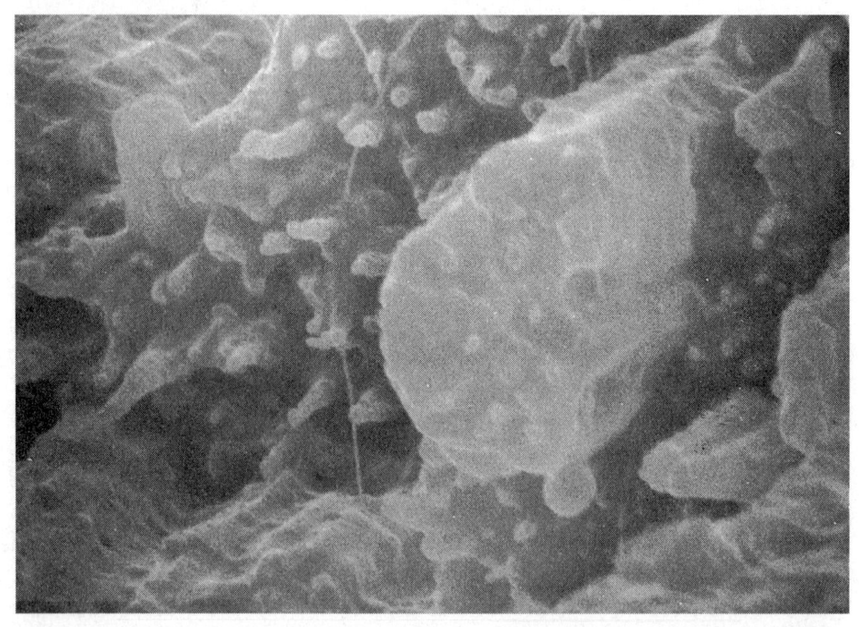

Figure 9.4 SEM of pure (COP) sputtered with Au. Fracture surface of a sample extruded at 245°C. The micrograph shows one of the fiber tips at a higher magnification compared with Figure 9.3.

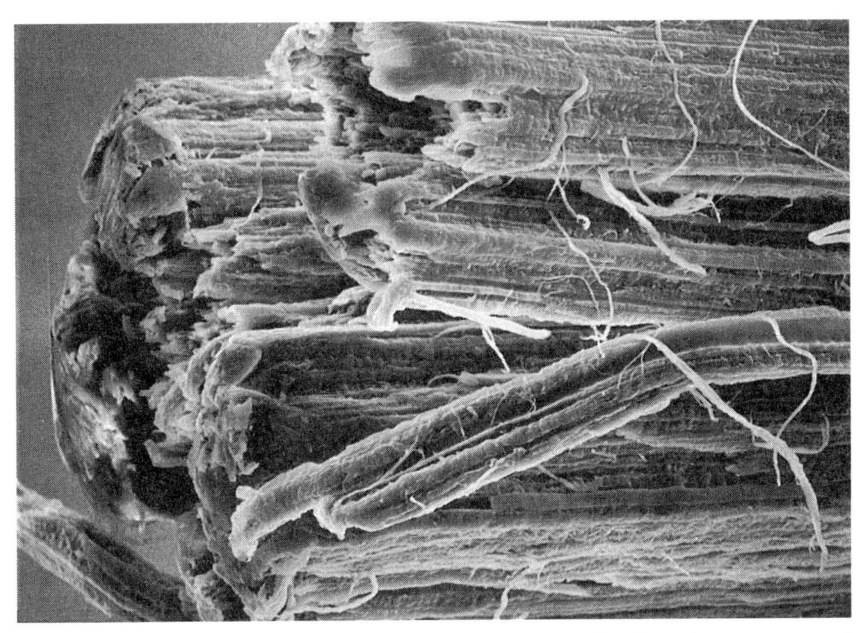

Figure 9.5 SEM micrograph of a fracture surface of a blend consisting of COP + 20% PC; extrusion at 245°C. The samples were sputtered with Au.

The PLC fibers become thinner with increasing amount of PC in the blend. In blends containing for example 83% PC, numerous PLC spheres can be observed together with fibers (Figure 9.8).

The coexistence of fibres and spheres may be caused by the fact that most of the spheres are rather small – less than 2 μm – and hence may not contain enough material to undergo fiber formation. The fibers merge above the fracture plane standing free. In cases where fibers have been pulled out they leave clear-cut holes indicating poor adhesion to the matrix. The PLC spheres are localized in spherical cavities with a larger diameter than the spheres themselves. They seem to stay free in these holes. The reason is that the expansivity of COP is about twice that of PC so that the cavities open during cooling down from the melt.

A direct analysis of the blend morphology is possible with a polarizing microscope. Independent of the method of preparation there are always two phases observable, except in the temperature and composition range where the phases i, j and l exist (see Figure 9.25 in section 9.11, where a complete phase diagram is discussed). In all other cases the components do not or only partially mix. Size and volume of the coexisting phases can be influenced by the composition and the preparation.

Blends prepared with less than 10% COP showed the same

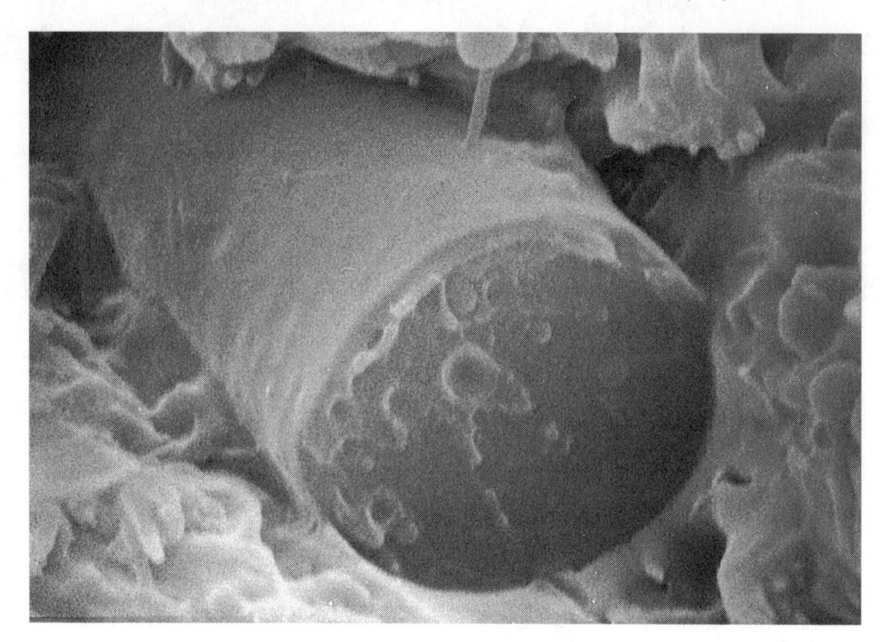

Figure 9.6 SEM micrograph of a fracture surface of a blend consisting of COP + 20% PC; extrusion at 245°C. The samples were sputtered with Au. The micrograph shows one of the fiber tips at a higher magnification compared with Figure 9.5.

appearance whether prepared with pressure or without: small spheres less than 2 μm in diameter dispersed in the matrix quite homogeneously. The size distribution of the spheres, however, is broad (Figure 9.9).

Blends with a composition between 10 mass% and 58 mass% COP can have a variety of different morphologies depending strongly on the preparation: Without pressure there are spheres dispersed in the continuous PC matrix. The spheres grow with increasing amount of COP – 14% COP corresponds to 7 μm diameter and 33% COP corresponds to 20 μm. The size distribution becomes broader in parallel. Beyond 30% COP the PLC phases start to contact and form larger, deformed regions; Figures 9.10 and 9.11 show examples.

The application of pressure to a sample placed between two microscope slides changes the picture drastically: there are no longer distinct PLC regions with a sharp boundary. Both phases seem to interpenetrate each other resulting in a better mechanical mixing of the two phase system (Figures 9.12 and 9.13).

Blends between *c.* 58 and *c.* 67% show a phase inversion without the application of pressure. With the application of pressure there is no phase inversion visible. There is so much COP that PC cannot be recognized as a separate phase, although it is known from other

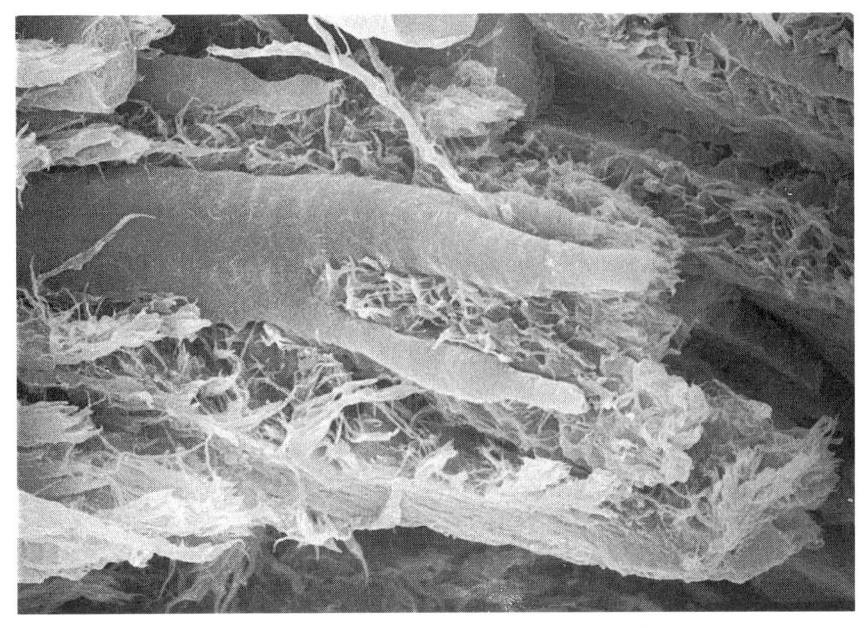

Figure 9.7 SEM micrograph of a fracture surface of COP + 33% PC; extrusion at 245°C. The samples were sputtered with Au.

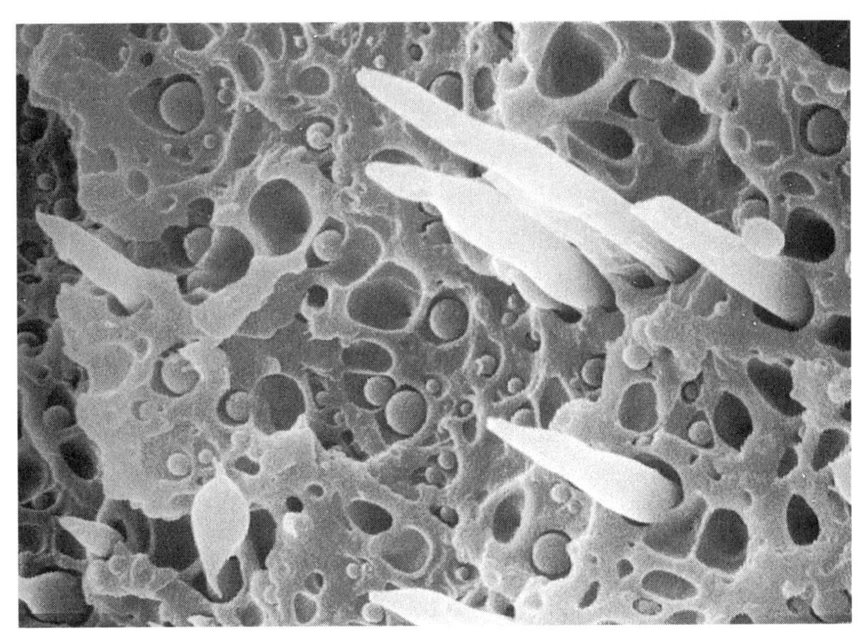

Figure 9.8 SEM micrograph of a fracture surface of COP + 83% PC; extrusion at 245°C. The samples were sputtered with Au.

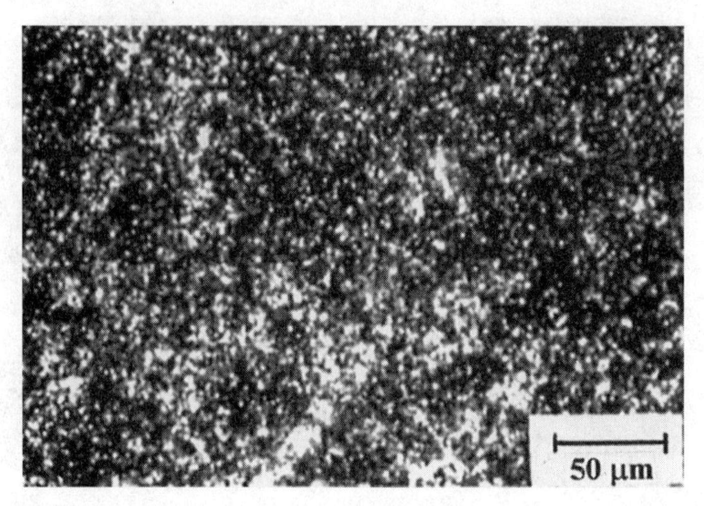

Figure 9.9 Polarization microscopy at room temperature of COP + 91% PC; 5 min at 230–240°C in the molten state, subsequently quenched with ice/water.

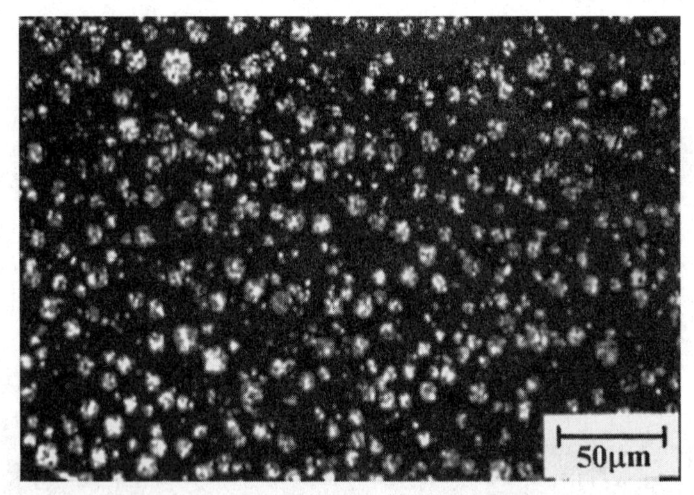

Figure 9.10 Polarization microscopy at room temperature of COP + 86% PC; 5 min at 230–240°C in the molten state, subsequently quenched with ice/water.

experiments, e.g. differential scanning calorimetry (DSC), that there are still two phases.

The morphology of blends with more than *c.* 67 mass% COP is almost indistinguishable from that of pure COP independent of whether pressure was applied or not.

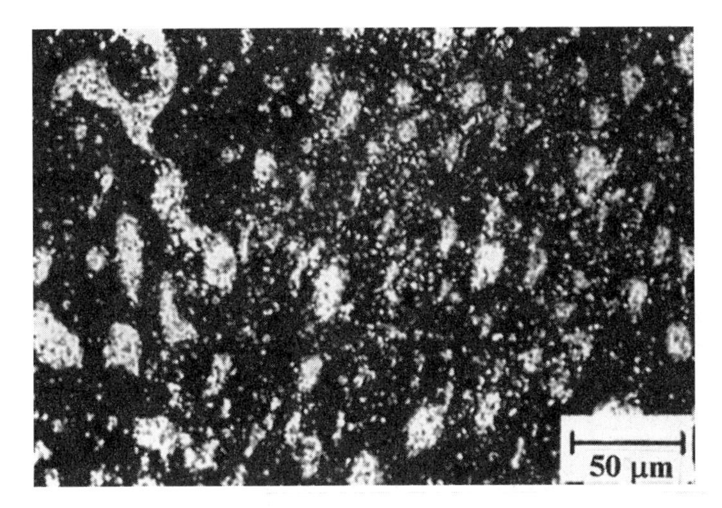

Figure 9.11 Polarization microscopy at room temperature of COP + 67% PC; 5 min at 230–240°C in the melt, subsequently quenched with ice/water.

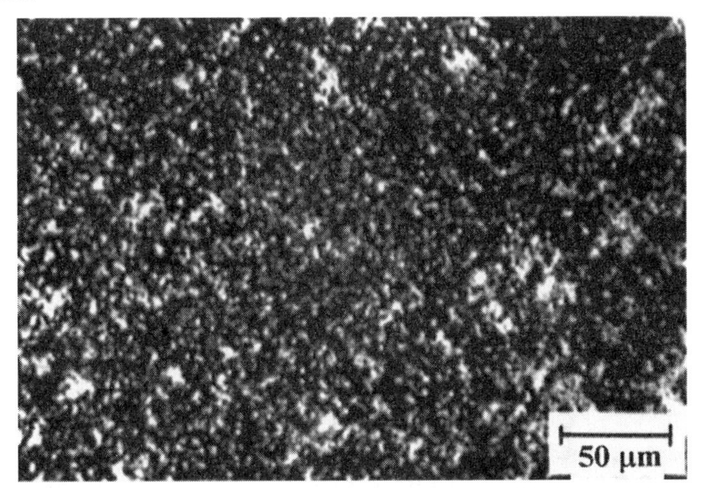

Figure 9.12 Polarization microscopy of COP + 86% PC; 5 min at 210–220°C and 4 MPa, subsequently quenched with ice/water.

9.4 X-RAY DIFFRACTION

Specimens that were simply pressed in a mold and not extruded showed no preferred orientation; there were rings in the X-ray diffraction patterns. Blends between *c.* 40% and *c.* 80% PC containing predominantly PC crystallites which were caused by the presence of the COP phase. They were not found in the pure PC. Good mechanical mixing turned out to be a precondition for effective nucleation. Bad mechanical mixture

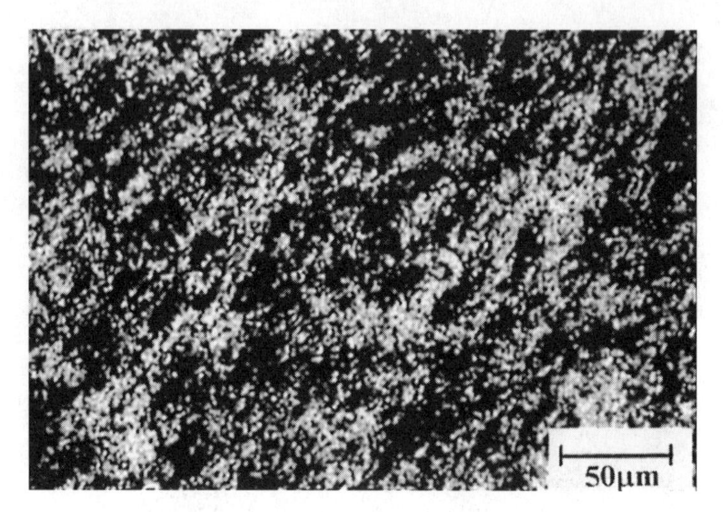

Figure 9.13 Polarization microscopy of COP + 67% PC; 5 min at 210–220°C and 4 MPa, subsequently quenched with ice/water.

may prevent any nucleation for the crystallization of PC by COP. PLC-rich blends showed both PHB and PC crystals after annealing. The presence of crystalline PET was also proved in certain phases (Table 9.2 in section 9.11; Figure 9.14).

Small angle X-ray scattering (SAXS) experiments give information about the radius r_c of a scattering superstructure. The boundary phase distribution function (BPDF), which can be calculated from the scattering curve [57], gives information about the size of the amorphous and crystalline regions.

Blends containing 50% PC (210–220°C/4 MPa/1 hr) were annealed without external pressure for 45 min at 70, 120, 175 and 215°C and subsequently quenched by liquid nitrogen. The results are shown in Table 9.1 (r_c) and Figures 9.15 and 9.16.

The temperatures were chosen so that they are between the transition temperatures. Blends that had been annealed at 70°C and 120°C showed no reflex so that the BPDF could not be calculated. However, the Guinier plot showed two scattering radii: r_{c1} and r_{c2}.

The situation changes after annealing at 175°C. There is a weak signal in the scattering curve; the corresponding BPDF is shown in Figure 9.16. At the same time a third scattering radius (r_{c3}) appears.

The values of d_a and l are obtained directly from the trace of BPDF: the first relative maximum gives d_c and the second one, which only appears as a shoulder in Figure 9.14, gives d_a. The sum of both is the long period l. The differences between the sum of d_c and d_a and the measured value of l (Table 9.1 and Figure 9.16) are within experimental

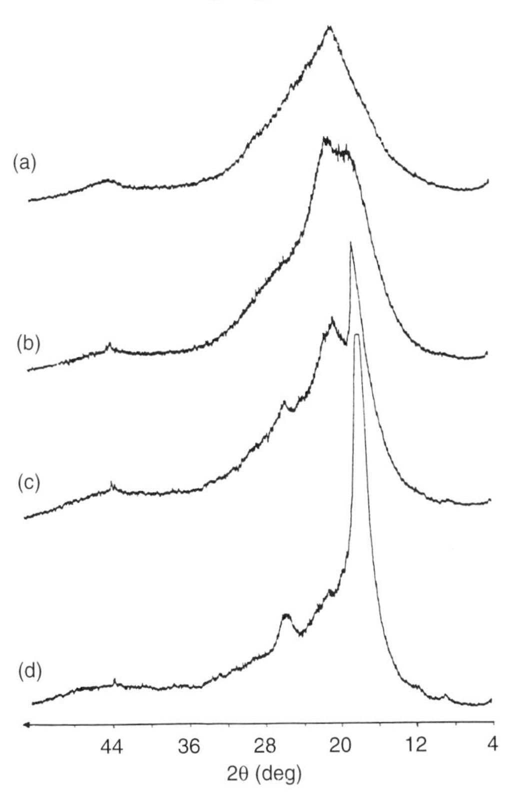

Figure 9.14 WAXS pattern at 25°C, $\lambda = 154.18$ pm (Cu Kα). The figure shows the development of PC crystallites in the presence of COP upon annealing. Pure PC does not crystallize under these conditions in the solid state. (a) Pure COP, 10 min at 210–220°C and 4 MPa; (b) COP + PC 67%, 5 min at 210–220°C and 4 MPa; (c) COP + PC 67%, 3 h at 210–220°C and 4 MPa; (d) pure PC, crystallized from $CHCl_3$ solution.

Table 9.1 r_{c1}–r_{c3} are the radii of the different scattering structures observed; d_c is the length of the crystallites; d_a is a measure of the size of the amorphous regions; l is the long period. Sample: COP + PC (50% mass) (210–220°C/4 MPa/1 h). For the pure COP only one r_c = 13.4 nm was obtained

Time (min)	Temperature (°C)	r_{c1} (nm)	r_{c2} (nm)	r_{c3} (nm)	d_c (nm)	d_a (nm)	l (nm)
45	70	2.9	–	23.9	–	–	–
45	120	2.6	–	20.3	–	–	–
45	175	2.6	4.7	19.1	3.2	7.4	11.2
45	215	3.0	2.3	19.9	3.6	8.2	11.2

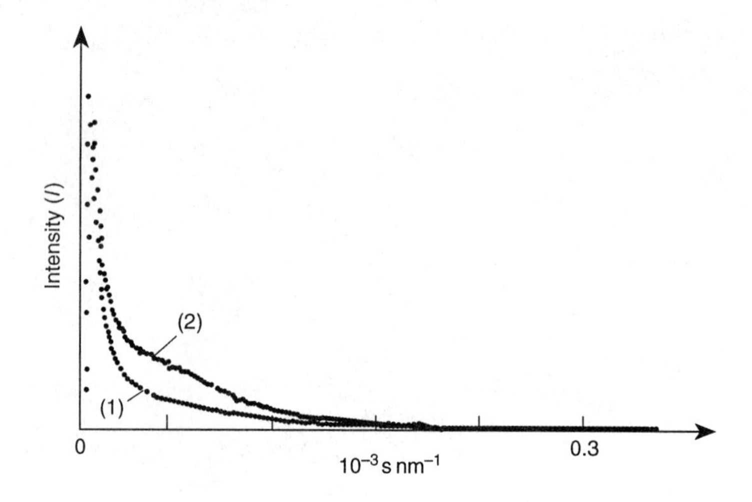

Figure 9.15 SAXS pattern of COP + PC 50% at 25°C, $\lambda = 154.18$ pm (Cu Kα). (1) after 45 min at 75°C, (2) after 45 min at 175°C. *s* is the scattering vector.

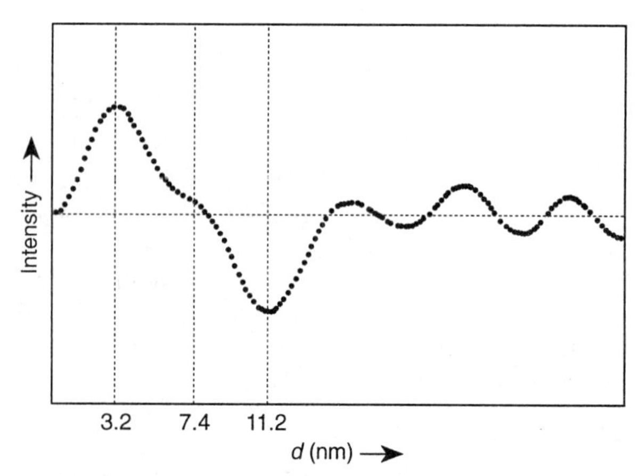

Figure 9.16 Boundary phase distribution function (BPDF) [125]. Small angle X-ray diffraction (SAXS) was recorded with a Kratky camera at the same wavelength as the WAXS experiments. For further explanation see text.

error. At 215°C there are no significant changes in the results, but the reflexes become weaker.

Because of the complexity of the system it was not possible to correlate the different r_c values unequivocally with identified morphological structures. However, it seems to be reasonable to attribute r_{c2} – which appears after annealing above the glass transition of PC – to small PC crystals. The growth of these crystals was proved by isothermal dynamic

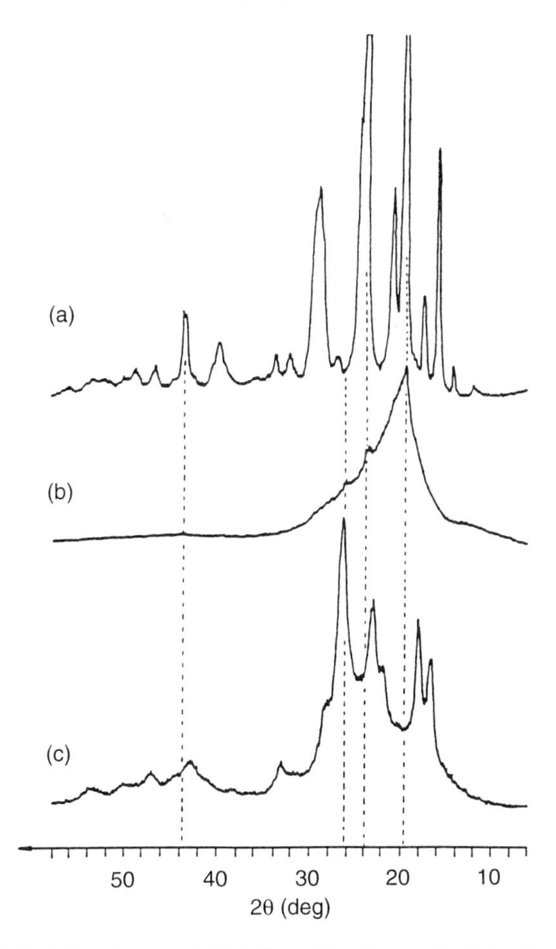

Figure 9.17 WAXS pattern at 25°C, $\lambda = 154.18$ pm (Cu Kα). Correlation of the WAXS peaks with the components in poly(*p*-hydroxybenzoic acid-co-ethylene terephthalate): (a) poly(*p*-hydroxybenzoic acid), a highly crystalline material; (b) COP, 18 h at 210–220°C and 4 MPa; (c) PET, 60 min at 210–220°C and 4 MPa.

mechanical testing and time-dependent scattering experiments with synchrotron radiation. Therefore, the d_c values correspond to the lengths of PC crystals. The origins of r_{c1} and r_{c3}, which do not show a clear tendency with temperature, are still unclear.

Scattering experiments with synchroton radiation at $\lambda = 154$ pm showed only small annealing effects proving the poor crystallizability of the COP; the well known block structures of PET and PHB, however, were clearly identified (Figure 9.17). Pure poly-PHB is highly crystalline and shows a slight increase of the lattice distances with increasing temperature (thermal expansivity) with almost no additional crystallization.

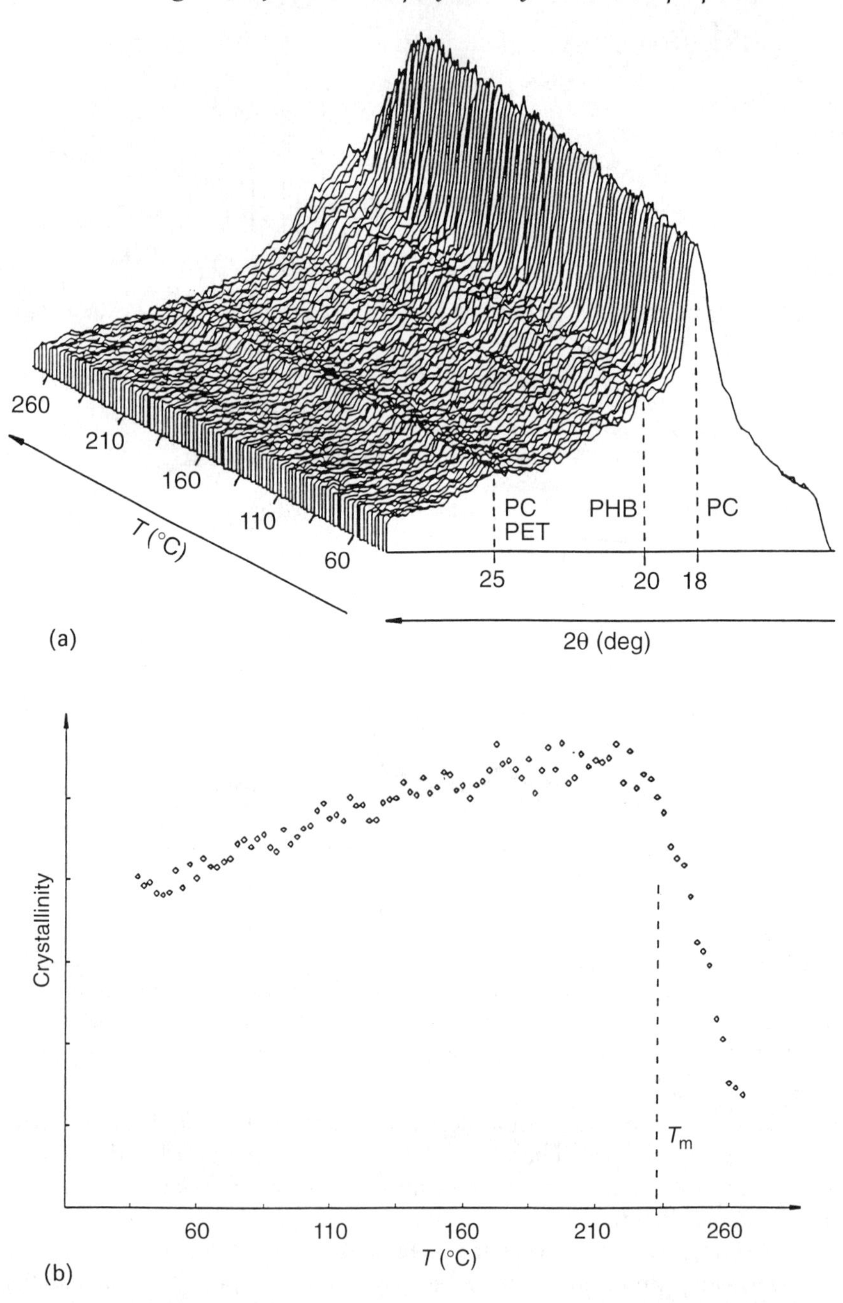

Figure 9.18 (a) Temperature dependent WAXS at a heating rate of 5°C min^{-1}; synchrotron radiation experiment; λ = 154 pm; PC 80 mass% + COP; 16 h at 210–220°C and 4.0 MPa. (b) Temperature dependence of the crystallinity; heating rate 5°C min^{-1}; Synchrotron radiation experiment; λ = 154 pm; PC 80 mass% + COP; 16 h at 210–220°C and 4.0 MPa.

Pure crystalline PC is also an almost temperature independent scatterer. In a blend with COP, however, crystallinity increases upon annealing mainly because the PC crystallizes (Figure 9.18).

9.5 THE CONCEPT OF HIERARCHICAL STRUCTURES

In the commonly used EPs, a multiphase structure is a quite common feature (semicrystallinity), and is a generally observed phenomenon in PLCs. These phases must not be restricted to equilibrium structures only because this would exclude the almost ubiquitious glassy phases. Brostow and coworkers [20] have worked out a model of hierarchical structures according to the basic work of Sawyer and Jaffe [51] (Figure 9.19).

A closer view of these hierarchical structures resulted in the definition of a series of rules governing these important structures [52]. These rules provide a guideline to understanding the complex cooperation of different structures and substructures.

- The complexity of hierarchical structures is related symbatically to the number of types of building entities (building blocks) which are not homeomorphic with respect to one another.
- Each level of a hierarchy is defined by the constituting (non-homeomorphic) types of building blocks, and by relations between the types. In materials the relations include (but are not limited to) connectedness by primary covalent bonds, hydrogen bonds, dispersion interactions and interactions between phases such as adhesion forces.
- Ascension in a hierarchy consists in defining relations such that the entities at the level h are divided into subsets. An entity at the level $h + 1$ corresponds to each subset, while each subset can consist of elements of one or more homeomorphic types. As a corollary, descent involves a relation between each level h entity and a subset of entities at the level $h + 1$.
- The structure of the smaller entity, for example the size or the shape of a smaller molecule, determines the size, shape and structure of a larger entity, for example an LC phase. Since macroscopic properties are determined through ascent within the hierarchy, they depend on entities and their interactions at lower levels.
- By assembling entities in a specific way one can achieve properties which a system of unassembled entities does not have.

A multiphase structure in polymer systems is not unusual. This is only true for polymer blends but also for pure polymers where microphase demixing in copolymers is a well known effect and where glassy regions are observed together with crystalline phases. The glassy

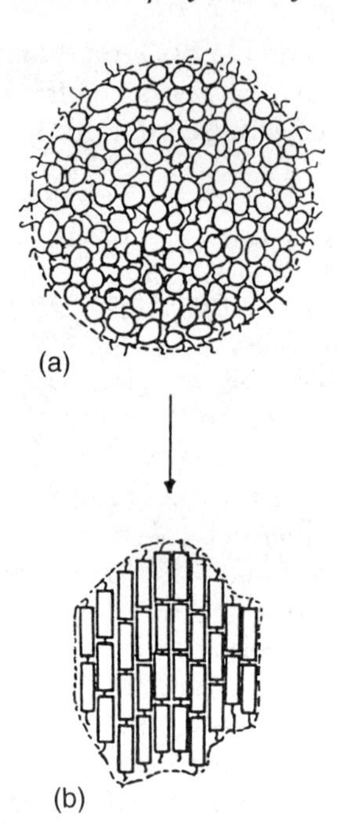

Figure 9.19 Hierarchical model of LC-rich islands. (a) Island, with diameter between 1.0 and 1.4 μm. Small unfilled regions inside with approximately spherical shapes represent LC crystallites. Lines between the crystallites represent some of the flexible sequences in chains. (b) LC crystallite, with a diameter of *c.* 12 nm. (According to [11], with permission of Elsevier Science Publ.)

regions, however, are not stable phases in the thermodynamic sense; they are 'frozen' non-equilibrium states that, at a given temperature below the glass transition temperature T_g, still tend to relax, although this relaxation is almost infinitesimally slow if the sample is more than 20 K below T_g. Consequently, it is obvious that 'intelligent processing' [51] is necessary to obtain maximal success in producing high performance polymer blends. Intelligent processing means that first the phase diagram has to be determined, and the processing conditions are subsequently adapted to the boundary conditions given by the phase diagram. This procedure is valid not only for polymer blends but also for pure PLCs if the concentration θ of liquid crystalline sequences is independently variable in a series of copolymers [58], but it is even more important in blends with EPs.

9.6 REASONS FOR STUDYING PHASE DIAGRAMS

There are different types of diagrams in physical chemistry which are called 'phase diagrams'. The most important are plots of the coexistence pressure p vs. the corresponding temperature T at constant composition x_i where x_i is the molar fraction of one component, and plots of the transition temperature T vs. the corresponding composition x_i at constant pressure p. The first type of diagram is most important for those who deal with the processing of a plastic material in an injection molding machine, an extruder, etc. This topic cannot be covered by this chapter although the processing parameters can influence the properties of the resulting product significantly. This has been stressed in section 9.1. The influence of the processing parameters on the product properties is strongly dependent on the system under consideration, the reactivity of the components, the composition, the equipment used, etc. Discussion of these correlations in more detail is beyond the scope of this chapter, which focuses mainly on the first steps of development, that is the investigation of the $T-x_i$ phase diagram.

The evaluation of the $T-x_i$ phase diagram gives information about the types of phases that can exist and/or coexist with other phases formed by components of the system. The most important phases in polymer systems are formed in the glassy state, the crystalline state, the melt and occasionally different types of liquid crystalline states. In contrast to the usual procedure in thermodynamics, it is important to take non-equilibrium states such as the glassy state into consideration. The glassy state is the most common state in polymers. Although it is a non-equilibrium state, the temperature determines whether or not the kinetics of phase transition can influence the properties of the material during its lifetime. As a rule of thumb it is found that the rate of phase transition is negligible in the majority of cases provided the temperature of use is at least 20 K below the glass transition temperature. However, experimental verification is self-evident for a reliable practicable application.

The investigation of the thermal behavior of a polymer not only shows that there is a phase transition, it may also provide, depending on the analytical method used (section 9.9), information about the type of the transition (melting, crystallization, glass transition, etc.). With additional information about the compositional dependence of these transitions it is possible to determine roughly the application range of the material by taking into consideration only the temperatures where drastic changes in mechanical properties, for example the modulus, are observed. Since the crystallinity of a polymer influences its applicability, for example with respect to its flexibility, it is important to know if cold crystallization is observed or if recrystallization takes place which also can cause changes in mechanical behavior.

The dependence of the transition temperatures on concentration shows, as has already been described, whether there is miscibility in multicomponent systems. Partial miscibility in a glassy or crystalline phase may be useful for compatabilization of two components. Eutectic blends may be desired to obtain a certain morphology of the crystalline regions of a blend.

It is obvious from the examples described above that the knowledge of the $T-x_i$ phase diagram provides the initial information about the principal usability of a certain combination of polymers. Investigation of the time dependence of the phases observed reveals whether there are chemical changes in the system under particular conditions or whether certain physical processes occur. The physical processes have been mentioned above. Chemical processes, for example transesterification, transamidation, formation of allophanate structures, etc., change the components of the blend itself, resulting in different molecular structures with desired or undesired properties. Knowledge of these types of processes together with information on the thermal stability of the components is important for polymer processing, in combination with the $P-T$ phase diagram which is now investigated for those compositions that promise the desired product properties. Knowledge of the phase diagrams and the means to influence them enables the development of polymer materials that are properly adjusted to their application. It is a precondition for the purposeful development of a product that avoids trial and error procedures.

9.7 INFORMATION FROM THERMODYNAMICS

Classical thermodynamics gives the criteria for miscibility with equations (9.5) and (9.6). The most important property that affects miscibility of polymer systems compared with other systems is the large molar mass of the components. The Gibbs free energy of mixing is given by

$$\Delta G_{\text{mix}} = \Delta H_{\text{mix}} - T(\Delta S^{\text{c}}_{\text{mix}} + \Delta S^{\text{e}}_{\text{mix}}) \tag{9.7}$$

where ΔH_{mix} is the enthalpy of mixing, $\Delta S^{\text{e}}_{\text{mix}}$ is the excess entropy of mixing and $\Delta S^{\text{c}}_{\text{mix}}$ is the combined entropy of mixing. It is the sum of the right hand terms in equation (9.7) that governs the miscibility of polymer systems; again the criteria are given by equations (9.5) and (9.6). On the whole, however, there is not always sufficient information about polymer systems to calculate a phase diagram. Moreover, almost every polymer system contains metastable glassy states under certain conditions. This is a non-equilibrium state so that Gibbs thermodynamics do not apply. The glass transition is not a second order process according to Ehrenfest. Kinetic effects have to be taken into account and more than one order parameter is necessary to describe this frozen-in state properly [59]. *Ab initio* calculations of phase diagrams in polymer systems are therefore still exceptional.

At present, the most common way to gain information about the phase behavior of polymer systems is first to obtain experimental data about the temperature and composition dependence of phase transitions and then to interpret them by applying thermodynamic principles. Following this semi-empirical method, the observation of glass transition(s) is commonly used to study miscibility in amorphous and semicrystalline mixtures.

The existence of a single glass transition between the glass transition of the pure components is probably the most unambiguous criterion for polymer–polymer miscibility. As will be pointed out later, (section 9.9), the different analytical methods that can be used have different limits of spatial resolution. In the first place, NMR spectroscopy can provide information about homogeneity on a molecular (nanometer and subnanometer) scale, but calorimetric measurements and mechanical spectroscopy are routine techniques and less expensive. Sharp glass transitions at the temperatures of the pure individual components indicate macroscopic phase separation. A broadening of the temperature range of the glass transition process may indicate microheterogeneity caused by additional fluctuations that are facilitated in the blend compared with the situation in an individual component. Broad transitions combined with shifts of T_g of the individual components towards each other indicate that the conditions for miscibility are close.

There are several equations to describe T_g-composition dependence in miscible blends, for example the Gordon–Taylor, Fox, Kelley–Bueche or Kanig equations [60–63]; however, there are systems that do not fit to any of them. The Gordon–Taylor equation may serve here as an example of the relations mentioned above:

$$T_{g,b} = \{T_{g,01} + (KT_{g,02} - T_{g,01})\phi_2\}/\{1 + (K - 1)\phi_2\} \qquad (9.8)$$

$T_{g,b}$ is the glass transition temperature of the blend, $T_{g,01}$ is the glass transition temperature of the pure polymer 1, $T_{g,02}$ is the glass transition temperature of the pure polymer 2, K is the ratio of the difference between the expansivities above and below the glass transition of polymer 2 and polymer 1, and ϕ_2 is the volume fraction of polymer 2.

Hierarchical systems were introduced in section 9.5. It is possible to treat these systems quantitatively [58, 64, 65]. George and Porter [64], for example, have mixed a monomolecular liquid crystal (MLC), i.e. a low molar mass molecule, with a thermotropic copolyester of the Kuhfuss and Jackson type [10]. They found that it was possible to describe the melting behavior of their system with reasonable accuracy by equation (9.9), although that has been derived by Flory [66] for 'simple', i.e. non-PLC, polymers only:

$$l/T_{mb} - 1/T_{mo2} = (R/\Delta H_m)(V_m/V_1)(\phi_1 - \chi\phi_1^2) \tag{9.9}$$

Here, T_{mb} is the melting point of the blend; T_{mo2} is the melting point of the pure polymer (100% crystallinity), component 2; R is the gas constant; ΔH_m is the molar transition enthalpy of the repeat unit of the polymer; V_m is the molar volume of the repeated unit; V_1 is the molar volume of the MLC or a diluent, component 1; ϕ_1 is the volume fraction of component 1 and χ is the Flory–Huggins–Staverman polymer–solvent interaction parameter.

Equation (9.9), however, only approximately describes the internal dilution of a liquid crystalline copolymer [64]. Here 'internal' means that the diluent is part of the chain and not a second independent component. Nishi and Wang [67] have derived equation (9.10) which describes polymers diluted by polymers. Their extension of the Flory–Huggins–Staverman theory gives the melting point depression:

$$T_{mo2} - T_m = -T_{mo2}(V_{2m}/\Delta H_{2m})B\phi_1^2 \tag{9.10}$$

where

$$B = RT(\chi/V_1) \tag{9.11}$$

A linearizing plot of the melting point depression vs. ϕ_1^2 theoretically starts at the origin. The Flory–Huggins–Staverman interaction parameter χ, which depends principally on the temperature and the composition, can be obtained from the slope of equation (9.10).

Only a few studies have concentrated on the phase transitions in blends containing EP and PLCs. Zachmann and coworkers [68], for example, have published experimental results from blends of a poly(ethylene naphthalene-2,6-dicarbonate) and a copolymer with PHB. Their paper again proves the complexity of these types of blends and the validity of the concept of hierarchical structures.

A rigorous theoretical treatment of copolymers consisting of rigid segments diluted by flexible comonomers as an internal diluent has been conducted by Jonah *et al.* [58], (section 9.10).

9.8 INFORMATION FROM STATISTICAL MECHANICS

There are several different statistical approaches to describe polymer–solvent and polymer–polymer miscibility phenomena:

- Flory's equation of state theory [69–72] and its extension by Jonah *et al.* [58];
- the Sanchez–Lacombe theory (lattice fluid theory) [73–75];
- other approaches, e.g. [76–81].

The prediction of miscibility requires knowledge of the parameters T^* (the characteristic temperature), p^* (the characteristic pressure) and V^* (the characteristic specific volume) of the corresponding equation of state which can be calculated from the density, thermal expansivity and isothermal compressibility. The isobaric thermal expansivity and the isothermal compressibility can be determined experimentally from $p-V-T$ measurements where these values can be calculated from $V(T)_p$ and $V(p)_T$. The characteristic temperature T^* is a measure of the interaction energy per mer, V^* is the densely packed mer volume so that p^* is defined as the interaction energy per V^*.

An important result of the Flory theory is that the interaction parameter χ for non-polar polymers is positive so that the conditions for miscibility favor low molar masses and similar characteristic temperatures T^*. $T_1^* > T_2^*$ and $p_1^* > p_2^*$ is a combination that is favorable with respect to miscibility while the combination $T_1^* > T_2^*$ and $p_1^* < p_2^*$ is quite unfavorable.

The lattice fluid theory predicts miscibility with a high probability if $V_1^*/V_2^* > 1$ and $T_1^*/T_2^* > 1$ while phase separation is favored if $V_1^*/V_2^* < 1$ and $T_1^*/T_2^* > 1$.

Jonah *et al.* [58] have extended Flory's lattice approach [82, 83] by considering PLCs consisting of random LC and flexible sequences with varying concentrations θ of LC segments. They obtain a partition function which is dependent on the concentration of LC sequences, the average length of these rigid parts of the polymer chain, the Hermans order parameter s [84, 85] and the thermodynamic temperature T (the order parameter s characterizes the degree of alignment of the principal molecular axis along a preferred direction $\boldsymbol{n}$ (director) in a mesophase). This enables the calculation, for example, of the dependence of the order parameter as a function of the temperature for a given polymer structure or of the LC–isotropic phase transition. Results from this approach will be discussed in more detail in section 9.10.

9.9 METHODS TO DETERMINE PHASE TRANSITIONS; DEFINITION OF PHASE AND MISCIBILITY

There are a variety of experimental methods to study phase transitions and related phenomena. The most convenient are

- Calorimetric investigations (DSC), e.g. [16]. DSC is a fast technique; only small amounts of material (in the milligram range) are necessary; the technique is not sensitive to secondary or higher relaxation phenomena because these processes usually do not have a high degree of cooperation [86].

- Dynamic-mechanical experiments (DMTA), e.g. [50, 87–89]. This technique is not only sensitive to the glass transition but also to secondary relaxation phenomena and can detect crank shaft motions, chain distortion, etc. [89]. This is important because these types of relaxation processes seem to be sensible to the dispersion of the phases in each other and interface effects. Mechanical quantities can be measured. DMTA seems to be more sensitive in the detection of small amounts of additional components compared with DSC and TMA [91, 92] (see also section 9.12.1).
- Thermomechanical analysis (TMA), e.g. [50]. The sensitivity of TMA is comparable with that of DMTA. In contrast to DSC and DMTA, the measurement is not the average over the whole sample but an average of the properties of a certain part of the sample that is limited by the shape of the tip of the probe. Direct measurements of the expansivity are possible which give access to the free volume and the relaxation behavior of the polymer chains [93–95].
- Optical transmission (OT), e.g. [87]. OT provides fast information if the size of the domains are of the order of magnitude of the wavelength of light and the refractive index of the phases under consideration differs significantly from each other.
- Optical microscopy (OM), e.g. [87, 96, 97] is useful if the samples can be prepared appropriately and the domain size is larger than the wavelength of light.
- Scanning electron microscopy (SEM), e.g. [98]. Unless there is a reasonable contrast, special sample treatment is necessary. The detectable domain size is in the 100 nm range.
- Tunnel microscopy (TM), e.g. [98, 99]. Depending on the technique, structures down to several 0.1 nm can be detected.
- Small angle X-ray scattering (SAXS), e.g. [100]. Structures in the nanometer range can be detected, as well as superstructures, e.g. layers.
- Wide angle X-ray scattering (WAXS), e.g. [97]. Structures in the subnanometer range can be detected, such as crystalline phases.
- Thermally stimulated depolarization (TSD), e.g. [101]. TSD is applicable to dielectric polymers that are electrets. Depolarization currents are monitored and these currents are related to the relaxation of electric charges in the sample as a function of the number of events that take place in the sample at the molecular and supermolecular level.
- Internal friction (IF), e.g. [102]. IF is a mechanical testing procedure which may be understood as a special technique of DMTA. The mechanical damping of the free oscillation is measured depending on the temperature at a constant frequency. The internal friction parameter is proportional to the natural logarithm of the ratio of two subsequent amplitudes of the oscillation and hence is also related to the damping factor $\tan\delta$, (see also section 12.1).

- Dielectric relaxation spectroscopy (DER), e.g. [103–105]. DER monitors the mobility of dipolar groups in the polymer and also of small dipolar molecules (e.g. water) that may be dissolved in the polymer system. Corresponding to mechanical measurements, the maxima of dissipated energy indicate phase transition processes.
- Dilatometry, pVT measurements, e.g. [50, 106]. These measurements unequivocally show a first order transition by a step in $V(T)$ and a bend if there is a glass transition. The important partial derivatives isobaric expansivity and isothermal compressibility can be derived from the corresponding measurements. The method is, however, quite time consuming and not widely used.
- Solid state nuclear magnetic resonance spectroscopy (NMR), e.g. [107–109]. This technique is sensitive to the local environment of certain nuclei, their mobility and orientation [108]. It provides information about the heterogeneity of polymer blends to *c.* 5 nm or less (spin diffusion experiments) or *c.* 0.3 nm in cross-polarization experiments, from which the direct (averaged) distance between two types of nuclei in a sample can be determined [107, 108]. Motions of moleuclar groups in a polymer chain can be analyzed and correlations with dispersion areas in the mechanical spectra may be possible [109]. Solid state NMR is not a standard technique at the present time but it is becoming increasingly important.

Discussion of phases and miscibility requires a proper definition. The general thermodynamic definition of a phase states that a phase is a region of constant or continuously changing physical and chemical properties, the state of which can be described by thermodynamic relations.

Phases with continuously changing properties can occur at the interface of neighboring phases and thus build an interphase. From the thermodynamic definition it is clear that the size of a phase is limited because the spatial continuity becomes lost when approaching molecular dimensions. Some of the analytical methods described above, however, go down to these dimensions. Solid state NMR, for example, gives information about the environment of a nucleus over short ranges – over distances of less than 0.6 nm (using chemical shift techniques, intermolecular cross-polarization, the nuclear Overhauser effect and rotational resonance) – and over long ranges – i.e. distances between 0.6 nm and 10 nm (using direct and indirect spin diffusion and dynamic nuclear polarization). This means that these techniques provide information about the neighborhood, and hence about the miscibility, of macromolecules or parts of them on a molecular scale. On a molecular scale there is no continuous change in physical properties. Besides the macromolecules themselves, there is also the free volume.

Being aware of this, an appropriate definition of a phase could be:

a region where the number of physical entities (particles, molecules) in the interior is several orders of magnitude larger than on the surface of the region. Thus the state of the region, i.e. the phase, is governed by the properties of the interior and not by the surface, and the total number of physical entities (particles, molecules) in this region has to be sufficiently large so that statistical methods can be applied. All parts of a finely dispersed phase together form a canonic ensemble. This distinction is necessary because the different analytical techniques are sensitive in different ways to the size of a phase, i.e. there is a lower size limit below which a certain technique is no longer able to defect the existence of a second phase properly. With NMR techniques [53, 110] it was shown above that it is possible to investigate the environment of only a few molecules, but a few molecules are certainly not a phase in the thermodynamic sense. The definition of a phase, however, must not depend on the method used to detect the phase. Consequently, the definition of miscibility must also be precise. Therefore, a system could be defined as miscible on the molecular scale if the statistics of the nearest neighbors of the smallest physical entity is the same as the statistics of the whole system (if the nearest neighbors of a certain molecule reflect the relative amounts of the components in the system).

The size of the phase domains influences the resolution of neighboring signals in a temperature dependent diagram obtained from the techniques described above. The position of the glass transition temperature and the signal widths is also influenced by the domain size if it is smaller than a certain limit [111, 112]. The domain size depends on the system, the preparation of the blend, processing parameters and thermal history. Most of the data allow a correlation of DMA and electron microscopy [113]. The values for the smallest detectable domain size using DMA differ and range from 1.5 nm and 15 nm [114–117]. This is the range where SAXS (1–100 nm) and WAXS (0.1–2 nm) overlap. Scanning electron microscopy (SEM), for example, can have a resolution down to 10 nm. The smallest domain size which is detectable by the other techniques is still not exactly known; however, it seems that a larger domain size is necessary in the case of DSC and DER compared with DMA and TMA. With the customary instrumentation it is difficult to separate two glass transitions with DMA if the temperature difference is smaller than 20°C. This is even more difficult with DSC if the first derivative of the DSC trace is not used. Solid state NMR, in contrast to all the other techniques mentioned above, is a direct probe for the microscopic structure in a polymer system and does not only provide information about transitions. At present, no other technique gives such a detailed view of the microscopic structure. Many of these NMR techniques are highly sophisticated and have not yet become routine procedures.

9.10 PHASE DIAGRAMS OF PURE POLYMER LIQUID CRYSTALS

A theoretical treatment of the general phase diagram of a pure copolyester consisting of rigid segments that under certain conditions can cause the existence of mesophases connected by flexible chains has been published by Jonah *et al.* [59]. They applied statistical thermodynamics to this type of system and obtained a partition function depending on the average length $\bar{\eta}$ and the concentration θ of the rigid segments, the order parameter s and the thermodynamic temperature T. It is well known from statistical thermodynamics that the Helmholtz function A is given by the sum of the natural logarithm of the combinatorial and the orientation partition function. The Helmholtz function of two neighboring phases is identical if there is internal equilibrium. Hence equation (9.12) describes the equilibrium situation at the transition from the isotropic to the anisotropic state of a liquid crystalline system. Of course this is identical to the reverse transition if there is true thermodynamic equilibrium.

$$A_{\text{iso}} = A_{\text{aniso}} \tag{9.12}$$

This leads to a system of nonlinear equations which is solved by iterative processes. Given, for example, the average length $\bar{\eta}$ of a rigid segment, the order parameter s (section 9.8), the transition temperature T_i to (or from) the isotropic state can be calculated with the concentration θ of the rigid segments as a parameter. For more details see reference 59. Figure 9.20, for example, shows the dependence of the order parameter s on $\bar{\eta}$ and θ at T_i.

The behavior of the system is strongly depending on θ. Assuming a constant temperature, the onset in a plot of $\bar{\eta}$ vs. s near $\theta = 0.2$ indicates a strongly increasing alignment of the rigid segments.

The transition from the anisotropic state to the isotropic state characterized by $T_{\text{lc-iso}}/T^*$, where T^* is the characteristic temperature in the equation of state of the system, shows an increase for all concentrations of the rigid segments depending on $\bar{\eta}$ (Figure 9.21).

The temperature range within which the mesophase exists grows symbatically with θ. The ascending parts of the curves are related to the athermal limit $T^{-1} \rightarrow 0$ for $\theta = 1$. The dependence on $\bar{\eta}$ of the transition temperature to the isotropic state shown in Figure 9.21 seems to be related to the assumption of having the same number of flexible sequences and rigid sequences. Since longer rigid sequences at a fixed concentration θ also means longer flexible sequences, the former therefore have a stronger capability of reinforcing any alignment.

The discussion of the statistical equations can also provide a prediction of $T_{\text{lc-iso}}$; however, a precise knowledge of the characteristic temperature T^*, θ and $\bar{\eta}$ is necessary to obtain reliable predictions. An

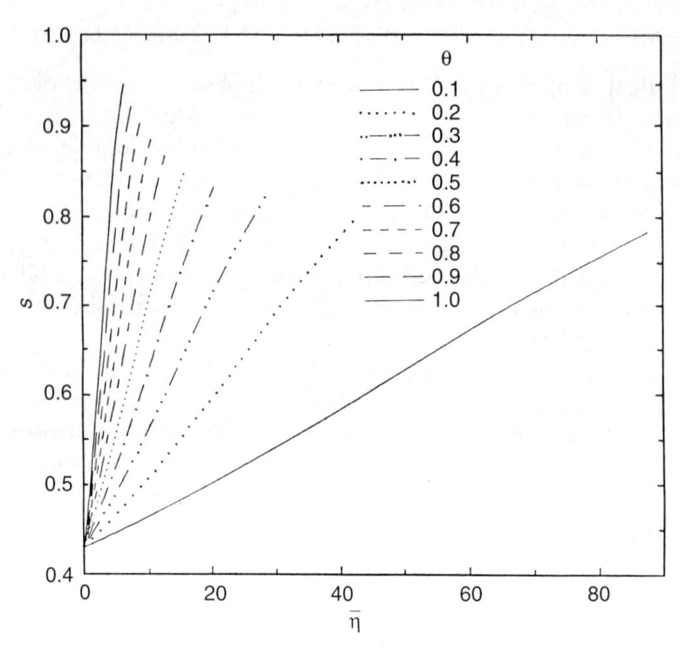

Figure 9.20 Dependence of the order parameter s on the average length $\bar{\eta}$ of rigid sequences in the PLC chain for selected values of the concentration θ of rigid segments. (According to [58] with permission of The American Chemical Society.)

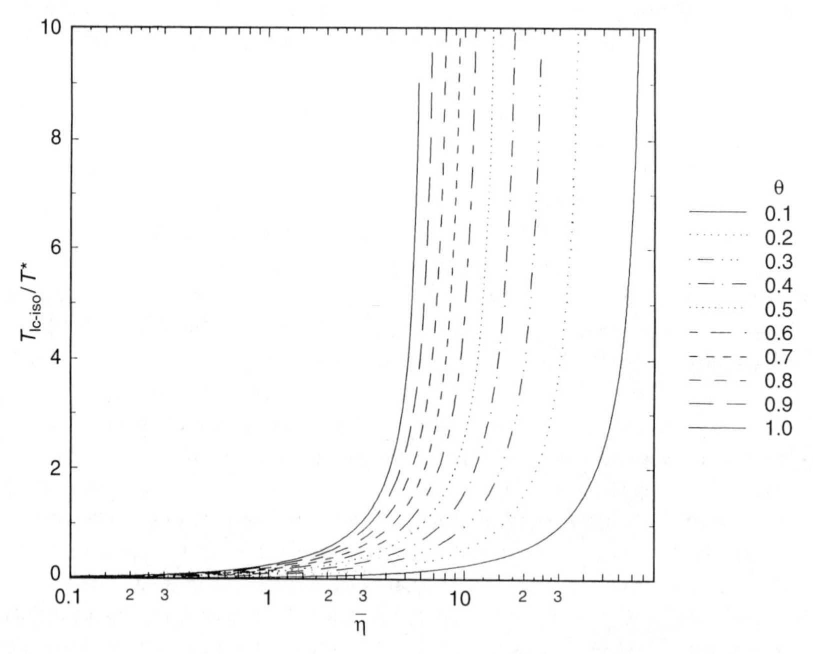

Figure 9.21 Dependence of T_{lc-iso}/T^* on the average length $\bar{\eta}$ of rigid segments in the PLC chain for selected values of the concentration θ of rigid segments. (According to [58] with permission of The American Chemical Society.)

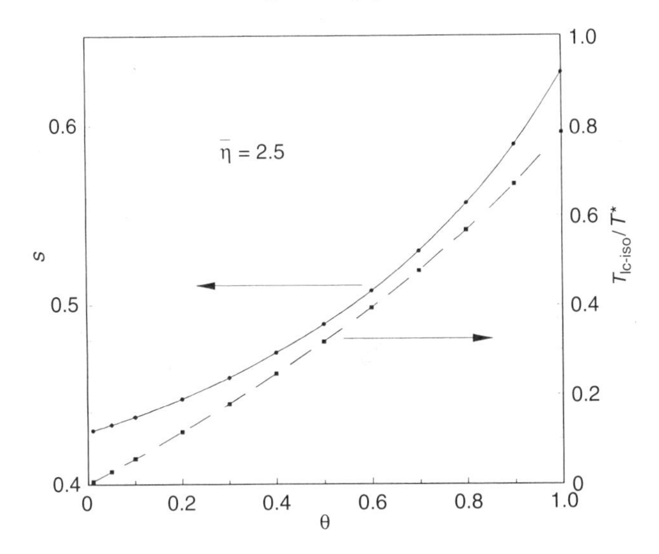

Figure 9.22 Dependence of the order parameter s and the anisotropic–isotropic transition temperature $T_{\mathrm{lc-iso}}/T^*$ on the concentration of rigid segments θ in a series of LC copolymers; $\bar{\eta} = 2.5$ is valid throughout the diagram. (According to [58] with permission of The American Chemical Society.)

example is given in Figure 9.22. It was also shown by Jonah *et al.* [58] that in any copolymer of this type a minimal concentration $\theta_{\mathrm{lc\ limit}}$ is necessary to cause LC-rich phases (Figure 9.23).

The phase diagrams of ternary systems consisting of a thermotropic copolyester as described above, a random coil polymer and a solvent have been treated by Blonski *et al.* [65], who also applied statistical thermodynamics. These systems are of interest because they show an isotropic mixture in a certain concentration range (Figure 9.24).

It is important that the thermotropic component is not completely rigid but contains flexible joints. Their presence has a strong influence on the phase structure of the mixtures as shown in the Gibbs' triangles in Figure 9.24. A fast 'jump' from the isotropic phase at the top of the triangle into the phase separated region below can result in a spinodal demixing and a 'frozen' structure of the isotropic state. Thus, if the change of the state of the isotropic mixture to the unstable state is fast enough, it is also possible to preserve the microstructure of isolated rigid segments, which exist in the isotropic liquid and in the precipitated polymer blend. In a dilute isotropic mixture the rigid segments are separated from each other and do not interact with each other. A slow change of state, for example by slow evaporation of the solvent, leads to a phase separation and precipitation of the polymers. The polymers

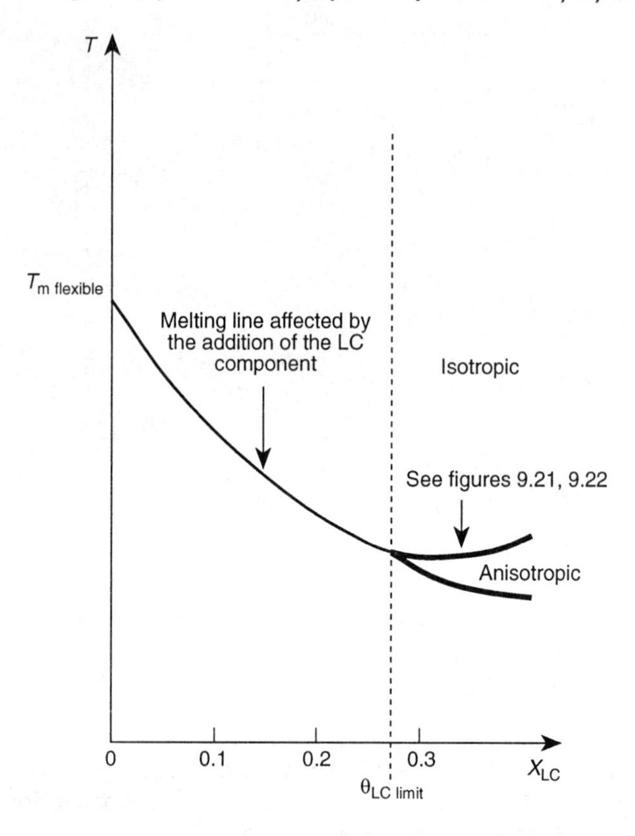

Figure 9.23 A section of a schematic phase diagram starting from the left at the pure flexible polymer with its melting temperature $T_{m, flexible}$. Mole fraction of the liquid crystalline component X_{lc} increases towards the right. An LC phase or phases exist to the right of $\theta_{lc\,limit}$. (According to [58] with permission of The American Chemical Society.)

will generally form two separate phases. In such a polymer blend there are islands of PLC dispersed in a matrix of coil polymers if the concentration is properly chosen. These structures are relatively weak compared to their mechanical properties because mechanical tensions can be created between the neighboring rigid segments. A much better reinforcing effect of the rigid segments in the blend is to be expected if the rigid segments are dispersed in the matrix and are separated from each other. In this way, optically transparent films (blends) from thermodynamically immiscible polymers were obtained by a fast quenching of the isotropic state, while in contrast a slow evaporation of the solvent resulted in a turbid film with a macroscopically visible phase separation of the two polymers.

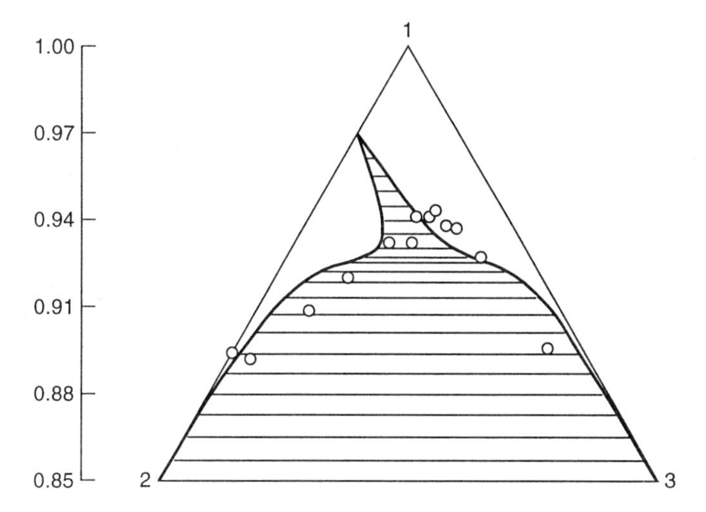

Figure 9.24 Comparison of experimental and calculated phase diagram. The circles are from experimental cloud point measurements. Lines were calculated for the system with $\theta = 0.27$, $r_2 = 600$, $r_3 = 1200$, $T^* = 600\,\mathrm{K}$ at a temperature of $T = 295\,\mathrm{K}$. The r_i values correspond to the respective degree of polymerization of the components. Component 1 is the solvent, 2 is the PLC and 3 is the coil polymer. (According to [65] with permission of The American Chemical Society.)

9.11 PHASE DIAGRAMS OF BLENDS

Combining the results of all techniques involved in the analysis of the phase transitions which were observed in the blends, the diagram shown in Figure 9.25 was obtained. It also contains non-equilibrium structures, and the lines were drawn in these cases during a closely controlled thermal history (Figure 9.25; Table 9.2).

For most of the results discussed below, the following is valid concerning COP, PC and mixtures if not otherwise stated. The thermotropic copolyester derived from *p*-hydroxybenzoate and poly(ethylene terephthalate) [10] containing 60 mol% PHB was used exclusively as the PLC component because this composition has the best mechanical properties [10, 118] of these copolymers. It was obtained from Eastman Kodak, Kingsport, TN, and had an average molar mass estimated from solvent viscosity of about 19000 g mol^{-1}. The sequence distribution was calculated from ^{13}C-NMR as described by Lenz *et al.* [119] and was nearly random: the statistical parameter Ψ which describes the randomness of the copolymer is $\Psi = 1$ for a block copolymer and $\Psi = 0$ for a completely random copolymer. The copolymer used for the experiments had $\Psi = 0.15$.

Polycarbonate from bisphenol-A served as the EP. It was obtained from the Bayer A.G., Leverkusen, Germany. The number-average molar mass was determined by vapor pressure osmosis: $\langle M_n \rangle = 18\,700\,\mathrm{g\,mol^{-1}}$. The mass average $\langle M_w \rangle$ was estimated by viscometry: $\langle M_w \rangle = 31\,000\,\mathrm{g\,mol^{-1}}$. The blends were prepared by co-precipitation from trifluoracetic acid/chlorofom (20/80 vol%) with acetone, if not otherwise stated.

The specimens were prepared by molding at 210–220°C (4.0 MPa) if not stated otherwise. Usually, the samples were cooled down to ambient temperatures over a period of 45 min. Under these conditions a reproducible sample preparation was possible. The melt viscosity was high enough to provide a macroscopic homogeneous sample, but the temperature was not high enough to cause transesterification or thermal degradation of the samples during molding.

The diagram is very complex with at least 17 phases. The main reason is the fact that the pure COP already shows a multiphase behavior [50]. According to Yoon and coworkers [105] there are two mesophases in the pure COP, SmE and SmB, which might also be observable in the blend. However, in the liquid state, especially at temperatures higher than 260°C, transesterification and/or degradation processes are observable, so that all phases claimed in this area are in some sense hypothetical because they could not be observed unequivocally due to the reasons outlined above, although their existence can be predicted from theoretical arguments.

The quasi-liquid state (ql) [50, 52, 120] only exists in polymers and is localized at the transition between the glass and the melt, where low mobility and a retarded response to external forces are observed. The low mobility distinguishes this state from a low molar mass (ordinary) liquid which flows much more rapidly under comparable conditions. The retarded mechanical response compares with the leathery state of an elastomer (e.g. Chapter 1 in reference 92). In the present case the ql phase originates from PET. This is also the state in which cold crystallization occurs. The temperature of maximum cold crystallization is indicated in Figure 9.25.

At about 50% (w/w), phase inversion takes place, and at COP concentrations beyond 50% the formation of LC-rich islands in an LC-poor matrix is observed. These islands have been described by several authors: in the COP system [118] and in a blend consisting of a rigid rod and a semiflexible coil polymer [121]. The formation of these islands is mainly due to differences in viscosity and surface tension of LC-poor and LC-rich phases and the tendency to achieve efficient packing by orientation and the lowering of lateral distances of the anisotropic parts of a polymer. The LC-poor (= PC-rich) matrix has a higher melt viscosity than that of COP. In a shear field or under

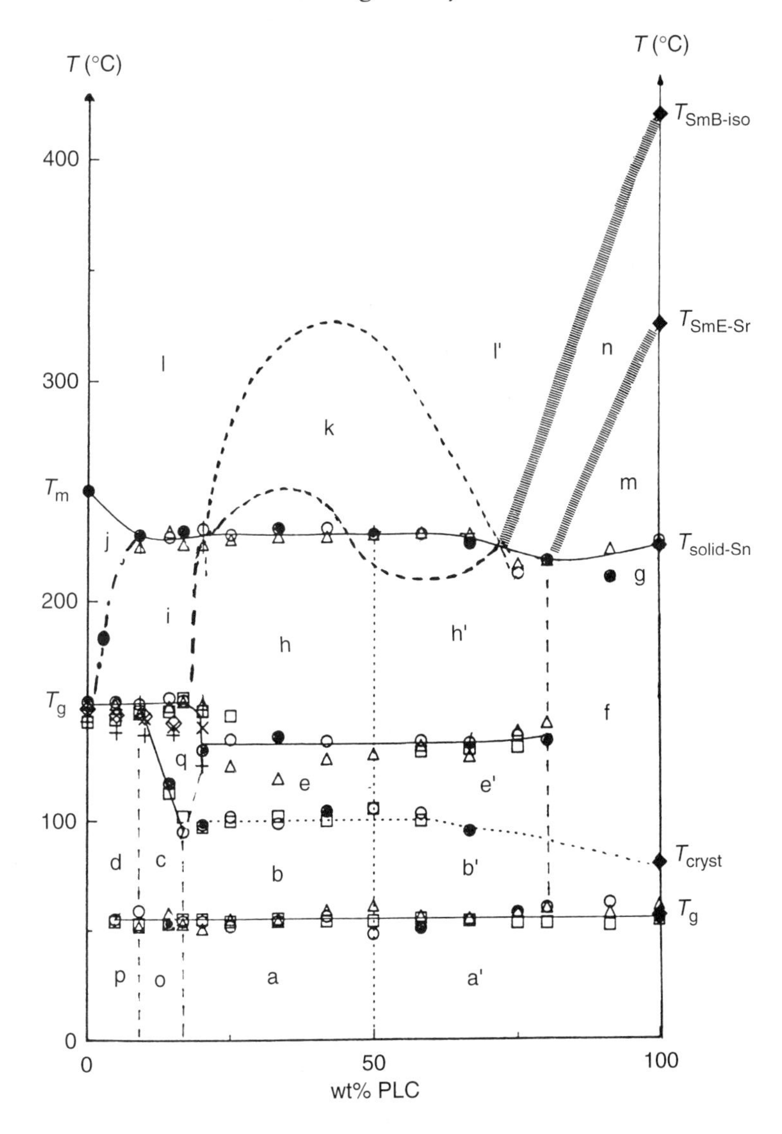

Figure 9.25 Phase diagram of the blend PC + COP. The identification of the phases is given in Table 9.2. The primes denote the PLC rich phases. For further explanation see text.

elongational flow the islands are deformed, first to an ellipsoidal shape and finally to fibers and fibrils which have already been considered in the hierarchical structure model of Sawyer and Jaffe [51]. Under elongational flow conditions there is no such deformation in the center of the conduit, therefore the islands are able to relax to some extent,

Table 9.2 Identification of the phases in Figure 9.25. ql means 'quasi-liquid', which is explained in the text

Phase	Structures present					
a	PET glass	PHB islands	PHB crystals	PC glass	PC crystals	PET/PC glass
b (meta-stable)	PET ql	PHB islands	PHB crystals		PC crystals	PET/PC glass
c	PET ql	PHB islands	PHB crystals	PC glass	PC crystals	PET/PC glass
d	PET ql			PC glass	PC crystals	
e	PET ql	PHB islands	PHB crystals	PC ql	PC crystals	
f	PET ql	PHB islands	PHB crystals	PC ql	PC crystals	
g	PET ql	PHB islands	PHB crystals			
h	PET crystals		PHB crystals	Isotropic liquid	PC crystals	PET/PC ql
i			Isotropic ql			
j	PET ql			PC ql		
k		PHB-rich liquid				PET/PC-rich liquid
l			Isotropic melt			
m		Isotropic liquid/SmE/SmB				
n		Isotropic liquid/SmB				
o		PHB islands	PHB crystals		PC crystals	PET/PC glass
p	PET glass	PHB islands	PHB crystals	PC glass	PC crystals	PET/PC glass
q		PHB islands	PHB crystals		PC crystals	PET/PC ql

almost retaining the spherical shape forced upon them by the interfacial tension. Consequently, this results in a rather homogeneous distribution of the low-viscosity islands in the high-viscosity matrix, producing some kind of lubrication effect which has been demonstrated by measurement of rheological properties [24, 39, 118, 122, 123].

All blends were prepared from amorphous PC. If a semicrystalline PC had been inserted, then the indicated increase of the melting line towards the melting point of pure PC would be observed. In the case of amorphous PC the material softens along the dash–dotted line in Figure 9.25, separating the regions i and j.

The phase diagram as shown in Figure 9.25 was obtained from samples which had been compression molded at 210–220°C at 4 MPa for 60 min. A thermal history which differs in some features would cause a different phase diagram.

Figure 9.26 clearly shows that the application of pressure is necessary to obtain an optimal miscibility of COP in PC. After 60 min at a pressure of 4 MPa, no further change of the glass transition temperature (T_g) of PC was observed. Due to the fact that T_g of COP remains unaltered and that NMR spectroscopy proved that under the given conditions no significant transesterification was observed [53], it can be concluded that there is partial miscibility of COP (60%) in PC in the concentration range between 60% and 85% PC.

The dashed lines in the upper part of Figure 9.25 which show a liquid–liquid miscibility gap are calculated values. In the temperature range beyond ~230°C no reliable measurements are possible because of the rapid change of the chemical constitution of the blend caused by transesterification and thermal degradation. The thick dashed lines on the right side of Figure 9.25 display the guessed trace of the phase separations which are expected. They are caused by the different mesophases which have been described in that temperature range for pure COP by Yoon *et al.* [105]. It is clear that the features of the phase

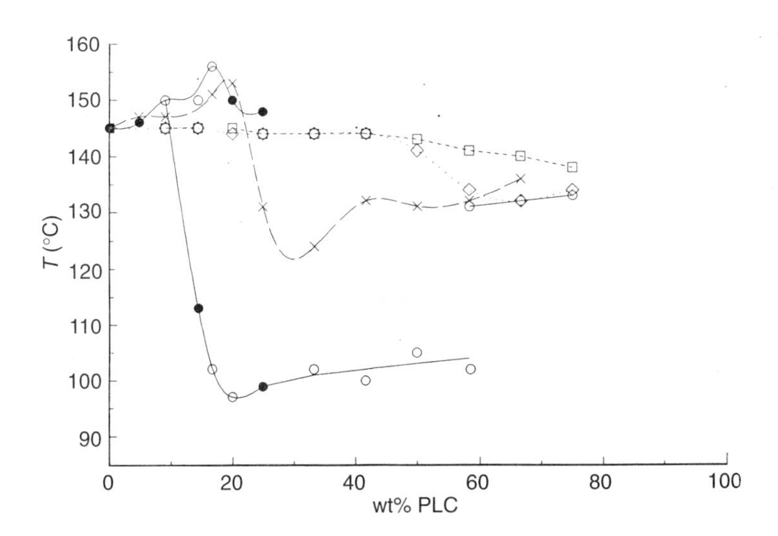

Figure 9.26 Effects of thermal history. Glass transition temperatures of PC + COP blends: ---- blends prepared in the solid state by compression molding for 10 min at 210–220°C at 4.0 MPa; ··· blends prepared by precipitation from solution and subsequent compression for 10 min at 210–220°C at 0.1 MPa; --- blends prepared by precipitation from solution and compression for 1 min at 210–220°C at 4.0 MPa; —— blends prepared by precipitation from solution and compression for 60 min at 210–220°C at 4.0 MPa.

diagram will be reflected by certain properties of the blends and that they are useful for an 'intelligent processing' [50].

In quasibinary blends A + B where one component is a copolymer, it can be advantageous to treat the copolymer itself as a quasibinary system (B + C) because the copolymer composition can be changed. A quasiternary diagram is obtained in these cases. B, for example, could represent the mesomorphic group in the copolymer (PHB in COP). A typical Gibbs phase triangle can be created in this way giving an isothermal section of the three-dimensional, prismatic diagram. Figure 9.27 shows a corresponding example for a hypothetical system.

Another suggestion, not only applicable to LC-containing systems but to any binary or quasibinary system, is shown in Figure 9.28. The corresponding amounts of crystalline phases for each component under specified conditions after a certain thermal history can be obtained directly from such a diagram.

Zachmann *et al.* [124] have taken advantage of these suggestions, describing not a polymer blend but binary and ternary copolyesters prepared from poly(ethylene-terephthalate), poly(ethylene-naphthalene-2,6-dicarbonoxylate), and poly(*p*-hydroxy-benzoate). This example is very instructive and clearly shows the practical value of phase diagrams in general and especially of the procedure suggested above.

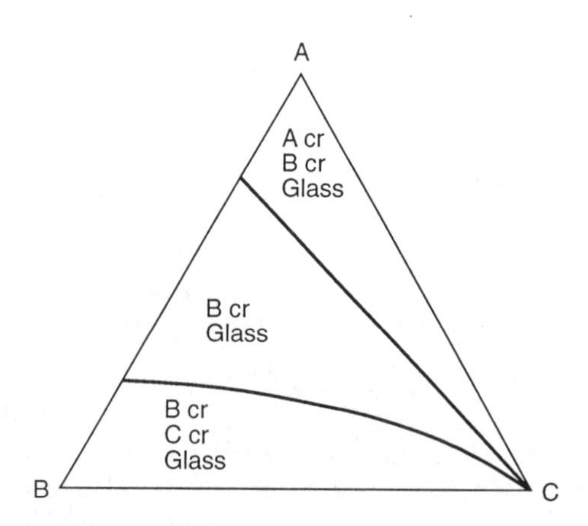

Figure 9.27 A binary A + BC system treated as ternary (hypothetical isothermal diagram). Component B is the mesomorphic component in the semirigid copolymer BC. The compositions along the axis are measured in molar fractions x. (According to [11] with permission of Elsevier Science Publ.)

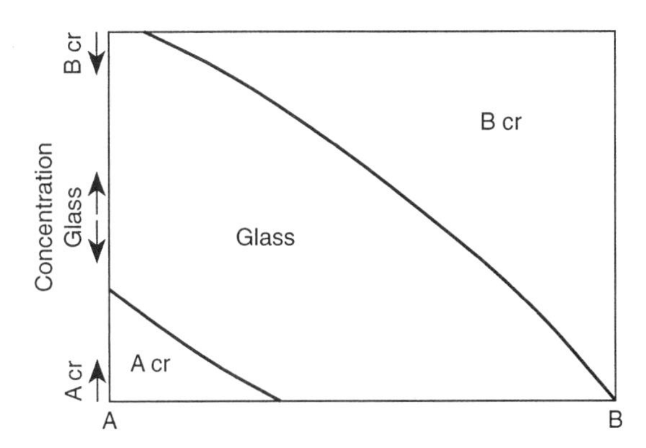

Figure 9.28 Hypothetical isothermal diagram for a binary system A + B showing the amounts of crystalline and glassy phases. (According to [11] with permission of Elsevier Science Publ.)

9.12 RELATION WITH MECHANICAL PROPERTIES

The phase morphology of the blends determines the mechanical properties. These correlations, which will be described in the following sections, have been investigated in different ways.

9.12.1 Torsion modulus

DMTA with a torsion pendulum not only shows thermal transitions but also provides mechanical data. From the decay of a free torsional oscillation at a fixed frequency the complex torsional modulus G^* is derived which consists of a real part, the storage modulus G', which is a measure of the energy stored reversibly in the system, and an imaginary part $i \cdot G''$, where G'' is the loss modulus, which is a measure of the energy dissipated by the system. The quotient G''/G' gives $\tan\delta$, where δ is the phase shift between the applied force and the deformation which is identical to the angle θ of the polar coordinates in the Gaussian plane of complex numbers. $\tan\delta$ is also called the 'loss factor'. It is well suited for the determination of thermal transitions as shown in Figures 9.29–9.31. At about the glass transition temperature there is a drop in G' because the polymer softens. The function $G''(T)$ shows a maximum because the amount of energy that is dissipated increases. The glass transition temperature can be defined from these measurements by the $\tan\delta$ or G'' maximum.

The figures show the effect of transesterification which was observed in COP/PC blends in the melt. The glass transition temperature of COP remains almost unaltered at about 55°C. The glass transition temperature of PC decreases with increasing annealing at about 215°C.

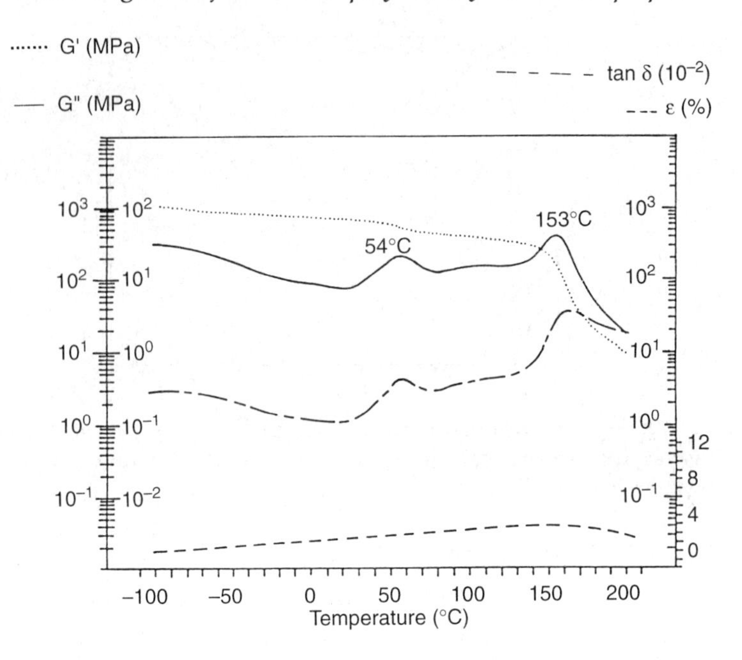

Figure 9.29 Pendulum DMTA (1 Hz) results for a blend containing COP 20 mass%, obtained from solution by acetone precipitation and kept under compression at 210–220°C and 4.0 MPa for 1 min [125] and [128]. Reproduced with permission of Elsevier Science Ltd. Torsion pendulum, Myrenne Company, Roetgen, Germany. Usually, the measuring frequency was 1 Hz, so that the measured values are close to equilibrium values. The heating rate was 10 K min^{-1}. The torsion pendulum was also equipped to measure sample elongation simultaneously.

This indicates that parts of the COP chains are incorporated in the PC phase, causing the observed decrease by chemical reaction between the different polyesters.

The thermal transition derived from DMTA at 1 Hz differ only slightly (*c.* 5°C higher) from DSC values (corrected for the heating rate) because DMTA is a dynamic method. In principle it was found that both methods gave the same transition temperatures.

Isothermal DMTA experiments revealed interesting results concerning the viscoelastic behavior of the blends. In blends containing relative small amounts of PC, G' decreases with the time while the strain ε increases until it reaches a plateau value (Figures 9.32 and 9.33).

The results may be interpreted as follows. At the given temperature of the experiment, the sinusoidal force of the deformation helps to surmount the surface tension so that the PLC islands can start coalescing more easily than they could without external force. A smaller number

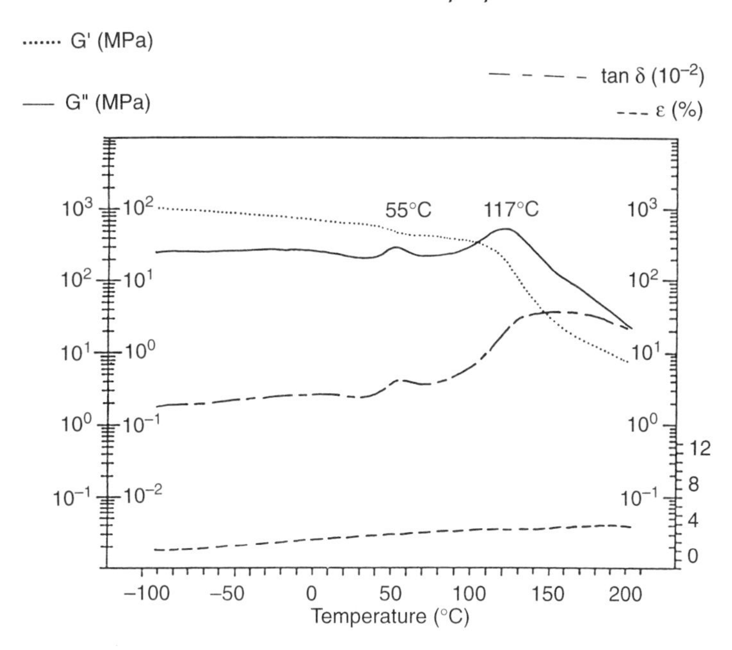

Figure 9.30 Pendulum DMTA (1 Hz) results for a blend containing 20% COP obtained from solution by acetone precipitation and kept under compression at 210–220°C and 4.0 MPa for 10 min [125] and [128]. Reproduced with permission of Elsevier Science Ltd.

of islands will consequently lead to a smaller storage modulus with a strain increasing to an upper limit. If, on the contrary, the PC concentration is relatively high, as shown for example in Figures 9.33 and 9.34, then the crystallization of PC (which has already been mentioned in section 9.11) governs the viscoelastic behavior, G' increases with time while the strain decreases (Figure 9.35). In the blend with a low PC concentration, formation of PC crystals is hindered while in PC-rich blends the presence of COP supports the formation of crystalline PC, which is also a hindered process in pure PC.

The influence of temperature and composition on G' are shown in Figure 9.36. Except for the highest isotherm, which is beyond all glass transition temperatures, all isotherms show a maximum in G' near a content of about 50 mass% PLC. As outlined above, this can partially be explained by the enhanced crystallinity of PC in these blends. The slope of the isotherms depends on the temperature. At temperatures below the glass transition of COP (*c.* 55°C) the shear modulus generally decreases with increasing PC concentration. This is caused by the lower G' of PC compared with COP. This situation changes at higher temperatures. While the COP phase becomes softer, the PC phase still

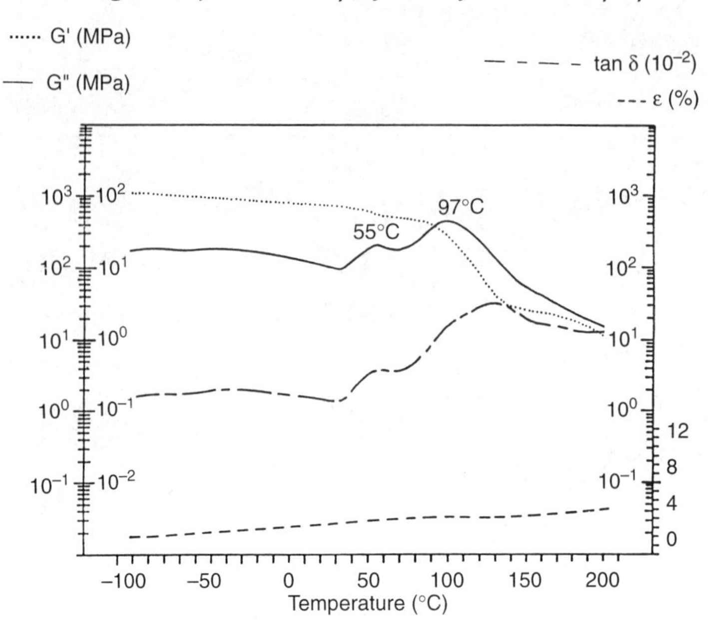

Figure 9.31 Pendulum DMTA (1 Hz) results for a blend containing 20% COP obtained from solution by acetone precipitation and kept under compression at 210–220°C and 4.0 MPa for 60 min [125] and [128]. Reproduced with permission of Elsevier Science Ltd.

remains rigid until it reaches its T_g. Therefore, an increase in the amount of PC leads to an improvement of the storage modulus. Starting from the pure PC (Figure 9.36) there are flat regions of the curves, particularly at low temperatures. These plateaus correspond to PLC concentrations which are smaller than $\theta_{PLC,limit}$ – the critical concentration of mesogenic groups which is necessary to form an LC phase – so that there are almost no reinforcing islands. The pure COP shows a $\theta_{PLC,limit}$ at about 30% PHB (Figure 9.23). Beyond $\theta_{PLC,limit}$ there is a gradual formation of islands which also contribute to the storage modulus. In the central part of the diagram the phase inversion takes place, combined with a decrease of the number of the PLC islands, and it is the number of the islands and not their volume that determines the reinforcement [126, 127]. The coalescence of the islands has been proved by SEM studies published by Brostow *et al.* [118]. The further increase of COP content results in a continuous increase of the modulus. Thus the viscoelastic behavior, in this special case the plateau region of G', its minimum and maximum value, clearly correspond with the phase diagram. It reflects the phase diagram and the morphology which is involved in it.

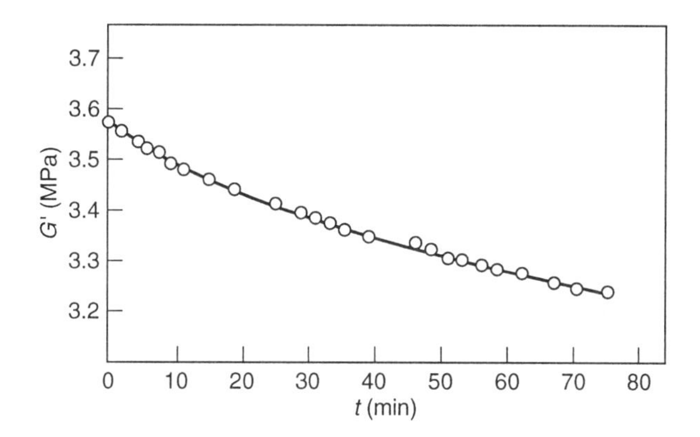

Figure 9.32 Isothermal (173°C) variation of the storage modulus G' with time. The blend contains 75% COP, obtained from solution by acetone precipitation and kept under compression at 210–220°C and 4.0 MPa for 1 min. Dynamic-mechanical thermal analysis (DMTA): Polymer Labs MKII (8.0 mm × 8.0 cm × 1.5 mm samples) with cantilever clamps [125] and [128]. Reproduced with permission of Elsevier Science Ltd. For explanation see text.

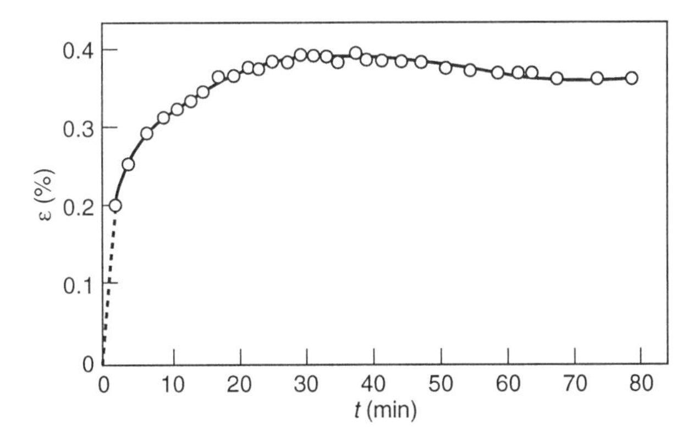

Figure 9.33 Isothermal (173°C) variation of strain with time. The blend contains 75% COP, obtained from solution by acetone precipitation and kept under compression at 210–220°C and 4.0 MPa for 1 min [125] and [128]. Reproduced with permission of Elsevier Science Ltd. For explanation see text.

9.12.2 Three-point bending

In a three-point bending experiment, a sample bar is placed on two supports separated by a certain distance. A normal force is applied on

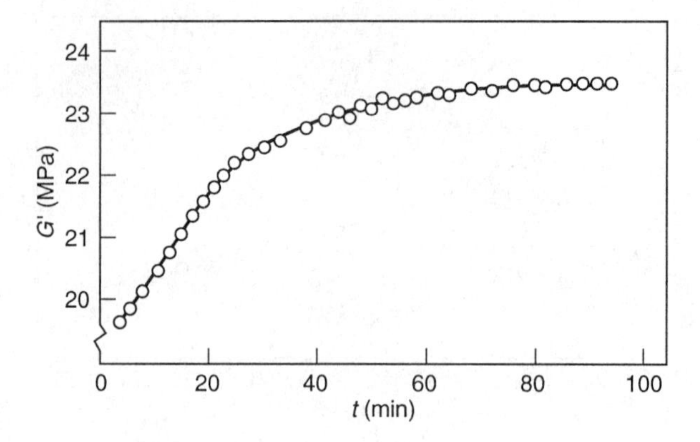

Figure 9.34 Isothermal (173°C) variation of the storage modulus G' with time. The blend contains COP 50 mass%, obtained from solution by acetone precipitation and kept under compression at 210–220°C and 4.0 MPa for 1 min [125] and [128]. Reproduced with permission of Elsevier Science Ltd. For explanation see text.

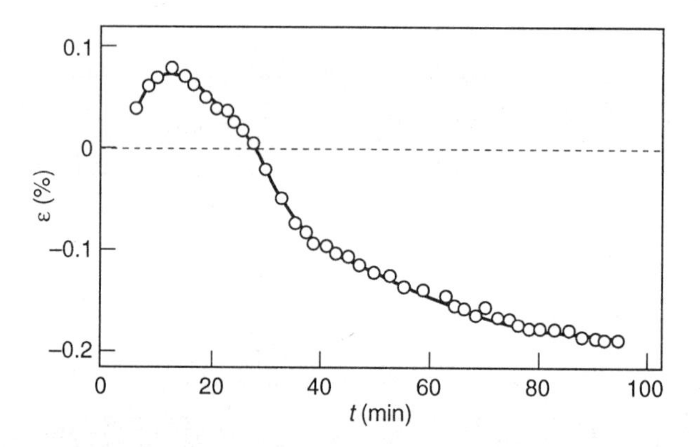

Figure 9.35 Isothermal (173°C) variation of the strain with the time. The blend contains COP 50 mass% , obtained form solution by acetone precipitation and kept under compression at 210–220°C and 4.0 MPa for 1 min [125] and [128]. Reproduced with permission of Elsevier Science Ltd. For explanation see text.

the sample by a ram with a curved tip which hits the sample bar in the middle between the two supports. The elastic modulus can be derived from the applied force, the sample dimensions and the bending of the sample. In the experiments described below the sample is not fixed to the supports.

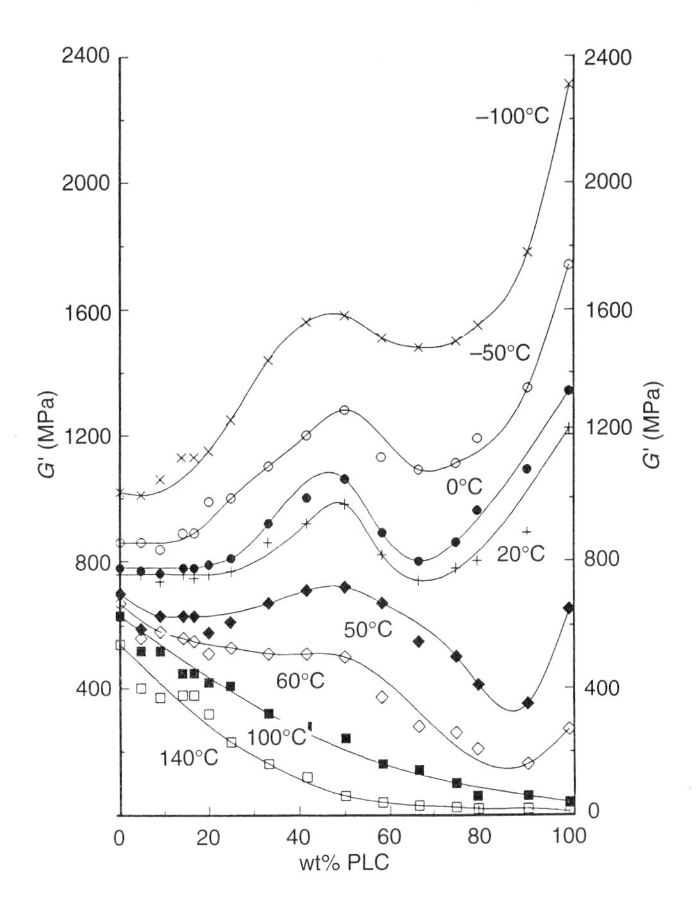

Figure 9.36 Storage modulus G' as a function of the composition at the temperatures indicated [125] and [128]. Reproduced with permission of Elsevier Science Ltd.

The elastic modulus E_b was studied at room temperature in a three-point bending experiment. The results are shown in Figure 9.37. From the behavior discussed above it is clear that annealing above the glass transition temperature for a certain time will increase E_b. This can be explained in terms of the crystallinity of PC in blends where there is enough PC present. The remarkable drop of E_b in blends containing more than 80 mass% PC is based on the hindered crystallization of PC, where there is more PLC necessary to support its crystallization. The sharp drop in E_b coincides with the change in the glass transition temperature and the miscibility as displayed in the phase diagram,

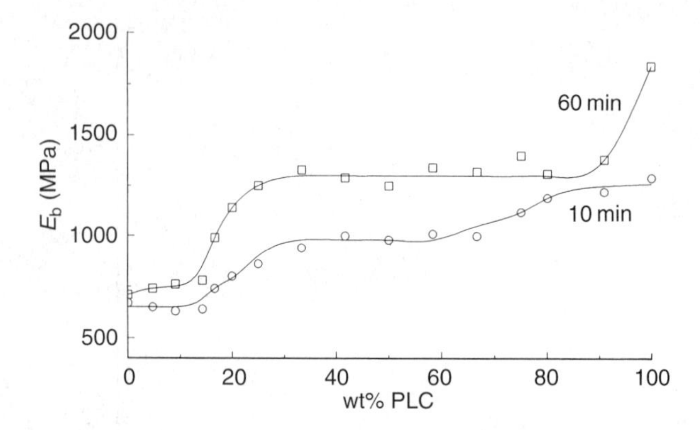

Figure 9.37 Three-point bending modulus of blends as a function of the concentration determined at 20°C [125] and [128]. Reproduced with permission of Elsevier Science Ltd.

Figure 9.25. The annealing temperature is important here, since the temperatures applied are below the onset of transesterification (*c.* 215°C) and thermal degradation (*c.* 280°C). It is important to note that in the PC + COP system there is a long plateau starting at about 18 mass% PLC up to about 80 mass% PLC (Figure 9.25), where the bending modulus, which is the measure of the reaction on a complex deformation, remains constant. Hence, only a relatively small amount of the expensive PLC is necessary in this particular system to achieve a reasonable reinforcing effect. Again, this effect can clearly be correlated with the phase diagram, in particular with the drop in the glass transition temperature of the PC-rich phase (borders of phase q in Figure 9.25). In other words, if a phase diagram indicates partial miscibility of a PLC with an EP, then mechanical reinforcement is to be expected. As already mentioned, miscibility can diminish or even prevent this effect [23, 28, 29]. A certain degree of miscibility, which is not too high, allows the formation of a second phase that may form fibrils under certain conditions depending on the processing conditions and the properties of the components of the blend. If there is a reasonable interfacial interaction that still preserves the fibrillar structure the result is a more effective adhesion. The fiber pull-out is reduced, and there is good transmission of applied forces through the blend by the fibrils. Hence these types of blends provide a wide variety of ways to influence morphology and thus to create tailored materials by intelligent processing.

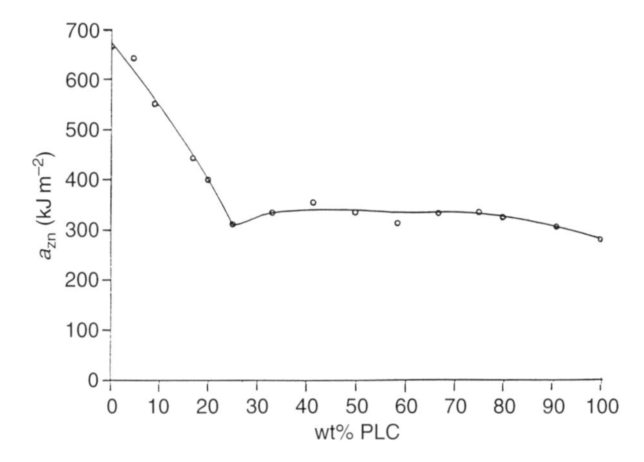

Figure 9.38 Impact resistance a_{zn} at 20°C of samples obtained from solution by acetone precipitation and kept under compression at 210–220°C and 4.0 MPa for 1 min [125] and [128]. Reproduced with permission of Elsevier Science Ltd.

9.12.3 Impact resistance

These studies were carried out at ambient temperatures; some results are shown in Figure 9.38. PC shows a much better impact resistance a_{zn} than COP because of its intense secondary phase transition at −100°C caused by the onset of the motions of the carbonate group. The flexibility enables the PC chain to absorb a certain amount of the impact energy and dissipate it. This fact stresses the importance of a phase diagram not only at ambient temperatures and beyond but also at quite low temperatures. The reason is that several dissipation processes which are attributed to the motions of smaller groups begin at low temperatures and they still act, of course, at higher temperatures. They may also be influenced by their starting temperature and strength by additional components.

Pure COP and partially crystalline blends with less than 80 mass% PC are rather brittle and show a consistently low impact resistance. Blends of this composition consist of a PC phase partially miscible with COP (Figure 9.25). The change in morphology is responsible for the sudden improvement of the impact resistance qualities in blends with more than 80 mass% PC. These particular products are almost amorphous and consist of dispersed COP spheres embedded in a tough PC matrix. The morphological and mechanical studies on COP + PC blends demonstrate the strong influence of sample preparation and blend composition on the morphological and mechanical behavior. Variation of the preparation conditions permits the production of blends with properties optimized for a certain application. There is no blend which could be regarded as the best one in its mechanical performance. A high shear modulus, for example, is always accompanied by a low impact resistance and vice versa.

ACKNOWLEDGEMENTS

Financial support was provided by the Deutsche Forschungsgemeinschaft, Bonn, Germany, and also by the Robert A. Welch Foundation, Houston, Texas, USA. (Grant B1203). An equipment grant from the Perkin-Elmer Corp., Newark, Connecticut, USA., is appreciated. Dr Frank Schubert, presently at Bakelite Gesellschaft mbH, Duisburg, Germany, participated in much experimental work at the Gerhard Mercator University, Duisburg, Germany. We also gratefully acknowledge the contribution of some specially processed samples by Dr Tomasz Sterzynski, Ecole d'Application des Hautes Polymères, Louis Pasteur University, Strasbourg, France. We further have to thank Messrs James M. Berry, Kelly R. Degler and Rogelio F. Rodriguez, all of University of North Texas, Denton, Texas, USA for their technical assistance. MH is pleased to thank the University of North Texas for inviting him as a visiting professor, Prof. H.G. Zachmann, University of Hamburg, Germany, for providing measuring time at the HASYLAB (Hamburger Synchrotron Laboratory) of the DESY (Deutsches Elektronen Synchrotron) and his crew for their technical assistance. Last but not least we wish to express our thanks to Witold Brostow, Department for Materials Science, University of North Texas, Denton, for providing equipment and support (not only financial). The fruitful discussions the authors had with him contributed to some of the results presented in this chapter.

REFERENCES

1. Halpin, J.C. and Kardos, J.L. (1976) *Polymer Eng. & Sci.*, **16**, 344.
2. Roviello, A. and Sirigu, A. (1975) *J. Polymer Sci. Lett.*, **13**, 455.
3. Hannell, J.W. (1970) *Polym. News*, **1**, 8.
4. Wilfong, R.E. and Zimmerman, J. (1973) *J. Appl. Polymer Sci.*, **17**, 2039.
5. Kwolek, S.L. (1971) US Patent 3600350.
6. Kwolek, S.L. (1974) US Patent 3819587.
7. Morgan, P.W. (1977) *Macromolecules*, **10**, 1381.
8. Kuhfuss, H.F. and Jackson, W.J. (1973) US Patent 3778410.
9. Kuhfuss, H.F. and Jackson, W.J. (1974) US Patent 3804805.
10. Kuhfuss, H.F. and Jackson, W.J. (1976) *J. Polymer Sci. Chem.*, **14**, 2043.
11. Brostow, W. (1990) *Polymer*, **31**, 979.
12. Jin, J.-I., Ober, C. and Lenz, R.W. (1980) *Br. Polymer J.*, **12**, 132.

13. Ober, C.K., Jin, J.-J. and Lenz, R.W. (1984) *Adv. Polymer Sci.*, **59**, 103.
14. Franek, J., Zbignew, J.J., Majnusz, J. (1992) in *Handbook of Polymer Synthesis*, Part B (ed. H.R. Kricheldorf), Marcel Dekker, New York, p. 1281.
15. Ciferri, A. (1982) in *Polymer Liquid Crystals* (eds A. Ciferi, W.R. Krigbaum and R.B. Mayer), Academic Press, New York, p. 63.
16. Wunderlich, B. and Grebowicz, J. (1984) *Adv. Polymer Sci.*, **60/61**, 3.
17. Wierschke, S.G. (1989) *Proc. Symp. Mater. Res. Soc.*, **134**, 313.
18. Brostow, W., Ertepinar, H. and Singh, R.P. (1990) *Macromolecules*, **23**, 5109.
19. La Mantia, F.P., Valenza, A., Paci, M. and Mangagnini, P.L. (1989) *J. Appl. Polymer Sci.*, **38**, 583.
20. Brostow, W., Dziemianowicz, T.S., Hess, M. and Kosfeld, R. (1989) *Polymer Prepr. Am. Chem. Soc.*, **30**, 542; (1990) *Mater. Res. Soc. Symp. Proc.*, **171**, 177.
21. Sukhadia, A.M., Done, D. and Baird, D.G. (1990) *Polymer Eng. & Sci.*, **30**, 519.
22. Dutta, D. and Weiss, R.A. (1991) *ACS Symp. Ser.*, **462**, 144.
23. Suenaga, J., Fujita, E. and Marutani, T. (1991) *Kobunshi Ronbunshu*, **48**, 573.
24. Hong, S.M., Kim, B.C., Kim, K.U. and Chung, I.J. (1991) *Polymer J.* (Tokyo), **23**, 1347.
25. Crevecoer, G. and Groeninckx, G. (1991) in *Integration of Fundamental Polymer Science and Technology* (eds P. Lemstra and L.A. Kleintjens), London, p. 251.
26. Pawlikowski, G.T., Dutta, D. and Weiss, R.A. (1991) *Ann. Rev. Mater. Sci.*, **21**, 159.
27. La Mantia, F.P., Valenza, A., Paci, M. and Mangagnini, P.L. (1990) *Mater. Eng.* (Modena), **1**, 223.
28. Classen, M.S., Schmidt, H.W. and Wendorff, J.H. (1990) *Polymer Adv. Technol.*, **1**, 143.
29. Seppala, J., Heino, M. and Kapanen, C. (1992) *J. Appl. Polymer Sci.*, **44**, 1051.
30. Akhtar, S. and Isayev, A.I. (1993) *Polymer Eng. & Sci.*, **33**, 32.
31. McMurrer, M. (1992) *Plast. Comp.*, **15**, 35.
32. Sukhadia, A.M., Datta, A. and Baird, D.G. (1991) *Proc. Ann. Techn. Conf. Soc. Plast. Eng.*, **49**, 1008.
33. Lee, W.C. and DiBennedetto (1993) *Polymer*, **34**, 684.
34. Heino, M.T. and Seppala, J.V. (1993) *Polymer Bull.*, **30**, 353.
35. Brostow, W. (1991) *Isr. Mater. Eng. Conf.*, **5**, 219.
36. Heino, M.T. and Seppala, J.V. (1993) *J. Appl. Polymer Sci.*, **48**, 1677.
37. Hsu, T.C., Lichkus, A.M. and Harrison, I.R. (1993) *Polymer Eng. & Sci.*, **33**, 860.
38. Michaeli, W., Brinkmann, T., Heidemeyer, P. *et al.* (1993) *Kunststoffe*, **83**, 312.
39. Heino, M.T., Hietaoja, P.T., Vainio, T.P. and Seppala, J.V. (1994) *J. Appl. Polymer Sci.*, **51**, 259.
40. Lee, S., Hong, S.M., Seo, Y. *et al.* (1994) *Polymer*, **35**, 519.
41. Baird, D.G., Bafna, S.S., De Souza, J.P. and Sun, T. (1994) *Polymer Comp.*, **14**, 214.
42. Gupta, B., Calundann, G., Carbonneau, L.F. *et al.* (1994) *J. Appl. Polymer Sci.*, **53**, 575.
43. Sun, L.-M., Sakamoto, T., Ueta, S. *et al.* (1994) *Polymer J.* (Tokyo), **26**, 953.
44. Engberg, K., Stromberg, O., Martinsson, J. and Gedde, U.W. (1994) *Polymer Eng. & Sci.*, **34**, 1336.
45. Shi, F. (1994) *Polymer Plast. Technol. Eng.*, **33**, 445.
46. Ide, Y. and Ophir, Z. (1983) *Polymer Eng. & Sci.*, **23**, 261.

47. Laivins, G. (1989) *Macromolecules*, **22**, 3974.
48. Blizard, K.G. and Baird, D.G. (1986) *ANTEC*, 311.
49. Kyotani, M., Kaito, A. and Nakayama, K. (1992) *Polymer*, **33**, 4756.
50. Brostow, W., Hess, M. and Lopez, B.L. (1994) *Macromolecules*, **27**, 2262.
51. Sawyer, L.C. and Jaffe, M.J. (1986) *J. Mater. Sci.*, **21**, 1897.
52. Brostow, W. and Hess, M. (1992) *Mater. Res. Symp.*, **255**, 573.
53. Kosfeld, R., Hess, M. and Friedrich, K. (1987) *Mater. Chem. & Phys.*, **18**, 93.
54. Roe, R.-J. and Rigby, D. (1987) *Adv. Polymer Sci.*, **82**, 104.
55. Wu, S. (1978) in *Polymer Blends*, Vol. 1, (eds D.R. Paul and S. Newman), Academic Press, New York, p. 243.
56. Wissbrun, K.F. (1985) *Faraday Trans.*, **79**, 161.
57. Ruland, W. (1977) *Colloid & Polymer Sci.*, **255**, 417.
58. Jonah, D., Brostow, W. and Hess, M. (1993) *Macromolecules*, **26**, 76.
59. Rehage, G. and Borchard, W. (1973) in *The Physics of Glassy Polymers* (ed. G. Haward), Applied Science Publ., London, p. 54.
60. Gordon, M. and Taylor, J.S. (1952) *J. Appl. Chem.*, **2**, 493.
61. Fox, T.G. (1956) *Bull. Amer. Phys. Soc.*, **1**, 123.
62. Kelley, N.F. and Bueche, F. (1961) *J. Polymer Sci.*, **50**, 549.
63. Kanig, G. (1963) *Kolloid Z.*, **190**, 1.
64. George, E.R. and Porter, R.S. (1988) *J. Polymer Sci. Phys.*, **26**, 83.
65. Blonski, S., Brostow, W., Jonah, D. and Hess, M. (1993) *Macromolecules*, **26**, 84.
66. Flory, P.J. (1953) *Principles of Polymer Chemistry*, Cornell University Press, Ithaca, New York.
67. Nishi, T. and Wang, T.T. (1975) *Macromolecules*, **8**, 909.
68. Buchner, S., Chen, D., Gehrke, R. *et al.* (1988) *Mol. Cryst. Liq. Cryst.*, **155**, 357.
69. Flory, P.J., Orwoll, R.A. and Vrij, A. (1964) *J. Am. Chem. Soc.*, **86**, 3515.
70. Flory, P.J. (1965) *J. Am. Chem. Soc.*, **87**, 1833.
71. Eichinger, B.E. and Flory, P.J. (1968) *Trans. Faraday Soc.*, **64**, 2035.
72. Flory, P.J. (1970) *Disc. Faraday Soc.*, **49**, 7.
73. Sanchez, I.C. and Lacombe, R.H. (1976) *J. Phys. Chem.*, **80**, 2352.
74. Lacombe, R.H. and Sanchez, I.C. (1976) *J. Phys. Chem.*, **80**, 2568.
75. Sanchez, I.C. and Lacombe, R.H. (1977) *J. Polymer Sci. Lett.*, **15**, 71.
76. Simha, R. and Somcynsky, T. (1968) *Macromolecules*, **2**, 342.
77. Somcynsky, T. and Simha, R. (1971) *J. Appl. Phys.*, **42**, 4545.
78. Nose, T. (1971) *Polymer J.*, **2**, 124.
79. Nose, T. (1971) *Polymer J.*, **2**, 427.
80. Curro, J.G. (1974) *J. Macromol. Sci.; Rev. Chem. C*, **11**, 321.
81. Beret, S. and Prausnitz, J.M. (1975) *Macromolecules*, **8**, 878.
82. Flory, P.J. (1956) *Proc. Royal Soc. A.*, **234**, 60.
83. Flory, P.J. (1956) *Proc. Royal Soc. A.*, **234**, 74.
84. Hermans, P.H. and Platzek (1939) *Kolloid Z.; Z. Polymerc*, **88**, 68.
85. Hermans, J.J., Hermans, P.H., Vermaas, D. and Weidinger, A. (1946) *Rec. Trav. Chim. Pays-Bas*, **65**, 427.
86. O'Reilly, J.M. and Karasz, F.E. (1965) *Polymer Prepr. Am. Chem. Soc.*, **6**, 731.
87. Sun, T. and Porter, R.S. (1990) *Polymer Commun.*, **31**, 70.
88. Benson, R.S. and Lewis, D.N. (1987) *Polymer Commun.*, **28**, 289.
89. Chen, D. and Zachmann, H.G. (1991) *Polymer*, **32**, 1612.
90. Schatzki, F. (1966) *J. Polymer Sci. Phys.*, **14**, 139.
91. Seefried, C.G. and Koleske, J.V. (1975) *J. Polymer Sci. Phys.*, **13**, 851.
92. Schurer, J.W., de Boer, A. and Challa, G. (1975) *Polymer*, **16**, 201.

93. Brostow, W. and Corneliussen, R.D. (1986) *Failure of Plastics*, Hanser Verlag, Munich.
94. Brostow, W. (1985) *Mater. Chem. & Phys.*, **13**, 47.
95. Brostow, W. and Macip, M.A. (1989) *Macromolecules*, **22**, 2761.
96. Viney, C. and Windle, A.H. (1982) *J. Mater. Sci.*, **17**, 2661.
97. Kricheldorf, H.R. and Schwarz, G. (1990) *Polymer*, **31**, 481.
98. Bonnell, D.A. (1993) *Scanning Electron Microscopy*, VCH, Weinheim.
99. Magonov, S.N. (1993) *Appl. Spectr. Rheol.*, **28**, 1.
100. Balta Calleja, F.J., Santa Cruz, C., Chen, D. and Zachmann, H.G. (1991) *Polymer*, **32**, 2252.
101. Brostow, W., Kaushik, B., Mall, S. and Talwar, I. (1992) *Polymer*, **33**, 4687.
102. Brostow, W. and Samatowicz, D. (1993) *Polymer Eng. & Sci.*, **33**, 581.
103. Gedde, U.W., Burger, D. and Boyd, R.H. (1987) *Macromolecules*, **20**, 988.
104. Takase, Y., Mitchell, G.R. and Odajima, A. (1986) *Polymer Commun.*, **27**, 76.
105. Yoon, D.Y., Masciocchi, N., Depero, L.E. *et al.* (1990) *Macromolecules*, **23**, 1793.
106. Walsh, D.J., Dee, G.T. and Wojtkowski, P.J. (1989) *Polymer*, **30**, 1467.
107. Veeman, W.S. (1993) *Makromol. Chem. Symp.*, **69**, 149.
108. Veeman, W.S., Maas, W.E.J.R. (1994) in *NMR, Basic Principles and Progress*, (eds P. Diehl, E. Fluck, H. Günther *et al.*, guest ed. B. Blümich), **32**, 127.
109. Schmidt-Rohr, K. and Spiess, H.W. (1994) *Multidimensional Solid State NMR and Polymers*, Academic Press, London.
110. Beckham, H.W. and Spiess, H.W. (1994) in *NMR, Basic Principles and Progress*, (eds P. Diehl, E. Fluck, H. Günther *et al.*, guest ed. B. Blümich), **32**, 163.
111. Bare, J. (1975) *Macromolecules*, **8**, 244.
112. Wetton, R.E., Moore, J.D. and Ingram, P. (1973) *Polymer*, **14**, 161.
113. Zhuang, P., Kuy, T. and White, J.L. (1988) *Polymer Eng. & Sci.*, **28**, 1095.
114. Stoelting, J., Karasz, F.E. and McKnight, W.J. (1970) *Polymer Eng. & Sci.*, **10**, 133.
115. Garton, A. (1982) *Polymer Eng. & Sci.*, **22**, 124.
116. Vinogradov, Y.L., Martinov, M.A. and Olshanik, G.A. (1975) *Polymer Sci. USSR*, **7**, 1605.
117. Hammer, C.F. (1971) *Macromolecules*, **4**, 347.
118. Brostow, W., Dziemianowicz, T.S., Romanski, J. and Werber, W. (1988) *Polymer Eng. Sci.*, **18**, 785.
119. Lenz, R.W., Jin, J.-I. and Feichtinger, K.A. (1983) *Polymer*, **24**, 327.
120. Brostow, W., Dziemianowicz, T.S., Hess, M. and Kosfeld, R. (1990) *Mater. Res. Soc. Symp.*, **171**, 177.
121. Krause, S.J., Haddock, T., Price, G.E. *et al.* (1986) *J. Polymer Sci.: Polymer Phys.*, **24**, 1991.
122. Kiss, G. (1987) *Polymer Eng. & Sci.*, **27**, 410.
123. Weiss, R.A., Huh, W. and Nicolais, L. (1987) *Polymer Eng. & Sci.*, **27**, 684.
124. Zachmann, H.G., Spies, C. and Thiel, S. (1993) *Physica Scripta*, **T49**, 247.
125. Schubert, F. (1989) Thesis, Gerhard Mercator University Duisburg, Germany.
126. Brostow, W. (1979) *Einstieg in die Moderne Werkstoffwissenschaft*, Carl Hanser Verlag, Munich–Vienna.
127. Brostow, W. (1979) *Science of Materials*, Wiley, New York–London.
128. Brostow, W., Hess, M., Lopez, B.L. and Sterzynski, T. (1996) *Polymer*, **37**, 1551.

10

Development and relaxation of orientation in pure polymer liquid crystals and blends

Ulf W. Gedde and Göran Wiberg

10.1 INTRODUCTION

Orientational order is a natural feature of polymer liquid crystals (PLCs) in all their mesomorphic states ranging from nematic to smectic. The early theoretical developments due to Onsager [1], Ishihara [2] and Flory [3] predicted the formation of an anisotropic phase in solutions of rigid rod molecules. The packing of molecules with an anisotropic shape, e.g. longitudinal molecules, is greatly facilitated if the molecules are oriented in a preferential direction, i.e. if they are close to parallel (Figure 10.1). The theory of Flory [3] predicts a transition from an anisotropic to an isotropic phase at a certain concentration of rigid rod molecules in a low molar mass solvent (lyotropics). The critical concentration of rigid rods for the formation of an anisotropic phase decreases with increasing length-to-width ratio of the rigid rod molecules. Principally the same effect can be achieved during the cooling of a rigid rod polymer melt. At high temperatures, the excessive volume allows the presence of non-parallel rods. At a certain lower temperature, at which the relative free volume is so small that the packing of the

Mechanical and Thermophysical Properties of Polymer Liquid Crystals
Edited by W. Brostow
Published in 1998 by Chapman & Hall, London.
ISBN 0 412 60900 2

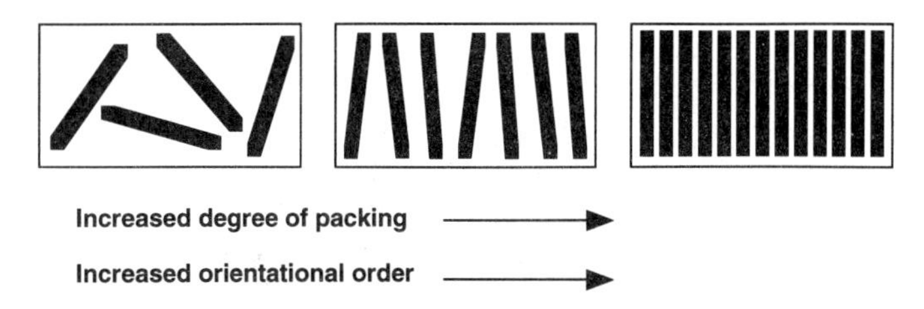

Figure 10.1 Illustration of the relationship between orientational order and degree of packing.

rods needs to be considerably more parallel, the anisotropic (LC) phase is formed. There has been further development of theory founded on the original work of Flory: rigid rods in solution considering also orientation dependent interactions [4], polydisperse rigid rods and quasi-spherical solvent molecules [5], rods connected via flexible units [6–8] and ternary systems consisting of PLC, flexible chain polymer and solvent [9] to mention a few important contributions. The theory of PLCs is discussed further by Brostow and Walasek in Chapter 1 in Volume 4 of this series, particularly from the point of view of competition between energetic and entropic effects.

This chapter deals exclusively with thermotropics (pure polymer systems). This review deals with longitudinal, one-comb and network PLCs and blends of longitudinal PLCs and conventional, flexible chain polymers (Figure 10.2). An excellent introduction to the subject including a presentation of definitions and classifications is given by Brostow [10]. It is important to emphasize that a given PLC at a given

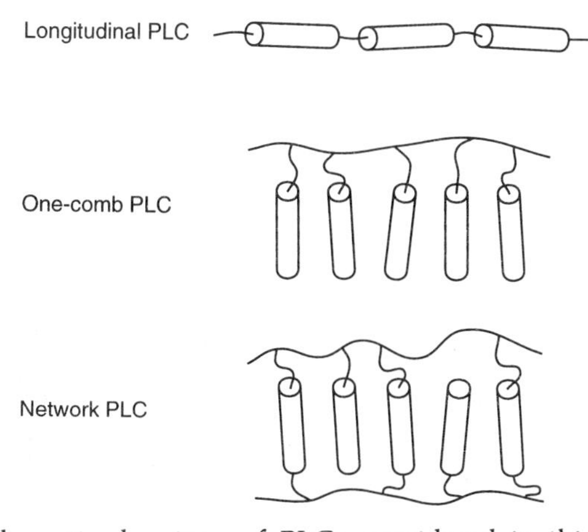

Figure 10.2 Schematic drawings of PLCs considered in this chapter. The mesogens are drawn as cylinders and the flexible and linking groups as thin curved lines.

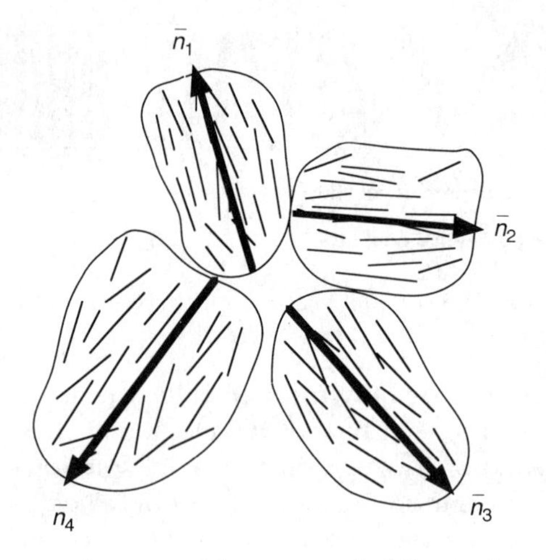

Figure 10.3 Schematic drawing of domains with different directors ($\boldsymbol{n}_i$) in a PLC.

temperature and pressure exhibits only a typical 'domain' or 'island' orientational order. The term 'island' was coined by Brostow *et al.* [11]. A 'domain' (island) constitutes a relatively small volume, of the order of a micrometer in diameter, which can be characterized by a director ($\boldsymbol{n}_i$), which corresponds to the average direction of the mesogens (Figure 10.3). It is typical for most nematics that they consist of many domains and that the average of all the domain directors is zero. The 'domain' concept should be considered only as a simplification because in most PLCs there is a continuous change in the director field within the structure. Domain boundaries are in many cases not readily found. The continuous change in the director field may be associated with a disclination line around which the director field rotates (Figure 10.4). The overall orientation for a whole macroscopic sample is in many cases negligible, although the domain orientation may be substantial (Figure 10.3).

In the next section we shall introduce concepts and quantities which

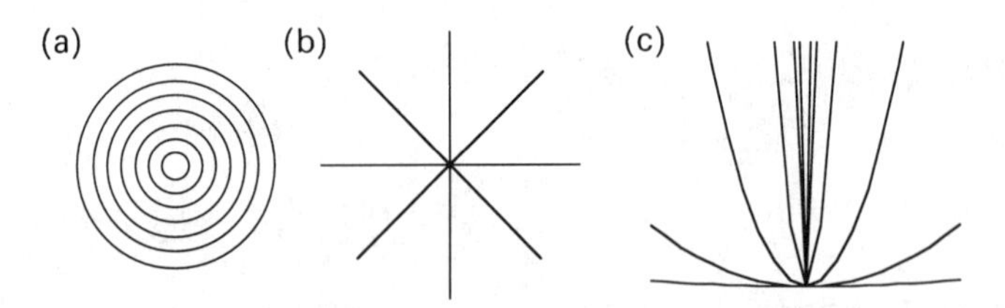

Figure 10.4 Illustration of disclinations of strength ± 1 ((a) and (b)) and $\frac{1}{2}$ ((c)). The lines indicate the director field orientation.

are used to describe orientational order in polymers in general and in PLCs in particular. A very brief description of methods which are used for the assessment of these quantities is also included in the first section. The next section deals with the development of orientational order in PLCs using shear flow, extensional flow, electric, magnetic and surface fields. The final section reviews the loss of orientational order in PLCs, a process strongly dependent on temperature and on the chemical and physical structure of the polymer.

10.2 CHAIN ORIENTATION: FUNDAMENTAL ASPECTS

Comprehensive reviews of chain orientation have been given by Samuels [12], Ward [13], Struik [14] and Gedde [15]. Chain orientation is a phenomenon unique to polymers. The unidimensional nature of the linear polymer chain makes it possible to obtain strongly anisotropic material properties. The anisotropy arises when molecules are aligned along a common director. The intrinsic properties of a polymer chain are strongly direction dependent. The strong covalent bonds along the chain axis and the much weaker secondary bonds in the transverse directions cause a significant anisotropy of any given tensor property. The concept of orientation would be meaningless if the chain-intrinsic properties were isotropic. The concept of orientation is also applicable to the one-comb molecules and in this case it may refer to the directional order of the mesogens. Orientation always refers to a particular structural unit which is determined by the experimental method used for the assessment. It may be a whole specimen or a macroscopic part of a specimen as is the case for the conventional X-ray diffraction methods, or it may be a microscopic part as in the case of electron diffraction. The director varies as a function of location in both types of sample, which emphasizes the fact that it is possible to find smaller 'domains' with a higher degree of orientation.

Figure 10.5 illustrates this with another example. The orientation of the segments of a molecule in a single crystal is indeed very high. More than 90% of the molecule is perfectly aligned provided that the

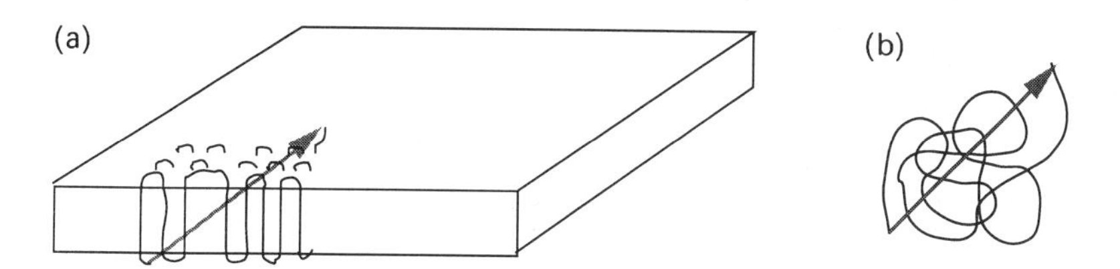

Figure 10.5 (a) A chain in a single crystal and (b) a Gaussian chain. (From Gedde [15] with permission from Chapman & Hall, London.)

crystal thickness is greater than 10 nm. The end-to-end vector is not, however, very long, i.e. the molecular draw ratio is not very large. The Gaussian chain shown in Figure 10.5 exhibits a similar end-to-end group separation but the segmental orientation is completely random. Segmental orientation is revealed by the measurement of optical birefringence and by wide angle X-ray diffraction and infrared spectroscopy. The orientation of the end-to-end vector has only more recently been assessed by neutron scattering.

The Hermans orientation function is probably the quantity most frequently used to characterize orientation. This orientation function was introduced by P.H. Hermans in 1946 and is part of an equation which relates optical birefringence (Δn) to chain (segmental) orientation:

$$f = \frac{\Delta n}{\Delta n_0} = 1 - \frac{3}{2}\langle \sin^2 \phi \rangle = \frac{3\langle \cos^2 \phi \rangle - 1}{2} \tag{10.1}$$

where f is the Hermans orientation function, ϕ is the angle between a chain segment and the director and Δn_0 is the maximum birefringence which is given by

$$\Delta n_0 = \frac{(\langle n \rangle^2 - 2)^2}{6\langle n \rangle} \frac{4\pi}{3}(p_1 - p_2) \tag{10.2}$$

where p_1 and p_2 are the polarizabilities parallel to and perpendicular to the chain axis and $\langle n \rangle$ is the average refractive index. The Hermans orientation function takes the value 1 for a system with complete orientation parallel to the director and it takes the value $-\frac{1}{2}$ for the same sample with the director perpendicular to the chain axis. Non-oriented samples have a value of f equal to zero (Figure 10.6). It is important to realize that the 'director' should here be considered as a vector pointing in any direction and not only as the average direction of mesogens or chains. If we however constrain the director so that it is only the average direction of the mesogens in a PLC, it will take values between 0 and 1.

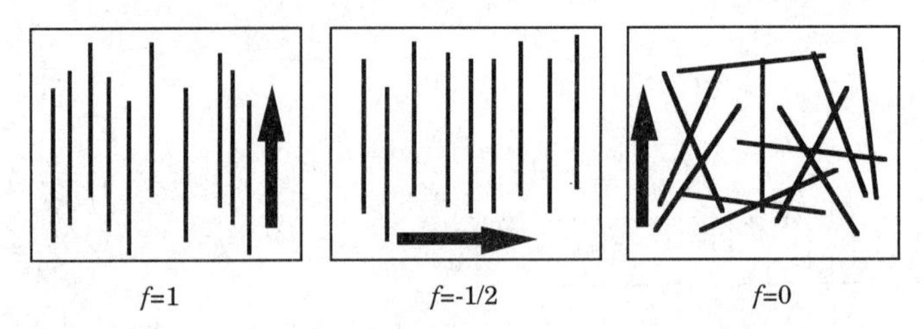

Figure 10.6 Values of the Hermans orientation function for three simple cases. The directors are marked with arrows.

The Hermans orientation function can be given a relatively simple interpretation. A sample with orientation f may be considered to consist of perfectly aligned molecules with mass fraction f and randomly oriented molecules with mass fraction $1 - f$. PLCs are often characterized by their order parameter (denoted s), a concept coined by Tsvetkov [16]. This quantity is basically equivalent to the Hermans orientation function. For nematics and smectics the director represents the average mesogen direction and the order parameter can thus only take positive values. For cholesteric phases, on the other hand, the director is chosen perpendicular to the layers and in this case the order parameter takes negative values.

The orientation can be viewed in more general terms in terms of an orientation probability function $f(\phi)$, where the angle ϕ is the angle between the director and the molecular segment. It is implicit in this treatment that the orientation in the plane perpendicular to the director is random, i.e. that the orientation is uniaxial (Figure 10.7). The in-plane orientation is the same in the zx and zy planes. It is possible to represent $f(\phi)$ by a series of spherical harmonic functions.

Polymers may exhibit a biaxial orientation. The segmental orientation function is in this case a function of two angular variables, i.e. $f(\phi, v)$, as shown in Figure 10.7. The in-plane orientation is different in the zx and zy planes (Figure 10.7). There are several methods commonly used to determine chain orientation: in-plane birefringence, wide angle X-ray diffraction, small angle X-ray diffraction, infrared spectroscopy and sonic modulus measurements. In the case of uniaxial orientation there is only

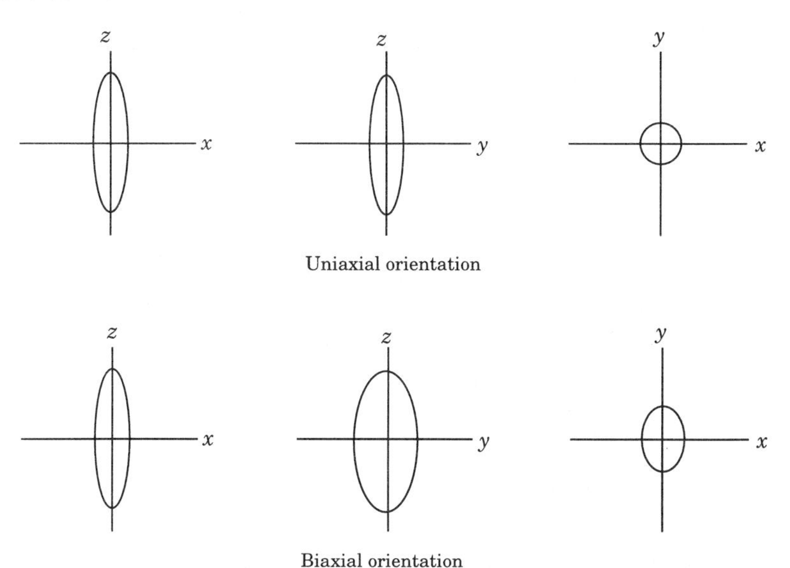

Figure 10.7 Illustration of uniaxial and biaxial orientations.

one unique direction, the z (3) direction. Orientation 'exists' only in the xz and yz planes. The uniaxial system appears to be isotropic in the xy plane.

It is possible to assess the Hermans orientation function by birefringence measurements, X-ray diffraction and infrared spectroscopy. The full description of uniaxial orientation $f(\phi)$ cannot thus be attained by a single birefringence measurement.

Birefringence measurements may involve the use of a polarized light microscope with the crossed polarizer–analyzer pair at an angle of 45° to the main optical axes of the sample (domain). A Babinet or tilt compensator, which is a birefringent component, is introduced between the sample and the analyzer. The compensator makes it possible to change the optical retardation of the vertically and horizontally polarized light components in order to compensate for the optical retardation introduced by the sample (domain). A comprehensive review of optical methods for the assessment of birefringence and chain orientation is given by Read *et al.* [17].

The in-plane orientation of (hkl) planes in semicrystalline polymers is revealed by wide angle X-ray diffraction. Comprehensive reviews on the subject are given by Samuels [12] and Baltá-Calleja and Vonk [18]. X-ray diffraction provides information about the structure and orientation of PLCs (Figure 10.8). An unoriented smectic polymer shows concentric rings for the intermesogen wide angle and smectic layer small angle reflections (Figure 10.8a). An oriented polymer shows pronouncedly azimuthal angle-dependent reflections (Figure 10.8b). The smectic layer reflection is concentrated to the meridian and the intermesogen reflection to the equator. The orientation, expressed by the mean square of the cosine of the angle ϕ between the scattering planes (smectic layers, s) and (intermesogenic, i) and the director, can be obtained from

$$\langle \cos^2 \phi_x \rangle = \frac{\int_0^\pi I_x(\phi) \cos^2 \phi \sin \phi \, d\phi}{\int_0^\pi I_x(\phi) \sin \phi \, d\phi} \tag{10.3}$$

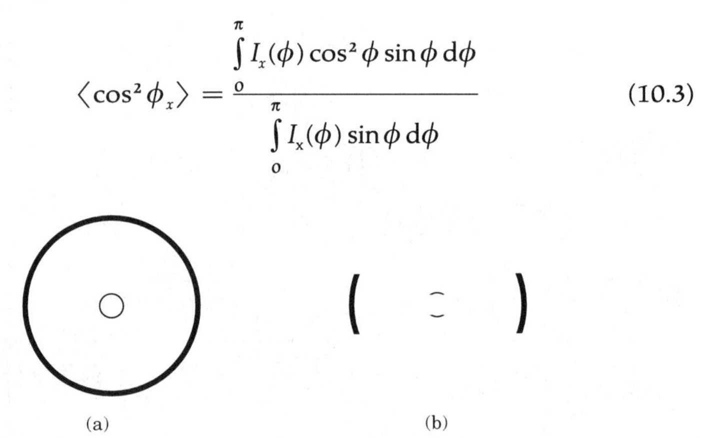

Figure 10.8 X-ray diffraction patterns from (a) unoriented smectic A and (b) oriented smectic A.

where $I_x(\phi)$ is the scattered intensity of the x reflection at the azimuthal angle ϕ. It is possible to calculate an order parameter for the smectic layers by inserting the result of equation (10.3) in equation (10.1). The Hermans orientation function (order parameter) can also be calculated for the mesogenic groups on the basis of $I_i = f(\phi)$ according to

$$f = -2f_i \tag{10.4}$$

One of the main attractions of the X-ray method is that the full orientation function $f(\phi)$ can be obtained in the case of uniaxial orientation and the function $f(\phi, \nu)$ in the case of biaxial orientation.

IR spectroscopy is very useful for the assessment of chain orientation. Recent excellent reviews on the subject have been given by Bower and Maddams [19] and Koenig [20]. The measurement of IR dichroism requires the use of IR radiation with polarization parallel to and perpendicular to a selected reference direction. The fundamental principle underlying the dichroism is that the absorbance A is proportional to the square of the cosine of the angle between the transition moment vector and the electrical field vector. The IR radiation is strongly absorbed when the electric vector of the light and the transition moment vector are parallel. No absorption of the IR radiation occurs however if the two vectors are perpendicular to each other. The dichroic ratio R is defined as

$$R = \frac{A_{\parallel}}{A_{\perp}} \tag{10.5}$$

where $A_{\parallel}$ is the absorbance of the radiation polarized parallel to the director and $A_{\perp}$ is the absorbance of radiation polarized perpendicular to the director. The Hermans orientation function f is given by

$$f = \frac{(R-1)(R_0+2)}{(R+2)(R_0-1)} \tag{10.6}$$

where R_0 is the dichroic ratio for a sample with perfect uniaxial orientation and is dependent on the angle between the transition moment vector and the chain axis (ψ):

$$R_0 = 2\cot^2\psi \tag{10.7}$$

Absorption bands with $\psi = 54.76°$ show no dichroism ($R_0 = 1$). Absorption bands associated with a perpendicular transition moment vector have $R_0 = 0$ and the Hermans orientation function is given by

$$f = 2 \cdot \frac{(1-R)}{(R+2)} \tag{10.8}$$

One of the attractions of the IR technique is that the dichroism of the absorption bands assigned to different groups can be determined and that the orientation of different groups of the repeating unit can be determined.

NMR spectroscopy to characterize thermotropic PLCs was first reported by McFarlane *et al.* [21] and Calundann and Jaffe [22], although the first actual measurements of the nematic order parameter (i.e. the domain order parameter of the nematic polymer; s_n) and the fraction of nematic phase (in the biphasic region) was first made by Blumstein and coworkers [23–25] and Samulski [26] using ^1H and ^2D-NMR. The nematic order parameter was calculated on the basis of the dipolar and quadrupolar splitting of the NMR spectrum when cooling from the isotropic phase.

The domain order parameter characteristic of the LC phase depends on the nature of the phase, i.e. whether it is nematic or smectic, and on the temperature. The order parameter of glassy nematic PLCs is in the range 0.45–0.65 whereas smectic PLCs show larger domain order parameter values, between 0.85 and 0.92 [27]. Nematic MLCs show higher domain order parameters than their polymeric analogs whereas smectic MLCs and PLCs exhibit similar domain order parameter values [27]. This effect has been found e.g. for one-comb PLCs based on acrylate, methacrylate and siloxane backbones [28, 29]. It may be concluded that the linkage of a mesogen to a backbone (one-comb PLC) or to another mesogen (longitudinal PLC) reduces the nematic order parameter.

The values quoted above for the domain order parameter refer to the alignment of the mesogenic groups. It is known, for instance, that the order parameter of the spacer methylene units in smectic one-comb PLCs is only 0.3–0.4 whereas the mesogenic group order parameter is close to 0.9 [30]. Finkelmann *et al.* [31] reported the following data on the order parameter of absorbing dichroic dyes, chemically bonded to a nematic, one-comb PLC with a siloxane backbone, close to the isotropization temperature T_i: 0.48 ($T/T_i = 0.90$); 0.40 ($T/T_i = 0.95$); 0.30 ($T/T_i = 0.99$); 0.2 ($T/T_i = 1.0$).

The overall order parameter s can, according to Donald and Windle [32], be obtained from the domain order parameter s_d and the distribution of the domain directors s_{dd} according to the expression

$$s = s_d \cdot s_{dd} \tag{10.9}$$

Equation (10.9) provides an elegant and simple view of the order parameter concept with the 'separation' of the overall order parameters into the two parameters s_d and s_{dd}. We may consider the development of orientation in a PLC as being primarily the increase of s_{dd} from zero to a value closer to unity. We shall in the remaining part of this chapter use the concept 'order parameter' for the description of orientation and it should be remembered that the order parameter is completely equivalent to the Hermans orientation function.

10.3 DEVELOPMENT OF ORIENTATION IN PLCs

10.3.1 General aspects

PLCs may be considered to consist of a great many 'domains', each domain characterized by an order parameter s_d. The directors of the domains can be more or less randomly distributed. Alignment of the domain and thus of the molecular directors to a common, global director may be accomplished by the application of mechanical, electric or magnetic fields or under the influence of a surface field. Different PLCs require different methods for their alignment. Mechanical action (shear or extensional flow) is effective in orienting longitudinal PLCs, whereas one-comb PLCs can be aligned in electric, magnetic, mechanic or surface fields. The surface methods are less effective in orienting longitudinal PLCs. The following sections make a brief presentation of these methods for the orientation of PLCs, while Chapter 9 reviews some data of the effect of drawing on the mechanical properties of PLCs.

10.3.2 Alignment by shear or extensional flow

Ide and Ophir [33] and Viola *et al.* [34] reported that extensional flow was more efficient than shear flow to orient nematic longitudinal PLCs. Cold drawing of longitudinal PLCs leads to a slippage of domains past each other without any significant increase in the overall degree of orientation [35].

Graziano and Mackley [36] reported a very extensive study on oscillatory sheared, nematic, longitudinal PLCs based on a series of random 50/50 copolyesters of chlorophenylene terephthalate and bis-phenoxyethane carboxylate. They observed a variety of textures in the polarized microscope and also noticed a pronounced difference between low and high molar mass samples.

The low molar mass sample displayed a collection of thin (1 μm) and thicker threads at zero and low shear rates. Thicker threads were observed only in the thicker films. Both types of threads formed mostly closed loops. Threads which were not attached to the surfaces were mobile and they moved with the 'bulk' of the film. The threads that were 'pinned' to the opposite surface broke up on shearing into dark lines which elongated along the flow direction. The entire field was 'covered' by a great many elongated black lines on prolonged shearing. This black line texture broke up into short, curled black entities (so-called 'worms'; the texture being denoted 'worm texture') on further increase of the shear rate. The worms flow along the flow field while continually changing their shape. The density of worms increased with increasing shear rate. At even higher shear rates, an ordered texture extinguishing light with the polarizer–analyzer pair parallel to or at an angle of 90° to the flow direction and with a birefringence of 0.03 ± 0.02. This oriented texture showed rapid relaxation during the first 10 s, first by the formation of thin threads in the form of close loops and at a later stage by the shrinkage and annihilation of the thin threads. The quiescent high molar mass sample exhibited only the worm texture without the appearance of thin threads. Shear led to appreciable orientation with a birefringence of 0.11 ± 0.03. The relaxation of the oriented texture is dealt with in this section, although only one of three relaxation types led to the formation of bands aligned perpendicular to the prior flow direction. Graziano and Mackley [36] argued that in a shear gradient small molecules will readily slip past each other and that the molecules will align in the flow direction and reduce their viscous drag. As the molar mass increases, the independent motion of the molecules becomes more hindered by their length. The molecules can no longer flow past each other and they are forced to restructure in the flow field. Structural changes created by shear should thus occur more readily in polymers with a high molar mass.

A follow-up study using the same oscillatory shear apparatus was reported in the following year by Alderman and Mackley [37] including both Vectra (poly(*p*-hydroxybenzoic acid-co-2,6-hydroxynaphthoic acid)) with 73 mol% *p*-hydroxybenzoic acid) and X7G (poly(*p*-hydroxy-benzoic acid-co-ethylene terephthalate)) with 60 mol% *p*-hydroxybenzoic acid). Both polymers were polydomain-like prior to shear and exhibited sharp schlieren textures with disclinations. With increasing angular velocity (increasing shear rate), the onset of an extensive disclination multiplication was observed starting at a certain critical shear rate. At even higher shear rates the birefringence became uniform and the structure within the melt disappeared. Alderman and Mackley [37] proposed that the disclination multiplication progressed to a point where the structures became smaller than the resolution of the microscope. Kulichikhin [38], on the other hand, suggested that the formation of the uniformly birefringent texture at high shear rates is due to the destruction of disclination rather than to the formation of new ones. He considers the polydomain morphology appearing at the lower shear rates to be viscous due to the many disclinations which he considers to be 'crosslinks'. Donald and Windle [32] summarize the status with the relevant questions: what is the nature of the structure of the system at high shear rates – is it uniformly oriented or not?

So-called banded structures, shown in Figure 10.9, are formed in PLCs. The banded structures do not appear in monomer liquid crystals

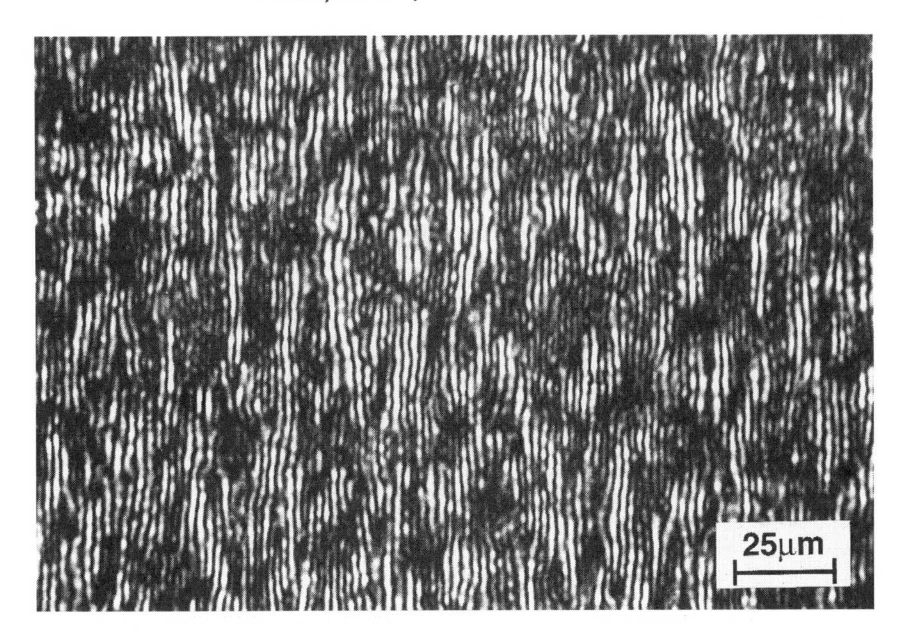

Figure 10.9 Polarized photomicrograph of oriented Vectra A950 (poly (hydroxybenzoic acid-co-hydroxynaphthoic acid)). The arrow indicates the shear direction.

(MLCs) [39]. In a very wide range of longitudinal PLCs, both rigid and semiflexible, thermotropic and lyotropic, shear causes the formation of the banded structure. It is universal in the appropriate range of shear rate and viscosity, controlled by molar mass and temperature. The bands are always perpendicular to the prior shear direction. In general, the bands form after the cessation of shear, presumably as a result of a relaxation process, but there is some evidence that under some circumstances they may appear during the flow process itself [32]. The bands are associated with a periodically varying molecular director about the flow axis (Figure 10.9). Early work by Dobb *et al.* [40] on Kevlar showed that the bands in this particular polymer correspond to a pleated sheet. The banding in thermotropic copolyesters correspond according to Viney *et al.* [41] to a more gentle serpentine trajectory of the molecules. Zachariades and Logan [42] studied the effect of shear flow of poly(*p*-hydroxybenzoic acid-co-ethylene terephthalate) in a plate-to-plate rheometer. Shearing at $240\,s^{-1}$ at temperatures below 328°C led to only a moderately oriented material probably due to the presence of crystalline poly(*p*-hydroxybenzoic acid). Shearing at 330°C under similar conditions led to the formation of highly oriented and banded textures, with a periodicity of 5 to 10 μm. Similar early observations were reported by Flory [43] and Viney *et al.* [44] on other longitudinal PLCs.

Attempts have been made to attribute the formation of shear bands to flow instabilities [45] and negative normal stresses [46]. It was argued by Zachariades *et al.* [47] that the Reynolds number for poly(*p*-hydroxybenzoic acid-co-ethylene terephthalate) with 60 mol% *p*-hydroxybenzoic acid, showing extensive banding in a plate-to-plate rheometer at 265 to 295°C, was under these conditions only $10^{-3}-10^{-5}$, which is very low for flow instability to occur. Zachariades *et al.* [47] based their explanation on the domain theory of Asada *et al.* [48] and Onogi and Asada [49] and they argued that the flow behavior may be divided into three types. At very low shear rates, the domains flow with significant rotation and slippage. As the shear rates increase, the domains deform, and at high shear rates they transform into a monodomain structure. For poly(*p*-hydroxybenzoic acid-co-ethylene terephthalate) it seems that the band structure is developed in the low to intermediate shear rate range (regions I and II according to the nomenclature of Onogi and Asada). The relaxation behavior of the band structures and of the domains after the cessation of shear is similar in that both relax faster as the shear rate increases. This provides an explanation of the fact that the band spacing decreases with increasing shear rate. At very high shear rates a monodomain structure is obtained.

Very little research has been conducted on the rheology of one-comb PLCs. The early studies of Fabre and Veyssie [50] and of Zentel and Wu [51] concluded that the flexible backbone dominates the rheological behavior, suppressing any significant manifestation of the LC character of the polymers. A more recent study by Kannan *et al.* [52a] showed the occurrence of limited, flow-induced alignment near the clearing point in a nematic one-comb PLC. The alignment in these PLCs originated from the coupling of the mesogens to the backbone. Similar perpendicular (to the flow field) alignment of the mesogens was very recently reported for a smectic one-comb PLC [52b].

Injection molded bars of longitudinal PLC (poly(*p*-hydroxybenzoic acid-co-ethylene terephthalate); 'X7G') showed a layered morphology [53]. Thinner specimens, of the order of 3 mm thick, showed three different layers symmetrically around the central plane of the specimen: an oriented skin ($s \sim 0.5$; Young's modulus at 23°C ~ 15 GPa), an oriented layer beneath the skin ($s \sim 0.5-0.6$; Young's modulus at 23°C ~ 15 GPa) and an isotropic core ($s = 0$; Young's modulus at 23°C = 2.8 GPa. Thicker specimens (6 mm) showed a more complex morphology with five different layers. The relative thickness of the isotropic core increased with increasing thickness of the specimen, which resulted in a thickness-dependent Young's modulus of the specimen, principally in accordance with data of Jackson and Kuhfuss [54]. The surface skin of injection molded bars of another longitudinal PLC, poly(*p*-hydroxybenzoic acid-co-2,6-hydroxynaphthoic acid) (Vectra A950), exhibited an even higher order parameter ($s \sim 0.7$) [55].

Blends based on longitudinal PLCs and flexible chain polymers have been the subject of many reports. A comprehensive list is not given here, but emphasis is placed on a few of the early papers [56–63]. The components of the blends should phase separate and the PLC are oriented during the processing and form a reinforcing, often fibrous component. The low viscosity of the nematic PLC components makes the blends readily processable.

Crevecoeur and Groeninckx [64] prepared blends of longitudinal PLCs, Vectra A950 and Vectra B950 (copolyester of 2,6-hydroxynaphthoic acid, terephthalic acid and 4′-hydroxy acetanilide) with polystyrene and a miscible blend of polystyrene and poly(2,6-dimethyl-1,4-phenylene ether). They assessed the crystalline Hermans orientation function from the dependence on azimuthal angle of the (110) X-ray reflection of the PLC component. The order parameter of the PLC component in melt-spun blend fibers increased with increasing PLC content from 0.6 at low PLC contents to close to 0.9 in pure PLC [64]. Similar results were obtained by Jung and Kim [65] for fibers of polycarbonate and poly(*p*-hydroxybenzoic acid-co-ethylene terephthalate). Chung [66], on the other hand, reported a composition-independent orientation of the PLC component in drawn blends of polyamide-12 and Vectra.

Engberg *et al.* [67, 68] prepared blends based on Vectra A950 and several flexible chain polymers. The order parameter of the PLC component in injection molded blends of polyethersulfone (PES) and Vectra, referring to the whole cross-section (thickness of specimen 1.5 mm), was according to X-ray scattering data close to 0.4 and also independent of the composition of the blend [67]. The lower chain orientation found in these samples reflects the fact that shear flow is less effective in orienting PLCs than extensional flow.

PLC elastomers are a special class of lightly crosslinked one-comb PLC which were first made by Finkelmann *et al.* [69]. A permanently oriented PLC may be achieved by mechanical means by stretching a lightly crosslinked polymer, and the orientation is made permanent by further crosslinking in the distorted state. The crosslinks are attached to the backbone chains. The end product is crosslinked polydomain PLC which can be further aligned by drawing.

10.3.3 Alignment in electric fields

Electric fields can be used to align permanent dipoles present in PLCs and, if the dipoles are 'rigidly' linked to the mesogens, the latter can be oriented along a common director. Centrosymmetrical molecules can only form a structure with quadrupolar orientation whereas molecules lacking centrosymmetry may in an electric field form a structure with dipolar orientation [32]. The general picture is that longitudinal polymers are difficult to align in an electric field, whereas one-comb polymers are effectively aligned in a 'suitable' electric field. Krigbaum *et al.* [70, 71] were unable to orient poly(*p*-hydroxybenzoic acid-co-ethylene terephthalate) with 60 mol% *p*-hydroxybenzoic acid in an electric field. A few years later Martin and Stupp [72] showed that crystallization was promoted by the electric field in a longitudinal copolyester. There is a demand for such extensive voltages that discharges and electrical degradation will occur prior to the orientation of the longitudinal PLC. Finkelmann *et al.* [73] reported early that the response of one-comb PLCs to an electric field was similar to that of MLCs but with the difference that PLCs exhibit longer response times. Williams and coworkers [74–79] showed that one-comb PLCs with a siloxane backbone can be aligned using directing electric AC or DC fields at temperatures both above and below the clearing temperature. Keller *et al.* [80] reported that photochromic one-comb LC polyacrylates were difficult to align by any of the available electrical methods due to the substantial low frequency conductance of the sample.

It is useful to consider the dielectric permittivity parallel to ($\varepsilon_{\|}$) and perpendicular to ($\varepsilon_{\perp}$) the mesogen and to bear in mind that they are dependent on both the frequency f of the applied AC field and the temperature. If $\Delta\varepsilon = \varepsilon_{\|} - \varepsilon_{\perp}$ is positive, then the long axis of the mesogens will be parallel to the applied field, whereas if $\Delta\varepsilon$ is negative, the mesogens align perpendicular to the applied electric field. Let us consider the system, a one-comb PLC, at a certain temperature (isothermal conditions) with a possible variation in the frequency of the applied AC field. Dipoles oriented perpendicular to the mesogen director are more readily reoriented than dipoles along the director, the latter requiring a rotation of the whole mesogen about the backbone chain. Hence the transverse relaxation process occurs at considerably higher frequencies than the longitudinal relaxation. At a frequency between ω_1 and ω_2 the mesogens will have a planar orientation, whereas at frequencies less than ω_1 they will display a homeotropic orientation (Figure 10.10). This is known as the 'two-frequency addressing' principle [81]. The transitional frequency (ω_1) is commonly referred to as the 'critical crossover frequency'.

The rotation of the mesogen about the backbone chain is referred to as the δ process. The mesogen director reverses its direction as a result of the δ process. This process also occurs in MLCs and exhibits almost a single relaxation time with an activation energy of 50–70 kJ mol^{-1} [82, 83]. The δ process is very narrow in one-comb polysiloxanes with

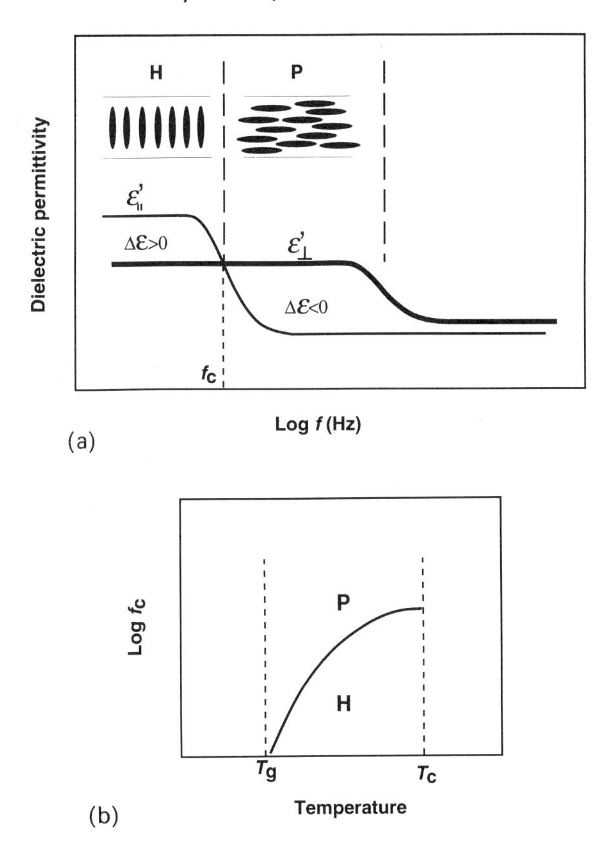

Figure 10.10 (a) Schematic relationship between isothermal parallel ($\varepsilon'_{//}$) and transverse ($\varepsilon'_{\perp}$) dielectric permittivities as a function of the logarithm of the frequency f. The resulting alignments of the mesogens, homoeotropic (H) and planar (P) orientations are shown. (b) Temperature dependence of the crossover frequency f_c.

their extraordinarily flexible backbone chain, and is rather similar to that observed in MLCs [74–79]. The δ process is broadened in polyacrylates and polymethacrylates, indicating that the backbone imposes constraints on these polymers. Kresse *et al.* [84] reported that the δ process was about 1000 times slower in smectics than in the corresponding nematics.

Talrose *et al.* [85] reported that the response for the alignment of one-comb PLCs was faster at higher temperatures (nematics) and that the response time was proportional to the reciprocal of the square of the applied voltage, a trend which indeed was the same as for MLCs. Ringsdorf and Zentel [86] found that the response time decreased with increasing difference between the actual and glass transition temperatures.

Polymers such as polysiloxanes with flexible backbones are therefore ideal for the purpose of achieving short response times [87–89].

The fastest response times reported have been of the order of 100 ms at temperatures close to the isotropization temperature or in the biphasic region [88–94]. For this reason, many researchers align their samples during a slow cooling through the isotropic–nematic phase transition.

The crossover frequency has a strong temperature dependence and the δ process undergoes a critical slowing down when the glass transition temperature is approached. Another important problem when using an AC field close to the crossover frequency is that the dielectric loss may be substantial, and this causes substantial heating of the sample [95].

Pranoto and Haase [96] reported the existence of a threshold voltage for alignment which is dependent on $(k_{11}/\Delta\varepsilon)^{1/2}$, where k_{11} is an elastic constant and $\Delta\varepsilon$ is the dielectric anisotropy, whereas Attard and Williams [92] observed alignment without any sign of a threshold voltage.

When the directing field is a DC field, the motion of ions leads to flow and electrodynamic instabilities, which are well known for MLCs. This is the reason why good alignment of PLCs may not be achieved in this case [32]. This problem can be eliminated by so-called electrical cleaning, a process first reported by Osaki *et al.* [97]. The procedure is as follows: the PLC is held at a temperature 1–2 K above the clearing temperature and a DC voltage of 20–50 V (100 μm^{-1}) is applied. The ions then migrate towards the electrodes and give rise to a small current, initially of the order of a few milliamperes. The current decreases with time and finally reaches a value of a few microamperes. The voltage is then increased stepwise to a few hundred volts per 100 μm. The DC voltage causes a homeotropic alignment. Cooling of the sample to its mesomorphic state leads to a more stable situation, because the mesomorphic state permits much less migration of ions than the isotropic phase. The electrical cleaning procedure greatly reduces the low frequency loss and enables further alignment in an AC field.

One major problem with using a DC field to align dipoles is the injection of charges which damage the materials, and it is common for DC field-induced alignment to be carried out at temperatures near the glass transition temperature. Conductivity effects can thus also be greatly reduced by electrical cleaning.

In simple terms, the alignment caused by the application of an electric field may be thought of as a removal of the domain boundaries and the overall order parameter should approach the order parameter value of the single domain. However, Shibaev [88] reported that electric field alignment of a one-comb PLC caused a lowering of the *trans* content in the spacer group. NMR data on one-comb LC poly(vinyl ether)s did not reveal any differences in the *trans* content between aligned and non-aligned specimens [98].

10.3.4 Alignment in magnetic fields

Any molecule will become oriented in a magnetic field as long as its magnetic susceptibility (χ) is anisotropic: $\Delta\chi = \chi_{\parallel} - \chi_{\perp} \neq 0$, where the indices '$\parallel$' and '$\perp$' denote the susceptibilities parallel to and perpendicular to the molecule. In general terms, 1 T is equivalent to an electric field of 100 V mm^{-2} in orienting power [32]. Magnetic fields can however be applied more readily than electric fields because there is no risk of partial discharges and breakdown. Very strong magnetic fields, greater than 10 T, can be obtained in superconducting magnets. A large number of reports on magnetic field alignment of one-comb and longitudinal PLCs were published in the early 1980s [28, 99–108]. In the case of longitudinal molecules with *p*-phenylene groups in the main chain, the polymer molecules align parallel to the magnetic field. This effect is most pronounced in liquid crystalline phases with their reduced thermal randomizing effect. The polarization is induced by the magnetic field rather than being present in the molecules as dipoles, and therefore only quadrupolar orientation may arise [32].

Anwer and Windle [109, 110] reported the kinetics of the alignment of poly(*p*-hydroxybenzoic acid-co-2,6-hydroxynaphthoic acid) s of different molar masses ranging from 4600 to 36000 g mol^{-1} in a magnetic field applied at elevated temperatures in the nematic state. The order parameter of a specimen held in a magnetic field increased rapidly with time and finally reached a plateau value. The data obtained fitted the following equation:

$$s = s_{max}\left[1 - e^{-\frac{t}{\tau}}\right] \tag{10.10}$$

where s_{max} is the plateau value and τ is the characteristic retardation time for the alignment process. The retardation time increased strongly with decreasing magnetic field strength B; $\tau \propto B^{-1}$, as shown for a low molar mass polymer ($\bar{M}_w = 6000$ g mol^{-1}) in Figure 10.11. The retardation time increased strongly with increasing molar mass M of the polymer; $\tau \propto M^{-2}$ (B = constant). Figure 10.11 shows the ultimate order parameter for the same polymer as assessed by X-ray scattering as a function of the strength of the applied magnetic field. s_{max} reaches impressively high values under the influence of stronger magnetic fields.

Unoriented poly(*p*-hydroxybenzoic acid-co-2,6-hydroxynaphthoic acid) exhibited smoothly wandering director fields in three dimensions. Alignment with a 1.1 T magnetic field for 30 min at 300°C transformed this structure to domains with an anisotropic shape within which the polymer was highly oriented, and the global order parameter amounted to 0.85 [110]. Boundaries were of the splay-bend type and involved a 180° director rotation. At lower field strengths, the domains were less

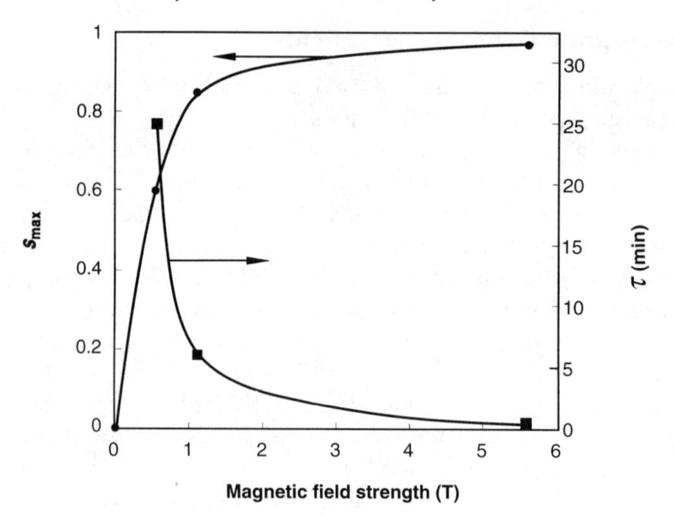

Figure 10.11 The maximum degree of orientation s_{max} and the retardation time τ as a function of the magnetic field strength for a longitudinal copolymer based on hydroxybenzoic acid and hydroxynaphthoic acid ($\bar{M}_w = 6000\ \mathrm{g\ mol^{-1}}$) magnetically aligned at 300°C. (After data of Anwer and Windle [110].)

elongated and the boundaries were wider. Vectra samples aligned by a 5.6 T field exhibited a very highly oriented texture with an overall value of $s_{max} = 0.97$ and with no boundary network, although the discrete boundaries were very sharp indeed. The boundary width was inversely proportional to the strength of the applied field [110]. Anwer and Windle [110] proposed that the maximum orientation can be accounted for by the simple assumptions that all domains are perfectly aligned and that the boundaries are completely unoriented. The observed increase in order parameter with increasing field strength is thus due to a reduction in the width of the domain boundaries with increasing field strength.

10.3.5 Alignment in surface fields

Rubbed polyimide films efficiently orient MLCs to a quadrupolar state but are considerably less efficient in orienting PLCs. There are a substantial number of papers reporting the preparation of highly oriented one-comb and network PLCs by polymerization of aligned MLCs. Broer and coworkers [111–114] aligned mesomorphic mono- and bifunctional acrylates using the surface technique and performed subsequent photopolymerization to obtain highly oriented (quadrupolar) one-comb PLCs. One-comb and network PLCs with very high values of the order parameter ($s \sim 0.90$) were prepared by photopolymerization of mono- and bifunctional vinyl ethers [30, 115, 116]. Further details on these topics are presented by R.A.M. Hikmet in Chapter 5 of Vol. 4 of this series.

Surface alignment may also have a positive effect on the alignment on side chain PLCs, particularly in combination with other alignment methods [86, 94, 117].

10.4 LOSS OF ORIENTATION IN PLCs

10.4.1 Background and general aspects

This section deals with thermotropic PLCs. The story starts, however, with lyotropic PLCs for which the long structural relaxation times of PLCs, compared to those of conventional flexible chain polymers, were first realized from rheo-optical studies [118]. Figure 10.12 presents the fundamentals of the important early discovery: a specimen first oriented by shear flow which is suddenly switched off exhibits a rapid stress relaxation within a period of seconds. The decay of birefringence is however slower and it takes a very significant period of time.

It was later confirmed by Ide and Chung [119] that these observations were also valid for thermotropics. Small-diameter rods of a longitudinal PLC (a copolyester based on 2,6-hydroxynaphthoic acid, terephthalic acid and *p*-aminophenol) were annealed above its crystal melting temperature in the nematic state for different periods of time. The dependence of the Young modulus E at room temperature on the

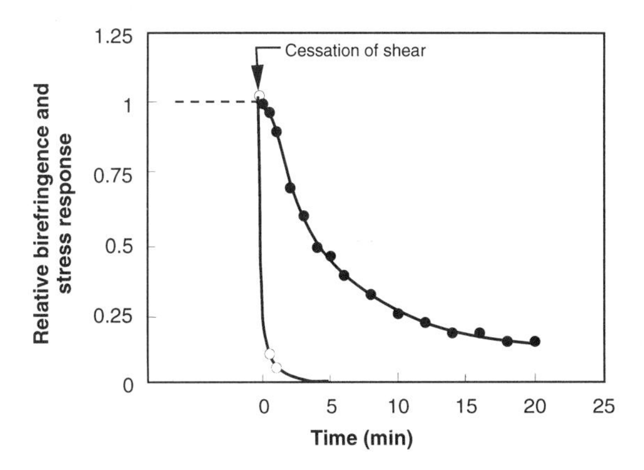

Figure 10.12 The evolution of shear stress (○) and birefringence stress (●) with time after cessation of shear for a 55% hydroxypropyl cellulose solution. The data are normalized with the steady state values obtained during shear. (After data of Onogi *et al.* [118].)

annealing time obeyed the following expression:

$$E = E_0 e^{-t/\tau} \tag{10.11}$$

where E_0 is a constant, t is the annealing time and τ is the relaxation time. The relaxation times for these longitudinal PLCs were 36 min at 295°C and 11 min at 305°C [108]. Similar studies on conventional thermoplastics gave relaxation times of the order of seconds [119]. More recently, the opposite and unexpected behavior has also been reported in a lyotropic system [120]. Hongladarom and Burghardt observed a considerable increase in orientation and also a sudden change in the dynamic modulus accompanying the switching off of the shear (Figure 10.13). The cause of this behavior remains unknown. The presentation of the relaxation behavior is subdivided into three sections covering longitudinal PLCs, one-comb polymers and network PLCs.

The structural relaxation of PLCs, the time dependence of the order parameter, can be recorded by a number of different experimental techniques: X-ray scattering, IR spectroscopy, NMR spectroscopy, rheo-optical methods, rheological methods and small angle light scattering (SALS). Some details of these methods are given in section 10.2. A distinction can be made between direct (*in situ*) observation and the assessment of orientation in a specimen after quenching to a low temperature. The *in situ* methods are for obvious reasons preferred. However, the long relaxation times characteristic of PLCs suggest that it should be possible in some cases to make a 'high temperature' structure permanent by quenching. It should be emphasized that this is not always true. Most of the aforementioned methods can provide direct

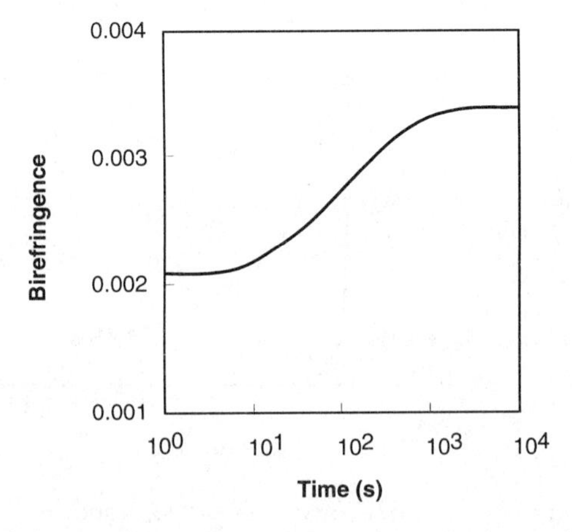

Figure 10.13 The evolution of birefringence with time after the cessation of shear ($1\,s^{-1}$) for a 13 wt% solution of polybenzylglutamate in *m*-cresol. (After data of Hongladarom and Burghardt [120].)

data, although a substantial number of the reports rely on data from quenched samples.

10.4.2 Relaxation of longitudinal PLCs

The relaxation behavior of nematic, longitudinal PLCs is greatly complicated by their multiphase character. The presence of both crystallites and a coexisting isotropic component has been reported [32]. Vectra A (poly(*p*-hydroxybenzoic acid-co-2,6-hydroxynaphthoic acid)), currently one of the more widely used and studied commercial polymers, exhibits, despite that fact that it is a statistical copolymer, a sizeable amount (10–60%) of crystallinity, either of the non-periodic layer type [121] or of a truly statistical type according to the model of Biswas and Blackwell [122]. It has become clear that the relaxation of oriented Vectra is to a substantial degree controlled by the thermal behavior of its tiny crystallites.

Lin and Winter [123] reported recrystallization of Vectra A at 290°C into poly(*p*-hydroxybenzoic acid)-rich crystallites with a melting point of 320°C. Annealing of an initially fully nematic polymer at 290°C for 200 min resulted, due to the ongoing crystallization, in an increase in the storage modulus by three orders of magnitude. The rate of modulus growth was higher for oriented samples. A stable nematic melt could only be achieved by first heating the sample to a temperature above 320°C to remove all crystal nuclei. The nematic melt treated in this way was also stable at lower temperatures down to 290°C. A considerable recrystallization primarily involving the growth of existing crystallites took place at temperatures between 250 and 270°C [123].

Another less attractive feature of these polymers is that they may undergo transesterification reactions at the high temperatures used in the studies [67]. Recommended reading on transesterification reactions in polyesters is Porter *et al.* [124].

Lin *et al.* [125] studied the relaxation of orientation in Vectra A950 (poly(*p*-hydroxybenzoic acid-co-2,6-hydroxynaphthoic acid)). Samples were squeezed at different constant temperatures between lubricated disks resulting in a thickness reduction of 90% and were then cooled at a rate of 20°C min^{-1} or faster. The order parameter decreased with increasing squeezing temperature and increasing cooling rate after squeezing. Quenching resulted in a higher degree of orientation but at the expense of reduced crystallinity. In a study by Lin and Winter [126], Vectra A900 samples (poly(*p*-hydroxybenzoic acid-co-2,6-hydroxynaphthoic acid)) were first heated to 320°C to remove all crystal nuclei, then cooled to different temperatures between 255 and 285°C, and sheared at that particular temperature to 'steady state' conditions. The strain recovery, also associated with a relaxation of orientation according to X-ray scattering data, was recorded after the cessation of shear, and the recovered strain increased with increasing shear temperature, showing the stabilizing action of the crystallites formed at the lower temperatures [126]. Samples sheared close to the melting point of the crystallites randomized state almost instantaneously when the shear was stopped. Viola *et al.* [34] studied another thermotropic copolyester and reported a higher final orientation for samples that were sheared and cooled at the same time compared to samples where shearing was stopped in the melt state and thereafter the samples were cooled.

Engberg and Gedde [55] reported order parameter data obtained by X-ray scattering on annealed, oriented Vectra A950. Annealing led to a very sudden and strong decrease in the order parameter at temperatures between 275 and 293°C. This very rapid relaxation process, the relaxation times of which were of the order of seconds, was suggested to be due to melting of the crystallites, causing a rapid randomization of nearby structural regions and a reduction in domain order parameter s_d. The subsequent (and parallel) randomization of the domain directors s_{dd} is a considerably slower process and required minutes or hours at the temperatures used in this study. Alderman and Mackley [37] reported slower textural relaxation after intense shearing of Vectra A950 and X7G (poly(*p*-hydroxybenzoic acid-co-ethylene terephthalate)) from uniformly birefringent to polydomain visible through the formation of closed disclination loops.

Graziano and Mackley [36] reported three types of relaxation from the ordered, oriented texture of nematic longitudinal copolyesters of chlorophenylene terephthalate and bis-phenoxyethane carboxylate after cessation of shear: (1) relaxation to worms which remain intact for a long time or which slowly relax; (2) relaxation to worms which then rapidly relax leaving a background with the optical axis oriented in the prior flow direction and a low density of bright-field entities; and (3) relaxation to a texture characterized by bands aligned perpendicular to the prior flow direction. Above 240°C, generally the formation of worm structures was accompanied by a simultaneous loss of birefringence within a few seconds. However, a significant amount of birefringence (orientation) remained. After shearing at higher shear rates, the threads grew together forming essentially perpendicular banding. Shearing and relaxation at 220°C led to the formation of the banded texture. The banded texture is thus a 'low temperature, high shear rate phenomenon'.

Samples of oriented Vectra A950 (poly(hydroxy-benzoic acid (73%) −co-hydroxy-naphthotic acid (27%))), obtained from the surface skin of injection molded specimens, were annealed at different temperatures and the order parameter was assessed by IR spectroscopy and X-ray scattering in a recent study of Wiberg and Gedde [127]. They found that the extent of orientational relaxation depended on the thermal history and the clamping procedure of the sample, whether the specimen was under restraint at constant length or free to shrink. Figure 10.14 shows schematically the change in order parameter in freely shrinking specimens which were rapidly heated (200°C min^{-1}) to a constant temperature at which they are annealed. Annealing at temperatures higher than 270°C led to a rapid decrease (within 6 s) in the order parameter in accordance with data of Engberg and Gedde [55]. At low temperatures (< 270°C) there is a progressive but retarding increase in orientation as a function of annealing time. Note the logarithmic time scale in Figure 10.14. Extensive (18 h) annealing at 270°C led to only marginal changes in the order parameter.

Specimens annealed while being restrained at constant length showed only an increase in order parameter, even after annealing at very high temperature (Figure 10.15a). Slow and gradual heating favors recrystallization and the formation of thermally more stable crystallites. These form new physical crosslinks which prevent shrinkage of the sample. The recorded increase in chain orientation during slow and gradual heating (Figure 10.15b shows one example) must be due to a perfection of the material between the crystallites. This process involves

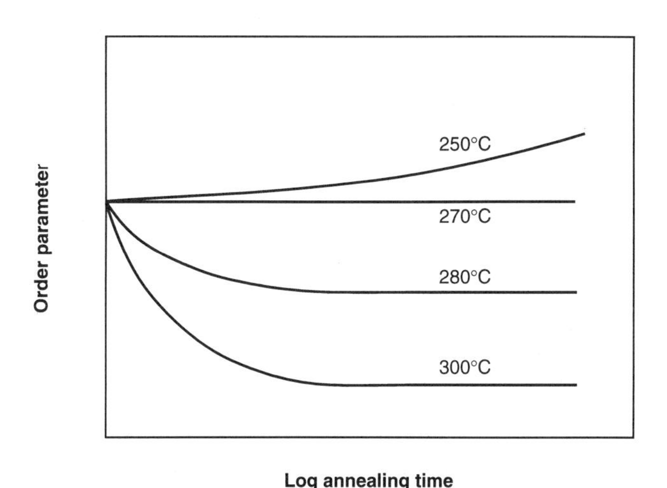

Figure 10.14 The order parameter of oriented Vectra A950 as a function of the annelaing time at different annealing temperatures as shown in the graph. The samples were heated at a rate of 200°C min^{-1} from room temperature to the annealing temperature and the samples were free to shrink during the thermal treatment. (Schematically drawn after data of Wiberg and Gedde [127].)

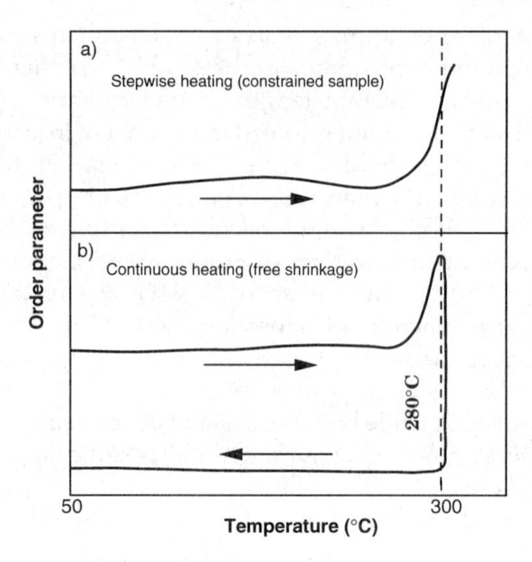

Figure 10.15 The evolution of the order parameter of oriented Vectra A950 as a function of temperature of specimens given different treatments as shown in the graph. (Schematically drawn after data of Wiberg and Gedde [127].)

improved packing of the chains in general and possibly also annihilation rather than generation of disclinations. In the example shown in Figure 10.15b, the final melting of the crystals occurred at 300°C which then caused a rapid relaxation of the oriented system.

10.4.3 Relaxation of one-comb PLCs

The relaxation of orientation of one-comb PLCs is reported in only a limited number of papers. Most of the papers deal with the temperature dependence of the mesogen order parameter. There are to our knowledge no papers dealing with the time dependence (at constant temperature) of the order parameter. Buerkle *et al.* [128] presented order parameter data by IR spectroscopy (dichroism of the CN-stretching band) of nematic one-comb polyacrylates with terminal cyano groups and two or six carbon atoms in the oligomethylene spacer. The order parameter of the PLC with a longer spacer was considerably higher (0.7) than that of the polymer with an ethylene spacer (0.5) at temperatures (50°C) well below the clearing point. The higher order parameter of the PLC with the longer spacer group is due to the fact that the decoupling of the mesogen from the backbone chain is more complete in the polymer with the longer spacer group. Heating of both oriented polymers led to a continuous and accelerating decrease in the order parameter approaching a zero level at the clearing point (Figure

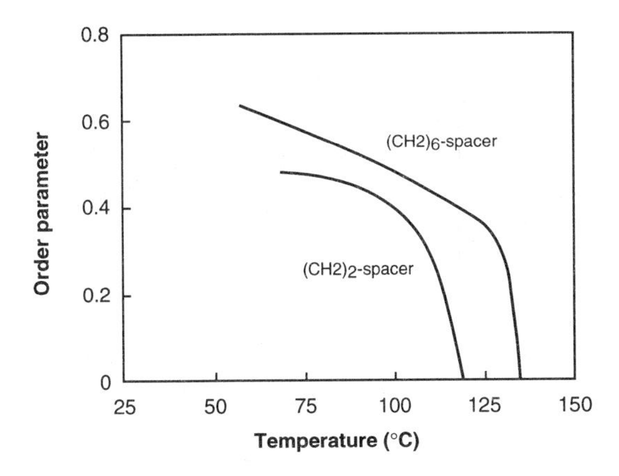

Figure 10.16 The order parameter of two oriented, nematic one-comb PLCs with different spacer groups as shown in the graph as a function of temperature. The samples were continuously heated and the order parameter was obtained by IR spectroscopy. (After data of Buerkle *et al.* [128].)

10.16). The high initial order parameter indicates that the samples were close to monodomain and the deterioration of orientation is mostly caused by a decrease in the domain order parameter.

Similar data were presented by Broer *et al.* [111] on polyacrylates which, in addition to a nematic phase, also exhibited smectic mesomorphism and a solid crystalline phase at lower temperatures. There was a stepwise decrease in the order parameter accompanying both first order transitions, i.e. the solid crystal–smectic and the smectic–nematic transitions. The order parameter of the smectic PLC was considerably higher than that of the nematic polymer. Figure 10.17 shows that the data for the relaxation of the nematic phase were indeed similar to the data of Buerkle *et al.* [128]. Furthermore, as expected, the order parameter of the nematic PLC was lower than that of the corresponding MLC at the same reduced temperature (Figure 10.17). Sahlen *et al.* [129] assessed the order parameter in an oriented one-comb PLC and recorded a constant order parameter below the glass transition temperature and a substantial decrease in order parameter accompanying both the glass transition and the transition between smectic B and smectic A states.

Hempel *et al.* [130] presented order parameter data by ^{13}C-NMR for individual groups in a nematic one-comb PLC (Figure 10.17). The terminal ring shows the highest order parameter. The innermost ring shows, particularly at the higher temperatures, a considerably lower order parameter. There are significant differences in the order parameter for the different carbon atoms of the spacer. The carbon atom closest to the mesogen exhibits the highest order parameter and the carbon

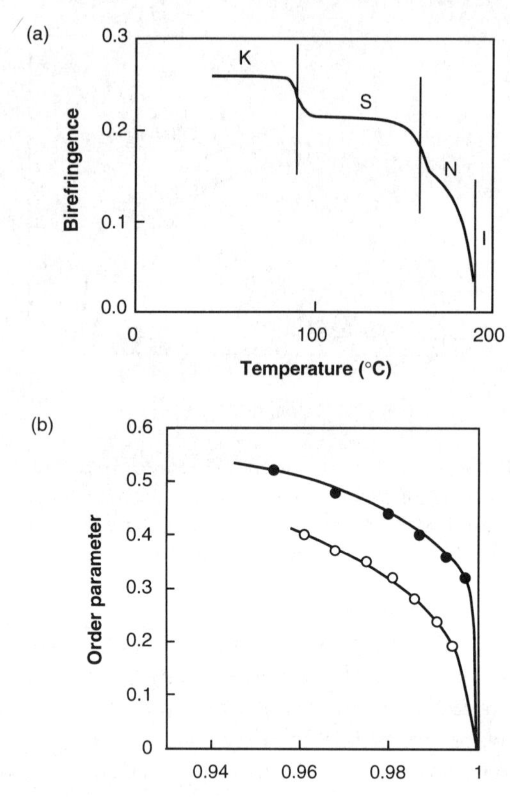

Figure 10.17 (a) The birefringence of an oriented one-comb PLC (polyacrylate obtained by photopolymerization of a surface aligned MLC) as a function of temperature. The various phases (K, crystalline; S, smectic; N, nematic and I, isotropic) are shown in the graph. (b) The order parameter of the same polymer (open symbol) in its nematic state as a function of reduced temperature, T/T_i, where T_i is the isotropization temperature. Data for the monomer (filled symbol) are shown in the same graph. ((a) and (b) from data of Broer *et al.* [111].)

atom closest to the backbone shows interestingly negative values, i.e. it is aligned perpendicular to the mesogens.

10.4.4 Relaxation of PLC networks

Several classes of crosslinked PLCs have been prepared:

- Dense PLC networks (thermosets) are often prepared from mixtures of mono- and bifunctional monomers resulting in a one-comb polymer-like structure with crosslinks.

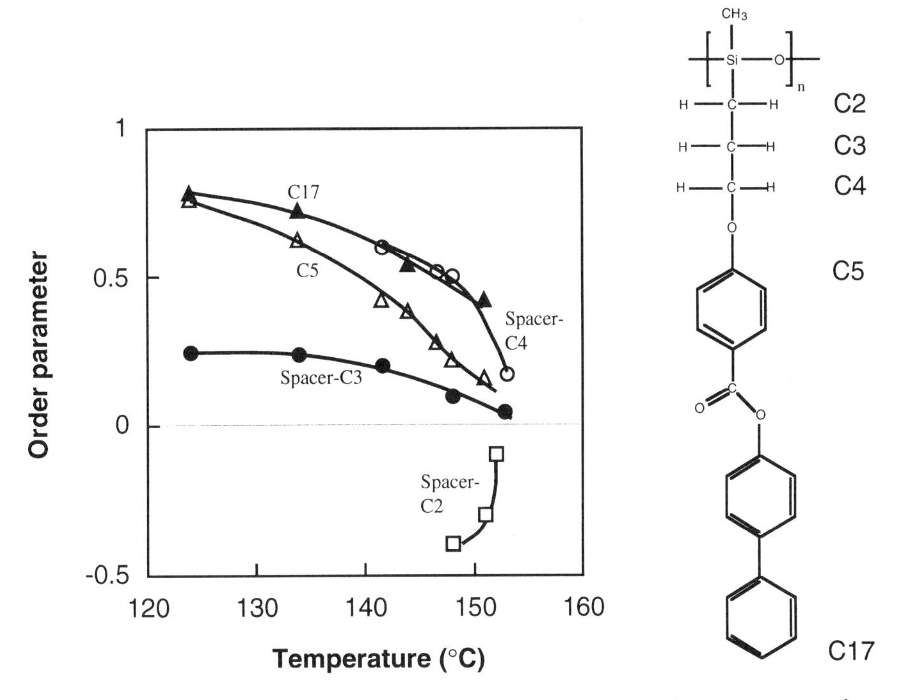

Figure 10.18 The temperature dependence of the order parameter for different moieties (explained in the graph) in a nematic, one-comb PLC with a polysiloxane backbone. (After data of Hempel *et al.* [130].)

- Anisotropic PLC gels consisting of a PLC network and a sorbed MLC liquid component.
- Lightly crosslinked PLC networks (elastomers).

Broer *et al.* [131] made oriented, dense PLC networks using diacrylates and the decrease in birefringence was very moderate (Figure 10.19). The birefringence behavior was completely reversible on heating to 240°C. Exposure to higher temperatures led to thermal degradation of the polymer [131]. A slight shift to lower birefringence values was observed on increasing the polymerization temperature (Figure 10.19).

Andersson *et al.* [116] showed that an oriented smectic PLC network based on mono- and bifunctional vinyl ethers (90/10) lost orientation when heated to 200°C, but regained close to 90% of its original order parameter after cooling. Further temperature cycling was completely reversible with regard to the order parameter. The small loss of orientation during the first heating and cooling cycle was due to unreacted monomer species postreacting in the isotropic state at high temperatures. Similar results were reported by Hikmet *et al.* on

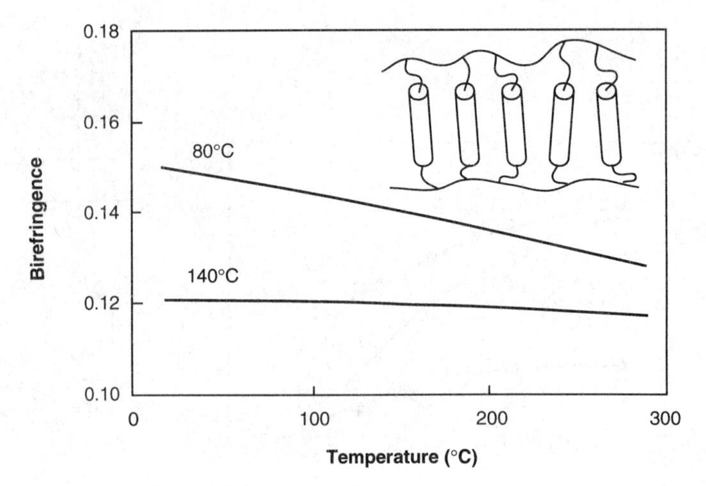

Figure 10.19 The birefringence of an aligned polydiacrylate network as a function of temperature with the photopolymerization temperature shown as a parameter. A schematic diagram of the network structure is inserted. (After data of Broer *et al.* [131].)

polyacrylates [132] and poly(vinyl ether) s [133], which after crosslinking exhibited only a minor decrease in order parameter on heating from 20°C to 140°C. Hikmet and coworkers [132, 134] showed that both the order parameter and the thermal stability increased with increasing crosslink density for a specific polymerization temperature.

In a follow-up study to the paper of Andersson [116], Sahlén *et al.* [129] reported the temperature dependence of the order parameter for a wider range of PLC networks based on mono- and bifunctional vinyl ethers. Figure 10.20 shows the evolution of the order parameter during the first heating and cooling cycle. The order parameter was not fully recoverable in the first heating–cooling scan due to postpolymerization of residual monomer. Further temperature cycling led to no further change in the order parameter–temperature curve. The one-comb PLC based on the monofunctional vinyl ether showed smectic B, smectic A and isotropic phases in order of increasing temperature. The introduction of the bifunctional monomer (crosslinking) led to a suppression of the phase transitions, and the PLCs with 20% of bifunctional units or more exhibited only smectic A mesomorphism [129].

Oriented PLC thermosets based on epoxy chemistry and with the mesogens in the backbone chain have been prepared by aligning an oligoether prepolymer either mechanically or in a magnetic field (13.5 T) prior to and during curing [135]. The temperature associated with a drop in order parameter increased with increasing crosslink density. Samples with few crosslinks showed a drop in order parameter at 130°C

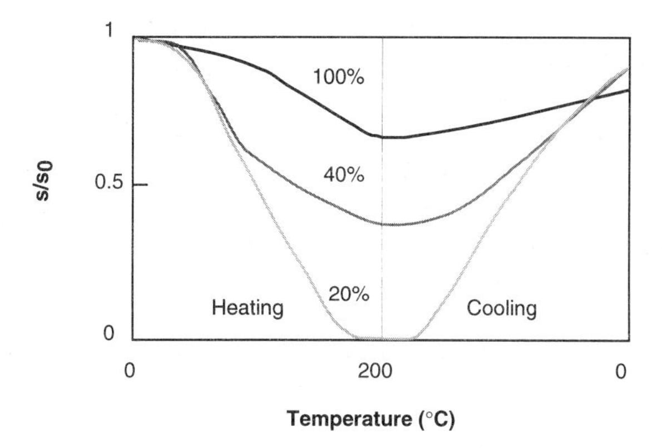

Figure 10.20 Normalized order parameter (s_0 is the original order parameter) of aligned PLC networks with different molar contents of crosslinks as shown in the graph as a function of temperature during the first heating and cooling scans. The order parameter was assessed by IR spectroscopy. (After data of Sahlén *et al.* [129].)

whereas more densely crosslinked PLCs showed only a minor decrease in order parameter after heating to 250°C and then cooling down to room temperature [135]. It should be noted that the less crosslinked samples reached higher levels of mesogen group orientation ($s \approx 0.6$) than the most densely crosslinked network ($s \approx 0.4$).

Polydiepoxide networks have lately received some attention owing to their superior thermal stability and adhesive properties compared to the polydiacrylates [136, 137]. The polymers, prepared by photocrosslinking of surface-aligned monomer films, showed very high virgin order parameters ($s \sim 0.8$) and it was concluded on the basis of refractive index measurements that the order parameter remained constant up to 140°C [136]. Jahromi *et al.* [137] reported that postcuring of the PLC network further increased the thermal stability.

Anisotropic PLC gels are blends of a PLC network and a miscible MLC, and the first reports appeared in the early 1990s [138–140]. Systems of this kind are also referred to as liquid crystal–liquid crystalline network composite systems.

Hikmet [138] prepared anisotropic PLC gels using polydiacrylates as the PLC network component and observed a small drop in the birefringence at the isotropization temperature of the MLC. The birefringence at high temperatures was however higher than could be expected from the network anisotropy. IR spectroscopy confirmed that the MLC also possessed orientational order above its isotropization temperature. Hikmet [138] suggested two principally different types of behavior of the MLC: MLCs strongly bound to the PLC network will show no nematic–isotropic transition, whereas MLCs interacting less with the PLC network should exhibit a clear isotropization like a conventional MLC. Braun *et al.* [139] also recorded a sudden decrease in birefringence in association with a nematic–isotropic transition of the MLC. Anisotropic gels containing up to 28 wt% dispersed chiral molecules showed good thermal stability [140].

It is possible to align the MLC differently to the PLC network in an electric field which will cause the system to scatter light. When the electric field is switched off, the alignment of the MLC will adapt to the alignment of the PLC component. The size of the electric field strength required to realign the MLC component increases with increasing PLC content [138, 140].

Lightly crosslinked PLC networks based on polysiloxanes were oriented by mechanical means as described in section 10.3.2. Schätzle and Finkelmann [141] showed that the decrease in order parameter on heating was similar to that of a corresponding uncrosslinked PLC. IR spectroscopy data showed that the mesogens were orientated perpendicular to the direction of the applied stress. Küpfer and Finkelmann [142] were able to apply a stress on a lightly crosslinked PLC elastomer and lock in the orientation by further crosslinking. The orientation remained after the removal of the stress. The studied polymers were of the polysiloxane type and are referred to as 'liquid single crystal elastomers' because of their optical similarity to conventional single crystals. The order parameter was essentially unchanged in the mesophase temperature region but decreased strongly near the isotropization temperature [142]. The original order parameter was regained when the polymer was cooled past the isotropization temperature.

10.5 SUMMARY

PLCs are readily oriented under the influence of mechanical, electrical, magnetic and surface fields. It is possible to align the domains (regions with a common director) reaching a 'monodomain' state. The state of orientation is adequately described by the order parameter although this quantity does not differentiate between polar and quadrupolar orientation. The transformation of a polymer with a polydomain morphology to a material with a monodomain structure is complex and includes transitional states. The loss of orientation is slow in PLCs compared with flexible chain polymers. The alignment of mesogens within a single domain depends strongly on the nature of the mesomorphic state and the temperature and it adapts rapidly to a temperature-specific equilibrium value on changes in temperature whereas the loss of alignment of domain directors in PLCs is slow. PLC networks may exhibit recoverable states of orientation.

ACKNOWLEDGMENTS

This study has been sponsored by the Swedish Research Council of Engineering Sciences (TFR; Grant 91-245), the Swedish Natural Science Research Council (NFR; Grant K-KU 1910-305) and by the Royal Institute of Technology, Stockholm.

REFERENCES

1. Onsager, L. (1949) *Ann. NY Acad. Sci.*, **51**, 627.
2. Ishihara, A. (1951) *J. Chem. Phys.*, **19**, 1142.
3. Flory, P.J. (1956) *Proc. Royal Soc.*, **234A**, 73.
4. Flory, P.J. (1978) *Macromolecules*, **11**, 1138.
5. Matheson, Jr, R.R. and Flory, P.J. (1981) *Macromolecules*, **14**, 954.
6. Brostow, W. (1988) *Kunststoffe*, **78**, 411.
7. Witt, W. (1988) *Kunststoffe*, **78**, 795.
8. Jonah, D.A., Brostow, W. and Hess, M. (1993) *Macromolecules*, **16**, 76.
9. Blonski, B., Brostow, W., Jonah, D.A. and Hess, M. (1993) *Macromolecules*, **26**, 84.
10. Brostow, W. (1992) in *Liquid Crystal Polymers: From Structures to Applications* (ed. A.A. Collyer), Elsevier Applied Science, London–New York, p. 1.
11. Brostow, W., Dziemianowisz, T.S., Romanski, J. and Weber, J. (1988) *Polymer Eng. & Sci.*, **28**, 785.
12. Samuels, R.J. (1974) *Structured Polymer Properties: the Identification, Interpretation and Application of Crystalline Polymer Structure*, Wiley, New York.
13. Ward, I.M. (ed.) (1975) *Structure and Properties of Oriented Polymers*, Applied Science Publ., London.
14. Struik, L.C.E. (1990) *Internal Stresses, Dimensional Instabilities and Molecular Orientation in Plastics*, Wiley, Chichester–New York.
15. Gedde, U.W. (1995) *Polymer Physics*, Chapman & Hall, London.
16. Tsvetkov, V.N. (1942) *Acta Physicochim. USSR*, **16**, 132.
17. Read, B.E., Duncan, J.C. and Meyer, D.E. (1984) *Polymer Testing*, **4**, 143.
18. Baltá-Calleja, F.J. and Vonk, C.G. (1988) *X-ray Scattering of Synthetic Polymers, Polymer Science Library*, **8**, Elsevier, Amsterdam.
19. Bower, D.I. and Maddams, W.F. (1989) *The Vibrational Spectroscopy of Polymers*, Cambridge University Press, Cambridge.
20. Koenig, J.L. (1992) *Spectroscopy of Polymers*, American Chemical Society, Washington, DC.
21. McFarlane, F.E., Nicely, V.A. and Davis, T.G. (1976) *Contemp. Topics in Polymer Sci.*, **2**, 109.
22. Calundann, G.W. and Jaffe, M. (1982) Anisotropic polymers, their synthesis and properties, in *Proceedings of the Robert Welch Conference on Chemical Research, XXVI, Synthetic Polymers*.
23. Blumstein, R.B., Blumstein, A., Stickles, E.M. *et al.* (1983) *Am. Chem. Soc., Polymer Prepr.*, **24**, 275.
24. Martins, A.F., Ferreira, J.B., Volino, F. *et al.* (1983) *Macromolecules*, **16**, 279.
25. Volino, F., Alloneau, J.M., Giroud-Godquin, A.M. *et al.* (1984) *Mol. Cryst. Liq. Cryst.*, **102**, 21.
26. Samulski, E.T. (1985) *Polymer*, **26**, 177.

27. Platé, N.A. and Shibaev, V.P. (1987) *Comb-shaped Polymers and Liquid Crystals*, Plenum Press, New York–London.
28. Geib, H., Hisgen, B., Pschorn, U. *et al.* (1982) *J. Am. Chem. Soc.*, **104**, 917.
29. Wassmer, K., Ohmes, E. and Kothe, G. (1982) *Makromol. Chem. Rapid Commun.*, **3**, 282.
30. Gedde, U.W., Andersson, H., Hellermark, C. *et al.* (1993) *Progr. Coll. & Polymer Sci.*, **92**, 129.
31. Finkelmann, H., Benthak, H. and Rehage, G. (1983) *J. Chimie Physique*, **80**, 163.
32. Donald, A.M. and Windle, A.H. (1992) *Liquid Crystalline Polymers*, Cambridge University Press, Cambridge.
33. Ide, Y. and Ophir, Z. (1983) *Polymer Eng. & Sci.*, **23**, 261.
34. Viola, G.G., Baird, D.G. and Wilkes, G.L. (1985) *Polymer Eng. & Sci.*, **25**, 888.
35. Shih, H.H., Hornberger, L.E., Siemens, R.L. and Zachariades, A.E. (1986) *J. Appl. Polymer Sci.*, **32**, 4897.
36. Graziano, D.J. and Mackley, M.R. (1984) *Mol. Cryst. Liq. Cryst.*, **106**, 73.
37. Alderman, N.J. and Mackley, M.R. (1985) *Faraday Discuss.*, **79**, 149.
38. Kulichikhin, V.G. (1988) *Mol. Cryst. Liq. Cryst.*, **169**, 51.
39. Kiss, G. and Porter, R.S. (1980) *J. Polymer Sci. Phys.*, **18**, 361.
40. Dobb, M.G., Johnson, D.J. and Saville, B.P. (1977) *J. Polymer Sci. Phys.*, **15**, 2201.
41. Viney, C., Donald, A.M. and Windle, A.H. (1983) *J. Mater. Sci.*, **18**, 1136.
42. Zachariades, A.E. and Logan, J.A. (1983) *Polymer Eng. & Sci.*, **23**, 797.
43. Flory, P.J. (1984) *Adv. Polym. Sci.*, **59**, 1.
44. Viney, C., Donald, A.M. and Windle, A.H. (1985) *Polymer*, **26**, 870.
45. Marucci, G. (1985) *Pure & Appl. Chem.*, **57**, 1545.
46. Chaffey, C.E. and Porter, R.S. (1985) *J. Rheol.*, **29**, 281.
47. Zachariades, A.E., Navard, P. and Logan, J.A. (1984) *Mol. Cryst. Liq. Cryst.*, **110**, 93.
48. Asada, T., Muramatsu, H., Watanabe, R. and Onogi, S. (1980) *Macromolecules*, **13**, 867.
49. Onogi, S. and Asada, T. (1980) in *Rheology*, Vol. 1 (eds G. Astarita, G. Marrucci and L. Nicolais), Plenum Press, New York, p. 127.
50. Fabre, P. and Veyssie, M. (1987) *Mol. Cryst. Liq. Cryst. Lett.*, **4**, 99.
51. Zentel, R. and Wu, J. (1986) *J. Makromol. Chem.*, **187**, 1727.
52a. Kannan, R.M., Kornfield, J.A., Schwenk, N. and Boeffel, C. (1993) *Macromolecules*, **26**, 2050.
52b. Wiberg, G., Skytt, M.-L. and Gedde, U.W. (1997) *Polymer Commun.*, in press.
53. Hedmark, P.G., Rego Lopez, J.M., Westdahl, M. *et al.* (1988) *Polymer Eng. & Sci.*, **28**, 1248.
54. Jackson Jr, W.J. and Kuhfuss, H.F. (1976) *J. Polymer Sci. Chem.*, **14**, 2043.
55. Engberg, K. and Gedde, U.W. (1992) *Prog. Coll. Polymer Sci.*, **87**, 57.
56. Siegmann, A., Dagan, A. and Kenig, S. (1985) *Polymer*, **26**, 1325.
57. Joseph, E.G., Wilkes, G.L. and Baird, D. (1985) *Polymer Eng. & Sci.*, **25**, 377.
58. James, S.G., Donald, A.M. and MacDonald, W.A. (1987) *Mol. Cryst. Liq. Cryst.*, **153**, 491.
59. Blizard, K.G. and Baird, D.G. (1987) *Polymer Eng. & Sci.*, **27**, 653.
60. Isayev, A.I. and Modic, M. (1987) *Polymer Comp.*, **8**, 158.
61. Weiss, R.A., Huh, W. and Nicolais, L. (1987) *Polymer Eng. Sci.*, **27**, 684.
62. Acierno, D., Amendola, E., Carfagna, C. *et al.* (1987) *Mol. Cryst. Liq. Cryst.*, **153**, 533.
63. Brostow, W. (1988) *Kunststoffe*, **78**, 411.

64. Crevecoeur, G. and Groeninckx, G. (1990) *Polymer Eng. & Sci.*, **30**, 532.
65. Jung, S.H. and Kim, S.C. (1988) *Polym. J.*, **20**, 73.
66. Chung, T.S. (1987) *Plast. Eng.*, October, 39.
67. Engberg, K., Strömberg, O., Martinsson, J. and Gedde, U.W. (1994) *Polymer Eng. & Sci.*, **34**, 1336.
68. Engberg, K., Ekblad, M., Werner, P.-E. and Gedde, U.W. (1994) *Polymer Eng. & Sci.*, **34**, 1346.
69. Finkelmann, H., Koch, H. and Rehage, G. (1981) *Makromol. Chem. Rapid Commun.*, **2**, 317.
70. Krigbaum, W.R., Lader, H.J. and Ciferri, A. (1980) *Macromolecules*, **13**, 554.
71. Krigbaum, W.R., Grantham, C.E. and Toriumi, H. (1982) *Macromolecules*, **15**, 592.
72. Martin, P.G. and Stupp, S.I. (1987) *Polymer*, **28**, 897.
73. Finkelmann, H., Naegele, D. and Ringsdorf, H. (1979) *Makromol. Chem.*, **180**, 803.
74. Attard, G.S. and Williams, G. (1986) *Liq. Cryst.*, **1**, 253.
75. Attard, G.S., Araki, K., Muora-Ramos, J.J. and Williams, G. (1988) *Liq. Cryst.*, **3**, 861.
76. Araki, K., Attard, G.S., Kozak, A. *et al.* (1988) *Faraday Trans. II*, **84**, 1067.
77. Kozak, A., Moura-Ramos, J.J., Simon, G.P. and Williams, G. (1989) *Makromol. Chem.*, **190**, 2463.
78. Nazemi, A., Keller, E., Williams, G. *et al.* (1991) *Liq. Cryst.*, **9**, 307.
79. Williams, G., Nazemi, A., Karasz, F.E. *et al.* (1988) *Macromolecules*, **24**, 5134.
80. Keller, E.J.C., Williams, G., Grongauz, V. and Yitzchaik, K. (1991) *J. Mater. Chem.*, **1**, 331.
81. Schmidt, H.W. (1989) *Angew. Chem. Engl.*, **28**, 940.
82. de Jeu, W.H. (1980) *Physical Properties of Liquid Crystalline Materials*, Gordon and Breach, New York.
83. Böttcher, C.J.F. and Bordewijk, P. (1978) *Theory of Electric Polarisation*, Vol. 2, 2nd edn, Elsevier, Amsterdam.
84. Kresse, H., Kostromin, S.G. and Shibaev, V.P. (1982) *Makromol. Chem.*, **3**, 509.
85. Talrose, R., Kostromin, S.G., Shibaev, V.P. *et al.* (1981) *Makromol. Chem. Rapid Commun.*, **2**, 305.
86. Ringsdorf, H. and Zentel, R. (1982) *Makromol. Chem.*, **183**, 1245.
87. Haase, W. and Pranoto, H. (1984) *Progr. Colloid & Polymer Sci.*, **69**, 139.
88. Shibaev, V.P. (1988) *Mol. Cryst. Liq. Cryst.*, **155**, 189.
89. Ujiie, S., Koide, N. and Imura, K. (1987) *Mol. Cryst. Liq. Cryst.*, **153**, 191.
90. Simon, R. and Coles, H.J. (1984) *Mol. Cryst. Liq. Cryst. Lett.*, **102**, 43.
91. Simon, R. and Coles, H.J. (1986) *Polymer*, **27**, 811.
92. Attard, G.S. and Williams, G. (1986) *Polym. Commun.*, **27**, 66.
93. Attard, G.S., Araki, K. and Williams, G. (1987) *J. Mol. Electr.*, **3**, 1.
94. Findlay, R.B. and Windle, A.H. (1991) *Mol. Cryst. Liq. Cryst.*, **206**, 55.
95. Schadt, M. (1981) *Mol. Cryst. Liq. Cryst.*, **66**, 319.
96. Pranoto, H. and Haase, W. (1983) *Mol. Cryst. Liq. Cryst.*, **98**, 299.
97. Osaki, S., Uemura, S. and Ishida, Y. (1971) *J. Polymer Sci. Phys.*, **9**, 585.
98. Hellermark, C., Gedde, U.W., Hult, A. and Richardson, R.M. (in press) *Macromolecules*.
99. Liebert, L., Strzelecki, L., van Luyen, D. and Levulut, A.M. (1981) *Eur. Polymer J.*, **17**, 71.
100. Noël, C., Monnerie, L., Achard, M.F. *et al.* (1981) *Polymer*, **22**, 578.

101. Volino, F., Martins, A.F., Blumstein, R.B. and Blumstein, A.L. (1981) *J. Physique Lett.*, **42**, 305.
102. Hardouin, F., Achard, M.F., Gasparoux, H. *et al.* (1981) *J. Polymer Sci. Phys.*, **20**, 975.
103. Maret, G. and Blumstein, A.L. (1982) *Mol. Cryst. Liq. Cryst.*, **88**, 295.
104. Piskunov, M.V., Kostromin, S.G., Stroganov, L.B. *et al.* (1983) *Makromol. Chem. Rapid Commun.*, **3**, 443.
105. Achard, M.F., Sigand, G., Hardouin, F., Weill, C. and Finkelmann, H. (1983) *Mol. Cryst. Liq. Cryst.*, **92**, 111.
106. Boeffel, C., Hisgen, B., Pschorn, U. *et al.* (1983) *Isr. J. Chem.*, **23**, 388.
107. Casgrande, C., Veyssie, M., Weill, C. and Finkelmann, H. (1983) *Mol. Cryst. Liq. Cryst.*, **92**, 49.
108. Sigaud, G., Yoon, D.Y. and Griffin, A.C. (1983) *Macromolecules*, **16**, 875.
109. Anwer, A. and Windle, A.H. (1991) *Polymer*, **32**, 103.
110. Anwer, A. and Windle, A.H. (1993) *Polymer*, **34**, 3347.
111. Broer, D.J., Finkelmann, H. and Kondo, K. (1988) *Makromol. Chem.*, **189**, 185.
112. Broer, D.J., Mol, G.N. and Challa, G. (1989) *Makromol. Chem.*, **190**, 19.
113. Broer, D.J., Boven, J., Mol, G.N. and Challa, G. (1989) *Makromol. Chem.*, **190**, 2255.
114. Broer, D.J., Hikmet, R.A.M. and Challa, G. (1989) *Makromol. Chem.*, **190**, 3201.
115. Jonsson, H., Andersson, H., Sundell, P.-E., Gedde, U.W. and Hult, A. (1991) *Polymer Bull.*, **25**, 641.
116. Andersson, H., Gedde, U.W. and Hult, A. (1992) *Polymer*, **33**, 4014.
117. Findlay, R.B., Lemmon, T.J. and Windle, A.H. (1991) *J. Mater. Res.*, **6**, 604.
118. Onogi, Y., White, J.L. and Fellers, J.F. (1980) *J. Non-Newtonian Fluid Mech.*, **7**, 121.
119. Ide, Y. and Chung, T.S. (1984/1985) *J. Macromol. Sci. Phys.*, **B23**, 497.
120. Hongladarom, K. and Burghardt, W.R. (1992) *Theoretical and Applied Rheology*, Proc. XIth Int. Congr. on Rheology, Brussels (eds P. Moldenears and R. Kennings), Elsevier, p. 537.
121. Windle, A.H., Viney, C., Golombeck, R. *et al.* (1985) *Faraday Discuss.*, **79**, 55.
122. Biswas, A. and Blackwell, J. (1988) *Macromolecules*, **21**, 3146.
123. Lin, Y.G. and Winter, H.H. (1988) *Macromolecules*, **21**, 2439.
124. Porter, R.S., Jonza, J.M., Kimura, M. *et al.* (1989) *Polymer Eng. & Sci.*, **29**, 55.
125. Lin, Y.G., Winter, H.H. and Lieser, G. (1988) *Liq. Cryst.*, **3**, 519.
126. Lin, Y.G. and Winter, H.H. (1988) *Liq. Cryst.*, **3**, 593.
127. Wiberg, G. and Gedde, U.W. (1997) *Polymer*, **38**, 3753.
128. Buerkle, K.R., Frank, W.F.X. and Stoll, B. (1988) *Polymer Bull.*, **20**, 549.
129. Sahlén, F., Andersson, H., Hult, A. *et al.* (1996) *Polymer*, **37**, 2657.
130. Hempel, G., Wobst, M., Israel, G. and Schneider, H. (1993) *Makromol. Chem. Symp.*, **72**, 161.
131. Broer, D.J., Gossink, R.G. and Hikmet, R.A.M. (1990) *Angew. Makromol. Chem.*, **183**, 45.
132. Hikmet, R.A.M., Lub, J. and Maassen van der Brink, P. (1992) *Macromolecules*, **25**, 4194.
133. Hikmet, R.A.M., Lub, J. and Higgins, J.A. (1993) *Polymer*, **34**, 1736.
134. Hikmet, R.A.M. and Higgins, J.A. (1992) *Liq. Cryst.*, **12**, 831.
135. Barclay, G.G., McNamee, S.G., Ober, C.K. *et al.* (1992) *J. Polymer Sci. Chem.*, **30**, 1845.
136. Broer, D.J., Lub, J. and Mol, G.N. (1993) *Macromolecules*, **26**, 1244.
137. Jahromi, S., Lub, J. and Mol, G.N. (1994) *Polymer*, **35**, 622.

138. Hikmet, R.A. (1991) *Liq. Cryst.*, **9**, 405.
139. Braun, D., Frick, G., Grell, M. *et al.* (1992) *Liq. Cryst.*, **11**, 929.
140. Hikmet, R.A. (1991) *Mol. Cryst. Liq. Cryst.*, **200**, 197.
141. Schätzle, J. and Finkelmann, H. (1987) *Mol. Cryst. Liq. Cryst.*, **142**, 85.
142. Küpfer, J. and Finkelmann, H. (1991) *Makromol. Chem.*, **12**, 717.

11

Flow induced phenomena of lyotropic polymer liquid crystals: the negative normal force effect and bands perpendicular to shear

Gabor Kiss and Roger S. Porter

11.1 INTRODUCTION

We first observed the negative normal force effect in 1975. At the time we were investigating the effect of concentration and shear rate on the viscosity of lyotropic polymers, subsequent to early reports [1, 2] that the onset of anisotropy due to volume filling and packing considerations [3, 4] caused a reduction in viscosity as concentration increased. Although a regime of decreasing viscosity with increasing concentration was unprecedented for polymer solutions, this behavior was in fact not unexpected, since the analogous decrease in viscosity with decreased temperature at the isotropic–nematic transition had already been observed with low molecular weight liquid crystals (referred to in this book as monomer liquid crystals, MLCs) [5, 6]. In addition to the intriguing rheology, there was an association of thought with the anisotropic polyparabenzamide spinning dopes that led to incredibly strong fibers, and the attendant ultrahigh performance applications such

Mechanical and Thermophysical Properties of Polymer Liquid Crystals
Edited by W. Brostow
Published in 1998 by Chapman & Hall, London.
ISBN 0 412 60900 2

as bulletproof vests, composite aircraft or even fiberoptic cables! At that time it was an intriguing and glamorous field of study.

We expected to find interesting science, but we were not expecting to wander into a phenomenon which was so unexpected, so counterintuitive that we ourselves assumed it was an artifact. We learned that, indeed, inertia is a source of potential artifact which can produce a positive-to-negative normal force transition in isotropic polymer solutions, as shown in Figure 11.1 [7].

No doubt, many a knowledgeable rheologist, upon hearing of our negative first normal stress difference (N_1) observations assumed that we had fallen into this trap. Needless to say, all of the data we published in the late 1970s was already corrected for the inertial contribution. In any case, the second sign change in N_1, from negative to positive again, could not have been so caused.

We felt privileged, therefore, to participate in the discovery of previously unknown territory in the field of rheology and to draw the first crude maps. Our contribution was to be in the right place at the right time, to keep an open mind and to make careful observations. The intellectual content of our contribution was not titanic, but we were willing to take risks, such as increasing the shear rate to beyond

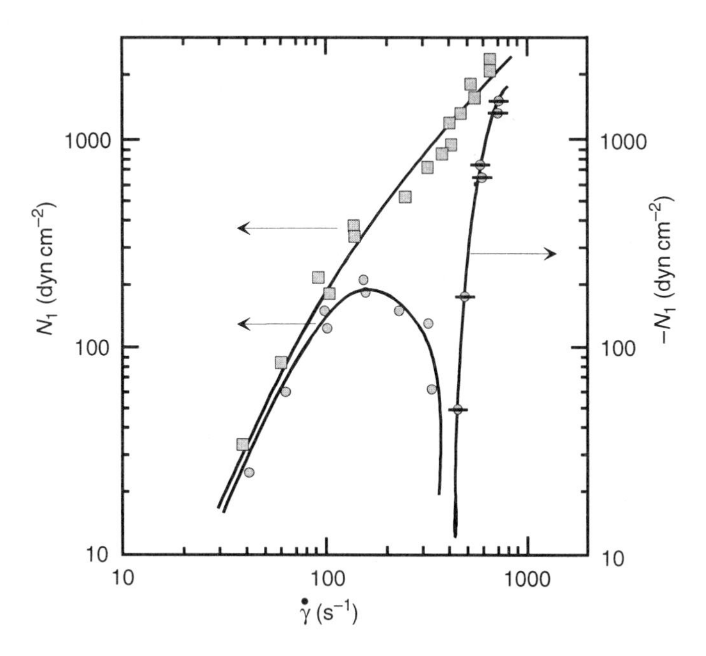

Figure 11.1 Inertial normal force correction (Reproduced from [7].)

the point at which sample ejection occurs (resolved by extrapolating back to the time of initiation of shear). We also used electronic filtering to remove large amplitude, high frequency oscillations. We ended up with much better data than we might have expected, and the data told an amazing story, that under easily accessible conditions of solute, concentration, molecular weight and solvent, one can **in a single experiment** of increasing shear rate observe normal forces that first increased to a maximum, then suddenly become negative, then reach a negative maximum and then suddenly become positive again. We conjured some 'handwaving' explanations, but essentially we had no clue and merely observed this phenomenon.

In the intervening years, the essential observation has been thoroughly confirmed specifically both for *m*-cresol solutions of poly(γ-benzyl glutamate) (PBLG) and for aqueous solutions of hydroxypropylcellulose. Section 11.4 amplifies this with some details of the negative N_1 phenomenon which have not been explored as thoroughly. The confirmation is gratifying, but not unexpected, since we had satisfied ourselves that the phenomenon was genuine. Even more gratifying, and much less expected, is the fact that subtle and insightful theoretical work, discussed in section 11.5, which has been accomplished by several sets of researchers, has completely explained the normal force observations. Indeed, the theory predicts a similar series of transitions in N_2, the second normal stress coefficient, which has been borne out experimentally: a major accomplishment.

Despite the numerous confirmations of the negative N_1 phenomenon, it has still been widely stated that the flow of all polymer systems exhibits only positive primary normal forces (i.e. a positive N_1, the first normal stress difference) [8, 9]. Even subsequent reviews and research papers on the specific subject of lyotropic main chain liquid crystal polymers have not mentioned the confirmed negative N_1 effect [10], and even equivalent shear measurements on the identical solutions did not report the negative N_1 effect [11].

At the same time that we were confronted with the negative normal force phenomenon, we also discovered an odd morphological phenomenon, that of the shear-induced band structure. Many polymer liquid crystal systems have been seen to exhibit striations perpendicular to the shearing direction upon cessation of shear. To our knowledge, the formation of striations or bands was first mentioned in passing by Aharoni [12], who did not specify the shearing direction. This phenomenon has also been confirmed repeatedly and has also now been satisfactorily explained, as discussed in section 11.6.

In 1980 we published polarized light micrographs taken during shear with stroboscopic illumination which appeared to show perpendicular striations. The occurrence of these striations during shear seemed to correspond to the shear rate at which the negative normal force occurred. This observation has not been confirmed and in fact some evidence suggests that it could not be correct [13]. We suspect now that this was in fact an artifact, as will be discussed in section 11.6.4.

In perspective, this chapter is not a thorough review so much as it is a retrospective, a narrative, and an homage to the great research of others which came after. In some places we have occasion to quote verbatim from other works. When the fragment is short, several words or so, it will be set off by quotation marks. When the fragment is extended, it will be set off by indentation. This device is not just an indication of our desire for accuracy. In many cases we acknowledge where they said it first, and said it best. By quoting verbatim we pay tribute.

While drafting the first version of this chapter in the year 1995, 20 years later, and the year of the death of Jerry Garcia of the Grateful Dead, we look back and echo his words: 'what a long, strange trip it's been!'.

11.2 VISCOSITY

One class of rod-like molecules that has received major attention in rheological studies is the synthetic polypeptide poly(γ-benzyl glutamate) (PBG), usually the L enantiomer (PBLG). The polypeptide has long been obtainable in narrow molecular weight distributions [14]. When a single enantiomer is dissolved in an appropriate solvent, the helices all have the same optical rotary sense, resulting in a cholesteric liquid crystal. However, for equimolar mixtures of both L and D enantiomers, a nematic liquid crystal is obtained [15]. For the two chiral enantiomorphs PBLG and PBDG solutions of each at rest form an anisotropic cholesteric phase.

Under shear, the cholesteric structure evolves into a nematic state. The flow properties of cholesteric PBLG and PBDG solutions have been found to be qualitatively the same as those of their nematic racemic mixtures [16–18]. A nematic–cholesteric transition reversion in PBG solutions occurs upon cessation of shear [19]. The cholesteric–nematic transition is not expected to have a major effect on the rheological results [20], as confirmed by an investigation on a racemic mixture of PBLG and PBDG by Berghmans [21]. The most frequently studied and addressed rheology of PLC solutions is their viscosity change as a function of PLC concentration. The equation given by Flory [3],

$$\phi = 8/p(1 - 2/p) \tag{11.1}$$

predicts a critical volume fraction ϕ, later called C_n^*, for formation of an anisotropic liquid crystal phase for polymer rods of length/diameter (aspect ratio) of p. Even this simplest evaluation revealed remarkable behavior in the early paper of Hermans, who studied the rheology of concentrated solutions of PBLG. This work revealed the dramatic viscosity maximum in capillary flow and its dependence on shear rate and molecular weight [2]. A useful model was offered but elastic effects were not considered. Iizuka [1] used solutions of PBLG in CH_2Br_2 and in dioxane, and made steady shear and oscillatory measurements using a cone-and-plate rheometer.

The remarkably simple and insightful two-dimensional Flory theory appears to underestimate slightly the critical concentration for the viscosity maximum of the anisotropic phase (Table 11.1). This concentration may be after the mesophase occurs and only when that single phase is dominant. As indicated in our studies of PBLG, for blends with PBDG, and found for other PLC lyotropics by Aharoni [12], the viscosity maximum is close, but not identical, to the concentrations for onset of a nematic state. A shift in this direction with concentration would also be expected if the helices were not perfectly rigid [22]. Doi has provided a valuable theoretical treatment of trends near the viscosity maximum [23].

Shoulders on the low concentration side of the viscosity–concentration curves suggest the transition from biphasic to a single mesophase (Figure 11.2a). The peak is identified with the concentration sufficient for a single LC phase. The actual peak maximum moves to lower shear rates and is vastly reduced in magnitude with increasing shear rate, as seen in Figure 11.2b. At the high shear limit, the polymer rods are aligned in shear over the entire concentration range, from dilute to high concentration, and the peak is almost completely suppressed. This

Table 11.1 Indicators of liquid crystalline order for helical polypeptides + *m*-cresol

Indication of liquid crystalline order	Concentration (wt%)		
	PBG ($M_w = 335\,000$)	PBG ($M_w = 150\,000$)	PCBZL ($M_w = 200\,000$)
Optical anisotropy (birefringence)	9.9	10.5	25.3
Maximum in zero shear limit of steady viscosity	11.0	11.2	23.6
Maximum in zero frequency limit of dynamic viscosity	11.0	11.4	23.2
Maximum G'	9.9	11.1	22.8
Extrema in N_1	8.1	11.2	25.2
Shear stress at sign change of N_1	11.0	12.6	27.6
Flory theory (wt%)	6.9	14.7	11.9

behavior has been reported for poly(benzyl glutamate) (single enantiomers as well as for racemic mixtures, PBLG + PBDG), poly(ε-carbobenzyloxy-L-lysine) (PCBZL) and poly-*p*-benzyl-L-aspartate (PBA), all in the helicogenic solvent, *m*-cresol.

Figure 11.2a shows that the low frequency limit of dynamic viscosity is very similar to the low shear limit of steady shear viscosity throughout the concentration sequence of isotropic to biphasic to anisotropic. At higher frequencies the dynamic viscosity peak is reduced in magnitude [24] for lyotropic cellulose esters + DMAC, just as viscosity is reduced at higher shear rates (Figure 11.2b). However the concentration of the peak was reported not to decrease with frequency. Thus it appears that dynamic shear, unlike steady shear, does not drive the thermodynamic transition for formation of an anisotropic phase to a lower concentration.

The effect of increased molecular weight, and therefore increased aspect ratio, is to decrease C^*. Increasing the flexibility of the rod effectively decreases the aspect ratio and therefore increases C^*. For example, a PCBZL of molecular weight 200 000 and nominal axial ratio of 85 had $C^* = 23.6\%$, while a PBLG of molecular weight 150 000 and nominal axial ratio of 68.5 had $C^* = 11.2\%$ [17]. The value of C^* calculated from the Flory theory was very close to the observed C^*, indicating that PBLG behaves essentially like a rigid rod, while the calculated C^* for PCBZL was much lower (9.2% vs. 23.6%), suggesting pronounced flexibility. Hydroxypropyl cellulose (HPC) in aqueous solution is also quite flexible, with $C^* = 25\%$ for a molecular weight of 90 000 [25]. HPC + *m*-cresol of molecular weight 50 000 had C^* of 22%, and C^* was seen to increase with temperature, that is with flexibility, to 26% at 50°C [26].

An interesting phenomenon which appears to be related to rod flexibility is that of a renewed viscosity increase upon increasing concentration well above C^*. This was observed by us for both PBLG and PCBZL, by Asada *et al.* [27] for PBLG + *m*-cresol, for HPC + *m*-cresol [26], and for HPC in aqueous solution [25, 28]. This phenomenon is not predicted by the Doi theory (discussed in section 11.5.1(b)) which is successful in predicting negative N_1 [29]. The behavior of the local minimum with increased shear rate depends on rod flexibility. For the rigid PBLG, shear has been observed not to affect concentration at the local minimum. For the more flexible PCBZL [17] and HPC [26], shear decreases the concentration of the local minimum, apparently due to effective stiffening of the rod molecules.

We observe that the transition from the isotropic to liquid crystalline state at a critical concentration of the rod-like helical polypeptide solute is manifested dramatically in each of the six properties that were examined [16, 17] (Table 11.1).

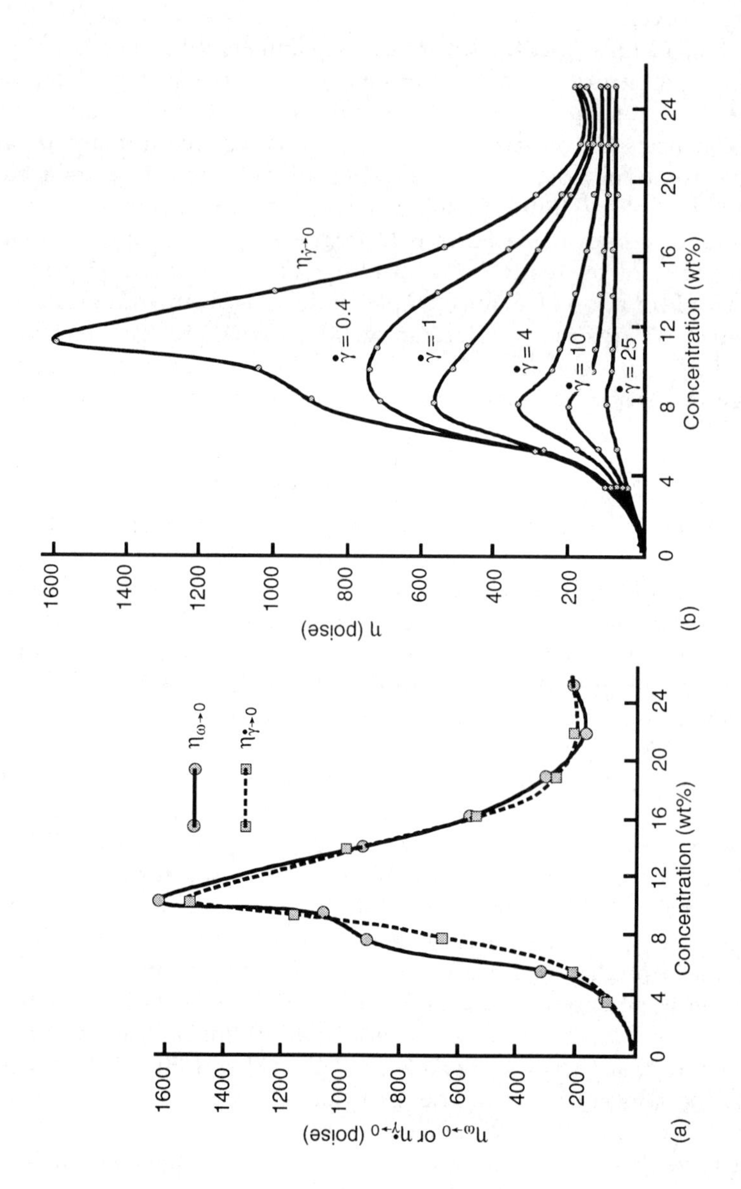

Figure 11.2 Viscosity vs. concentration for PBG (M_w = 335 000) + *m*-cresol. (a) Zero shear/frequency limit of steady shear and dynamic viscosity. (b) Steady shear viscosity at various shear rates.

As shown in Table 11.1, liquid crystalline order may be manifested at slightly different concentrations by various measured properties, possibly corresponding to different ratios of anisotropic to isotropic phase. The Flory prediction of the critical concentration for formation of an anisotropic phase is lower than the indications of any of the observed properties. Birefringence would be expected to be the most sensitive to a small amount of anisotropic phase coexisting with a large amount of isotropic phase.

PCBZL + *m*-cresol at room temperature is known to be in a random coil, and undergoes an inverse coil–helix transition as the temperature is raised [30]. The observation of lyotropic behavior in PCBZL + *m*-cresol solutions at room temperature [17] may represent a peculiar case of a conformational transition driven by the opportunity to form a liquid crystalline phase.

11.3 RELATIONSHIP OF VISCOSITY TO NORMAL FORCE

We would now like to glance at the viscosity behavior in the light of the dramatic normal force phenomena which we will presently discuss explicitly. The viscosity–shear rate curves for the anisotropic solutions which exhibit a negative N_1 are in general not remarkable. They tend to exhibit a low shear Newtonian plateau followed at higher shear by power law shear thinning. We were surprised by how little impact the simultaneously measured and dramatic normal stress phenomena apparently had on the shear stress response of these solutions (see, for example, the typical data of Figure 11.5). In concentrated solutions ($\sim$ 40%) a hysteresis loop was seen in viscosity vs. shear rate [29]. Upon shearing at a critical rate, a structural change in the fluid caused an increase in viscosity which was reversed upon resting for a day or more; however this change was not manifested in N_1 [29].

Nevertheless, we did find some indication of an unexpected viscosity feature of a 'second Newtonian plateau' [17] in some solutions. This feature has also been noted by other authors, who call it a 'dip' or 'hesitation' (it should not be confused with a 'region 0' Newtonian plateau at very low shear rates; section 11.5.1(b)). While we did not necessarily associate this feature with the sign change in N_1, others have done so. Driscoll *et al.* made this connection explicitly for a thermotropic HPC derivative and for PBLG + *m*-cresol [31]. They suggested that the presence of such a hesitation might be used as an indicator for the presence of negative N_1, even if it was experimentally inaccessible in a particular system. Magda *et al.* found that the magnitude of the hesitation was greater for lower concentration solutions of PBLG + *m*-cresol and that changes in convexity of the flow curve which result in the dip occur at the same shear rates as the relative maxima and minima in the N_2 shear rate curve (see section 11.4.2 for more discussion of N_2) [32].

An extremely pronounced hesitation, in fact a slight increase, in viscosity at intermediate shear rates was reported by Asada *et al.* [27] for racemic PBG + *m*-cresol for concentrations above 15% and moderate shear rates (see their Figure 8, p. 235). These authors did not comment on the presence of negative N_1, though it would have been expected to occur since they used a rotational rheometer.

Another interesting relationship between viscosity and normal stress was noted by Berry [33]. He used a quantity analogous to viscosity, which was termed 'normal viscosity' (introduced by Doi [23]), defined as $\eta_n = N_1/(d\gamma/dt)$, and showed that for 'slow flows' η_n should tend to η_0. He shows this behavior occurring in his Figure 5 (p. 350) by reanalyzing data taken from [17]. This figure also shows a region of convergence of $-\eta_n$ to η_0 at intermediate shear rates (where N_1 is negative and a viscosity 'dip' or 'hesitation' is observed), as well as another convergence of η_n to η_0 at the highest shear rates.

We take this opportunity to replot some of our early data from this perspective, as shown in Figure 11.3 and we note that these three observations generally hold.

A generalized 'three-region' flow curve has been proposed by Onogi and Asada [34] for flow curves of liquid crystal polymers, in which a low shear rate, shear thinning regime (I) is followed by a Newtonian plateau (II), and then by a shear rate, shear thinning regime (III). Larson *et al.* have pointed out that if accessible data for a particular system only show one shear thinning regime, it is not possible to say whether this is region I or region III; however, Doi theory predicts a sign change in N_1 to negative as one enters regime III. Thus the two shear thinning regimes can be identified if viscosity data are supplemented with N_1 data [29].

11.4 NORMAL FORCE PHENOMENA

The section above has already introduced observations of negative normal force in the form of a 'normal viscosity'. In this section we will discuss in much greater detail the behavior of the first normal stress difference N_1 (also referred to as ν_1 and defined as $\tau_{11} - \tau_{22}$). Figure 11.4 [16] shows the negative N_1 effect in the format usually seen, as log N_1 plotted against log $d\gamma/dt$, for a 16.4 wt% solution of racemic 350 000 M_w PBG + *m*-cresol.

The obviously notable features of N_1 with increasing shear rate are the sequence:

- an increase at low shear rate with an initial log–log slope of 1.0 ± 0.2;

- a maximum;
- a sharp transition to negative values (i.e. cone and plate pulled together during shear);
- a negative maximum;
- a sharp transition to positive values again;
- linear increase with a slope > 1.0 to the limits of the apparatus.

After a brief background, we will return to a detailed discussion of the features noted above, as well as other subtle observations.

The early developments in this field have been reviewed [6]. The earliest mention of negative N_1 that we have encountered for any system was by Hutton [35], who measured normal stresses in lubricating greases. He concluded that this 'strange behavior' was a manifestation of a directional effect, involving the gap-setting servomechanism. When this mechanism was switched off, 'a more regular type of result was obtained'.

Normal stress measurements for some MLC nematics was reported to be consistent with that of a second-order fluid, that the low frequency limit of G'/ω^2 equaled the low shear limit of $N_1/(d\gamma/dt)^2$ [36]. Coleman and Markowitz demonstrated that for a second-order fluid in slow Couette flow, the viscoelastic contribution to the normal thrust must have a sign opposite to the inertial contribution on thermodynamic grounds [37]. A textbook by Walters stated that the measurements of first normal stress difference have invariably led to a positive quantity except for one case which was later found to be in error [38]. Adams and Lodge reported the possible observation of a negative value for N_1 for solutions of polyisobutylene + decalin [39]. This result was obtained by a combination of $\tau_{11} + \tau_{22} - 2\tau_{33}$, obtained from radial variation in normal stress in a cone-and-plate rheometer, with measurements of N_2 (defined as $\tau_{22} - \tau_{33}$) that were reportedly of uncertain accuracy. In a discussion of this point, they concluded that negative values of N_1 would be curious but not impossible.

Another observation of negative normal stress which had been reported prior to our work was observed upon reversal of the direction of shear and was thus transient. In a study of rheological properties of aqueous lecithin, positive normal thrust was observed in steady cone-and-plate shear, but on reversal of the rotation direction, the normal thrust became negative and eventually increased to the positive steady state value [40]. This effect was attributed to incomplete normal stress relaxation upon cessation of flow. Time-dependent negative normal thrust was reported on melts of an SBS copolymer [41]. In this case a negative normal thrust was generated upon initiation of shear, but decayed to zero after about 75 s. This effect was thought to be due to a small volume decrease caused by shearing.

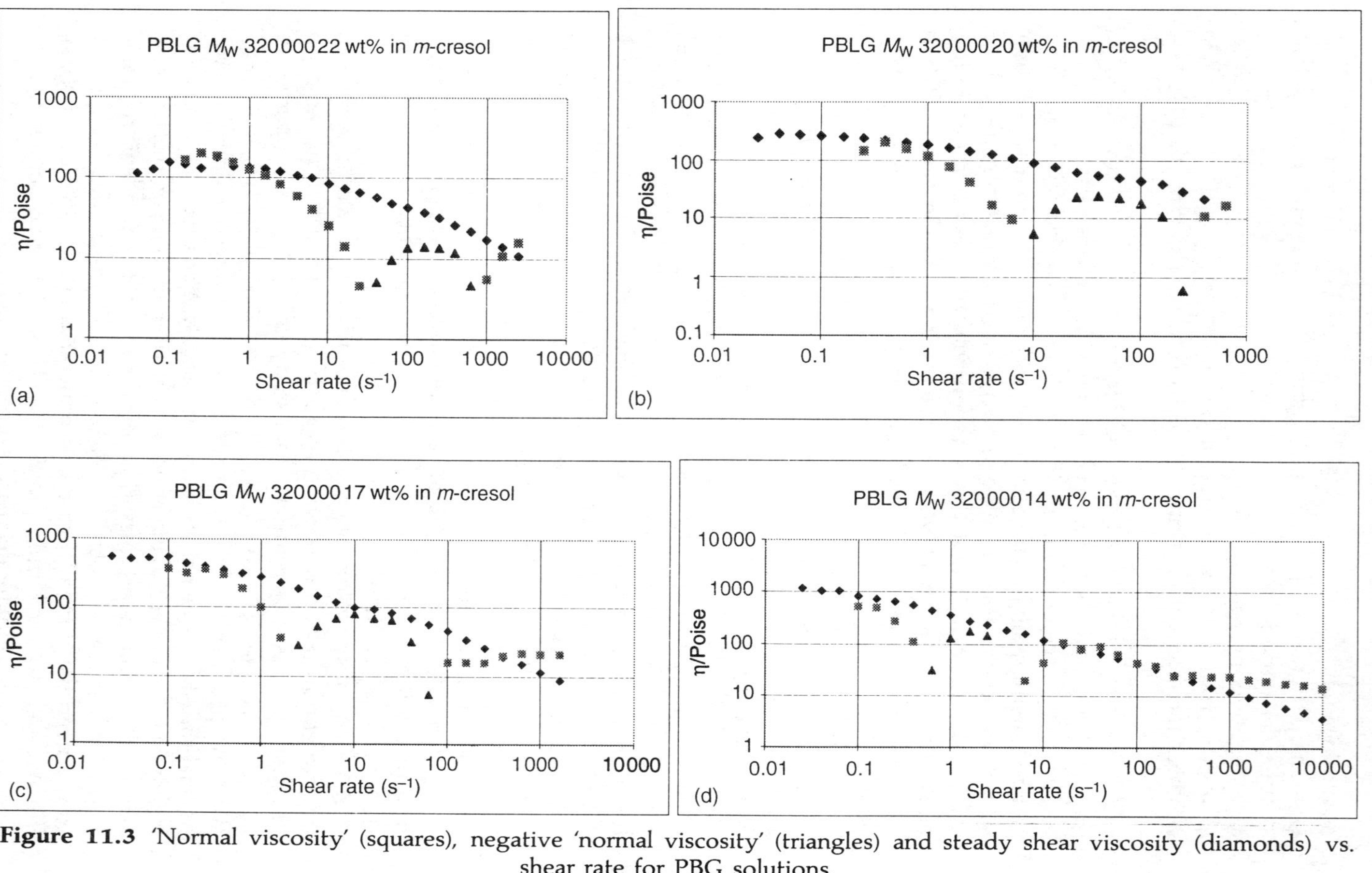

Figure 11.3 'Normal viscosity' (squares), negative 'normal viscosity' (triangles) and steady shear viscosity (diamonds) vs. shear rate for PBG solutions.

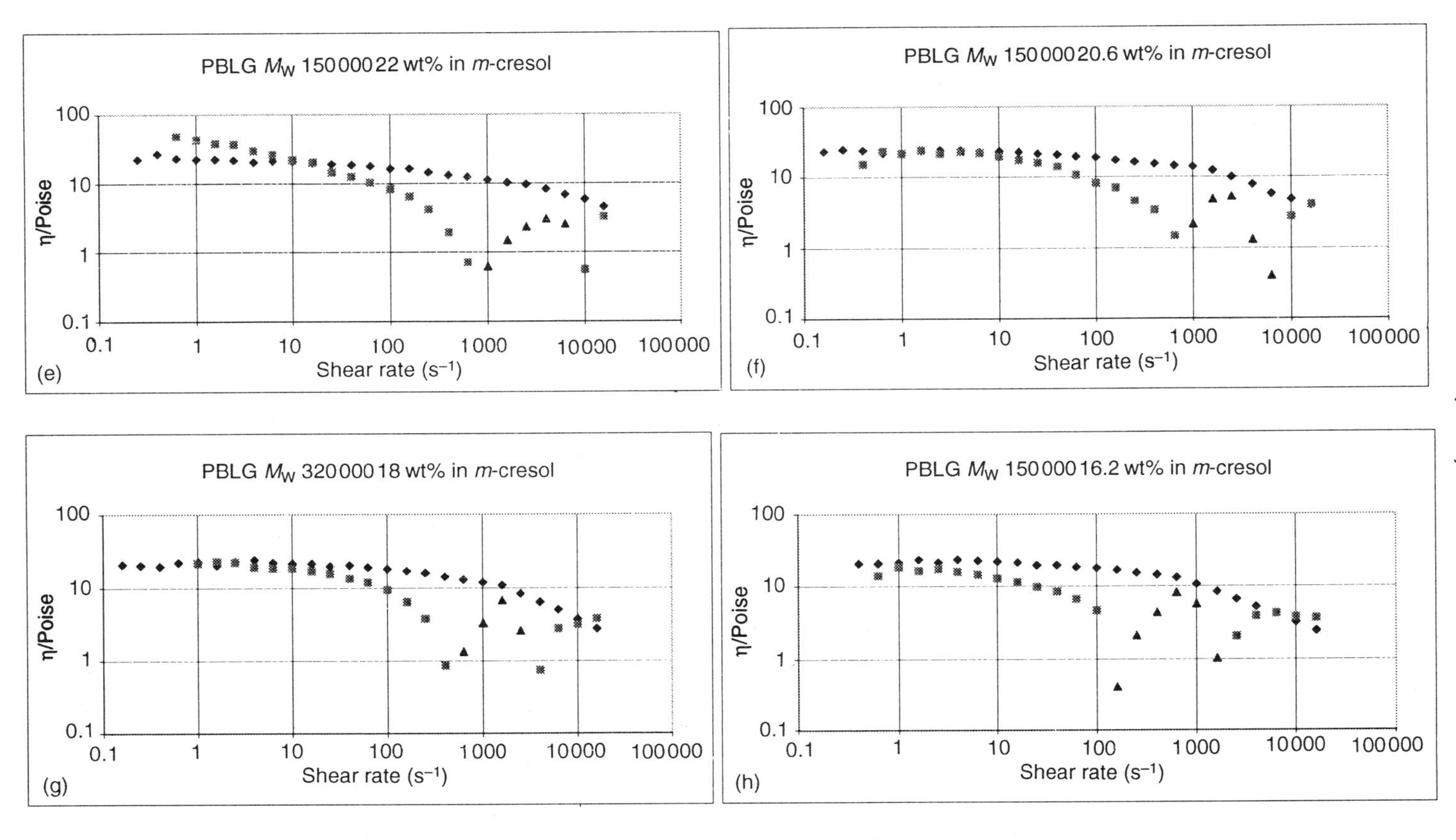

Figure 11.3 (Continued)

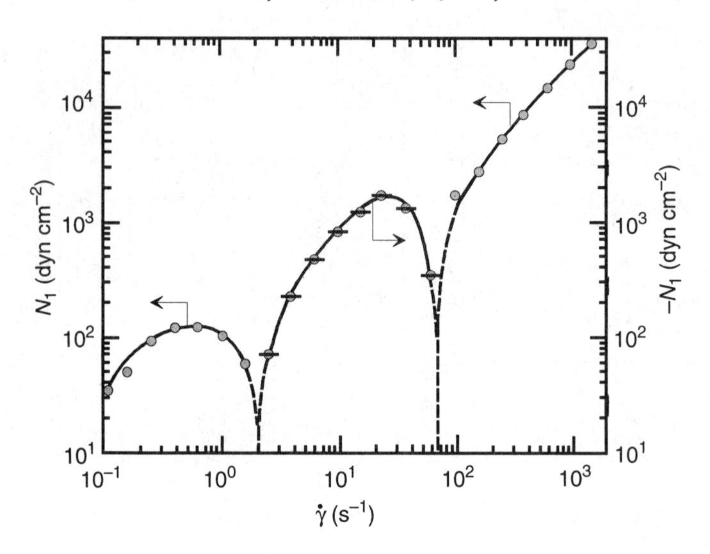

Figure 11.4 The negative N_1 effect: first normal stress difference vs. shear rate for 16.4 wt% PBG (M_w = 335 000) + *m*-cresol. (Reproduced from [16].)

Normal stress measurements on concentrated solutions of helical polypeptides were conducted by Iizuka [1, 42]. However he used these to calculate extinction angles, from which the rotary diffusion constant was deduced, and thence an apparent particle size from tables given by Scheraga [43]. In a personal communication to Kiss and Porter, Iizuka commented that he had observed negative normal stresses in solutions of PBLG + Ch_2Br_2 with concentrations of greater than 10% (i.e. probably liquid crystalline); however he ascribed this to the 'adhesive force' of the solution (E. Iizuka, personal communication, April 1977).

We see then that, interestingly, at the time that we began our exploration of the rheology of lyotropic polypeptide solutions, the possibility of the occurrence of negative N_1 had been discussed. Some scattered observations had in fact been reported, and generally dismissed. This state of affairs made us, if anything, more inclined to suspect an artifact than if we had stumbled onto a hitherto unsuspected phenomenon. The magnitude of our skepticism at the time is indicated by the fact that our first publication was on the correction for the negative inertial contribution to the normal thrust observed in cone-and-plate shear [7]. Incredibly, there was indeed a potential artifact which could have caused an erroneous observation of a negative normal thrust (Figure 11.1). However the contribution proved to be small, and in any case could only explain a single sign change from positive to negative. Nevertheless, it led to an emotional roller coaster, thinking we had made an important observation, then realizing that it might be due to artifact, then realizing that the artifact was insignificant and that we may have made an important observation after all.

In researching this chapter we have found a large number, approaching 300, of citations to the work we described in three papers [16–18] dealing with rheology and rheo-optics of lyotropic polypeptide solutions, including much of issue 5 of the *Journal of Rheology* (1994). In many cases, particularly before repeated confirmations of the phenomenon, the language used to describe it indicates a certain incredulity on the part of the authors. We find phrases such as 'unusual behavior', 'peculiar indeed', 'dramatic', 'an enigma', etc. We did not encounter the term 'ludicrous', but would not have been surprised if we had! Well after our discovery and confirmation of the negative N_1 effect, it was still widely stated that the flow of all polymer systems exhibit only positive primary normal forces [8, 9].

We wish to acknowledge here that in 1981 we learned orally from Professor Arthur B. Metzer that, at the University of Delaware, he with Professor J. Mewis had already obtained, but had not yet published, verification of our reports in 1978 and 1980 of negative normal stresses on equivalent PBLG + *m*-cresol systems. A contemporary report [11] on PBLG + *m*-cresol did not find a negative N_1, but rather that 'normal stresses are very small and can only be observed at the highest concentrations'.

In 1986 the first published reports of confirming measurements of a negative N_1 region on similar PBG + *m*-cresol solutions appeared in the papers of Mewis and Moldenaers [19, 44]. Also Navard published a confirming observation in a totally different system, that of aqueous hydroxypropylcellulose (HPC) solutions [45]. The HPC solutions were more concentrated, 30–50 wt%, and only the first positive region and the onset of the negative region were observed: the negative maximum and the second positive region were not accessible 'due to the high rotation speed involved' (presumably sample ejection).

To the present, few well defined lyotropic PLCs have been well characterized by rheology and in uniform shear geometries (which excludes capillary flow). Baek and coworkers provided an excellent summary in 1993 [46]. Table I in that paper indicates that only two systems have been unambiguously found to demonstrate negative N_1, i.e. PBG and HPC. Much outstanding experimental work has been conducted by e.g. Moldanaers, Burghardt, Mewis, Larson, Baek, Navard and Grizzutti on PBG + *m*-cresol and HPC solutions. Steady state negative N_1 in PBT solutions in 'fast flows' has also been reported [47].

Work by Suto *et al.* [26] demonstrated that HPC + *m*-cresol forms a lyotropic solution with a peak in viscosity vs. concentration at around 20 wt%; however no indication was given as to the presence of negative N_1. Navard and Hauden [28] examined HPC + acetic acid and found the characteristic peak in viscosity vs. concentration, but not negative N_1. Solutions of ethyl cellulose + *m*-cresol [48] were also found to be lyotropic, however rotational rheometry was only performed at a single low shear rate. Other manifestations of elasticity were studied using capillary measurements (entrance pressure drop, Bagley correction factor, die swell) but negative N_1 was not discussed in this work [48].

Table I in Baek *et al.* [46] does not mention the observation of negative N_1 for the system PCBZL + *m*-cresol which we reported in [17]. This is a very similar system to PBG + *m*-cresol in that both polymers are helical polypeptides. However they are very different in their flexibility. Using the composition of solvent at which the helix–coil transition occurs as a criterion of helix stability, and hence flexibility, Fasman *et al.* [49] determined that PBLG helices are more rigid than PCBZL helices. PBG has a persistence length of about 1200 Å [50] compared to estimates of around 100–200 Å for PCBZL [51].

The observation that two polymers of such different flexibility both demonstrate negative N_1 would lead one to think that this is an easily produced phenomenon and would be commonly observed. It seems odd therefore that so few other systems have been found. It is further extremely odd that the other frequently reported system, HPC in water, differs greatly from PBG + *m*-cresol in both polymer and solvent characteristics. What is the factor or balance of factors which appears to be so difficult to achieve, and yet is readily achieved in these two extremely disparate systems? Why does HPC show negative N_1 in aqueous solutions but not in *m*-cresol or acetic acid? These are questions which do not yet appear to have been addressed.

Table I of Baek *et al.* [46] also indicates that negative N_1 may have been seen in thermotropes, but they conclude that these observations are not definitive nor unambiguous. There is a report of negative N_1 in blends of thermotropic polyester and PET [52]. We hope to return to the issue of negative N_1 in thermotropes in a future paper, possibly attempting to correlate with observations of negative die swell, but at this point we state merely that the case for negative N_1 in thermotropes is tenuous.

In contrast to the paucity of systems which exhibit negative N_1, Baek *et al.* [46] indicate that the phenomenon of 'banding' perpendicular to the direction of shear (or 'transverse striations') appears to be universal or nearly so. We will return to this banding phenomenon in section 11.6, and in particular to an examination of the merit of our report in 1980 of a correlation between banding and negative N_1, which now appears to have been erroneous.

Despite the large body of work contributed by several active groups, there remain numerous voids in the literature. For example, there is insufficient rheological data to address the temperature dependence of the negative N_1 region, as it might be interestingly affected by changes in flexibility of rod polymers due to differences in temperature and solvent and others. The predicted large effect of molecular weight distribution [53] on the negative N_1 shear region for lyotropic solutions is only beginning to be evaluated [54].

The following cogent paragraph is taken directly from Hongladarom *et al.* [55]:

> There is no question that the model system of poly(benzyl glutamate) (PBG) + *m*-cresol solutions is, rheologically, the most well understood liquid–crystalline polymer solution. As one of the first materials for which negative N_1 were observed, as well as for which tumbling was directly confirmed, it has naturally served as the prototype for detailed rheological and structural studies. A great deal of the progress of the past five years ultimately amounts to the ability to explain phenomena that are most consistently observed in PBG. Since the theoretical foundation of this progress has been the Doi model, which assumes only rigid rod molecular structure and an anisotropic excluded volume potential, one might have expected that the success in describing behavior in PBG solutions would translate directly to other PLC systems that share these underlying characteristics. However, a recent summary by Baek and co-workers (1993) [46] of the level of understanding in diverse materials reveals a frustrating lack of universality in behavior.

One exception has been another model system, HPC solutions in water. HPC solutions have been shown to show sign changes in N_1 with increasing shear rate, like PBG solutions [28]. Moreover, the transition shear rates vary as a function of concentration [56] and molecular weight [54] in a manner consistent with the predictions of the Doi model, again in agreement with PBG [32]. Since HPC forms a cholesteric phase, formation of stable uniformly oriented monodomains is impossible in this system, so that direct confirmation of tumbling in HPC has to date been impossible. Nevertheless, HPC solutions show many 'signatures' of tumbling behavior at low shear rates, such as oscillatory responses in transient flows [57, 58], so that it may be reasonably assumed that HPC and PBG share this characteristic as well. There are, however, some differences in the rheological behavior of these two model systems.

11.4.1 Review of experiments

The remarkable results of experiments have drawn attention to the need for the development of theory, which has now been extensive, as discussed in section 11.5. Indeed, the value of such unexpected observations is that they are a demanding test of theory. Assuming, of course, that the measurements are an accurate reflection of 'what really happens' and not due to artifacts or wishful thinking, as in the case of pathological science, they represent a boundary condition for a proposed theory. Any proposed theory must be able to reproduce the observed complex features. This attitude is well expressed in a sentence taken from a paper by Mewis and Moldenaers dealing with the transient responses of PBLG + *m*-cresol solutions to step changes in shear rate [19]: 'The objective of this work is to identify some of the characteristics of polymeric liquid crystals which should be described by rheological constitutive equations'.

We were willing to pursue the conformation of most unusual observations. We did this with clear eyes and a constant awareness of the possibility of artifact. In the end, the results vindicated the use of data which some may have considered questionable and which were certainly contrary to expectation of all prior experiments and theory. Indeed, we might have ventured further. We were in a position to attempt measurements of values of N_2 by combining our cone-and-plate tests with parallel-plate measurements. Having previously published a rare N_2 study on conventional polymers [59], we knew of this possibility. We elected not to pursue these additional measurements. Our contribution was confined to revealing the existence of a regime of negative N_1, an unprecedented observation, and the existence of the transversely banded structures produced by shearing of PLCs.

(a) *Measurement technique, sample geometry and edge condition*

The Rheometrics Mechanical Spectrometer we originally used was a standard laboratory instrument, as is the Weissenberg Rheogoniometer, which could also have been used at that time. The polymers were also readily available, the solvent was not exotic or toxic, and measurements were made at room temperature with no particular precautions. The three techniques which we employed to stretch the shear rate range were:

- a low pass filter to extract the signal of interest from within a very large oscillation;
- use of rotational speeds high enough to eject material, so that the measurement needed to be extrapolated to zero time;
- use of many cone/plate combinations of widely varying cone angle and diameter.

The design of the Rheometrics Mechanical Spectrometer of the era was helpful, as it allowed rapid reloading and trimming of excess sample.

We first published this negative N_1 feature in 1978 for an *m*-cresol solution of the most definitively studied lyotropic PLC, i.e. PBLG [16].

We confirmed the observation in a lower molecular weight PBG and found it also for the more flexible PCBZL polymer in 1980 [17]. We found this distinct negative N_1 region to be independent of cone angle, as seen in Figure 11.5, which also shows the necessary independence of viscosity.

The occurrence of negative N_1 was also seen to be independent of the nature of the edge condition, as demonstrated by performing experiments in a Couette geometry using a conical bottom surface on the cylindrical bob, and by performing experiments in an 'infinite sea', that is a small cone and a large plate, as seen in Figures 11.6a and 11.6b. Comparison of these two figures also shows that:

- both of the changes in sign of N_1 (first positive to negative, and negative to second positive) are essentially independent of edge condition;
- the response of cholesteric PBLG solutions is essentially the same as nematic solutions of the racemic PBLG–PBDG mixture.

Another confirmation that the observed negative N_1 behavior represents a true material function is given by Magda *et al.* [32] by comparing negative N_1 values derived from total thrust with those derived from the radial normal stress distribution as obtained by small

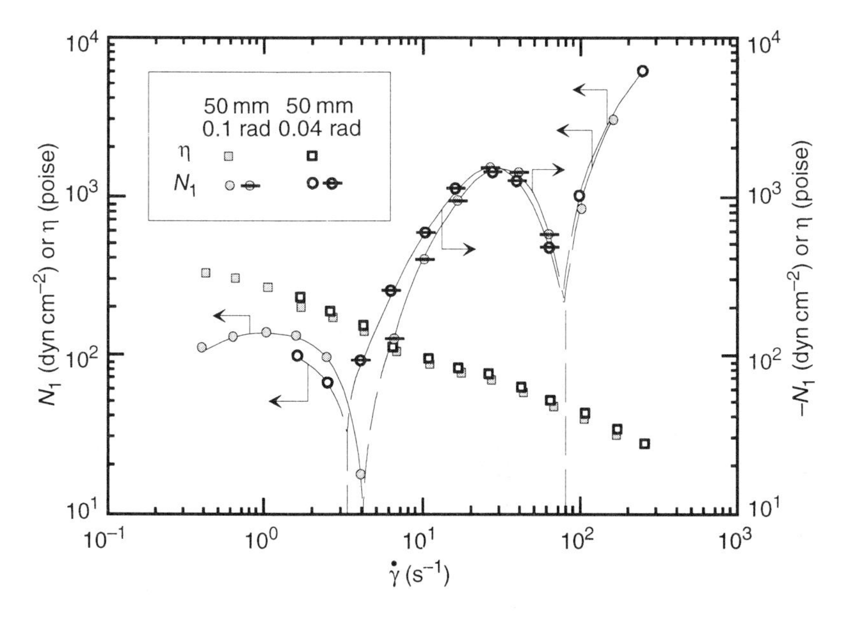

Figure 11.5 N_1 vs. shear rate of 17 wt% PBLG ($M_w = 350000$) + *m*-cresol for two cone angles. (Reproduced from [17].)

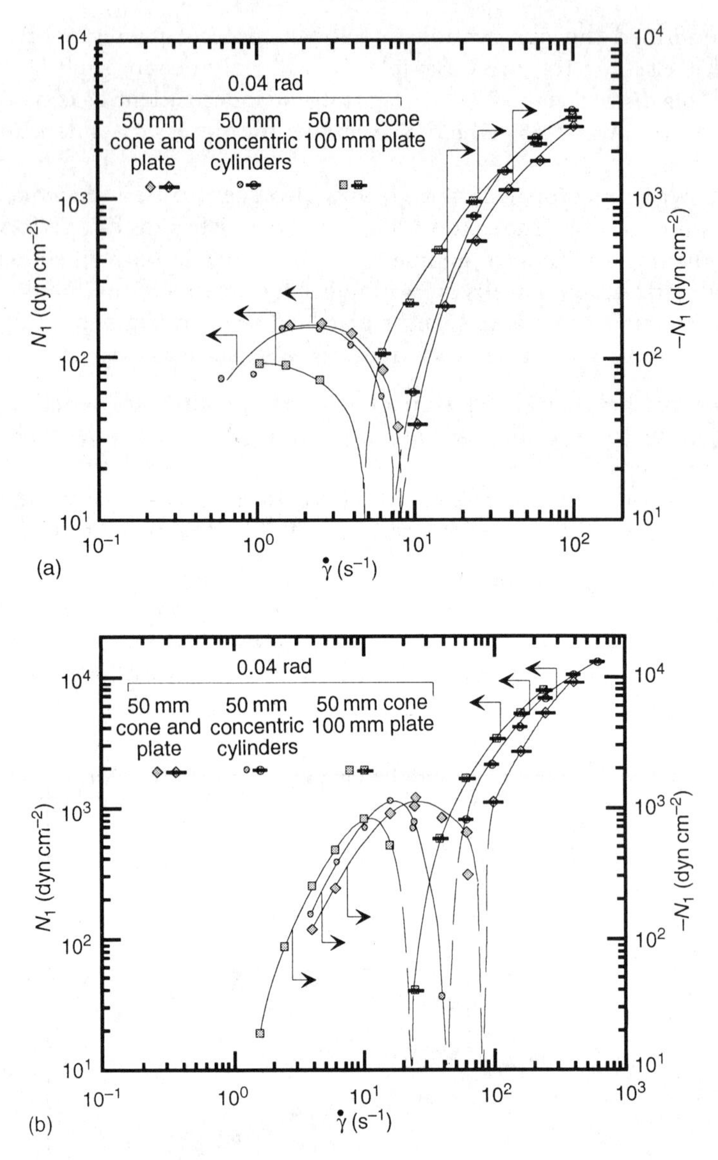

Figure 11.6 N_1 vs. shear rate for several edge conditions in (a) solution of racemic mixture 16.0 wt% PBG ($M_w = 335\,000$) + m-cresol and (b) cholesteric solution of single enantiomer 17.0 wt% PBLG ($M_w = 350\,000$) + m-cresol. (Reproduced from [17].)

flush-mounted transducers. The values given by the two methods agree precisely in Figure 11.5 of that reference. This same fine work makes a further unprecedented observation in the slope of the stress distribution.

Continuum mechanics requires that a semilogarithmic plot of the stress profile should be linear, but does not dictate the slope. A negative slope is normal for isotropic fluids and also for isotropic solutions of rod-like polymers. In the region of negative N_1 it may not be surprising that the slope is positive (i.e. atypical) but the surprising result is that even for shear rates for which N_1 is positive, this slope remains positive (i.e. atypical). This was considered to be due to the observation that N_2 is comparable in magnitude to N_1 for these solutions and hence both contribute significantly to the stress profile. In conventional fluids N_2 is much lower than N_1 and often negligible.

(b) Effect of shear rate and concentration on N_1

Although we have seen earlier that the peak in viscosity vs. shear rate is much reduced ('washed out') at high shear (cf. Figure 11.2), this is not the case for the negative N_1 effect. Figures 11.7 and 11.8 show that plotting N_1 vs. concentration gives a very pronounced peak at both low and high shear rate for both PBG and PCBZL.

These two shear rates were selected to fall in the first and second positive N_1 regions for all the concentrations, i.e. they bracket the negative N_1 region. In the high shear flow regime, where viscosity is nearly independent of concentration, N_1 varies over a wide range. This demonstrates that even at high shear rate, packing constraints still have

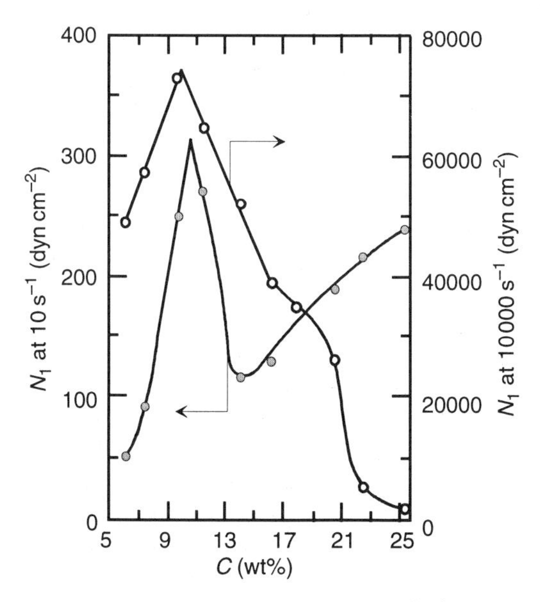

Figure 11.7 N_1 vs. concentration for a low and high shear rate in PBLG + *m*-cresol. (Reproduced from [17].)

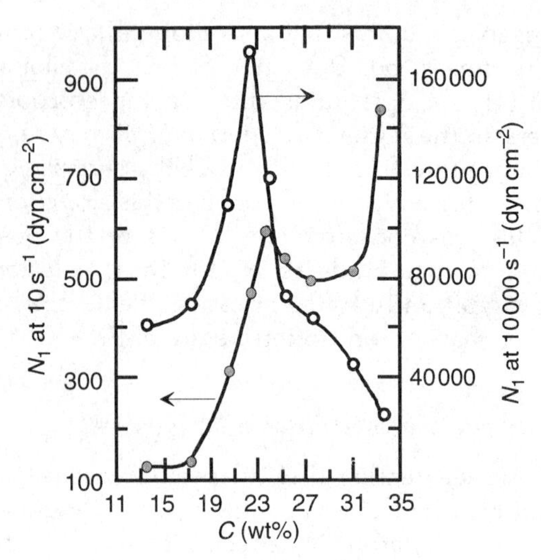

Figure 11.8 N_1 vs. concentration for a low and high shear rate in PCBZL + *m*-cresol. (Reproduced from [17].)

a significant effect on fluid structure, and that normal stress is a much more sensitive probe of this process than shear stress.

Interestingly, the low shear rate curves in Figures 11.7 and 11.8 show a minimum followed by renewed increase at higher concentration, as was also seen in viscosity. The high shear rate curves have an inflection point, but no minimum. Rather, they show monotonic decrease with higher concentration beyond the peak. Again, this behavior is seen for both polymers. Moreover, the concentration at which the peak occurs is not reduced at high shear rate, as was the case for viscosity, but is slightly increased.

Various other views can be 'mined' from the measurements to give further insight into the nature of the negative N_1 effect. Figure 11.9 shows the shear stress present within the fluid at the shear rates corresponding to sign changes in N_1.

In this view, plotting shear stress on a linear scale shows (for both PBG and PCBZL) that the boundaries for the three regions existing above C^* are straight lines which intersect at slightly above C^*. The three regions will be seen later to correspond to the director tumbling regime (first positive N_1), the wagging regime (negative N_1) and finally the flow-aligning regime (second positive N_1). Below C^* the first two regimes are seen to disappear, leaving only the flow aligning regime in isotropic solutions. The slope of the boundaries is the same for both PBG and PCBZL. Predicting the values of these slopes is a test of theory, which to our knowledge has not yet been attempted.

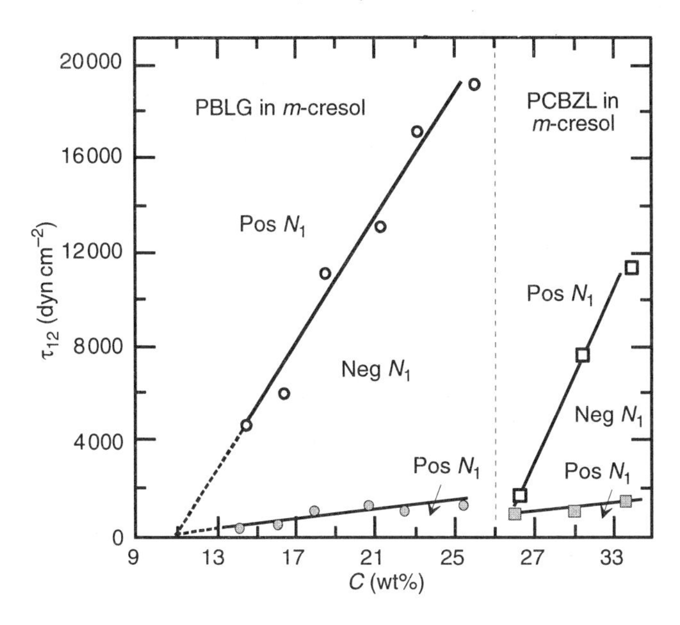

Figure 11.9 Shear stress at shear rates corresponding to sign changes in N_1. (Reproduced from [17].)

A somewhat similar data treatment is seen in Figure 11.10 in which the value of the first normal stress difference at the maximum first positive N_1 and maximum negative N_1 are plotted against concentration. As in Figure 11.9, the boundaries of the three regimes are approximately linear and intersect at or slightly above C^*, marking the disappearance of the director tumbling and wagging regimes. As before, the same slopes are seen for PBG and PCBZL, and again, existing theories have not been tested against this observation.

A closer look at this concentration dependence of the first positive maximum N_1, as well as the shear rate at which this occurs, was conducted by Baek *et al.* [46]. Both were seen to be monotonically increasing functions up to concentrations of 40%. They demonstrated qualitative agreement with the predictions of Doi theory with the Hinch–Leal closure and the Maier–Saupe potential. They also note that the ratio of shear rate at which N_1 becomes negative to the shear rate of the first positive maximum N_1 remains constant at about 3.5. (Our data from 1978 and 1980 yielded an average ratio of 3.2 with a standard deviation of 0.9 for nine PBG solutions and an average ratio of 2.2 with a standard deviation of 0.3 for three PCBZL solutions.) They determined that the rapid increase in $(d\gamma/dt)_{max}$ with concentration cannot be attributed to a decrease in molecular relaxation time with

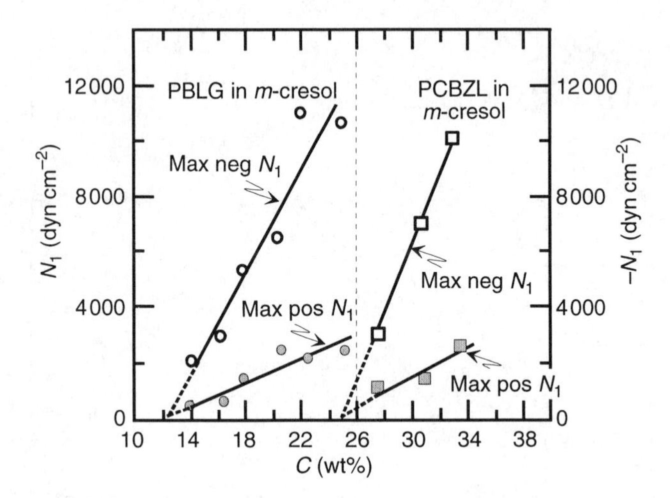

Figure 11.10 Maximum positive and negative values of N_1 vs. concentration. (Reproduced from [17].)

concentration, since the rotary diffusion constant D_r is only a weak function of C. They state 'as a result, N_1 does not have its maximum at a fixed value of the Deborah number De; instead, De_{max} increases with increasing U.' [46].

(c) *Molecular weight dependence of first maximum*

In a 1993 paper Baek *et al.* [46] state that according to Doi theory the shear rate at which the first (positive) maximum occurs in solutions just above C^* should scale with molecular weight to the -5th power. A series of solutions of PBG with different molecular weight was found to obey this prediction [60] as shown in Figure 11.11.

The diamonds are data taken from reference 60 and the line has a slope of -5. In addition, we plot here data taken from [17]. We see that our lower molecular weight datum falls on the same line. The higher molecular weight datum is slightly below this line and would fall on it if the molecular weight is taken as 400000 rather than the nominal 320000.

(d) *Relationship of G' and G'' to negative N_1*

An examination of dynamic storage and loss moduli for concentrated PBLG + *m*-cresol solutions (13% to 37%) is described in [29]. They observe that empirically shifting G' and G'' to cause them to superimpose allows one to estimate characteristic relaxation times τ. They find that

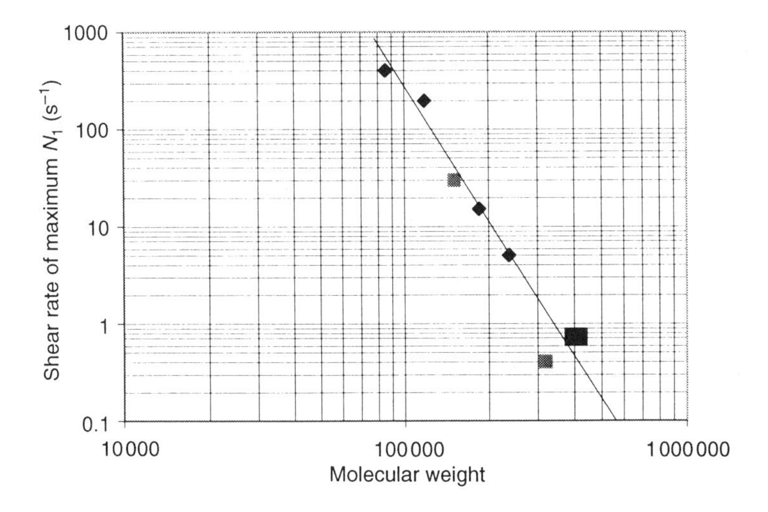

Figure 11.11 Shear rate of maximum N_1 in first positive region vs. molecular weight. (Diamonds from [59] with additional data from [17].)

the shear rate at which the second sign change in N_1 (positive to negative) occurs is given by $d\gamma/dt \sim 1/\tau$ for the range of concentrations studied. Since the crossover frequency $\omega = 1/\tau$ is roughly the rate of molecular relaxation, the shear rate $d\gamma/dt = 1/\tau$ is rapid enough to overcome molecular relaxation processes and N_1 changes sign. This paper suggests that if the correlation of N_1 with G' and G'' is general, then easily accessible dynamic measurements may be used to estimate the shear rate for the occurrence of the negative N_1 effect in other PLCs.

11.4.2 Measurement of second normal stress coefficient N_2

One of the rewards of writing a retrospective review such as this is the opportunity to marvel at the careful, subtle and elegant work which has been done in some sense as a follow-on to one's own work. An example of this is the remarkable work by Magda *et al.* [32]. Using small flush-mounted pressure transducers to measure the normal stress profile in the radial direction, combined with measurements of total normal thrust, they were able to obtain excellent meaurements of N_2, which they compared to the predictions of the Doi theory as extended to multidomain fluids [61–63].

The behavior of N_2 vs. shear rate for isotropic solutions of PBLG + *m*-cresol was negative and monotonically increasing in magnitude for all shear rates. This was the first measurement of N_2 for solutions of rod-like polymers. The ratio of $-N_2/N_1$ was constant at 0.29, which is consistent with the prediction of the Doi theory of 0.287 and with observations on entangled solutions of flexible polymers [64].

The ratio of N_2/N_1 for anisotropic solutions of PBLG was above 0.5, as surmised earlier based on the unexpected positive slope of the normal force profile (section 11.4.1(a)) even for positive N_1 values. N_2 was found to behave similarly to N_1 in that it exhibited two sign changes for a fully liquid crystalline solution of concentration 14%. In fact, at first glance, the N_2 vs. shear rate curve looks almost like the mirror image of the N_1 curve: at shear rates for which N_1 is positive, N_2 is negative and vice versa. The paper states (p. 4466) 'Since measurement of a positive N_2 value for *any* fluid appears to be unprecedented, we took great pains to ensure that the data ... cannot be attributed to experimental artifacts.' (italics in the original.) The value of N_2 calculated from the slope of the stress profile was also positive, so they concluded that the positive maximum of N_2 represents a true material property.

A solution of 12.5% (still fully liquid crystalline) had a very similar shape, but was shifted to lower shear rates as predicted by the Doi molecular theory. However, positive values of N_2 were not observed: the maximum of N_2 did not exceed zero. In fact, the maximum value of N_2 at intermediate shear rates dropped systematically as the concentration was reduced from 15 to 14 to 13 to 12.5% (Figure 12 of [32]).

The shear rates at which the relative minima and maxima N_2 occurred were found to be significant in two other ways. They correlated to the two changes in convexity which comprise the 'dip' or hesitation in the flow curve which has been seen in lyotropics which exhibit a negative N_1. Also, they report the existence of large, time dependent and periodic fluctuations in N_1 at shear rates between these two points. They speculated that this may be due to coherent wagging of the director in various domains. (Note: we also had observed, but did not report, large periodic fluctuations in N_1 for lyotropic solutions, which necessitated the use of low-pass filtering to extract the N_1 data. We saw that these were periodic with the rotation rate of the cone and had reproducible fine structure. We attributed them to imperfections of machining of the cone-and-plate tooling.) All three phenomena (maximum in N_2, fluctuations in N_1 and a dip in viscosity) correlated similarly with concentration, and were considered to be attributes of a particular regime of flow behavior.

An interesting experimental observation was noted by Magda *et al.* [32] regarding the effect of smoothness of the rheometer plates. For a solution of flexible polymers, two sets of plates gave identical linear normal stress profiles, as required to extract N_2 from a combination of point normal force measurements and total thrust measurements. However for the lyotropic solutions, only the smoother plate yielded a linear profile. They concluded that although it is possible to impose a homogeneous shear field on polydomain lyotropic PBLG, and to make point normal stress measurements using flush-mounted transducers, more care is required to avoid significant experimental errors due to roughness than for conventional polymer solutions.

The comparison with Doi theory described in [32] shows excellent agreement, as seen in Figure 11.12.

In addition to the obvious visual similarity, in fact they describe six points of detailed agreement (which will be easier to digest if the reader refers to the original figures), reproduced here verbatim:

1. The plots of N_1 and N_2 versus shear rate intersect twice, both times at negative values of the stress; at the second intersection the stress is more negative than at the first and at both intersections the stress becomes more negative as the concentration increases.
2. The width of the region of negative N_1 is approximately one decade in shear rate, and the range of shear rates over which it occurs shifts with concentration to an extent that agrees with theory.
3. At the highest shear rates, N_1 increases with a slope that is 2–3 times steeper than the slope of the decrease in N_2.
4. The maximum values of N_2 occurs at roughly the same shear rate as the minimum value of N_1 (particularly at lower concentrations).
5. The depth of the minimum in N_1 increases by slightly less than a factor of 2 when the concentration is increased from 12.5% or U is increased from 10.67 to 12. [Note: U is a dimensionless parameter proportional to concentration appearing in the Doi theory. $U = 10.67$ at the lowest concentration for which a lyotropic system is fully crystalline.]
6. For the 12.5% solution ($U = 10.67$), the maximum value of N_2 is approximately 4 times smaller than the negative minimum value of N_1.

However three points of disagreement were noted as well, reproduced here verbatim also:

1. As the concentration increases, the experimental maximum value of N_2 drops into the negative region, while the predicted maximum increases.
2. The maximum in N_1 at low shear rates is much broader and occurs at a lower shear rate than predicted.
3. At the smallest shear rates, the experimental absolute values of N_1 and N_2 decrease toward zero much less quickly than the predicted values.

Such a wealth of detailed points of agreement argues compellingly that the Doi theory is a rather faithful description of 'what really happens' and not merely a self-consistent mathematical construct. The authors conclude that 'stresses can be calculated from the Doi theory

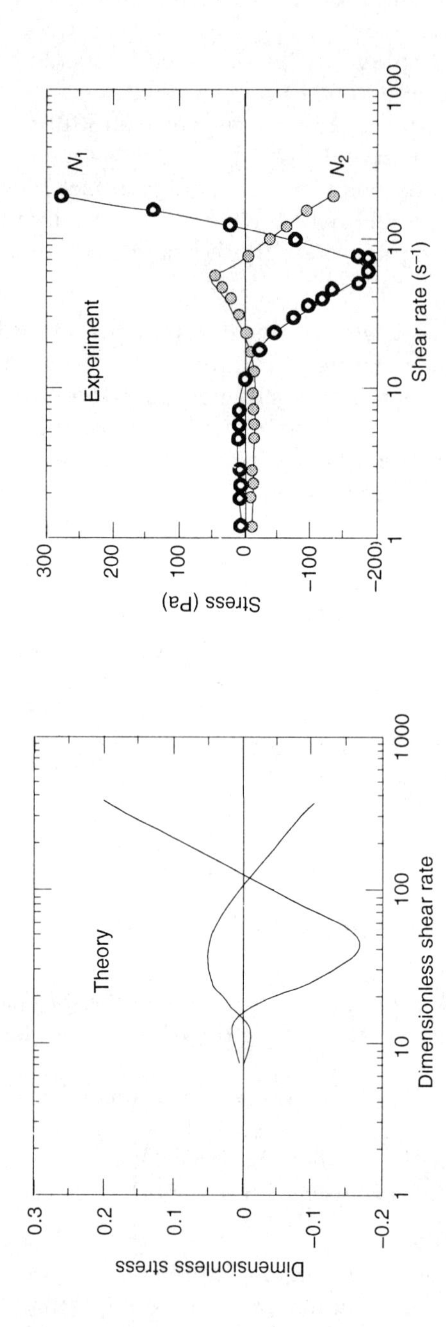

Figure 11.12 Comparison between Doi theory prediction and experimental data for N_1 and N_2 vs. shear rate. (Adapted from [32].)

without considering interdomainal interactions, *except at low shear rates in the first positive region of N_1*' (emphasis as in the original).

Similar N_2 behavior was observed in HPC + *m*-cresol solutions [65].

11.4.3 Behavior at high concentrations

A motivation for studying the rheology of liquid crystalline polymer solutions, besides of course curiosity and intellectual challenge, is the hope that deeper understanding of lyotropics could in some way lead to technologically and commercially useful advances in high strength and high modulus materials processed from thermotropic melts. A logical strategy would then be to study lyotropics of ever increasing concentration, to uncover insights which would extend into the thermotropic regime.

PBLG + *m*-cresol solutions at high concentrations were studied by Larson *et al.* [29] and Baek *et al.* [46], and were found to pose experimental difficulties. The sign change in N_1 was found to shift to higher shear rates as the concentration was increased, as expected from Doi theory, and by 30% was beyond the accessible shear rate range. Therefore negative N_1 was not observed. At concentrations between 37% and 40%, the shear rate dependence of viscosity abruptly changed to give 'region I' shear thinning at the lowest accessible shear rates. At concentrations above 40%, a 'gel-like' phase formed which gave non-reproducible results and proved impossible to study.

A more useful model system for the study of high concentration is HPC, since solutions are lyotropic and bulk HPC is thermotropic, melting into a liquid crystalline state at about 160°C [25]. In the liquid crystalline state the shear rate dependence of the steady state viscosity and normal stress difference of HPC solutions are qualitatively similar to those of PBLG solutions. Unlike PBLG, the concentration of HPC + *m*-cresol can be increased significantly beyond 40% and remain rheologically tractable. They also studied a semiflexible thermotropic polyester designed to give rather low transition temperatures, to avoid degradation, transesterification and crystallization. In fact, this particular polyester is also capable of forming a lyotropic 60% *m*-cresol solution, so that this work spans a continuum for moderate concentration through high concentration to bulk melts.

A complicating feature of rheological measurements of HPC + *m*-cresol at high concentrations is the onset of edge fracture at rather low shear rates. However it was shown in this paper [25] that, for this lyotropic system at least, edge fracture does not significantly affect the values of N_1 measured and, in particular, the sign of N_1 is not affected.

We have seen that the behavior predicted by Doi theory, and generally observed, is that the negative N_1 region shifts to higher shear rates as concentration is increased. A new and surprising finding in this paper [25] is that at concentrations above 35% this trend is reversed and the negative N_1 region occurs at lower shear rates as concentration is increased further. The value of the maximum N_1 in the first positive region $N_{1\,max}$ continues to increase, however. This is evidence of a flow mechanism unaccounted for by Doi theory.

As concentration is increased even farther, to 55%, the negative N_1 disappears entirely, leaving a local minimum in positive N_1. Thus N_1 vs. shear rate for the 55% HPC + *m*-cresol was non-monotonic but positive at all shear rates. Furthermore, increasing temperature from 25° to 37°C caused the regime of negative N_1 to reappear.

The Doi theory assumes that the solution is dilute, in that separations between rods is large compared to the rod diameter and that intermolecular contacts are rare. This paper [25] discusses a number of ways in which the Doi theory might break down at high concentration such as rotational and translational rod jamming and polymer–polymer drag interactions. Although quantitative predictions of these effects are not attempted, it is shown that if the rotational diffusion coefficient D_r increases more rapidly with concentration than predicted by Doi theory, then a 'reverse' shift of the negative N_1 region to lower shear rates with concentration can result. Based on the observed flow activation energy (higher than that of solvent alone), it is deduced that polymer–polymer friction plays a significant role.

It is interesting that the authors consider the cases of PBLG + *m*-cresol (persistence length *c.* 90 nm) and HPC + *m*-cresol (*c.* 10 nm), and show that they behave qualitatively differently at high concentrations, but do not consider PCBZL + *m*-cresol, which is probably intermediate in flexibility [17].

There is a discussion in this paper [25] of the relationship between tumbling behavior, required for negative N_1, and polymer flexibility. It is possible that slight flexibility enhances tumbling, and greater flexibility suppresses it. Indeed, it may be that negative N_1 is impossible in thermotropic polymers since any polymer flexible enough to melt at temperatures below degradation may be too flexible to exhibit tumbling. The authors note also that tumbling that occurs due to the nearby presence (in temperature) of a smectic phase [66] need not produce a negative N_1. There is a qualitative difference between tumbling as a transition from wagging to flow aligning (which is the scenario which produces negative N_1) and tumbling due to the formation of 'cybotactic clusters' near the transition to a smectic phase.

11.5 THEORY OF NEGATIVE N_1

Theories which have attempted to explain the occurrence of a negative N_1 can be categorized into three broad classes: molecular, domain and continuum.

11.5.1 Molecular theories

An early molecular model of the origin of negative N_1 was presented by Chaffey and Porter [52]. This theory considers the balance of torques present on individual rigid rod dumbbell molecules due to the interplay of Brownian motion, intermolecular potential and shear stress. They examined the behavior of the orientation distribution function of individual molecules and calculated the macroscopic stress state from the microscopic state using microrheological theory. Molecules are considered to be present initially with a distribution of alignments. Hydrodynamic torques will try to restore alignment with minimum rotation and intermolecular interactions will try to align a molecule with its neighbors. If a given molecule is oriented between 0 and 90° to the streamlines, hydrodynamic and intermolecular torques will add, and the molecule will rapidly rotate to alignment. If, however, a given molecule is oriented between 90 and 180° to the streamlines, then the two torques are opposed and a non-aligned orientation can persist during steady flow. The misaligned molecule has a smaller projected length in the streamline direction than the aligned molecule, so the net effect is shortening or buckling of the layer by shear flow. This compressive stress along the streamline corresponds macroscopically to negative N_1.

This model is able to predict the positive–negative–positive transitions which are observed. In addition, they show that plotting a dimensionless shear rate at which the negative extremum occurs, multiplied by the axial ratio cubed, vs. the volume fraction, caused data for two different molecular weight PBG solutions to superimpose, while data for PCBZL solutions fell on a different line of the same slope. The model proved to be non-quantitative in that it underestimates the range of shear rates over which N_1 is negative and also the magnitude of negative normal stresses. This may have been due to polydispersity of the polymers, which was not considered in the model.

The molecular approach which we will see eventually proved to be most successful in treating negative N_1 is based on the work of Doi [23]. Doi noted that the well established phenomenological theories for thermotropes (which he termed 'ELP' for Ericksen, Leslie and Parodi; [68]) which is successful in describing many dynamic phenomena in MLC nematics, is limited for polymeric liquid crystals in that it does not predict nonlinear viscoelasticity. Doi's approach determines the phenomenological coefficients from molecular parameters, so that the effects of, for example, molecular weight and concentration can be treated. He considers a single molecule (the 'test rod') and notes that as concentration increases, constraints on its motion are imposed by collisions with other rods. This constraint can be modeled as a tube within which the test rod is confined. Hence its rotary diffusion is severely restricted, but it is able to translate in the direction of its own axis. Moreover, since the tube has a finite diameter, as the rod partially escapes from the tube it is able to rotate through a small angle. By successive translations and small rotations, the test rod can eventually realize an end-to-end motion.

His theory is applicable to both isotropic and nematic phases, and so can be tested by the peak in viscosity with concentration, which it successfully predicts (see Figure 5 in [23]). Doi recognized a limitation in his theory, in that it assumes a spatially uniform system in which the director vector does not change with position. As will be seen in section 11.6, this is far from reality. Doi does not address negative N_1 in this work (it does not appear that he was aware of it), but notes that for small shear rates, normal stress is proportional to shear rate in isotropic solutions and to the square of the shear rate in the nematic phase.

A breakthrough in the understanding the negative N_1 phenomenon was the paper of Marucci *et al.* [68], who indeed titled a section of their paper 'Explanation of the negative normal stress phenomenon'. In the following discussion we shall again quote liberally from the original text. They note that the theory of Doi [23] does not predict negative N_1 in shear flow, due to a mathematical approximation. Another consequence of this approximation is 'stable shear flow with an aligned director, whereas calculations made without that approximation, which are possible in the linear limit, predict tumbling of the director due to shear.' They progress beyond the limitations of Doi by recourse to a two-dimensional rather than a three-dimensional analysis, and avoiding the Doi approximation. They did this consciously, noting 'We are aware, of course, that a change in dimensionality is not exempt of risks. ...The two-dimensional analysis proves to be, in fact, extremely rich in results, including the prediction of negative normal stresses in a range of shear rates.'

Figure 11.3 shows the two-dimensional distribution of orientation vectors of solutions of rod-like polymers in a variety of circumstances. Figure 11.3a shows the isotropic orientation distribution expected in an isotropic phase at rest. Figure 11.3b shows that the effect of shear on an isotropic phase is to distort the spherical (circular in two-dimensions) distribution into an elliptical one. The normal stress implication is that 'the tendency of the distorted distribution to relax back to the spherical shape implies a traction in the shear direction, or, equivalently, a compression orthogonal to it.' Elastic normal stresses associated with this situation are invariably positive. The situation for a nematic phase of rod-like polymers is shown in Figure 11.3c. The following three paragraphs are taken verbatim from [68]:

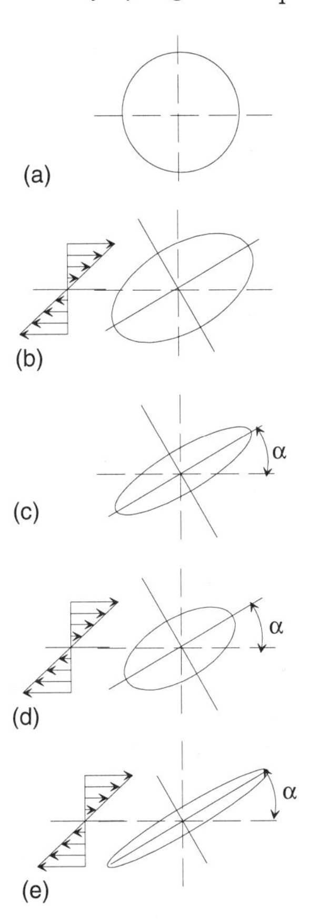

Figure 11.13 Two-dimensional distribution of orientational vectors of rod-like polymer solutions. (a) Isotropic, at rest; (b) isotropic, under shear; (c) nematic, at rest; (d) nematic, under shear, distribution blunted by shear, leading to negative N_1; (e) nematic, under shear, distribution sharpened by shear, leading to positive N_1. (Adapted from [68].)

The distribution of the unit vectors specifying rod orientation is already non-spherical at equilibrium. It is elongated in a direction (which is arbitrary at equilibrium) defining the so-called *director* of the nematic. The spread of rod orientations about the director is usually accounted for by a single scalar quantity, s, called the order parameter, a larger value of s implying a smaller spread. Shear flow will now give rise to two separate effects. On the one hand, there will be an effect on the director. We know already that either the director is made to tumble, or it is aligned to some small angle α above the shear direction. As a second effect, the shear can alter the spread of rod orientations about the director. Now, s can either increase with respect to equilibrium, i.e. the orientational distribution is sharpened by the flow, or decrease, i.e. the ellipsoid becomes blunter.

... assume now that the flow has oriented the director, as well as having created a distinction less peaked than that at equilibrium. There is no doubt that the elastic normal stresses associated with such a situation are negative. Indeed, the relaxation to equilibrium implies an expansion in the shear direction and a contraction orthogonally to it ... The plausibility that the flow can in fact generate a larger spread of rod orientations than that existing at equilibrium remains to be examined. In this regard we recall that shear flow exerts a torque on each rod which, by itself, would make the rod rotate clockwise. Thus rods that are oriented 'above' the director are pushed toward it, whereas rods sitting 'below' the director are pulled away from it. In turn the nematic potential opposes these tendencies, attempting to reestablish the equilibrium distribution. At steady state, a balance will be struck between the opposing effects, giving rise to some suitably 'distorted' distribution.

Notice now that, for rods above the director, the influence of the flow is that of sharpening of the distribution. Conversely, for those below, the tendency is that of increasing the spread. Which of these two opposite effects will prevail in a given case cannot be anticipated. There is no reason, however, to exclude *a priori* any one of the two possible outcomes. In other words, either positive or negative normal stresses can be generated by the same mechanism, depending on conditions. It can perhaps be expected that by varying those conditions, the system might switch from one situation to the other. It is fair to note that Chaffee and Porter had already indicated the conflict between shear and intermolecular potential as the possible source of negative normal stresses in liquid crystalline polymers.

A key insight required to understand this concept of the origin of negative normal stress is that in some circumstances shear can induce disorientation. This represents something of a paradigm shift relative to the behavior expected in isotropic fluids, to say the least! The fact that in extremely similar circumstances, shear can also induce orientation leads to the sign changes in N_1.

This fine study by Marucci and coworkers was extended in subsequent work to consider polydomain fluids and higher shear rates [62, 63]. The N_1 and viscosity predictions obtained are shown in Figure 11.14. Note that not only are all three regimes of N_1 well captured by the model, but the 'hesitation' or 'dip' in viscosity is also obtained at the transition from tumbling to wagging behavior.

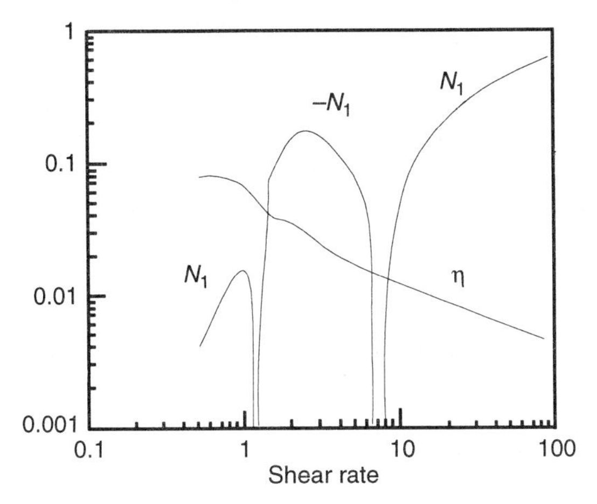

Figure 11.14 Predictions for N_1 and viscosity vs. shear rate. (Adapted from [63].)

Larson [61] was able to extend the work of Marucci and Maffettone into the third dimension to confirm the basic conclusions and remove some of their approximations. Larson also used the Onsager potential rather than the Maier–Saupe potential to model excluded volume interactions between rod-like molecules to improve quantitative predictions. Larson's major finding is that in between the tumbling and steady state regimes, there is a 'wagging' regime in which the birefringence axis wags around a fixed direction. This wagging has also been referred to as 'frustrated tumbling' [69]. The first sign change in N_1 corresponds to the transition from tumbling to wagging behavior. The wagging to steady state transition corresponds to the most negative extremum of negative N_1.

In addition, by virtue of being a three-dimensional model, Larson was able to make predictions about the second normal stress difference N_2. For fluids which are isotropic at rest, N_2 has always been found to be negative or zero, and much smaller in magnitude than N_1. For nematic solutions of rod-like molecules, Larson and coworkers find that N_2 is comparable in magnitude and opposite in sign to N_1 for most shear rates, hence at some shear rates N_2 is positive. Within a small range of shear rate near where N_1 changes sign, however, both N_1 and N_2 are found to be negative. This behavior is summarized in Figure 11.15.

The existence of domain tumbling, which is a key feature of the Doi/Larson/Marucci approach discussed above was conclusively demonstrated [70] by performing scanning-wavelength linear dichroism on monodomains of racemic PBG solutions in a flow cell. Director tumbling has also been observed experimentally [71].

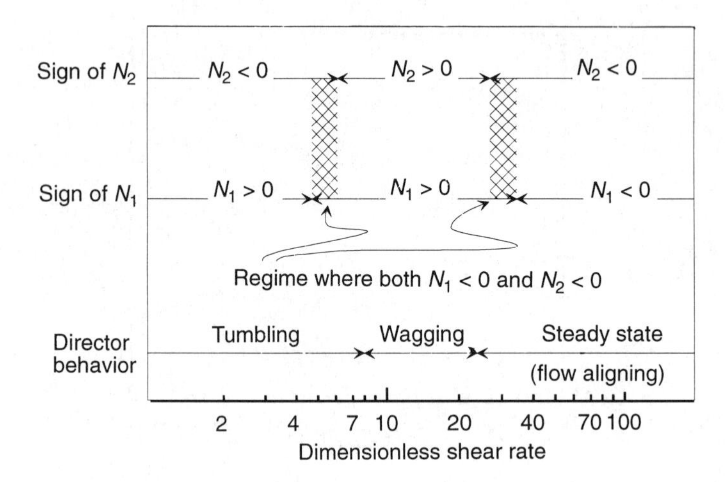

Figure 11.15 Normal stress and director behavior as a function of shear rate. (Adapted from [61].)

11.5.2 Domain theories

Doi [23] had already noted that his theory was restricted to a monodomain or textureless sample. The extension by Marucci and Maffettone [68] retains that restriction. This issue was addressed by Larson and Doi [72], who proposed a model for the rheology of textured lyotropic solutions in the tumbling regime. In the linear Larson–Doi polydomain model the response of the material is expressed in terms of a variable l proportional to the defect density. The defect density is proportional to the shear rate, so that texture refinement is a feature of this model. The steady state predictions for the order parameter S are independent of shear rate.

Hongladarom *et al.* [69] tested the theory quantitatively using birefringence under flow of textured solutions compared to quiescent monodomains. They found that birefringence in the low shear linear regime was relatively constant with shear rate and 53–63% of that of the monodomain. At higher shear rate, birefringence was again relatively constant with shear rate and around 90% of that of the monodomain. They found that the Larson and Doi tumbling model overpredicts the degree of orientation. However, 'the combined predictions of a constant degree of orientation and texture refinement may be taken to be a major success of the Larson–Doi model'. They discuss three possible reasons why director tumbling causes a greater reduction in the degree of orientation than expected, and indicate that two of them are unlikely. The most likely explanation, they feel, is that 'the qualitative idea of a continuous distribution of domain orientations may be too simplistic.' Rather, the director orientations may occur in two discrete populations oriented symmetrically about the flow direction. 'Crudely speaking, within any given "domain", the orientation flips back and forth between these two orientations as the director rotates in its tumbling orbit. ... Since orbits which pass through the flow direction ... do not appear to be present, this picture provides a natural explanation for a lower than expected macroscopic birefringence.' This picture is also extremely similar to the molecular configuration found in band structures, discussed in section 11.6.

The Larson–Doi model was further refined by Kim and Has [73]. They modified the Larson–Doi defect density evolution equation by including a domain size a_0 corresponding to the defect density L_0 of the quiescent state.They assumed that in the quiescent state nematic domains are surrounded by an isotropic phase consisting of solvent molecules and polymer molecules separated from the ordered domain regions. Upon inception of shear, the domains become smaller, as polymer molecules are 'partially released' from domains and free flow-aligned polymer molecules are generated. The steady state domain size a_s corresponds to the steady state defect density L_s. This model is illustrated in Figure 11.16. Thus inception of shear causes texture refinement, and cessation of shear causes texture coarsening.

Their modification prevents the indefinite coarsening which is found with the Larson–Doi model. The model shows region I shear thinning behavior and a region II Newtonian plateau (cf. Onogi and Asada [74]). In addition, for initial domains of very small size, the model shows another Newtonian plateau at extremely low shear rates, named 'region 0' to precede Onogi and Asada's three regions. This kind of plateau was reported by Sigillo and Grizutti [54] for aqueous HPC solutions. They were able to fit all four regions of the flow curve using two superposed Carreau-type power law models with two sets of plateau viscosities, relaxation times and and power law indices, identified with domains and molecules, respectively. In addition, N_1 was found to correlate with the molecular relaxation time through both sign changes.

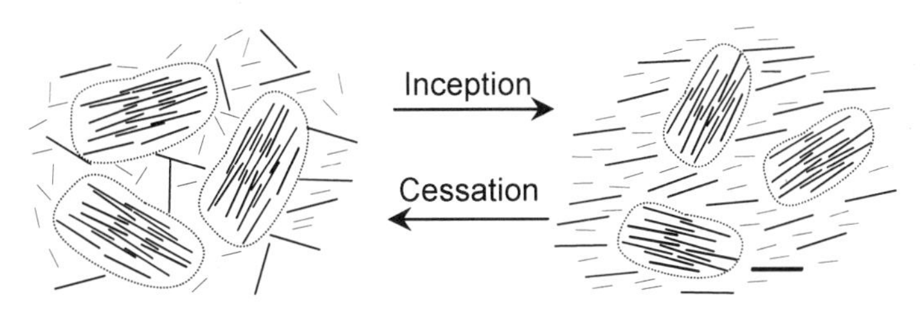

Figure 11.16 Polydomain structure during quiescence and steady shear. (Adapted from [73].)

Note also that reference [17] shows such a 'region 0' feature for the most concentrated PCBZL + *m*-cresol solution.

This model proposed by Kim and Has [73] shows N_1 proportional to $(d\gamma/dt)^2$ at the lowest shear rates and N_1 proportional to $d\gamma/dt$ at higher shear rates. However it does not seem to be applicable at yet higher shear rates, where N_1 reaches a maximum and the sign changes in N_1 occur.

11.5.3 Continuum theories

The following is not intended to be a complete review of continuum theories of polymeric nematics, but rather a taste to provide an entrée or starting point to the interested reader. We refer to the work of Currie for historical interest, and then summarize two recent works.

The first attempt at a theory of negative N_1 which we are aware of (we do not dignify our simple picture presented in [17] as a 'theory') was a letter from P.K. Currie communicated to us by K.F. Wissbrun (December 14, 1979). Currie's analysis was based on the Leslie–Ericksen theory for MLC nematics and is of debatable relevance to polymeric liquid crystals. Nevertheless, he does conclude that a negative N_1 is possible and would occur for a narrow range of boundary orientations. This analysis is available to interested parties from G. Kiss. In MLC nematics, N_1 may change from negative to positive as the shear rate increases [75].

Farhoudi and Rey [76] proposed a continuum tensorial theory. The following quoted extract describes the objectives of their work.

> The first objective of the present paper is to present a simple and tractable tensorial formulation that describes the isothermal flow of incompressible uniaxial nematic polymers; here we follow the earlier work of Hand [77] and the more recent work of MacMillan [78], giving emphasis to the simplest formulation that reduces to the well-known Leslie–Ericksen theory. Given the complexity in the modeling of nematic polymer flows, we restrict this work to uniaxial nematic polymers described by a [tensor order parameter] $\boldsymbol{Q}$ with two equal eigenvalues ($P = 0$) or equivalently by [director vector] $\boldsymbol{n}$ and [scalar order parameter] S. The distinction between thermotropics and lyotropics is not taken into account in this work except in the nondimensionalization step. The second objective of this paper is to present the reduction of the tensor theory corresponding to the simple shear flow of uniaxial nematic polymers displaying at most temporal variations of $\boldsymbol{n}$ and S. Here we wish to simulate the complex and unusual steady phenomena, observed experimentally in PLCs such as changes in sign of normal stress differences [32]. Lastly we use bifurcation theory to present a full characterization of the complex

orienting mode and the orientational transitions present in the simple shear flow of uniaxial nematic polymers, already predicted by the mesoscopic theories [79].

Their theory predicts two modes, a simple aligning mode and a complex mode which has three regimes: tumbling (director rotates, S oscillates), oscillatory (director and S oscillate), and stationary (stable steady state). The theory captures the sign changes in N_1 and predicts that the first sign change occurs at the end of the tumbling regime, and the second sign change occurs after the oscillatory regime and within the stationary regime. This is in agreement with the mesoscopic theories. The sign changes in this model are a direct consequence of the nonlinear dependence of the order parameter and the director orientation angle on shear rate. Another point of agreement with the mesoscopic theories is that for values of the nematic potential close to the nematic–isotropic transition, the two transient regimes of the complex mode are not observed, and only the stationary mode is observed. This is in fact the observed behavior, since the sign changes in N_1 are only observed in fully liquid crystalline solutions, somewhat beyond the critical concentration for formation of the anisotropic phase.

Andrews *et al.* [80] give an analysis of the dynamic behavior of polymeric liquid crystals in terms of a second-rank order parameter tensor describing the orientational state of the local microstructure. They credit Edwards *et al.* [81] with demonstrating that phenomena such as sign changes in N_1 and tumbling and wagging behaviors could be generated with continuum-level equations. The following quoted extract describes the objectives of their work.

> The first purpose of this article is to demonstrate that a rigorous analysis of the continuum dynamical equations for liquid-crystalline materials reproduces the essential features of the more complex distribution function equations. It is reasonable to expect that some information may be lost upon restricting attention to the continuum level theories: however, we believe that the lost information is entirely quantitative and not qualitative. Nevertheless, the continuum equations offer the nontrivial benefit that they are simple to solve using standard numerical algorithms, requiring very little computational time. Moreover, if one wishes to incorporate more interesting phenomena, such as nonhomogeneities arising from Frank distortion effects, the continuum equations offer a viable alternative to working in terms of distribution function equations, which are filled with computational pitfalls. The second purpose of this article is to examine how the shear flow of a polymeric liquid-crystalline material is affected by simultaneous application of a magnetic field (or an electric field, as if there is no difference between the two conceptually).

They note that Tse and Shine [82] are studying the problem of simultaneous shear and electric fields, using a two-dimensional distribution function approach similar to Marrucci and Maffettone. However, there are no experimental results at present, so this area is another in which fascinating experiments remain to be performed.

Although at the conclusion of this work Andrews *et al.* [80] state that they have demonstrated that a continuum theory of the dynamic response of polymeric liquid crystals provides a computationally efficient approximation to the complex distribution function theories, they acknowledge that not all of the rheological features are reproduced. Specifically, although the positive–negative sign change in N_1 is obtained, the negative–positive one is not, so the second positive region of N_1 is lacking. (This is in contrast to the theory of Farhoudi and Rey [76], discussed above, which obtains both sign changes.) They state:

> We see no evidence of this second sign transition, even for extremely high shear rates. Furthermore, the second normal-stress difference is positive for all shear rates, in contrast to the distribution function calculations which found that, generally, the second difference had the opposite sign to N_1. It is difficult to determine the source of the discrepancies noted above; however, we note that the orientation angle for the steady-state solutions was always negative, whereas those of the distribution function calculation actually passed from negative to positive in roughly the same shear rate regime where N_1 changed from negative to positive. ... Conceptually, there are two main differences between the types of effects incorporated in the two theories. First, Larson [61] used a specific functional form for the rotational diffusivity, dependent on the system orientation, whereas we essentially used a constant form. ... The second difference between the two sets of calculations is that the aligning potential is different in the two cases. Larson [61] used the Onsager potential, whereas we used a simple Landau/de Gennes expansion...

Finally, Andrews *et al.* [80] conclude that their continuum theory gives adequate representation of the orientational response even under simultaneous application of shear and an electric field, based on a comparison with results from Tse and Shine [82].

11.6 BAND STRUCTURES

11.6.1 Background

In the course of studying the negative N_1 phenomena, we also observed a phenomenon which consisted of the formation of transverse striations, or bands perpendicular to the direction of shear, at a range of shear rates [18]. We nicknamed this structure 'row-nucleated' due to the superficial similarity to structures formed during the stress-induced crystallization of polyethylene [83], and to emphasize that the direction of overall orientation is in fact perpendicular to that of the shear which created the structure. This nomenclature has not been generally adopted and these structures are generally known as 'bands', as distinguished from 'stripes', which are oriented parallel to flow. The observation of these bands constitutes our second major surprise while pursuing the investigation of flow of lyotropic PLCs. Examples of these bands are shown in Figure 11.17.

When one first encounters these bands, it is natural to assume that they orient in the direction of shear. Indeed, a reported observation by Aharoni preceded ours [12], in which lovely photographs of sheared solutions of polyisocyanates appear (his Figures 6 and 8). However the direction of shear was not identified, and cannot be inferred from other features in the photographs, as for example in Figure 3E of [18], which shows an air bubble elongated in the direction of shear. It should be noted that another prior mention of this phenomenon can be found from 1970 when Toth and Tobolsky [84] reported that slight mechanical shearing of 15% PBLG + $CHCl_3$ produces 'parallel multi-colored bands whose long direction is perpendicular to the direction of shear'. Band structures were found by Fincher [85] in PBA solutions when samples were first subjected to magnetic fields in one direction, and then to fields in a second, orthogonal direction.

It was of course very tempting to associate the two phenomena, that of negative N_1 and band structures, and by using a shearing stage and stroboscopic illumination, we obtained evidence that in fact at the shear rate at which negative N_1 was observed, transverse bands existed during shear. This observation has not been corroborated by other investigators, who have been unable to locate steady state bands during shear. We feel now that our observation was an artifact of imperfect alignment of the plates, so that steady state conditions were never achieved. We shall see shortly that transverse bands during shear transients have been observed.

In addition to the imperfect bands during shear 'frozen' by stroboscopic illumination, we observed that much more perfectly organized bands could be obtained by using a volatile helicogenic solvent and allowing the sheared solutions to dry. Most of the work which has been conducted on band structures since then has involved elucidation of kinetics and morphology of band structures formed after cessation of shear.

It is striking that these two phenomena, which both appear to be related to the organization of rigid rod molecules and how they respond to shear, vary so widely in generality. As pointed out by Baek *et al.* [46] (see their Table 2), observation of negative N_1 is limited to a very

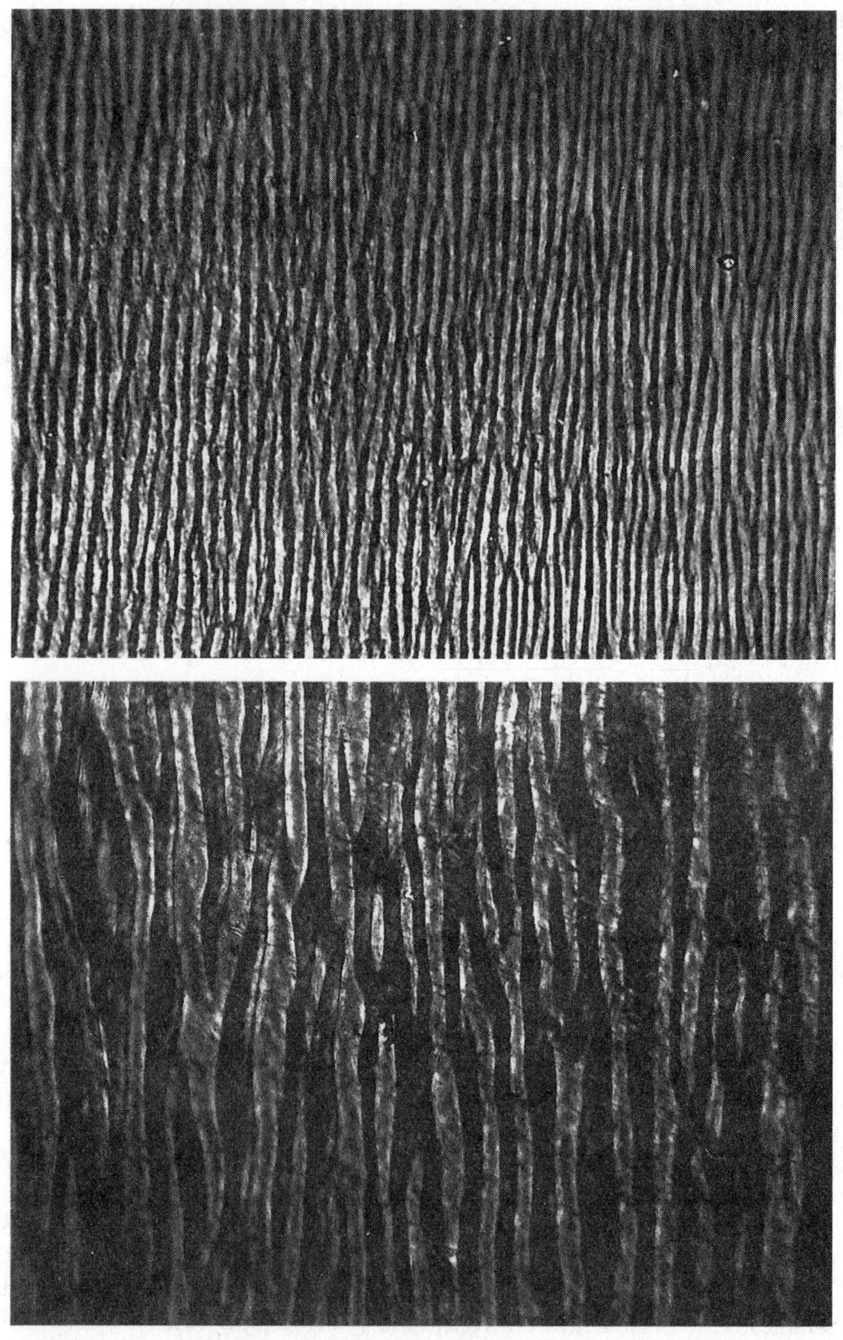

Figure 11.17 Dried films from 15 wt% PBLG ($M_w = 350000$) + dioxane at (a) high shear rate and (b) low shear rate.

few lyotropic systems (all the more remarkable given the extremely wide range of flexibility between PBG and HPC), while the formation of band structures is essentially universal in both lyotropes and thermotropes. Table 11.2 gives a partial listing of works in which transverse bands have been observed.

Even more highly structured variations on the band structure have been reported, such as the 'helical coarse rope' structure in PBG solutions [18] (Figure 11.18) or a 'torsad' structure in HPC solution [100] and a 'tractor' or 'herringbone' texture in thermotropic polyester [101].

Table 11.2 Observations of band structures perpendicular to shear

Polymer	Reference
Lyotropics	
Poly(γ-benzyl-glutamate) (PBG)	Kiss and Porter (1980) [18], Picken *et al.* (1992) [13], Gleeson *et al.* (1992) [60], Vermant *et al.* (1994) [86]
Hydroxypropylcellulose (HPC)	Navard (1986) [45], Marrucci *et al.* (1987) [87], Marsano *et al.* (1988) [88], Ernst and Navard (1989) [56], Vermant *et al.* (1994) [89]
Poly(*n*-hexyl isocyanate) (PHIC)	Marsano *et al.* (1988) [90]
Poly(*p*-phenylene-2,6-benzobisthiazole) (PBT)	Einaga *et al.* (1985) [91]
Poly(*p*-phenylene-2,6-benzobisoxazole) (PBO)	Einaga *et al.* (1985) [91]
Poly(*p*-phenyleneterephthalamide) (PPTA)	Horio *et al.* (1985) [92], Doppert and Picken (1987) [93]
Poly(*p*-benzamide) (PBA)	Marsano *et al.* (1989) [94]
Poly(*p*-anisole isocyanate), poly(50% butyl + 50% nonyl isocyanate)	Aharoni (1979) [12]
Cyanoethyl cellulose	Yong (1991) [95]
Thermotropics	
PET/PHBA copolyester	Zachariades *et al.* (1984) [96], Viney and Windle (1986) [97]
Chlorophenylterephthalate/bisphenoxyethane carboxylate copolyesters	Graziano and Mackley (1984) [98]
Trifluoroacetoxypropylcellulose	Navard and Zachariades (1987) [99]
PSHQ10 copolyester (cast from solution)	Kim and Han (1994) [73]

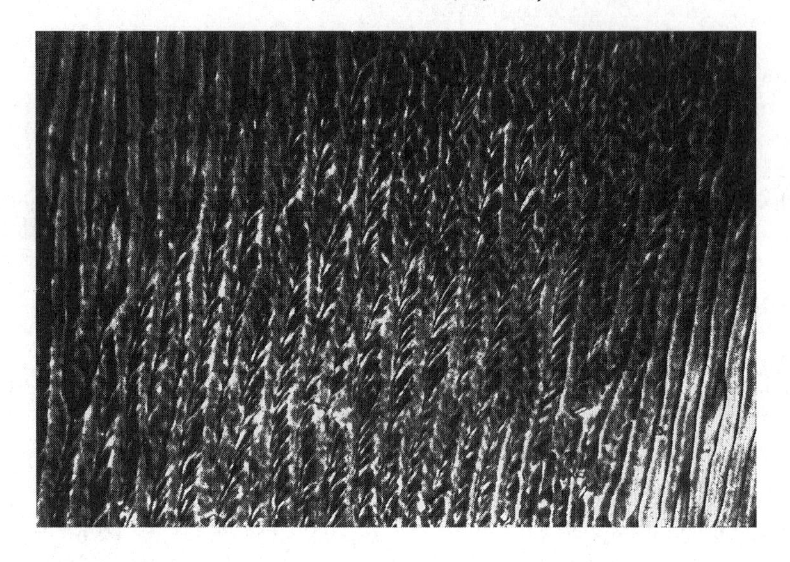

Figure 11.18 'Coarse rope' texture in dried films of sheared 15 wt% PBLG ($M_w = 350000$) + dioxane.

It has been noted earlier that negative N_1 can be explained by Doi theory, which is a molecular theory and does not consider the effect of domains. In others words, rheologically speaking, the presence of domains is irrelevant except at lowest shear rates. Thus it appears likely that negative N_1 and band structures are not linked. However, any practical application of liquid crystalline polymers is likely to involve shaping operations which result in textured materials. Some understanding of the evolution of textures and the resulting orientations on several scales is necessary to predict the result of shaping and processing, and in particular the strength and anisotropy of the resulting object.

11.6.2 Kinetics of formation upon cessation of shear

Oddly enough, although band structures have been reported in many systems, as indicated in Table 11.2, the most detailed and complete investigation has been performed on our old friends, PBG + *m*-cresol and HPC + water. The evolution phenomena of band structures in these two systems appear to be essentially the same, with the exception that the length scales of features are about five times smaller for HPC than for PBG [86]. There does not appear to have been a direct comparison of the evolution processes for lyotropes and thermotropes; however the ultimate morphologies appear to be the same.

The process of band formation is characterized by several parameters: the length of time during which the solution is sheared (t_s), the shear rate $(d\gamma/dt)_s$, the time required for band formation (t_b) and the length of time the bands persist before disappearing (t_d).

There appears to be some question as to whether band formation requires fully anisotropic solutions [56] or whether it can occur in biphasic solutions [102]. Most studies of band structures have been performed on solutions well into the fully anisotropic concentration regime.

The following observations are taken from a variety of sources, but are probably applicable to most band-forming nematic and cholesteric PLC systems.

- There is a critical shear rate below which bands do not form, regardless of how long one waits, i.e. t_b becomes infinite.
- This critical shear rate varies inversely with molecular weight.
- An upper critical shear rate beyond which bands no longer form has also been observed [89].
- The shearing time prior to cessation must be longer than a certain time which depends on the shear rate. However the band formation does not occur at a particular value of total shear, since a weak dependence on $(d\gamma/dt)_s$ remains [56], i.e.

$$\gamma_c (d\gamma/dt)_s^{0.3} = a \qquad (11.2)$$

  where a is large (of the order of 100) and varies inversely with molecular weight.
- If prior shear of sufficiently high rate is imposed for a sufficiently long time, then band formation occurs after a waiting time ranging from zero to several minutes. A t_b of zero leads to speculation that the bands may have been present during shear. This is addressed in detail in a subsequent section.
- After formation of the bands, they gradually begin to lose their parallelism, thicken by merging of adjacent bands (rather than widening of individual bands) and eventually disappear in tens of minutes, leaving a texture similar to that of the undeformed state. This t_d is extremely sensitive to molecular weight (higher M_w gives longer t_d) and the applied shear rate (higher resulting in more rapid disappearance), but is not sensitive to concentration or total shear.
- The fineness of the band texture increases as the rate of prior shear is increased. However, since band formation occurs more rapidly and is then followed by broadening, in fact the band spacing as a function of time is independent of the rate of prior shear. Bands are about five times finer for HPC solutions than for PBG solutions at a given rate of prior shear (of the order of 2–5 μm vs. 10–30 μm). This point is disputed by Vermant *et al.* [89] who find a weak effect.

The band structure seems to be energetically very stable. An interesting experiment described by Ernst and Navard [56] was to shear an HPC + water solution at a shear rate such that upon cessation, bands gradually formed. Then shear is renewed, but this time at $d\gamma/dt < (d\gamma/dt)_s$ such that bands do not form upon the second cessation. However, if this renewed shear is 'unwound' in the opposite direction, at the same rate, upon cessation a third time, bands are formed instantaneously! This process can be repeated several times, with instantaneous band formation each time, if the total shear upon cessation is the same as that at the first cessation, until the time for disappearance of the bands t_d (in the absence of renewed shearing) has elapsed.

In our rheo-optics paper of 1980 [18] we summarized the behavior of band structures formed after cessation of shear as follows: 'With increasing rate of prior shear, the row-nucleated texture forms more rapidly, with better definition, and thinner striations, and also degenerates more quickly.' A more sophisticated analysis described by Gleeson *et al.* in [60] yielded consistent results. This technique employed by Gleeson *et al.* was that of direct microscopic observation combined with digitization and analysis of video-recorded images, specifically fast Fourier transform (FFT). This strategy has a number of advantageous features. It allows the FFT to be conducted 'off-line' and thus tailored to enhance sensitivity to specific features, as opposed to scattering techniques which also provide Fourier transforms, but in a much less flexible way. A second advantage is that a correspondence between real space images and FFTs is retained, since in many cases it is impossible to identify shapes based only on Fourier transform information. Compare, for instance, Figures 4 and 12 of their paper [60], in which the real-space images are dramatically different yet the FFTs differ quantitatively, but subtly. Video analysis thus combines the strengths of scattering with those of microscopy.

Gleeson *et al.* [60] reported results in terms of shear strain, rather than shearing time, and found that the strain which needed to be imposed on a PBG solution to obtain bands increased with decreasing shear rate, independent of the gap. The decision as to whether or not bands form in a particular circumstance is of course subjective. Below a critical shear rate $(d\gamma/dt)_b$, bands were not formed at all, even after strain of at least 1400 units. This shear rate was a strongly decreasing function of molecular weight, ranging from about 40 s^{-1} for $M_w = 86000$ to about 0.25 s^{-1} for $M_w = 186000$. They noted also that this shear rate was at least an order of magnitude less than the shear rate at which N_1 reaches its positive maximum.

The digital analysis resulted in a power spectrum which is surprisingly undramatic. Although the eye can clearly see bands, the spectrum has no features which can be attributed to the bands. However by defining a characteristic wavelength λ^* which is the midpoint of the power spectrum, a single quantity was unambiguously obtained, the evolution of which could be easily followed. Note that in [60] $\lambda^*_{\parallel}$ is somewhat non-obviously the characteristic wavelength related to the width of the perpendicular bands, while $\lambda^*_{\perp}$ is the characteristic wavelength of the length of the bands. It was observed that the power spectrum evolved with time after cessation of shear, and power was lost in the short wavelength part, but gained in the long wavelength part, as the bands coarsened. A compact and powerful tool for characterizing the behavior of the bands was obtained by monitoring the evolution of $\lambda^*_{\parallel}$ and $\lambda^*_{\perp}$ with time after a cessation of shear.

Both $\lambda^*_{\parallel}$ and $\lambda^*_{\perp}$ were seen to increase with time after cessation of shear; however the rate of increase of $\lambda^*_{\perp}$ was much faster than that of $\lambda^*_{\parallel}$. In other words, the characteristic length scale in the vertical direction (the lengths of the bands) increased much faster than that in the horizontal direction (the widths of the bands). In fact, this process could be followed from the instant of cessation, well before band structures were formed, and indeed the formation of visible bands seemed to occur at the point at which the $\lambda^*_{\parallel}$ and $\lambda^*_{\perp}$ lines crossed, or had their closest approach. For example, with a prior shear of about $5\ s^{-1}$, the $\lambda^*_{\parallel}$ vs. time and $\lambda^*_{\perp}$ vs. time lines nearly intersect at 25 s, and bands were seen by eye at *c.* 20 s. Similarly, for prior shear of about $21\ s^{-1}$, these lines cross at *c.* 5 s and bands were visible at *c.* 3 s. The time required for band formation was roughly inversely proportional to the rate of prior shear.

The evolution of $\lambda^*_{\parallel}$ vs. time was essentially independent of prior shear rate, once visible bands were present, and increasingly a function of time. Before bands become visible, the data for various prior shear rates do not superimpose. If the prior shear is slow enough that bands never form, then not only do the $\lambda^*_{\parallel}$ vs. time data not superimpose, they are in fact decreasing functions of time. The $\lambda^*_{\perp}$ vs. time data, by contrast, do not superimpose for various rates of prior shear. Rather, the rise in $\lambda^*_{\perp}$ vs. time becomes much steeper as the rate of prior shear increases, reflecting an increase in the speed of band formation.

Another investigation into the kinetics of band formation more sophisticated than simple observation was reported by Picken *et al.* [13], using conoscopic small angle light scattering (CSALS) conducted in a shearing cell mounted in a polarizing microscope. As with the previous paper [60], this gave corresponding structural and real image information. This paper also confirms the observation we reported [18] that more rapid prior shear results in more rapid band formation for PBG + *m*-cresol, and also notes that a similar dependence was found for poly(*p*-phenyleneterephthalamide) (PPTA) + H_2SO_4 using time-resolved X-ray scattering. Again the finding is that the kinetics of band growth are independent of prior shear rate, once bands have been formed. (Again, since higher prior shear leads to earlier band formation, the initial spacing does depend on prior shear rate.) It was noted that the sign of N_1 during the prior shear does not affect the kinetics, nor does sample thickness. Another interesting observation was that the increase of average band thickness was due to a collapse or coalescence of adjacent bands, rather than broadening of individual bands. The question of lateral spacing or length of bands was addressed by determining the number of 'connected white regions' in an arbitrary surface area. In a less quantitative way than Gleeson *et al.* [60], Pickens *et al.* also observe an increase in lateral band size in the direction perpendicular to that of flow.

In a study of the kinetics of band formation for HPC + water [88], it was observed again that a minimum deformation is required for band formation, and that it is inversely related to the rate of applied prior shear. They observed, however, that two distinctly different behaviors were found, depending on the rate of prior shear. At lower rates the time for band formation reached a plateau value, so that beyond a certain total strain, the time required for band formation was constant. At higher rate of prior shear, this plateau dropped to zero, so that beyond a certain total strain applied at a high enough rate, bands were formed instantaneously. The texture which formed instantaneously was termed the 'precursor banded' texture, since it was poorly developed, and sharpened with time.

Marsano *et al.* [94] investigated HPC + water, poly(*n*-hexylisocyanate) (PHIC) + toluene, and poly(*p*-benzamide) (PBA) + DMAC + 3%LiCl. PBA + DMAC was slightly different from the other two systems in that at all shearing times studied, only the plateau value of t_b was obtained; in other words they did not see a region of t_b decreasing with total strain. Plotting the (presumably) plateau values of t_b vs. rate of prior shear, they found three distinct regions (four if one includes the low-shear region in which bands are not formed at all). Initially t_b decreased as $(d\gamma/dt)_s$ increased, then a region of instantaneous band formation upon cessation of shear was observed, and finally at even higher rates of prior shear, t_b was observed to increase again. They termed this the 'upturn' region or the 'reincrease' region. In HPC + water this was found to occur at relatively low shear, of the order of $50-100\ s^{-1}$. In PHIC and PBA solutions the upturn did not occur until much higher shear, around $2000\ s^{-1}$. They determined also that viscosity has opposite effects in the 'pre-upturn' regions and the 'upturn' regions. In the pre-upturn region, lower viscosity corresponds to larger t_b, whereas in the upturn region the re-increase of t_b is facilitated by higher viscosity. This behavior does not seem to have been documented elsewhere.

Vermant *et al.* [89] also observed two regimes of band formation in PBG solutions, depending on rate of prior shear. At low shear they found that both band width and band length increase with time after cessation. At higher shear they found that only band width increased, and band length instantaneously reached its final value upon cessation. The shear rate at which this transition in behavior occurred was found to correspond to a negative N_1, somewhat lower than the negative extremum. They found no effect of temperature on the band formation kinetics, however the shear rate of the transition from two-dimensional to one-dimensional behavior did vary with temperature. At even higher shear rate, within the flow aligning regime, they find that a persistent monodomain is induced, so that bands are not formed at all, in agreement with the findings of Larson and Mead [103].

11.6.3 Effect of sample thickness (gap)

A study of the effect of gap thickness on band formation in HPC + water solutions was conducted [87]. The apparatus used had a parallelism of about 20 μm over a plate length of about 10 cm. Note that this was a linear and not a rotary parallel plate apparatus. The gap was varied between 150 and 1000 μm. A rest time of 5–6 h was allowed between sample loading and initiation of shear. Since the time scale of changes in the band structure was slow, of the order of minutes, there was plenty of time for detailed examination.

By focusing the microscope at different depths, they found that in most cases the band structure did not fill the sample throughout the gap thickness. As has been observed by other authors, they noted that a threshold value of total shear needed to be applied in order to obtain bands, and that this value was a decreasing function of shear rate. However they noted also that 'A striking result is the strong effect of the sample thickness on threshold values. ... Up to one order of magnitude differences are found between the 150 μ and the 500 μ gap values. The direction is that increasing values of the thickness favour band formation.' They interpreted this observation to preclude the possibility that band formation reflects a bulk property, and that it might be in some way be due to an instability.

A study of band formation in thermotropes by Zachariades *et al.* [96] commented that deformation appeared to be non-uniform and dependent on the thickness of the sample, and that 'non-uniform deformation occurs because the domains close to the polymer–wall interface deform more effectively by shearing versus the domains in the bulk of the sample which may slip or rotate...' They found band structures after solidification of sheared thin samples, but not sheared thick samples. A personal communication from J.F. Fellers which reported a gap dependence was referenced in [56].

Vermant *et al.* [86] found that 'For both the two- and one-dimensional mechanisms a clear effect of the gap on the kinetics is observed, suggesting that the walls hinder band formation.'

Other authors have not found gap dependence. Ernst and Navard [56] used a transparent cone-and-plate rheometer to study aqueous HPC solutions. They found that sample thickness had no effect by scanning along the radial direction of the cone and plate, to observe different gaps, and observing simultaneous formation of the bands at all points. They also state that by focusing through the sample thickness they 'observe no modification of the formation of the band texture.' Fincher [104] used light scattering on HPC solutions and found that band width was independent of gap, shear rate and time after cessation of shear. Gleeson *et al.* [60] also observed no gap dependence in a study of PBG + *m*-cresol using a rotary parallel plate apparatus, neither for the width of bands nor for the time required for appearance after cessation of shear. They caution however that the time for appearance of bands is highly subjective and may be the source of discrepancy between different observers.

The consensus appears to favor the position that band formation is a bulk phenomenon and not due to instability or surface effects; however wall effects may produce gap dependence.

11.6.4 Bands during transient and steady shear

We reported in [18] that a negative N_1 is correlated with the presence of transverse bands during shear. Support for this observation can be inferred from the observation that under some circumstances, band textures formed instantaneously upon cessation of shear. Navard comments '... in the higher limit, the bands appeared almost immediately after, if not during, shearing. At higher shear rates, it appears that the bands were present during shearing' [28]. The question was open as recently as 1988, when Marsano *et al.* [88] stated 'So far only Kiss and Porter have reported band formation during shear. ... We have however no reason for questioning the correlation... between negative normal stress and band formation during shear.' They offered a tentative model to link negative N_1 with existence of bands during shear, postulating a state of flow instability which is now known not to be the source of a negative N_1.

More recently, the existence of transverse band structures during shear has been discounted. Ernst and Navard [56] used SALS of HPC + water and found no H_v patterns indicative of band structures, even at a shear rate when N_1 was negative. A confirming negative result was found by Picken *et al.* [13], who used conoscopic SALS and found no evidence of meridional reflections during steady shear.

The evolution of textures upon shear start-up, rather than steady shear or after shear cessation, was investigated by Vermant *et al.* [89]. An initial uniformly aligned sample was generated either by shear for extended time in a high rate, flow-aligning regime, or by shearing at a lower rate, and allowing bands to develop and then relax for a long rest period to a uniformly aligned condition. The sequence of structures was related to well defined start-up transients in the shear stress. Between 1 and 5 strain units, a transient band structure was observed, in which the bands display 'a rolling motion' and the conoscopic image is elongated in the shear direction. Between 5 and 15 strain units, the bands are broken up into smaller units, leading to a speckled appearance, but the conoscopic image has components both perpendicular and parallel to shear. After 15 strain units, these defects elongate in the shear direction, and by 25 strain units, very regular striations (stripes) parallel to shear are formed with a conoscopic image perpendicular to shear. After about 50 strain units, and then on into the steady state, only a weakly striated texture remains.

A similar observation was made by Gleeson *et al.* [60] that transient bands can sometimes be observed prior to the formation of stripes, which are present in the steady state. Similarly, Baek *et al.* [65] observed (Figure 16a in their paper) that for a low molecular weight thermotropic polyester OQO(OEt) 10, 'bands orthogonal to the flow direction quickly form at a strain of about 3. ... After a strain of 75, the perpendicular bands are gone and only the parallel stripes remain. ... After cessation of shear, bands again appear perpendicular to the previous direction of shear.'.

The observation of bands **during the start-up transient** is probably a key observation in resolving the discrepancy between our report [18] that negative N_1 is correlated with the presence of transverse bands during shear and the observations of other authors. Although some authors have agreed that this was possible, the observation had never been duplicated. Nevertheless, as recently as 1994 Kim and Han [73] left the possibility open: '... at present there is no experimental evidence, which proves that the formation of banded structure and negative values of N_1 in TPLCs are necessarily related to each other.'

We feel now that the most likely explanation is that, due to lack of parallelism in the plates of our primitive shearing stage, we never achieved a true steady state, and that in fact the shear rate at the point of observation oscillated significantly. A certain 'pumping' action could in fact be observed by eye, to the best of our recollection. It would be an interesting exercise to determine under what conditions of oscillating shear rate a quasi-steady state band structure could be formed. Either this will turn out to be relatively easy to do, or we were unbelievably lucky to observe it. Not so lucky, in fact, since it led to an apparently erroneous conclusion, but it seemed fitting at the time that the two phenomena should be linked.

An extensive study of the textures present during shear of lyotropic PBLG was conducted by Larson and Mead [103] in an experimental *tour de force*. Their parallel plate shearing stage had a gap of between 10 μm and a few centimeters with a precision of 5 μm and could give a shear rate of between 5×10^{-4} and $100\ s^{-1}$. With the range of PBG solutions they studied, they could achieve a range of Ericksen number between 1 and 10^7. They observed that, surprisingly, texture-free samples exist at both extremely low and at extremely high shear rates. In between, at the intermediate shear rates where most experiments are carried out, there is a sequence of transitions, or as they termed it, a 'cascade' of textures.

At relatively low shear rates the texture is controlled by the Ericksen number

$$Er \equiv \gamma_1 Vd/K_1 \tag{11.3}$$

where γ_1 is the rotational viscosity (in terms of Leslie–Ericksen viscosities $\gamma_1 = \alpha_3 - \alpha_2$), V is the velocity of the plates, d is the gap and K_1 is the Frank splay constant.

As the shear rate increases, large viscous stresses cause refinement of the texture until eventually individual molecules experience significant viscous torques, and the molecular order parameter can be distorted from its equilibrium value. This occurs when the Deborah number approaches or exceeds unity:

$$De \equiv \lambda(d\gamma/dt) \tag{11.4}$$

where λ is the average molecular relaxation time, given by the shear rate $d\gamma/dt_{max}$ at which N_1 has its local positive maximum. The sequence of texture transitions occur at certain values of Er at lower shear rates, and at certain values of De at higher shear rates, as shown in Table 11.3 (excerpted from [103]). Roll cells are structures analogous to Taylor vortices in Couette flow.

We have already concluded that our observation of a correlation between band structures during shear and negative N_1 is probably erroneous, and that band structures only appear during the start-up transient. Nevertheless, an interesting comment appears: 'For PBG198 ... the director also exits the shearing plane, but does so by producing a periodic "band" pattern, with bands oriented in the vorticity direction. These bands do not form during shearing of PBG118 ...' In fact a particular texture is associated with negative N_1: ... where we observed the distinct stripe texture ... N_1 is negative and typically quite noisy over much of this range of shear rates.'

Table 11.3 Flow regimes present for various Ericksen and Deborah numbers [103]

Er or *De* number	Flow regime
$Er \sim 1$	In-plane tipping
$Er \sim 10$	In-plane tumbling
$Er \sim 10$	Out-of-plane motion
$Er \sim 100$	Roll cell formation
$Er \sim 1000$	Roll cell refinement
$Er \sim 10\,000$	Director turbulence
$De \leqslant 1$	Tumbling
$De \sim 2$	Wagging
$De \geqslant 1$	Steady state (flow aligning)

Another observation made by Larson and Mead [103] is that in addition to a lower critical shear rate for band formation after cessation of shear, there is also an upper critical shear rate. 'For this sample at lower shear rates, $0.2 \leqslant d\gamma/dt \leqslant 30\,s^{-1}$, after cessation of shearing birefringent bands form that are orthogonal to the previous flow direction but for $d\gamma/dt = 100\,s^{-1}$, these bands usually do not form.' This is a rather moderate shear rate and it is odd that such an interesting observation had escaped the notice of other investigators. The reason might be found in the following comment:

> We should note, however, that in some runs, bands formed after high rates of shearing, we believe because the motor did not stop quite quickly enough in those particular runs.

Larson and Mead [103] state further:

> ... as the shear rate becomes much higher, and enters the steady-state regime, the average orientation increases, and should approach that of a monodomain. As this occurs, the tendency for molecular rearrangements upon cessation of flow should diminish. That is, if during flow the texture is defect-free and the orientation is that of a stationary monodomain, there is no reason for rearrangements at the molecular or texture level to occur when flow is suddenly halted. Thus the sample should remain the same when flow stops, which is what Mather [P. Mather, personal communication] and we observed. ... At lower shear rates, orientation is severely disrupted by shearing flow at the molecular and/or texture level. As a result, when flow suddenly stops, the molecules and/or texture must rearrange themselves to find a lower energy configuration. This rearrangement tendency creates an elastic stress that is no doubt responsible for the large recoverable strains observed.

As we have noted earlier, other investigators have associated a recoil with the formation of band structures. Vermant *et al.* [89] observed an upper critical shear rate for band formation and measured recoil in a controlled stress rheometer, noting that 'Bands indeed developed during recoil; they only become visible after the recoil was close to its final value.' An implication is that shearing beyond the upper critical rate should not produce recoil either. We seem to have identified yet another experiment waiting to be performed.

11.6.5 Morphology

Up to this point we have spoken loosely of the band texture as consisting of striations perpendicular to the shearing direction, with width determined by composition, rate of prior shear and time duration after cessation of shear. The disposition of the molecules within the bands has until now not been mentioned. The most detailed investigations have been carried out on dried films of sheared lyotropic solutions, since the structure is obviously stabilized and amenable to careful examination after evaporation of the solvent.

We reported in [18] that insertion of a first-order red plate at 45° to the shearing direction gave alternate blue and yellow bands, indicating that the molecules were oriented roughly parallel and perpendicular to the red plate, and thus at approximately $\pm 45°$ to the shearing direction. It is interesting, by the way, that Figure 3c and 3d of that paper, taken during shear at nominally 16 and 19 s^{-1}, shear rates at which N_1 was negative, shows alternately colored bands. At 16 s^{-1} the bands were blue and orange, at 19 s^{-1} they were blue and yellow. In addition to pursuing the possibility of creating bands during shear by means of oscillating the shear rate, it would be interesting to insert a first order red plate at various angles to the shear direction.

Our assessment of the angle of the molecules within the bands was semiquantitative at best. More properly, this angle α is the projection of the optical axis of the molecules onto the plane of the polarizer and analyzer [100]. Various attempts to quantify this angle α have given varying results. In HPC solutions angles of 15–24° were reported by Nishio *et al.* [105] and 8° was reported by Horio *et al.* [92]. Rather than using a first-order red plate, Yong [95] studied films from DMAC solutions of cyanoethyl cellulose by varying the direction of the crossed polars relative to shear. He found that bright and dark bands had equal width at an angle of 20°, indicating a sawtooth molecular pathway at $\pm 20°$. Takahashi *et al.* [106] deduced an angle of 40° using X-ray diffraction on dried HPC solutions. Fried and Sixou [100] also obtained a value of 40° on dried films of sheared HPC + acetic acid solutions.

The possibility of a shear rate dependence of this angle has been

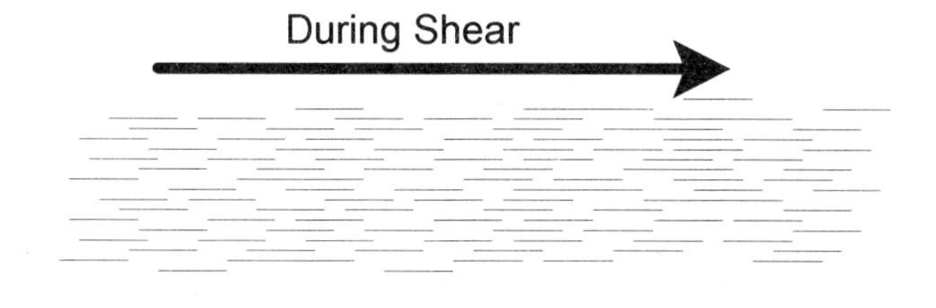

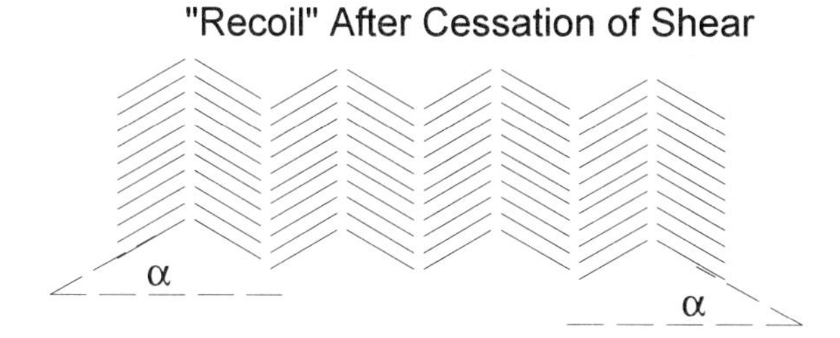

Figure 11.19 Molecular disposition during shear and after cessation.

explored. Yong [95] states with little supporting evidence that 'The higher the shear rate is, the larger the angle is, and the higher the orientation degree of the film is.' Vermant *et al.* [89] observed the evolution of this angle in sheared PBG solutions (not dried films) and found a variation from about 15–20° to a maximum of about 30–35° and then back to 10–15°. The maximum occurred at a shorter time for a higher rate of prior shear. A simple model of how this internal morphology is obtained when structures are formed is shown in Figure 11.19.

The disposition of the molecules has been described as a 'sawtooth' [95] and indeed, a 'recoil' phenomenon upon cessation of shear has been reported [107]. This picture is idealized, and departures from a perfect sawtooth or zigzag have been detected. Yong [95] examined grating diffraction from the bands and compared their observations with expectations if:

- the degree of orientation differed from band to band, but not the angle, and;
- if the angle varied from band to band, but not the degree.

He concluded that 'it is clear that the orientation direction between neighboring bands differs slightly and it plays an important role in grating diffraction. At the same time, although it is small, the difference of the orientation degree in the oriented film exists and plays a minor role in grating diffraction.' He reports also that the grating effect is present in cyanoethyl cellulose (CEC) solutions (presumably after cessation of shear).

Yong [95] also examined band structures via SEM in dried CEC films after etching with saturated aqueous $K_2Cr_2O_7$. He observed crystalline bands 2–3 μm wide perpendicular to the shear direction. Within the bands he observed fibrils 0.2 μm wide disposed in a zigzag about the shear direction.

A different departure from the idealized zigzag was proposed by Fried and Sixou [100], who investigated the change in angle at neighboring bands. They felt that 'the direction of the optical axis in these bands join in a smoother fashion than does a true zigzag. ... Near the apex of the zig-zag the optical axis probably follows a sine curve. So the path of the optical axis could be described by an intermediate way between the zig-zag and the serpentine figures.'

11.6.6 Theory of band formation

Referring to Figure 11.19, note that in order to form bands, an apparently highly oriented system in which the molecules are essentially parallel to the shear direction must in some sense degenerate into a system where the molecules have on the whole rotated away from the shear direction. In other words, the director vector which characterizes the overall local orientation of domains must rotate away from the direction of shear. One would have naively expected that once a nematic has become highly oriented, it would want to stay that way, even to the extent that Brownian motion might become ineffective. (Of course, the fact that director disorientation occurs does not necessarily preclude the possibility that local ordering may increase. Also, Larson and Mead [103] indicate that in fact this naive view is correct, and that after shearing at very high shear rates, no band formation is observed.)

This counterintuitive process must be driven by a source of energy. In some way, some of the boundless energy available from the rheometer drive while the fluid is being sheared must be stored in the viscoelastic ordered fluid as it is 'wound up' during the series of start-up transients (including transient bands perpendicular to shear). The fluid reaches a steady state when the distortion energy E_{tex} becomes proportional to the viscous energy $(d\gamma/dt)_0\eta$ [60]. The fluid then 'snaps back' upon cessation of shear, leading to the sequence of director reorientations which globally form the evolving band texture. The rate at which this happens would be determined by the ratio of the elastic forces available and the resisting viscosities.

Picken *et al.* [13] write:

> The elastic storage in a nematic is either in the molecules themselves and/or in the macroscopic texture of the nematic phase. A mechanism for the storage of elastic energy in the texture of the liquid crystal

> has been proposed in terms of an affine deformation of a 'disclination gas' type texture. In this model the elongational component of the applied strain leads to a stretching of the disclination gas and thus to an increase of the Frank elastic energy stored in the continuum surrounding the defect structure. ... We find it hard to envisage a reasonable molecular mechanism to explain this phenomenon.

They go on to derive a continuum mechanical model to describe the band spacing as a function of time. In their model, upon cessation of flow, an unspecified random perturbation of the oriented director field occurs. 'The relaxation of the perturbation is driven by the Frank elastic constants K_i and dissipated by the Miesowicz rotational viscosity γ_i.' This one-dimensional model is successful in duplicating some of the features seen experimentally, such as variation of the band spacing with time, the effect of prior shear rate on the fineness of the texture and the existence of a minimum rate of prior shear for band formation. However, being a one-dimensional model, it is incapable of describing the relaxation of the lateral spacing or length of the bands (what they refer to as 'connected white regions').

In the same paper Picken *et al.* [13] also describe a more general two-dimensional lattice model to deal with this additional aspect. In this model only splay and bend deformations are allowed to occur. The longitudinal relaxation (i.e. the band spacing) will be determined mainly by the bend elastic constant K_3, as used in the one-dimensional model, and the lateral relaxation is driven by the splay modulus K_1. They find the result of this model to be not completely satisfactory. '... it is clear that from the present model the aspect ratio of the bands $L_\perp/L_\parallel$ is expected to be a constant (and equal to $(K_1/K_3)^{1/2}$). This is not in agreement with the experimental results where the lateral correlation length seems to depend on the applied preshear rate,' (and the longitudinal correlation length does not). They speculate that 'Possibly, the applied shear rate influences the details of the initial texture that is formed upon cessation of flow.'

Picken *et al.* [13] further conclude that:

> ...the band texture phenomenon can be adequately explained by director relaxation. The rapid lateral relaxation is attributed to the large value of the splay constant in polymeric liquid crystals. This explains why the band texture is not observed in MLC nematics: the splay twist ratio K_1/K_3 is too small. ... The main disturbing feature of our model is the required high splay/bend ratio, which is not in agreement with some of the available literature. [108]

Gleeson *et al.* [60] agree that the only two possible sources of elastic energy which could drive the band formation phenomenon are molecular elasticity and texture elasticity:

> ... molecular elasticity is the energy stored in distortion of the local molecular order parameter from its equilibrium value. This molecular distortion occurs when velocity gradients are high enough that the degree of orientational order is made either higher or lower than it would be at equilibrium; the distortion becomes significant when the Deborah number becomes appreciable.

Note that 'molecular elasticity' does not refer to energy actually stored in the molecules' i.e. in bond distortions or non-equilibrium conformations.

> The second source ... is the elasticity stored in the texture, i.e. the micron-scale orientational inhomogeneities created or amplified by the previous shearing flow. Like molecular elasticity, the influence of texture elasticity can be quantified by a dimensionless number, in this case an Ericksen number.

Based on the observation that molecular elasticity relaxes quickly, in hundredths of a second, whereas band formation and evolution occurs in times of the order of minutes, it seems clear that texture elasticity is the direct energy source.

However Gleeson e.a. [60] note that molecular elasticity may play an indirect role, since below a threshold value of the Deborah number, bands do not form at all. For example, for a solution of PBG of molecular weight 186000 this critical shear rate was about $40\,s^{-1}$, while for a solution of PBG of molecular weight of 86000 bands were formed at a shear rate of $0.25\,s^{-1}$. For four solutions of widely varying viscosity and relaxation time, a critical value of $De \sim 0.01-0.04$ was obtained. 'Thus molecular elastic effects during shear seem to be necessary to produce the texture that in turn drives band formation after shearing ceases.'

Gleeson e.a. [60] do not find their model to be completely satisfactory either, however, stating '... we conclude that [the band spacing] b should decrease with increased viscosity η or shear rate $(d\gamma/dt)_0$ as $((d\gamma/dt)_0\eta)^{-1/2}$. However, we have found for PBG solutions that there is little change ... when $(d\gamma/dt)_0$ is changed by a factor of 10 or η is changed by a factor of 25' ... 'Thus the scaling for the band spacing b that we predict from the small gradient (or Frank) theory of texture elasticity is not satisfied.' A possible reason for this failure is that the texture is too fine for Frank theory to be valid. Although the size of the bands in PBG is large enough, of the order of 20 µm, the precursor striped texture has a broad distribution of length scales, ranging down to a few micrometers. In addition, the optical observations during shear only detect features in the x and y direction, not in the z (gap) direction. Based on the behavior of the coarsening process in the xy plane, they infer indirectly that the finest texture length scales in tumbling nematics are probably in the z direction, and approach the molecular persistence length of 0.1 μm (for PBG), so that Frank theory fails. Additional evidence that this may be the case is drawn from the observation that the initial band spacing in PBG solutions is 5–10 times larger than for HPC solutions, while perhaps not coincidentally, the persistence length of PBG is also 5–10 times larger than that of HPC.

Chan and Rey [109] proposed a model based on the Larson–Ericksen continuum theory, modified to accommodate static and moving defects [110], and the Landau–de Gennes nematic continuum theory. They performed a numerical study of pattern formation after cessation of shear of an incompressible uniaxial nematic polymer phase composed of rigid rod molecules, specifically noting the effect on time for formation of the bands and the orientation angle. They were able to derive a behavior which appears similar to the plots of time for band formation vs. shearing time, at a series of shear rates (compare their Figure 8 to Figures 2 and 3 in [88]). They also obtain a plausible explanation of the presence of a critical shear rate below which bands do not form. They present some of their results in the form of digitized optical light patterns of the director fields. These patterns are seen to evolve with time in that contrast develops with time (their Figure 7), and that the degree of contrast increases with rate of prior shear or time during which prior shear is applied (their Figures 11 and 12). However it appears that the band spacing does not evolve with time, as was experimentally observed.

In conclusion, we see that even at this late date the theories and models which have been put forward, to the degree that we understand them, appear to reproduce some but not all of the features of the band structure phenomenon. It is somewhat astonishing that the process of band formation should occur similarly in such widely differing systems as lyotropic solutions of both extremely rigid and somewhat flexible polymers, and in thermotropes. The theoretical work which has been done to date seems to be driven mostly by observations made on lyotropes. It would be extremely interesting to determine the ways in which the details of the band formation process change as concentration increases into the highly concentrated regime studied by Larson *et al.* [29] and on into bulk systems. HPC, which is both lyotropic and thermotropic, should be particularly interesting and amenable to such an investigation.

11.6.7 Band formation in elongational flow

The question of formation of band structures during or after cessation of elongational flow has not been extensively addressed. Marsano *et al.* [88] note that 'the texture should be present during elongational flow since bands perpendicular to the chain direction have been observed in the core of fibers' [111, 112]. A prior and somewhat more accessible work along the same line is that of Dobb *et al.* [113] which revealed banding in Kevlar 49 PPTA fibers using dark field microscopy. One diffraction spot gave an image of bands with 500 nm spacing, and another gave an image with spacing of 250 nm. Their Figure 3, which shows the more widely spaced bands, is extremely similar to those of band structures in dried lyotropic films and thermotropes. This paper did not speculate about the presence of banding in the lyotropic Kevlar PPTA dopes during spinning, which is an elongational process.

A brief mention occurs in Ernst and Navard [56] as follows: 'By varying the shear rate ... we observed ... no influence on the band spacing. In addition, the same behavior has been observed when bands are generated in elongational flow. These two later results show that the periodicity of the band texture is independent of the flow condition.' However, neither an external reference nor additional details are given. Presumably this was observed in their laboratory.

We have not seen additional discussion or description of experiments on this interesting topic of band formation in the literature.

There continue to be new developments in the field of shear-induced band structures in both PBG + m-cresol and HPC + water.

Müller e.a. [116] observed that the orientation of transient band structures upon shearing of an oriented monodomain was perpendicular to initial monodomain director, and independent of the direction of shear! They did not determine whether the orientation of band structures formed upon cessation of shear was perpendicular to the initial monodomain director or to the shearing direction. We suspect that it will prove to be the latter.

Fischer e.a. [117] proposed an entirely new origin for the band structures. They felt that it was due to 'buckling of uniaxially oriented birefringent entities'. They observed 'self-extension phenomena' which 'provide the axial compression which produces the buckling'. In addition they observed a novel related rheological phenomenon: a spontaneous stress build-up upon cessation of shear.

11.7 FINAL THOUGHTS

This chapter has been both a retrospective of our work and an overview of some studies which have been conducted since about 1980 on the negative N_1 effect during shear of lyotropic PLCs and the formation of band structures transverse to the shearing direction upon cessation of shear. The two phenomena are a manifestation of (1) liquid crystalline order, how this order is affected by forces imposed on molecules and domains during shear, and (2) how the stored energy relaxes after cessation of shear. The former effect seems to have been confirmed only in lyotropes, the latter in thermotropes as well. Neither has been reported in non-ordered fluids nor in MLCs. In the course of researching and preparing this retrospective and review, it became humblingly clear that since our initial publications, a veritable cornucopia of beautiful related science had been accomplished. This chapter provides a glimpse, a taste, and an entrée to this surprisingly large body of work. The reader is certainly encouraged to access the original sources directly, from which we have quoted liberally. In fact, the reader is specifically referred in many instances to certain figures and tables in the source material.

Both of these surprising phenomena now appear to be well understood. What was at first 'an enigma', 'a peculiar phenomenon', 'peculiar indeed', etc. has been amply verified and thoroughly explained. Neverthless a number of unanswered questions and intriguing experiments occurred to us, as we have noted. We enumerate those here.

1. The negative N_1 effect has been seen in very few systems. Oddly, the two most studied polymers, PBG and HPC, are extremely different in flexibility. What is the factor or balance of factors which appears to be so difficult to achieve, and yet is readily achieved in these two extremely disparate systems? Why is it that two such different polymers exhibit this phenomenon, yet it is not general? PCBZL + *m*-cresol exhibits the negative N_1 effect [17] and should be intermediate in flexibility, yet has been little studied. Poly(benzyl-L-asparate) (PBA) does not [114]. What about other helical polypeptides?
2. Negative N_1 has been seen for PBG only in *m*-cresol. We found *m*-cresol to be convenient because its low vapor pressure prevented evaporation and concentration changes from becoming an issue. Is this merely a coincidence, or is there a connection? Why does HPC show a negative N_1 in aqueous solutions and in *m*-cresol, but not acetic acid?
3. We have seen that the peak in the second positive region of N_1 (i.e. at high shear rate) vs. concentration persists and is only slightly shifted, whereas the peak in viscosity vs. concentration is suppressed at high shear rate. The solution is well into the flow aligning regime at such a high shear rate (the transition from wagging to flow aligning occurs at about the negative extremum of N_1). What change in the solution is occurring in this concentration regime at high shear rate that causes a peak in N_1?
4. Plots of shear stress at the sign change in N_1 vs. concentration (Figure 11.9) and maxima in the positive and negative N_1 vs.

concentration (Figure 11.10) have not been constructed for other systems, nor have the observed slopes been used to test theories.

5. How would changes in rod flexibility due to temperature changes and addition of non-helicogenic co-solvent impact the occurrence of negative N_1 and band structures?
6. What is the effect of molecular weight distribution on negative N_1?
7. What is the origin and nature of the large periodic fluctuations in normal force observed in lyotropic but not isotropic solutions of PBG + *m*-cresol?
8. What are the effects of simultaneous shear and electromagnetic fields on negative N_1 and formation of band structures?
9. It appears likely that we observed band structures during shear due to an inadvertant superposition of oscillatory and steady shear. An attempt should be made to reproduce a quasi-steady state band structure and to explore this phenomenon. Either this will turn out to be relatively easy to do, or we were unbelievably lucky (yet again!) to observe it. If this is accomplished, it would be interesting to insert a first order red plate at various angles to the shear direction.
10. Since there appears to be an upper shear rate for formation of band structures, shearing beyond this critical rate should not produce recoil.
11. It would be extremely interesting to determine in which ways the details of the band formation process change as concentration increases into the highly concentrated regime studied by Larson e.a. [29] and on into bulk systems. HPC, which is both lyotropic and thermotropic, should be particularly interesting and amenable to such an investigation.

Many other interesting possibilities for future work would no doubt present themselves upon further reflection.

The introductory section of many investigations of negative N_1 and band structures give as justification for the work the 'technological importance' of high strength and rigidity lyotropic (Kevlar) and thermotropic polymers. It is taken for granted that deeper understanding of orientation phenomena which occur during flow and deformation will somehow eventually translate into new products which would have been unobtainable in the absence of this understanding and knowledge or, at least, cheaper ways of making useful products.

Is this really so? Has this been simply an exercise in intellectual beauty, satisfying the curiosity, or can one hope for a practical result? Will the insights developed in explaining negative N_1 lead to an unexpectedly effective extrusion die design or processing technique? When one of us (G. Kiss) was a graduate student at the University of

Massachussetts, friends would sometimes (rarely) ask about my research. After a brief explanation, which always felt rather tepid ('... first the plates get pushed apart, then pulled back together, then pushed apart again!' followed by a question about what it is good for), I finally resorted, only half in jest, to the observation that 'What it's good for is getting a doctorate.'

In a review of the rheology of processing polymer liquid crystals, Muir and Porter [114] give no indication that the insights gained in determining the origins of negative N_1 and band structures have had a pronounced impact on the control of orientation and the physical properties of thermotropes. There may be some link to the observation of near-zero or even negative die swell, which would be 'very attractive to those interested in the design of dies and injection cavities, particularly for thin sections.' However the performance of thermotropic PLCs still lags that of lyotropes.

Orientation impacts other properties besides physical, so that control of orientation may be exploited by means of other properties. Control of the coefficient of thermal expansion (CTE) is critical for fiber optics applications, since optical fiber has an extremely low CTE and the waveguiding property of optical fiber is disrupted by bending. This is particularly the case at the longer wavelength (1550 nm as opposed to 1310 nm) which is becoming more widely used for high bit rate and AM video systems, due to the lower attenuation of optical fiber at this wavelength, and the availability of erbium-doped fiber amplifiers (EDFAs) which operate only at this wavelength. Thus thermotropic PLCs have found application in fiber optic components requiring extreme dimensional stability such as splices and connectors. They have also been considered for optical cable reinforcement members and buffer tubes. Another property which is extremely dependent on orientation is the diffusion rate. PLCs have extremely good barrier properties, even when incorporated into blends with isotropic polymers (i.e. *in situ* composites; R. Lusignea, Superex Corp., Boston, MA, personal communication).

By far the most important application area of liquid crystals *per se* is of course in display devices. Who can fail to be amazed at the technological triumph of an active matrix LCD flat projection panel capable of displaying 1024 × 768 pixels × 256 colors? Perhaps the insights derived from PLCs will have some impact on displays, or even modulators, nonlinear optical devices or data storage devices.

We hope that time will reveal that the first 20 years after our confirmation of the negative N_1 effect and band structures in lyotropic PLCs was simply the explanation phase, interesting as it may be to a small niche of specialists. We hope that the second 20 years will prove to be the exploitation phase.

REFERENCES

1. Iizuka, E. (1974) *Mol. Cryst. Liq. Cryst.*, **25**, 287.
2. Hermans, Jr, J. (1962) *J. Colloid Sci.*, **17**, 638.
3. Flory, P.J. (1956) *Proc. Royal Soc. London*, **A234**, 60.
4. Onsager, L. (1949) *Ann. NY Acad. Sci.*, **51**, 627.
5. Porter, R.S., Barrall, E.M. II and Johnson, J.F. (1966) *J. Chem. Phys.*, **45**, 1452.
6. Porter, R.S. and Johnson, J.F. (1967) in *Rheology: Theory and Applications*, Vol. IV (ed. F.R. Eirich), Academic Press, New York, p. 317.
7. Kulicke, W.M., Kiss, G. and Porter, R.S. (1977) *Rheol. Acta*, **16**, 568.
8. Bird, R.B. and Curtis, C. (1984) *Phys. Today*, January, 36.
9. Kiss, G. (1984) *Phys. Today*, May, 15.
10. Northold, M.G. and Sikkema, D.J. (1990) *Adv. Polymer Sci.*, **2A**, 119.
11. Aoki, H., White, J.L. and Fellers, J.F.J. (1979) *J. Appl. Polymer Sci.*, **23**, 2293.
12. Aharoni, S.M. (1979) *Macromolecules*, **12**, 94.
13. Picken, S.J., Moldenaers, P., Berghmans, S. and Mewis, J. (1992) *Macromolecules*, **25**, 4759.
14. Doty, P., Bradbury, J.H. and Holtzer, A.M. (1956) *J. Am. Chem. Soc.*, **78**, 947.
15. Robinson, C. and Ward, J.C. (1957) *Nature*, **180**, 1183.
16. Kiss, G. and Porter, R.S. (1978) *J. Polymer Sci. Symp.*, **65**, 193.
17. Kiss, G. and Porter, R.S. (1980) *J. Polymer Sci. Phys.*, **18**, 361.
18. Kiss, G. and Porter, R.S. (1980) *Mol. Cryst. Liq. Cryst.*, **60**, 267.
19. Mewis, J. and Moldenaers, P. (1987) *Mol. Cryst. Liq. Cryst.*, **53**, 291.
20. Larson, R.G. and Mead, D.W. (1989) *J. Rheol.*, **33**, 1251.
21. Berghmans, S. (1989) Eng. Thesis, Katholieke Universiteit Leuven, Belgium.
22. Frenkel, S.Y. (1974) *J. Polymer Sci.*, **C44**, 49.
23. Doi, M. (1981) *J. Polymer Sci. Phys.*, **19**, 229.
24. Dave, V. and Glasser, W.G. (1992) *Am. Chem. Soc. Symp. Ser.*, **489**, 144.
25. Baek, S.-G., Magda, J.J., Larson, R.G. and Hudson, S.D. (1994) *J. Rheol.*, **38**, 1473.
26. Suto, S., Gotoh, H., Nishibori, W. and Karasaa, K. (1989) *J. Appl. Polymer Sci.*, **37**, 1147.
27. Asada, T., Tanaka, T. and Onogi, S. (1985) *J. Appl. Polymer Sci.*, **41**, 229.
28. Navard, P. and Hauden, J.M. (1986) *J. Polymer Sci.: Phys.*, **24**, 189.
29. Larson, R.G., Promislow, J., Baek, S.-G. and Magda, J.J. (1994) in *Ordering in Macromolecular Systems* (eds A, Teramoto, M. Kobayashi and T. Norisuje) Springer-Verlag, Berlin–Heidelberg, p. 191.
30. Matsuda, M., Norisuye, T., Teramoto, A. and Fujita, H. (1973) *Biopolymers*, **12**, 1515.
31. Driscoll, P., Takigawa, T., Nakamachi, N. and Masuda, T. (1990) On negative first normal stress difference and a 'dip' in steady shear viscosity in PLCs, unpublished manuscript obtained from A. Metzner.
32. Magda, J.J., Baek, S.G., Devries, K.L. and Larson, R.G. (1991) *Macromolecules*, **24**, 4460.
33. Berry, G. (1988) *Mol. Cryst. Liq. Cryst.*, **165**, 333.
34. Onogi, S. and Asada, T. (1985) in *Rheology* (eds G. Astarita and G. Marrucci), Plenum Press, New York.
35. Hutton, J.F. (1975) *Rheol. Acta.*, **14**, 979.
36. Erhardt, P.F., Pochan, J.M. and Richards, W.C. (1972) *J. Chem. Phys.*, **57**, 3596.
37. Coleman, B.D. and Markowitz, H. (1964) *J. Appl. Phys.*, **35**, 1.
38. Walters, K. (1974) *Rheometry*, Halstead Press, New York, p. 88.
39. Adams, N. and Lodge, A.S. (1964) *Phil. Trans. Royal Soc.*, **256**, 149.

40. Duke, R.W. and Chapoy, L.L. (1976) *Rheol. Acta.*, **15**, 548.
41. Huang, T.A. (1976) PhD Thesis, Univ. of Wisconsin, Dept. of Engineering Mechanics, p. 93.
42. Iizuka, E. (1973) *J. Phys. Soc. Japan*, **35**, 1792.
43. Scheraga, H.A., Edsall, J.T. and Gadd, J.D. (1951) *J. Chem. Phys.*, **19**, 1101.
44. Moldenaers, P. and Mewis, J. (1986) *J. Rheol.*, **30**, 567.
45. Navard, P. (1986) *J. Polymer Sci. Phys.*, **24**, 435.
46. Baek, S.-G., Magda, J.J. and Larson, R.G. (1993) *J. Rheol.*, **37**, 1201.
47. Berry, G.C. (1991) *J. Rheol.*, **35**, 943.
48. Suto, S., Ohshiro, M., Nishibori, W. e.a. (1988) *J. Appl. Polymer Sci.*, **35**, 407.
49. Fasman, G.D., Idelson, M. and Blount, E.R. (1961) *J. Am. Chem. Soc.*, **83**, 709.
50. Tsvetkov, V.N., Riumtsev, E.I. and Shtennikova, I.N. (1978) in *Liquid Crystalline Order in Polymer Molecules* (ed. A. Blustein) Academic Press, New York, p. 58.
51. Daniel, E. and Katchalski, E. (1962) in *Polyamino Acids, Polypeptides, and Proteins* (ed. M.A. Stahmann), University of Wisconsin, Madison, p. 183.
52. Chaffey, C.E. and Porter, R.S. (1985) *J. Rheol.*, **29**, 281.
53. Lin, Y.G. ansd Winter, H.H. (1992) *Polymer Eng. & Sci.*, **32**, 773.
54. Sigillo, I. and Grizutti, N. (1994) *J. Rheol.*, **38**, 589.
55. Hongladarom, K. Secakusuma, V. and Burghardt, W.R. (1993) *J. Rheol.*, **38**, 1505, 1549.
56. Ernst, B. and Navard, P. (1989) *Macromolecules*, **22**, 1419.
57. Grizutti, N., Cavella, S. and Cicarelli, P. (1990) *J. Rheol.*, **34**, 1293.
58. Moldenaers, P. and Mewis, J. (1993) *J. Rheol.*, **37**, 367.
59. Sakamoto, K. and Porter, R.S. (1970) *J. Polymer Sci. B*, **8**, 177.
60. Gleeson, J.T., Larson, R.G., Mead, D.W. e.a. (1992) *Liq. Cryst.*, **11**, 341.
61. Larson, R.G. (1990) *Macromolecules*, **23**, 3983.
62. Marrucci, G. and Maffettone, P.L. (1990) *J. Rheol.*, **34**, 1217.
63. Marrucci, G. and Maffettone, P.L. (1990) *J. Rheol.*, **34**, 1231.
64. Ramachandran, S., Gao, H.W. and Christiansen, E.B. (1985) *Macromolecules*, **18**, 695.
65. Baek, S.-G., Magda, J.J. and Cementwala, S. (1993) *J. Rheol.*, **37**, 935.
66. Srinivasaro, M., Garay, R.O., Winter, H.H. and Stein, R.S. (1992) *Mol. Cryst. Liq. Cryst.*, **223**, 29.
67. De Gennes, P.G. (1974) *The Physics of Liquid Crystals*, Clarendon Press, Oxford.
68. Marrucci, G. and Maffettone, P.L. (1989) *Macromolecules*, **22**, 4076.
69. Hongladarom, K., Burghardt, W.R., Baek, S.G. e.a. (1993) *Macromolecules*, **26**, 772.
70. Burghardt, W.R. and Fuller, G.G. (1991) *Macromolecules*, **24**, 2546.
71. Srinivasaro, M. and Berry, G.C. (1991) *J. Rheol.*, **35**, 379.
72. Larson, R.G. and Doi, M.J. (1991) *J. Rheol.*, **35**, 539.
73. Kim, S.S. and Han, C.D. (1994) *J. Polymer Sci. Phys.*, **32**, 371.
74. Onogi, S. and Asada, T. (1980) in *Proceedings of the VIIIth International Congress on Rheology* (eds G. Astarita, G. Marrucci and L. Nicolais), Plenum Press, New York, p. 127.
75. Currie, P.K. (1981) *Mol. Cryst. Liq. Cryst.*, **73**, 1.
76. Farhoudi, Y. and Rey, A.D. (1993) *J. Rheol.*, **37**, 289.
77. Hand, G.L. (1961) *J. Fluid Mech.*, **13**, 33.
78. MacMillan, E.H. (1989) *J. Rheol.*, **33**, 1071.
79. Marrucci, G. (1991) in *Liquid Crystallinity in Polymers* (ed. A Ciferri), VCH, New York, p. 395.

80. Andrews, N.C., Edwards, B.J. and McHugh, A.J. (1995) *J. Rheol.*, **39**, 1161.
81. Edwards, B.J., Beris, A.N. and Grmela, M. (1991) *Mol. Cryst. Liq. Cryst.*, **201**, 51.
82. Tse, K.L. and Shine, A.D. (1995) *J. Rheol.*, **39**, 1021.
83. Keller, A. and Machin, M.J. (1967) *J. Macromol. Sci. Phys.*, **B1**, 41.
84. Toth, W.J. and Tobolsky, A.V. (1970) *Polymer Lett.*, **8**, 531.
85. Fincher, C.R. (1986) *Macromolecules*, **19**, 2431.
86. Vermant, J., Moldenauers, P., Pickens, S.J. and Mewis, J. (1994) *J. Non-Newtonian Fluid Mech.*, **53**, 1.
87. Marrucci, G., Grizutti, N. and Buonario, A. (1987) *Mol. Cryst. Liq. Cryst.*, **153**, 263.
88. Marsano, E., Carpento, L. and Ciferri, A. (1988) *Mol. Cryst. Liq. Cryst.*, **158B**, 267.
89. Vermant, J., Moldenauers, P., Mewis, J. and Pickens, S.J. (1994) *J. Rheol.*, **38**, 1571.
90. Marsano, E., Carpento, L. and Ciferri, A. (1988) *Liq. Cryst.*, **3**, 1561.
91. Einaga, Y., Berry, G.C. and Chu, S.G. (1985) *J. Polymer Sci. Phys.*, **17**, 239.
92. Horio, M., Ishikawa, S. and Oda, K. (1985) *J. Appl. Polymer Sci., Appl. Polymer Symp.*, **41**, 269.
93. Doppert, H.L. and Picken, S.J. (1987) *Mol. Cryst. Liq. Cryst.*, **153**, 109.
94. Marsano, E., Carpento, L. and Ciferri, A. (1989) *Mol. Cryst. Liq. Cryst.*, **177**, 93.
95. Yong. H. (1991) *Cellulose Chem. & Tech.*, **25**, 283.
96. Zachariades, A., Navard, P. and Logan, J.A. (1984) *Mol. Cryst. Liq. Cryst.*, **110**, 93.
97. Viney, C. and Windle, A.H. (1986) *Polymer*, **27**, 1325.
98. Graziano, D.J. and Mackley, M.R. (1984) *Mol. Cryst. Liq. Cryst.*, **106**, 73.
99. Navard, P. and Zachariades, A.E. (1987) *J. Polymer Sci. Phys.*, **25**, 1089.
100. Fried, F. and Sixou, P. (1988) *Mol. Cryst. Liq. Cryst.*, **158B**, 163.
101. Kwiatkowski, M. and Hinrichsen, G. (1990) *J. Mater. Sci.*, **25**, 548.
102. Viney, C. and Windle, A.H. (1987) *Mol. Cryst. Liq. Cryst.*, **148**, 145.
103. Larson, R.G. and Mead, D.W. (1993) *Liq. Cryst.*, **15**, 51.
104. Fincher, C.R. (1988) *Mol. Cryst. Liq. Cryst.*, **155**, 559.
105. Nishio, Y., Yamane, T. and Takahashi, T. (1985) *J. Polymer Sci. Phys.*, **23**, 1053.
106. Takahashi, J., Shibata, K., Nomura, S. and Kurokawa, M. (1982) *Seni Gakkaishi*, **38**, 375.
107. Picken, S.J., Aerts, J., Doppert, H.L. *et al.* (1991) *Macromolecules*, **24**, 1366.
108. Lee, S.-D. and Meyer, R.B. (1988) *Phys. Rev. Lett.*, **61**, 2217.
109. Chan, P.K. and Rey, A.D. (1993) *Liq. Cryst.*, **13**, 775.
110. Ericksen, J.L. (1989) *IMA Preprint Series*, No. 559.
111. Shouxi, C., Chenfen, L. and Liying, C. (1981) *Gaofenzi Tongxun*, **6**, 424.
112. Horio, H., Kaneda, T., Ishikawa, S. and Shimanura, K. (1984) *Sen I Gakkaishi*, **285**, 40.
113. Dobb, M.G., Johnson, D.J. and Saville, B.P. (1977) *J. Polymer Sci. Phys.*, **15**, 2201.
114. Kiss, G., Orrell, T.S. and Porter, R.S. (1979) *Rheol. Acta.*, **18**, 657.
115. Muir, M.C. and Porter, R.S. (1989) *Mol. Cryst. Liq. Cryst.*, **169**, 83.
116. Müller, J.A., Stein, R.S. and Winter, H.H. (1996) *Rheol. Acta*, **35**, 60.
117. Fischer, H., Keller, A. and Windle, A.H. (1996) *J. Non-Newtonian Fluid Mech.*, **67**, 241.

12

Creep and stress relaxation

Josef Kubát and Robert D. Maksimov

12.1 INTRODUCTION

Thermotropic polymer liquid crystals (PLCs) are a relatively new class of materials; they are under extensive investigation which is still growing in scope. Many studies have been reported concerning their rheological, dielectric, thermal and mechanical properties, and processing and morphology [1–14]. New areas of their potential applications continue to emerge. The PLCs have good physical and mechanical properties as well as better dimensional stability and chemical resistance than flexible polymers. They show ease of processing due to their low melt viscosity. Their properties are highly anisotropic in fabricated products; this is associated with the sensitivity of the stiff molecules to the flow fields causing a molecular orientation in the flow direction.

During the last few years, most attention has been paid to the blending of PLCs with less expensive thermoplastic engineering polymers (EPS). Addition of PLCs to such polymers not only enhances mechanical properties (strength and stiffness) of the resulting composites obtained due to the orientation of the PLC phase, but also improves their processing properties. Even relatively small amounts of a PLC may induce a reduction in the melt viscosity and thus improve the processability. In most cases, under appropriate processing conditions the dispersed PLC phase can be deformed into a fibrillar one. The

Mechanical and Thermophysical Properties of Polymer Liquid Crystals
Edited by W. Brostow
Published in 1998 by Chapman & Hall, London.
ISBN 0 412 60900 2

resulting self-reinforced materials are hence sometimes called *in situ* composites. Often the so-called skin–core morphology is found with the thin fibrillar PLC phase in the skin region and spherical or ellipsoidal PLC domains in the core. Improvements in the mechanical properties as well as dimensional and thermal stability have been revealed in blends of PLCs with EPs, such as polycarbonate [15–22], polypropylene [23–28], poly(ether-ether-ketone) [29, 30], poly(ether-imide) [29, 31], polystyrene [15, 32], poly(ethylene terephthalate) [15], polysulfone [33], polyethersulfone [21], poly(butylene terephthalate) [34], poly(ester carbonate) [21] and polyamide [35].

Despite this large volume of results, there are almost no papers on the thermoviscoelastic behavior of PLCs and their blends with EPs under conditions of creep or stress relaxation. At the same time, it is known that the time dependence of the mechanical behavior of these materials is significant, i.e. they are distinctly viscoelastic. Thus creep and stress relaxation studies can give important information for better understanding of the peculiarities of this type of polymer material and for the application conditions of PLCs as engineering materials.

The only data on creep and stress relaxation of PLCs and their blends available at present are those obtained by the present authors and their colleagues. In the following sections we analyze those results, noting possible general features.

12.2 CREEP OF A LIQUID CRYSTALLINE COPOLYESTER

12.2.1 Systems studied

In this and the following two sections we consider the data for a PLC and for different blends of that PLC with isotactic polypropylene (PP). The PLC used in this study was liquid crystalline copolyester PET/0.6PHB, where PET = poly(ethylene terephthalate), PHB = *p*-hydroxybenzoic acid and 0.6 is the mole fraction of PHB in the copolymer, manufactured by Unitika Ltd, Kyoto, Japan. The content of the PLC in the blends with PP was varied from 0 to 20 wt%. The samples were prepared by injection molding. PP, resin VB65 11B, was supplied by Neste OY, Finland.

12.2.2 Linear thermoviscoelastic creep behavior

In this section we analyze the results reported in [36]. The creep tests in the linear viscoelastic region were carried out at 14 temperatures in the range from 20°C to 160°C. In order for the creep behavior to stay in the linear region, it was necessary to reduce the stress gradually, together with the temperature increase. Thus, at 20°C the stress value was 20 MPa, while at 160°C it was only 0.2 MPa. In all cases, total strains at the moment of specimens unloading did not exceed 0.5%. The loading time was approximately 1 s; then creep curves were measured during 1 h. Afterwards the samples were completely unloaded and creep recovery curves were measured. It should be noted that 24 h after unloading the residual strains were negligible, i.e. the material under the investigation conditions exhibited an inverse viscoelastic deformation with time-recoverable creep.

The creep data were represented as creep compliance $D(t) = \varepsilon(t)/\sigma$, where $D(t)$ is the compliance, t is time, and ε and σ are the engineering strain and stress, respectively. To minimize scattering, three to five creep curves were measured on different samples at each temperature level. Standard deviations were less than 10%. Averaged experimental creep compliance data are shown by the dots in Figure 12.1. As may

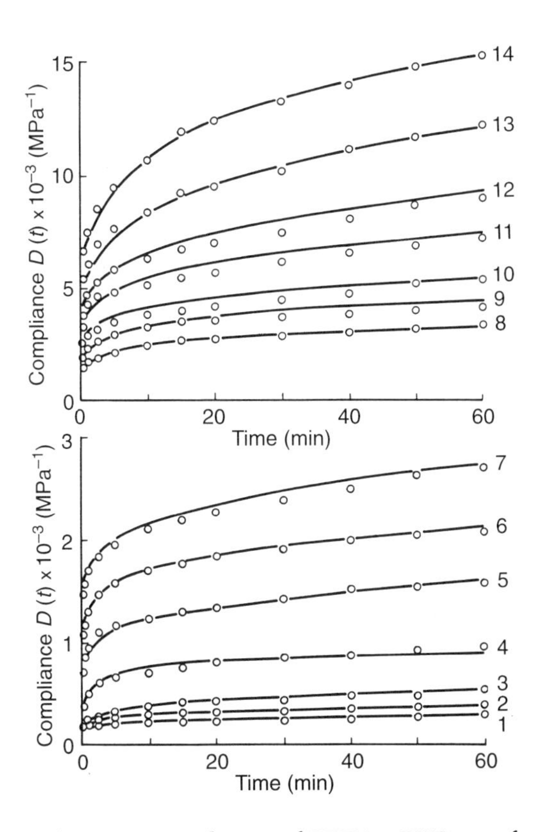

Figure 12.1 Tensile creep compliance of PET/0.6PHB as a function of time at different temperature levels: 20°C (curve 1); 40° (2); 50° (3); 60° (4); 70° (5); 80° (6); 90° (7); 100° (8); 110° (9); 120° (10); 130° (11); 140° (12); 150° (13); 160°C (14). Dots = experimental values; lines = calculated from an analytical expression. (After [36].)

be expected, temperature has quite a large effect on the creep of the PLC copolyester. The creep compliance after deformation during 1 h at 160°C is larger than that at 20°C by a factor of 65. However, the material examined reveals a marked creep even at room temperature. The compliance after creep during 1 h at 20°C is larger than the initial compliance (at the loading time) by a factor of 1.26.

The time-temperature correspondence principle is known to be widely applicable to polymers other than PLCs, e.g. [37] or [38]. It was therefore of interest to test the applicability of the principle to our PLC. For this purpose, the experimental creep data are presented in Figure 12.2 in logarithmic coordinates. It can be seen that the shape of these curves suggests that a master curve may be constructed by shifting experimental curves along the log time axis until they all superimpose on one curve for the chosen reference temperature. Figure 12.3 presents such a master curve for the reference temperature of 20°C. As can be seen in

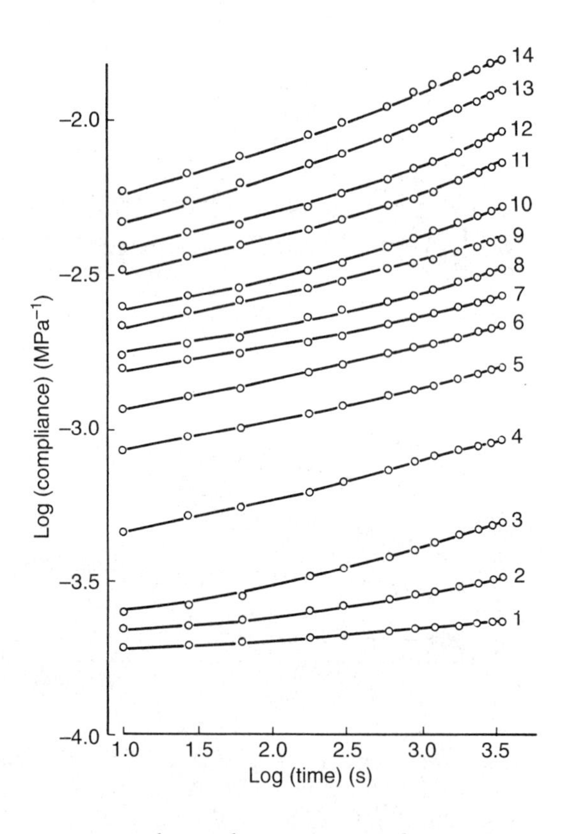

Figure 12.2 Experimental tensile creep compliance data in logarithmic coordinates at the same temperatures as in Figure 12.1. (After [36].)

Figure 12.3, the procedure of time–temperature superposition for obtaining a master curve can be performed without any difficulty. Thus the compliance curves at different temperature levels differ only by the time scale, that is $D(\log t, T) = D(\log t - \log a_T) = D[\log(t/a_T)]$.

In Figure 12.4 the shift factor $\log a_T$ has been plotted vs. temperature. There are two horizontal scales, since the first corresponds to 20°C and the second to the glass transition temperature of the PET-rich phase in the PLC, that is $T_{g\,\mathrm{PET}} = 62°\mathrm{C}$. Incidentally, the $T_{g\,\mathrm{PET}}$ evident in Figure 12.4 agrees well with values obtained by several other techniques [13, 39]. The broken line in the figure has been calculated from the Williams–Landel–Ferry (WLF) equation. The large deviation from experimental values (circles) was expected, since Ferry [37] states that their equation works well around $T_g + 50\,\mathrm{K}$, while here an attempt was made to use it below T_g. There are also other problems with the WLF equation, as discussed by Brostow in Chapter 10 of reference [38].

The continuous $a_T(T)$ curve in Figure 12.4 was calculated [36] using the general shift factor equation of Brostow [40] (see also Chapter 10 of reference 38):

$$\log a_T = A + \frac{B}{\tilde{v} - 1} \qquad (12.1)$$

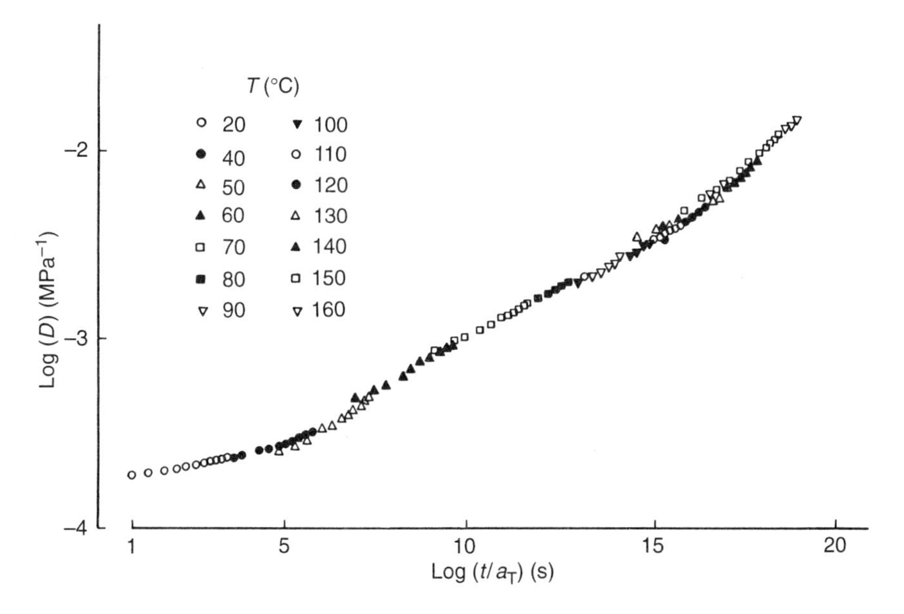

Figure 12.3 Creep compliance master curve for PET/0.6PHB at the reference temperature 20°C. (After [36].)

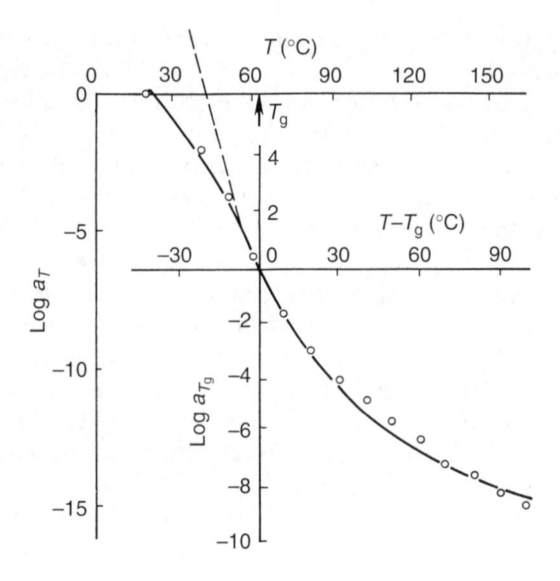

Figure 12.4 Temperature–time shift factor a_T; explanation in the text. (After [36].)

in conjunction with the Hartmann equation of state [41–43]

$$\tilde{P}\tilde{v}^5 = \tilde{T}^{3/2} - \ln \tilde{v} \tag{12.2}$$

Here B is the Doolittle constant, since Brostow derived his equation starting from the Doolittle formula relating viscosity and free volume [44]. The reduced quantities are

$$\tilde{P} = \frac{P}{P^*}; \quad \tilde{v} = \frac{v}{v^*}; \quad \tilde{T} = \frac{T}{T^*} \tag{12.3}$$

P, v and T are the actual pressure, specific volume and temperature while the values with asterisks are characteristic for a given material. Moreover, the free volume is defined as $v^f = v - v^*$. The Hartmann equation has been demonstrated to work well for both polymer solids and melts. Figure 12.4 shows that the combination of equations (12.1)–(12.3) provides the capability of prediction of the shift factor a_T values both *below and above* the glass transition temperature.

12.2.3 Nonlinear creep behavior

The stress dependence of the creep of PET/0.6PHB was also investigated in [36] at the room temperature of 20 ± 1°C. Experimental creep curves were measured for 2 h at nine stress levels in the range from 10 to 50 MPa at 0.5 MPa intervals. It should be noted that the maximal value of the applied stress is equal to 0.5 of the short-term strength of the material.

Averaged experimental creep curves at different stress levels are shown by dots in Figure 12.5. Figure 12.6 shows the isochronal stress–strain curves at $t = 0$ (the strain at the loading time) and at $t = 120$ min (total strain just before unloading). We see from the latter that the stress–strain dependence on loading time is almost linear. A small departure from linearity may occur as a result of a viscoelastic strain contribution at the moment of the first measurement. This effect apparently increases with increasing stress level. By contrast, the isochronous curve at $t = 120$ min is significantly nonlinear. The contribution of the nonlinear component to the total creep strain increases with increasing stress. At $\sigma = 50$ MPa the value of the creep strain is more than twice that in the linear range. Therefore, in an analytical representation of stress–strain–time relations we have to consider the nonlinearity of the stress dependence of the viscoelastic deformation.

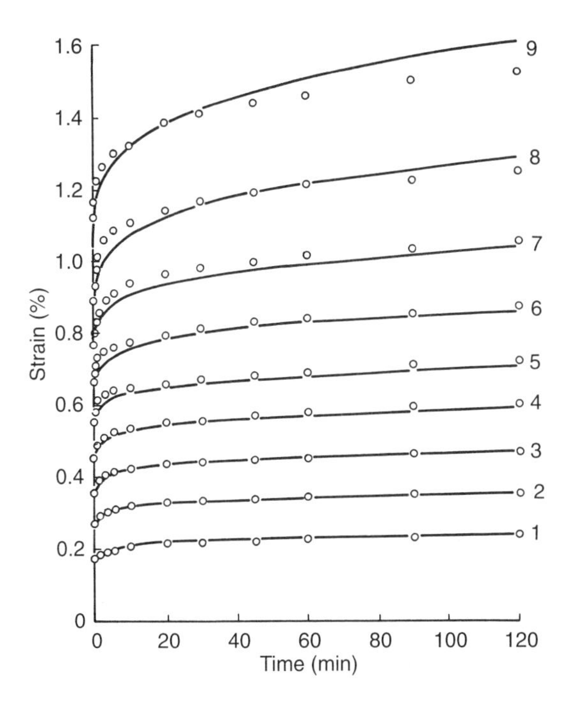

Figure 12.5 Tensile creep of PET/0.6PHB at temperature of 20°C for different stress levels: 10 (1); 15 (2); 20 (3); 25 (4); 30 (5); 35 (6); 40 (7); 45 (8); 50 MPa (9). Dots = experimental data; lines derived from equation (12.6). (After [36].)

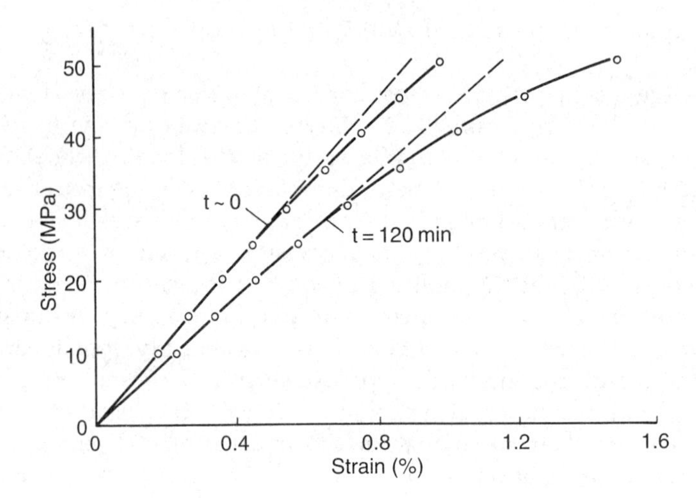

Figure 12.6 Isochronous stress–strain curves at $t = 0$ and $t = 120$ min for creep at 20°C (After [36].)

In the general case, the nonlinear creep behavior could be described by a multiple integral representation using several kernels. The practical application of such equations is limited by the absence of a clearly defined strategy of creep kernel determination as well as relations between the kernel and resolvent. Various simplified versions of this approach have also been proposed. Usually it is assumed that the creep kernel and relaxation time are independent of the stress value. At the same time, it is known that a good approximation to nonlinear creep may be obtained by using the following equation:

$$\varepsilon(t) = \frac{\sigma(t)}{E} + \frac{1}{E}\int_0^t K(t - s, \sigma(s))\,\sigma(s)\,ds \qquad (12.4)$$

where the kernel K is written as a function of the stress and of the time difference $(t - s)$; that is, it is assumed that the material properties are invariant with respect to the starting time of experiment and observation. It is assumed also that the initial strains correspond to the law of linear elasticity.

It was demonstrated in several papers [45–49] that increasing the stress leads to a shift of the time relaxation spectrum in a manner similar to increasing the temperature. Thus we have correspondence between time, temperature, frequency and also stress. This means, among other things, that the stress dependence of kernel K in equation (12.4) may be described by a function a_σ analogous to the temperature shift factor a_T. Then the kernel K expressed as a sum of exponential functions may be represented by

$$K(t-s,\sigma)=\frac{1}{a_\sigma}\sum_{i=1}^{k}\frac{b_i}{\tau_i}\exp\left(-\frac{t-s}{\tau_i a_\sigma}\right) \tag{12.5}$$

where b_i and τ_i are coefficients of discrete relaxation spectra at some reference stress σ_0.

Using equation (12.4) in conjunction with (12.5), we obtain the following expression for the compliance $D(t)=\varepsilon(t)/\sigma$ when $\sigma=$ constant:

$$D(t)=\frac{1}{E}\left\{1+\sum_{i=1}^{k}b_i[1-\exp(-t/a_\sigma\tau_i)]\right\} \tag{12.6}$$

The last equation implies that the compliance curves at different stress levels differ only by the time scale, that is $D(\log t,\sigma)=D(\log t-\log a_\sigma)=D[\log(t/a_\sigma)]$.

To test the stress–time superposition, the curve of creep strains shown in Figure 12.5 are presented as the dependence of creep compliance on log time in Figure 12.7. In the stress range from 10 to 30 MPa the difference between averaged compliance curves at different stress levels was less than the mean square error of the experimental data. Therefore, the dashed line in Figure 12.7 corresponds to the average compliance for stresses from 10 to 30 MPa. Thus, if the stress

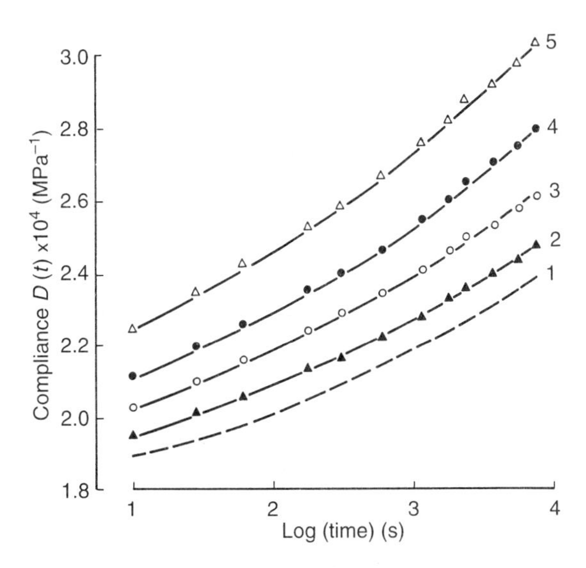

Figure 12.7 Creep compliance of PET/0.6PHB vs. log time at different stress levels: averaged data for the range 10–30 MPa (1); 35 (2); 40 (3); 45 (4); 50 MPa (5). (After [36].)

does not exceed 30 MPa, that is 0.3 of the short-term strength of the material, we may use the less complicated theory of linear viscoelasticity for the analytical representation of the creep results at room temperature.

The effect of nonlinearity increases gradually with increasing stress. It is seen from Figure 12.7 that the compliance curves at different stress levels are displaced relative to each other along the log time axis. These curves were shifted horizontally to produce the master curve consisting of several different symbols (see Figure 12.7) in Figure 12.8 [36]. The stress dependence of the stress shift factor is shown in an insert. The reference stress for the master curve is 30 MPa.

The master curve so obtained corresponds to the compliance in the region of linear viscoelasticity. Because of this, it is interesting to compare the master curve with that obtained using the temperature–time superposition (shown in Figure 12.3). The latter curve is shown in Figure 12.8 as the dashed line. It is seen that both master curves agree with each other fairly closely.

There is also good agreement between experimental and calculated creep curves in Figure 12.5. The calculated curves are obtained by the use of equation (12.6) with stress shift factors a_σ shown in the insert to Figure 12.8. Thus description and prediction of creep of our PLC

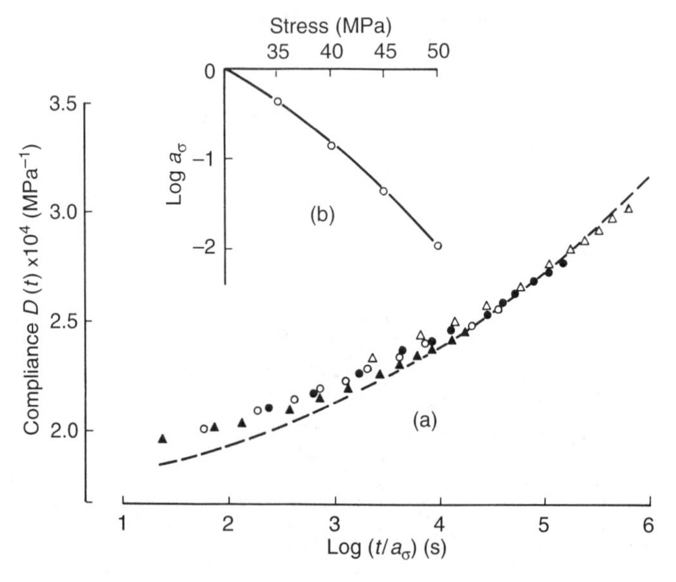

Figure 12.8 Creep compliance master curves (a) and stress–time shift factor (b). The master curve obtained using the stress–time superposition is indicated by short dashes; the point symbols are the same as in Figure 12.7. The master curve obtained by using the temperature–time superposition is shown by long dashes (b; cf. Figure 12.3); after [36].

examined in regions of both linear and nonlinear viscoelasticity are quite sufficient for engineering applications.

12.3 CREEP OF BLENDS OF A LIQUID CRYSTALLINE COPOLYESTER WITH POLYPROPYLENE

12.3.1 Introductory remarks

In this section we analyze creep results reported in [50] for PLC + PP blends, where the PLC is the same PET/0.6PHB discussed above, while PP is the isotactic polypropylene. In particular, thermal effects on the short-term creep in a wide temperature range have been investigated, as well as long-term (more than 1 year) creep. On this basis we discuss further the time–temperature correspondence and possibilities of long-term creep prediction from short-term tests.

The contents of PET/0.6PHB in the blends were varied in [50] from 0 to 20 wt% in 5 wt% intervals. The plaques were prepared by injection molding, and the samples were cut from the plaques along the major melt flow direction. Blending and sample preparation in [50] were the same as in [28]. We note that results of short-term tensile tests for this type of blends were reported in [28]. A significant influence of the addition of PLC on the mechanical properties of PP has been found. Generally, the modulus of elasticity increased with increasing PLC content. At the same time, the strain at yield and the total elongation during drawing (strain at break) decreased significantly.

12.3.2 Thermoviscoelastic creep behavior

The temperature dependence of the creep was investigated [50] at nine temperature levels in the range from 20°C to 100°C in 10°C intervals. These tests were performed for samples with PLC contents of 0, 10 and 20 wt% . The experiments were carried out in the linear viscoelasticity region. Thus it was necessary to reduce gradually the stress level with the temperature increase. The stress value was 4 MPa at 20°C while at 100°C it was only 1 MPa. The creep curves were measured during 1 h. In all tests the total strains at the moment of unloading of specimens did not exceed 0.5%. It is significant that 24 h after unloading the residual strains were negligibly small; that is, also in this case the materials exhibited a reversible creep behavior.

Averaged experimental creep compliance curves for the blends containing 20 wt% PLC are shown in Figure 12.9 as a series of curves $D = D(\log t, T)$. The respective diagrams for 0% PLC (= pure PP)

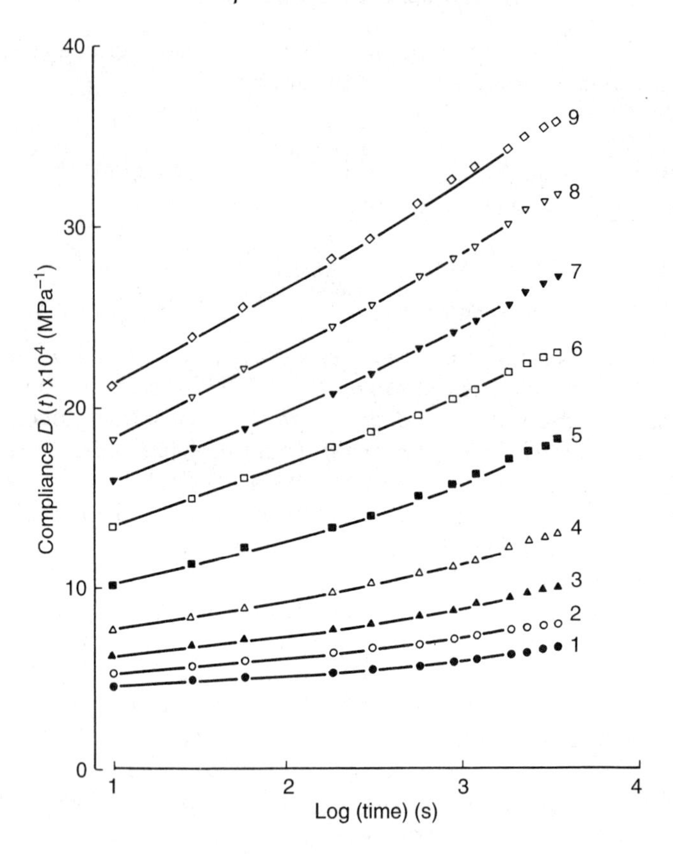

Figure 12.9 Tensile creep compliance of 80% PP + 20% PLC at different temperature levels; 20°C (1); 30° (2); 40° (3); 50° (4); 60° (5); 70° (6); 80° (7); 90° (8); 100°C (9). Point symbols = experimental values; lines = calculation results. (After [50].)

and 10 wt% are similar [50]. As expected, the effect of temperature on the creep of pure PP as well as blends containing the PLC is significant. The creep compliance after deformation lasting 1 h at 100°C for pure PP and blends with 10% and 20% PLC is larger than the compliance at 20°C by factors of 4.4, 5.1 and 5.3, respectively. Thus, addition of the PLC has enhanced the effect of temperature on the creep behavior. On the other hand, addition of the PLC leads to a *decrease* in creep at equal temperature levels.

We have discussed above the applicability of the time–temperature correspondence principle to pure PET/0.6PHB. Now of course the question of the applicability of the principle to PLC-containing blends

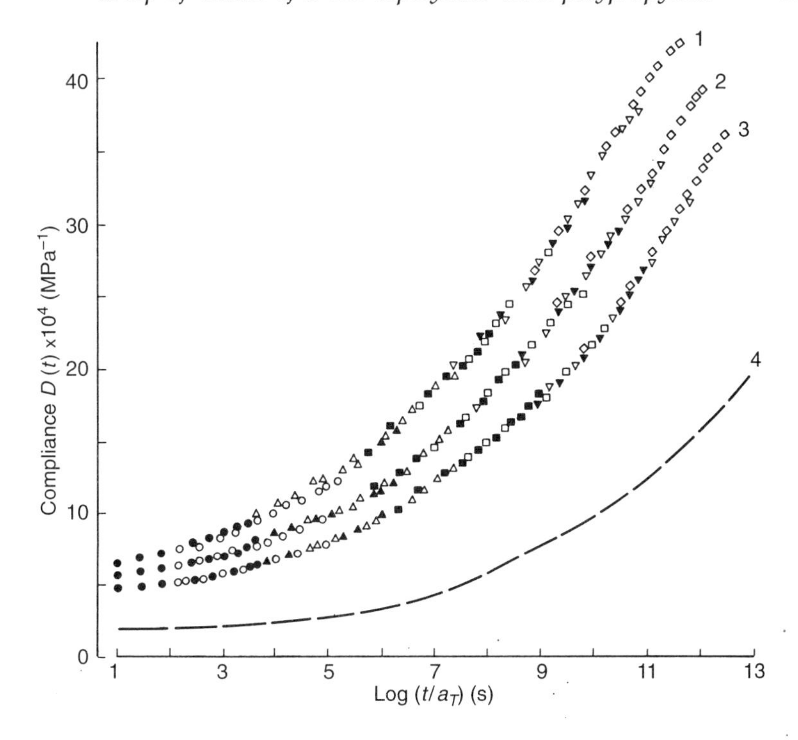

Figure 12.10 Creep compliance master curves for different materials: pure PP (1); a blend of 90% PP + 10% PLC (2); a blend of 80% PP + 20% PLC (3); pure PLC (cf. Figure 12.3) (4). The symbols are the same as in Figure 12.9. (After [50].)

arises. Figure 12.10 presents the respective master curves for 0, 10 and 20 wt% PLC. The reference temperature is 20°C. It can be seen from the figure that master curves for PP and PP + PLC blends can be constructed very satisfactorily using only a horizontal shift factor. For comparison the master curve for pure PLC, already discussed in section 12.2, is shown also. In Figure 12.11 we display $\log a_T$ plotted against temperature for pure PP and the blends. As can be seen, $\log a_T$ for pure PP is a monotonic function of T in the range from 20°C to 100°C. As the content of PLC increases, the curve of $\log a_T$ shows a noticeable change in slope. It is known [39] that for the PET-rich phase in PET/0.6PHB the values of T_g obtained by different methods are within the temperature range from 60 to 65°C. This glass transition is apparently reflected in the curves 2 and 3 in Figure 12.11.

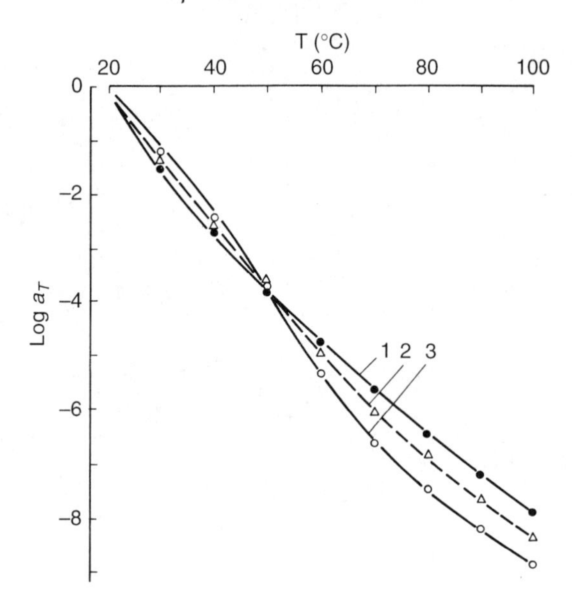

Figure 12.11 Temperature–time shift factors for PP (1) and the blends 90% PP + 10% PLC (2) and 80% PP + 20% PLC (3). (After [50].)

12.3.3 Long-term creep behavior

Long-term tests were carried out [50] for 10 000 h (14 months) at 20 ± 1°C in a room adapted specially for this purpose. No other experiments were performed in that room. The contents of PLC in the PP + PLC blends were 0, 5, 10, 15 and 20 wt%. In all tests the stress level was 5 MPa.

Averaged experimental creep compliance data are represented by dots in Figure 12.12 as a function of log time. The tests show that even *after 14 months the creep continued* and deformations kept increasing. After tests of 10 000 h the compliance of the materials examined exceeds the instantaneous elastic compliance by a factor varying from 3.8 (for pure PP) to 3.2 (for the 80% PP + 20% PLC blend). This shows that the inelastic strain of these materials manifests itself substantially; methods of accelerated testing and prediction have to be subjected to very detailed scrutiny.

Since we are now concerned with properties of blends, we show in Figure 12.13 the dependence of the creep compliance on the PLC concentration [50]. As expected, the relatively rigid LC component (here PHB) in the PLC lowers the compliance of PP.

In the present case it is possible to verify directly the applicability

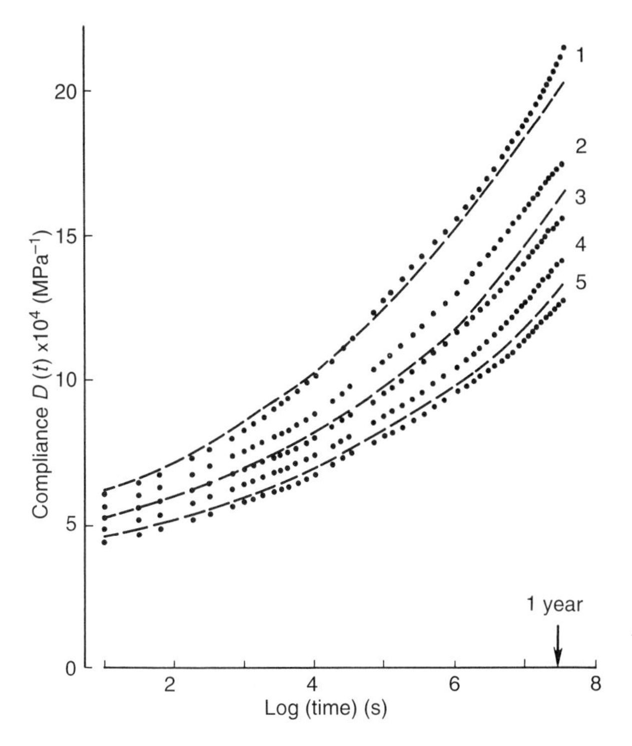

Figure 12.12 Long term (14 months) creep compliance curves (dots) for PP (1) and the blends 95% PP + 5% PLC (2), 90% PP + 10% PLC (3), 85% PP + 15% PLC (4) and 80% PP + 20% PLC (5). Dashed lines = master curves from short term testing at different temperatures for PP and the blends 90% PP + 10% PLC and 80% PP + 20% PLC (cf. Figure 12.10). (After [50].)

of the time–temperature correspondence principle. We have at our disposal long-term control tests with duration four orders of magnitude longer than short-term tests at several temperature levels. For this purpose the master curves corresponding to the temperature 20°C must be compared with curves from long-term tests. This comparison is made in Figure 12.12 where the master curves are shown as dashed lines. A comparative analysis of the statistical scattering of experimental points and errors of the master curves reveals that the predicted results are generally within the confidence intervals of the averaged experimental data. On the whole, the obtained accuracy of predicting the long-term creep of the examined materials may be considered as acceptable for engineering applications.

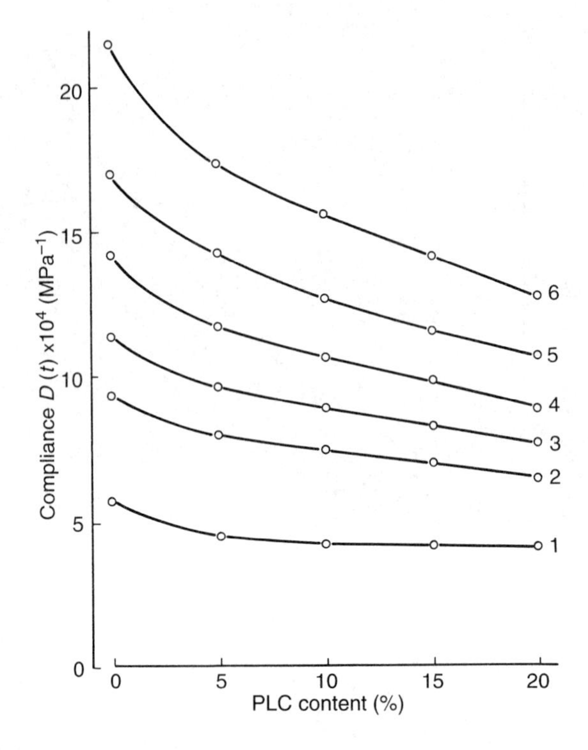

Figure 12.13 Isochronous creep compliance curves as a function of PLC content in the blends PP + PLC for different values of time: 0 (1); 1 h (2); 10 h (3); 100 h (4); 1000 h (5) and 10000 h (6). (After [50].)

12.4 STRESS RELAXATION AT DIFFERENT TEMPERATURES

12.4.1 Liquid crystalline copolyester

Stress relaxation of PET/0.6PHB was also studied [51]. Plates were prepared by casting and the specimens were cut from them along the major melt flow direction. The testing was carried out at ten temperature levels in the range from 20°C to 120°C in the region of the linear viscoelasticity. In all cases the strain was $\varepsilon_0 = \text{const.} = 0.5\%$. The stress relaxation curves were measured for 1 h. A first value of stress was measured 10 s after loading. To minimize scattering, three to five curves were measured on different samples at each temperature level. The results are shown in Figure 12.14, presented as $\log E$ vs. $\log t$.

In Figure 12.15 we show [51] the stress relaxation master curve at the reference temperature 20°C, presented as σ vs. $\log (t/a_T)$, and the corresponding temperature shift factor $\log a_T$ vs. T curve. For comparison, in Figure 12.16 we show also the values of $\log a_T$ from creep data

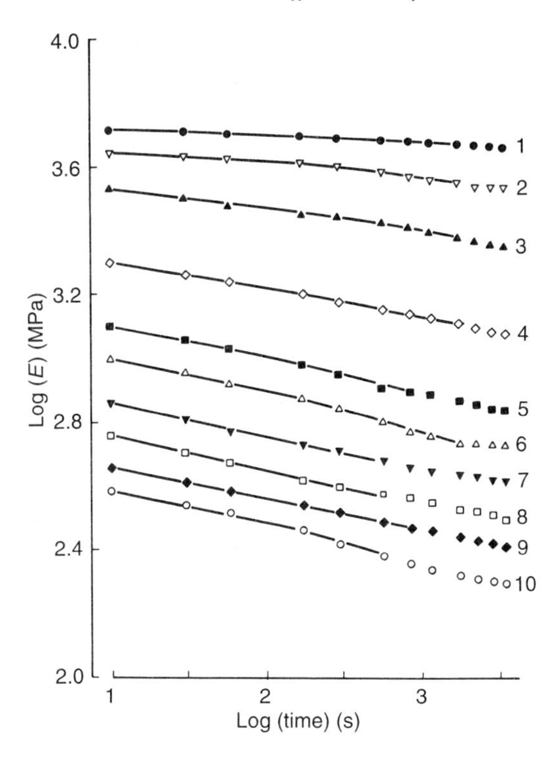

Figure 12.14 Relaxation modulus of PET/0.6PHB versus log t for different temperature levels: 20°C (1), 40° (2), 50° (3), 60° (4), 70° (5), 80° (6), 90° (7), 100° (8), 110° (9); 120°C (10). (After [51].)

discussed in section 12.2 and reported in [36] along with $\log a_T$ data from stress relaxation data. The two sets agree with each other perfectly below $T_{g\,PET}$ but only approximately above $T_{g\,PET}$.

Figure 12.17 shows a master stress relaxation curve, presented as $(\sigma - \sigma_i)/(\sigma_0 - \sigma_i)$ vs. log time. The value of σ_i is determined for $t \rightarrow \infty$ by an approximation to the experimental stress relaxation master curve. This type of plot has been recommended by one of us [52–61]. It brings out the common features of stress relaxation curves for metals, polymers and other materials. This is true not only for experimental but also for computer generated stress relaxation curves [60–62]. For a discussion of these common features see also [62]. The type of plot recommended first in [52] is also being used successfully for instance by Wortmann and coworkers [63–66]. They investigated a variety of materials including wool fibers and also very stiff aramid fibers such as Kevlar.

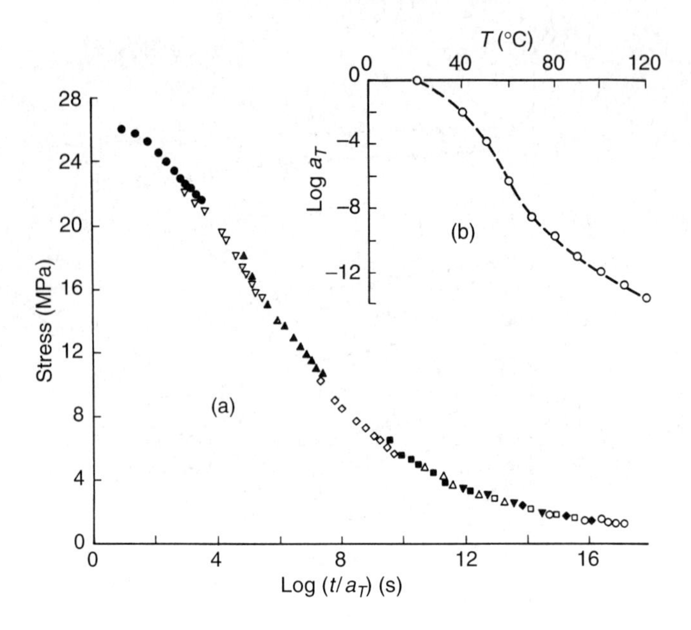

Figure 12.15 Stress relaxation master curve for PET/0.6PHB at the reference temperature 20°C (a) and temperature dependence of shift factor a_T (b). The symbols in the master curve are the same as in Figure 12.14. (After [51].)

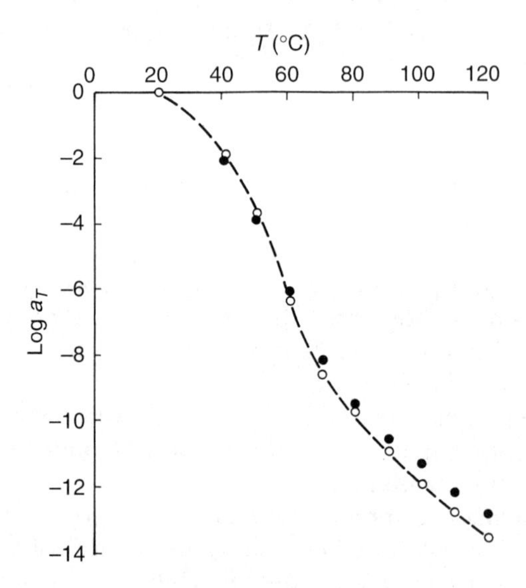

Figure 12.16 Temperature dependence of shift factor log a_T determined from experimental creep (●) and stress relaxation (○) results. (After [51].)

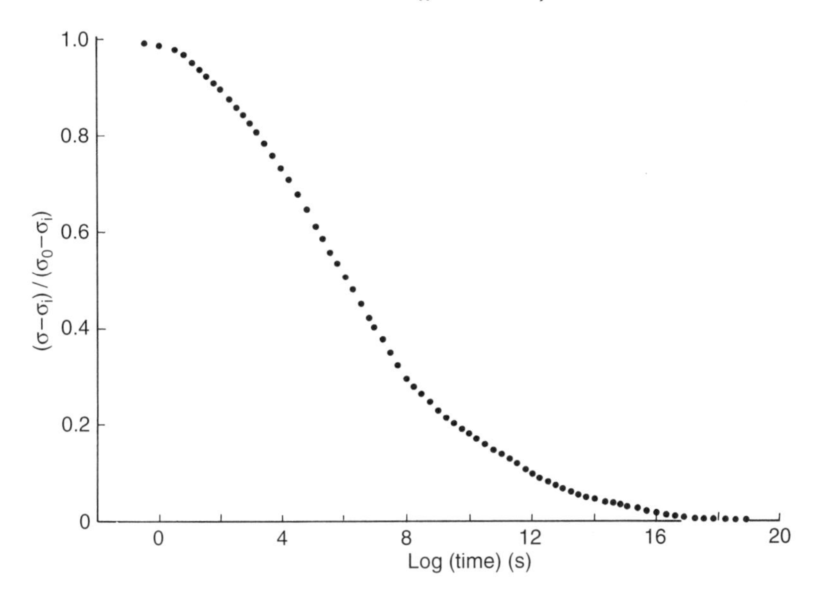

Figure 12.17 Stress relaxation curve for PET/0.6PHB, given as $(\sigma - \sigma_i)/(\sigma_0 - \sigma_i)$ vs. $\log t$. (After [51].)

12.4.2 Blends of a liquid crystalline copolyester with polypropylene

The results discussed in this section have been reported in reference 67. Stress relaxation was determined at nine temperature levels in the range from 20°C to 100°C in the region of the linear viscoelasticity. The contents of PLC in the blends were 0, 10 and 20 wt%. In all cases the strain was $\varepsilon_0 = \text{const.} = 0.5\%$. The stress relaxation curves were measured for 1 h, the first value of stress being measured 10 s after loading.

Stress relaxation curves, which represent averages of at least three tests, for 80% PP + 20% PLC blends are shown in Figure 12.18 as E vs. $\log t$. Figure 12.19 shows [63] the master curves at the reference temperature 20°C as σ vs. $\log(t/a_T)$, and the temperature shift factors $\log a_T$ vs. T for the same blend of 80% PP + 20% PLC. The curves for pure PP and for the blend with 10% PLC are similar.

The similarity is also preserved when we plot the master curves as $(\sigma - \sigma_i)/(\sigma_0 - \sigma_i)$ vs. log time (Figure 12.20). The initial stress σ_0 (0) is obtained as $\sigma_0 = E\varepsilon_0$, where $\varepsilon_0 = 0.5\%$. The values of the elastic modulus of these materials are known from [28]; the values of σ_i are determined for $t \to \infty$. We see that the PLC addition to PP increases the effective stress $\sigma - \sigma_i$.

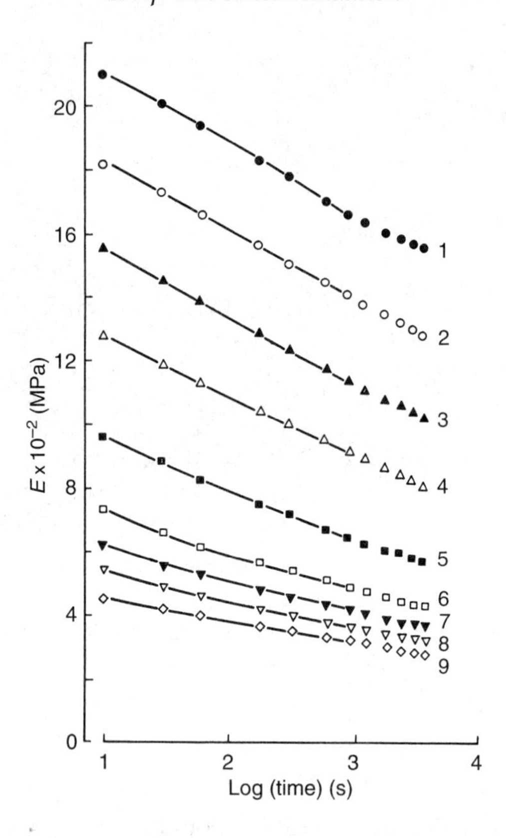

Figure 12.18 Relaxation modulus for 80% PP + 20% PLC presented as E vs. $\log t$ for different temperature levels: 20°C (1), 30° (2), 40° (3), 50° (4), 60° (5), 70° (6), 80° (7), 90° (8) and 100°C (9). (After [67].)

12.5 FINAL REMARKS

The liquid crystalline copolyester PET/0.6PHB for which data are available reveals a marked linear viscoelastic creep even at room temperature. As could be expected, the temperature effect on the creep is quite significant. The creep compliance after deformation lasting 1 h at a temperature of 160°C is 65 times larger than that at room temperature. It is particularly interesting to note that, in spite of the complex (multiphase) nature of the PLC-containing materials [8, 13], the time–temperature correspondence principle is applicable over a wide range of temperatures. The so-called master curves can be constructed very satisfactorily using only a horizontal shift factor. This is true also for the blends. Thus, for these materials the short-term creep tests at different temperatures may be considered as accelerated tests for long-term creep prediction.

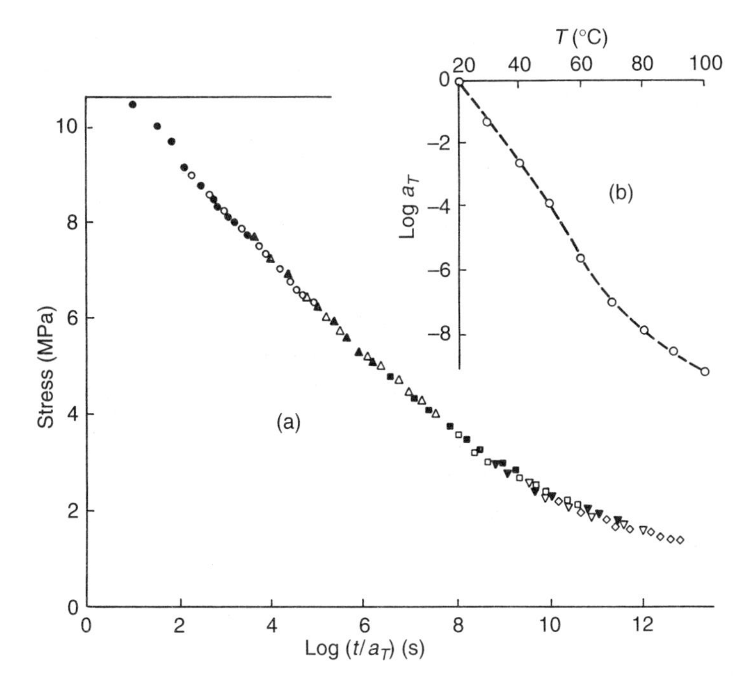

Figure 12.19 Stress relaxation master curve for the blend 80% PP + 20% PLC at reference temperature 20°C (a) and the temperature dependence of the shift factor log a_T (b). The symbols are the same as in Figure 12.18. (After [67].)

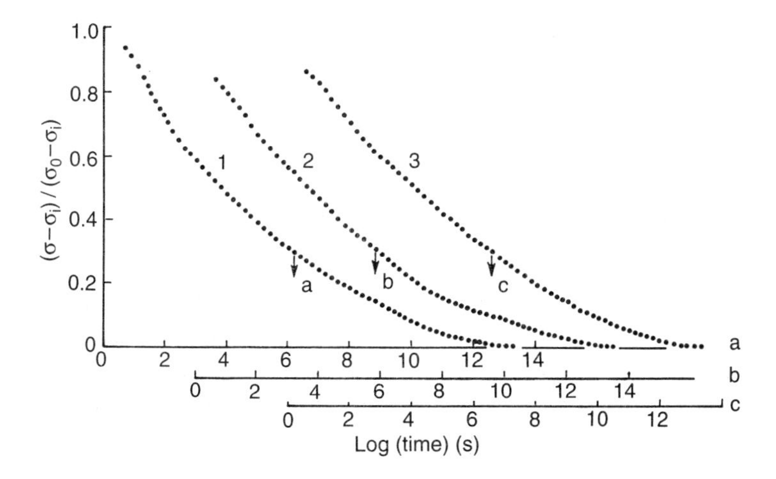

Figure 12.20 Stress relaxation curves, given as $(\sigma - \sigma_i)/(\sigma_0 - \sigma_i)$ versus $\log t$, for pure PP (1) and the blends 90% PP + 10% PLC (2) and 80% PP + 20% PLC (3).

The theory of linear viscoelasticity may be used for analytical representation of PET/0.6PHB creep behavior at relatively small stress levels only. The effect of nonlinearity increases gradually as the stress level increases. For a stress level of 0.5 of the short-term strength, the nonlinearity becomes significant. Increasing the stress, like increasing the temperature, accelerates the relaxation processes. Nonlinear creep data can be related by simple stress–time superposition. The master curve obtained by use of stress–time superposition in the nonlinear region is in a good agreement with that obtained by temperature–time superposition in the linear region.

The addition of the PLC to PP affects not only the strength, elastic modulus, strain at yield and strain at break, but also creep behavior. The creep compliance after deformation for 1 h for pure PP is 1.4 times larger at 20°C and 1.2 times more at 100°C than the respective values for the 80% PP + 20% PLC blend.

The effect of temperature on the creep of pure PP as well as blends containing the PLC is considerable. The creep compliance after deformation for 1 h at 100°C for pure PP and blends with 10% and 20% PLC is larger than the compliance at 20°C by factors of 4.4, 5.1 and 5.3, respectively.

The long-term tests demonstrate that the effect of time on the deformation of materials investigated is quite large. Let us again include actual data. After tests lasting 10000 h (up to 14 months) the creep compliance exceeds the instantaneous elastic compliance by a factor varying from 3.8 (for pure PP) to 3.2 (for the 80% PP + 20% PLC blend). The inelastic strain of these materials manifests itself substantially.

The effect of PLC content on the long-term creep is significant as well. After tests lasting 14 months the compliance of blends containing 5, 10, 15 and 20% PLC is respectively 1.22, 1.37, 1.52 and 1.70 times less than the compliance of pure PP. The creep compliance master curves obtained from short-term tests are in good agreement with the long-term control tests. Thus, for the examined materials the short-term tests at different temperatures may be considered as accelerated tests for long-term creep prediction.

We conclude that the addition of the PLC to PP strengthens the material in a predictable and quantifiable way. Surveying the above creep as well as stress relaxation, we find that the time–temperature correspondence principle is applicable here in spite of the multiphase character of the PLC. Hence the predictive capabilities, noted above for various specific instances, are quite extensive. It remains to be seen whether other PLC-containing blends behave similarly, although there is no *a priori* reason why they should behave differently.

ACKNOWLEDGEMENTS

This work has been supported by the North Atlantic Treaty Organization, Brussels, the Royal Swedish Academy of Sciences, Stockholm, and the Latvian Council of Science, Riga.

REFERENCES

1. Blumstein, A. (ed.) (1985) *Polymeric Liquid Crystals*, Plenum Press, New York.
2. Platé, N.A. (ed.) (1988) *Liquid Crystalline Polymers*, Chemistry, Moscow (in Russian).
3. Collyer, A.A. (ed.) (1992) *Liquid Crystal Polymers: From Structures to Applications*, Elsevier, London; Vol. 1 in this series.
4. Jackson, W.J. and Kuhfuss, H.F. (1976) *J. Polymer Sci. Chem.*, **14**, 2043.
5. Brostow, W. (1990) *Polymer*, **31**, 979.
6. Shibaev, V.P. and Beliaev, S.V. (1990) *Vysokomol. Soyed. A*, **32**, 2266.
7. Jakobson, E.E., Faitelson, L.A. and Kulichikhin, V.G. (1991) *Mech. Compos. Mater.*, **27**, 514.
8. Brostow, W. and Hess, M. (1992) *Mater. Res. Soc. Symp.*, **255**, 57.
9. Zülle, B., Demarmels, A., Plummer, C.J.G. *et al.* (1993) *Polymer*, **34**, 3628.
10. Plummer, C.J.G., Wu, Y., Davies, P. *et al.* (1993) *J. Appl. Polymer Sci.*, **48**, 731.
11. Plummer, C.J.G., Zülle, B., Demarmels, A. *et al.* (1993) *J. Appl. Polymer Sci.*, **48**, 751.
12. Buijs, J.A.H.M. and Vroege, G.J. (1993) *Polymer*, **34**, 4692.
13. Brostow, W., Hess, M. and Lopez, B.L. (1994) *Macromolecules*, **27**, 2262.
14. Kudryavtseva, S.E. and Kovriga, V.V. (1994) *Mech. Compos. Mater.*, **30**, 435.
15. Zhuang, P., Kyn, T. and White, J.L. (1988) *Polymer Eng. & Sci.*, **28**, 1095.
16. Nobile, M.R., Amendola, E., Nicolais, L. *et al.* (1989) *Polymer Eng. & Sci.*, **29**, 244.
17. Lin, Q., Jho, J. and Yee, A.F. (1993) *Polymer Eng. & Sci.*, **33**, 789.
18. Berry, D., Kenig, S. and Siegmann, A. (1993) *Polymer Eng. & Sci.*, **33**, 1548.
19. Lin, Q. and Yee, A.F. (1994) *Polymer Compos.*, **15**, 156.
20. Lin, Q. and Yee, A.F. (1994) *Polymer*, **35**, 3463.
21. Engberg, K., Strömberg, O., Martinsson, J. *et al.* (1994) *Polymer Eng. & Sci.*, **34**, 1336.
22. Turek, D.E., Simon, G.P. and Tiu, C. (1995) *Polymer Eng. & Sci.*, **35**, 52.
23. Heino, M.T. and Seppälä, J.V. (1993) *J. Appl. Polymer Sci.*, **48**, 1677.
24. Datta, A., Chen, H.H. and Baird, D.G. (1993) *Polymer*, **34**, 759.
25. Qin, Y., Brydon, D.L., Mather, R.R. *et al.* (1993) *Polymer*, **34**, 1196.
26. Qin, Y., Brydon, D.L., Mather, R.R. *et al.* (1993) *Polymer*, **34**, 1202.
27. Qin, Y., Brydon, D.L., Mather, R.R. *et al.* (1993) *Polymer*, **34**, 3597.
28. Maksimov, R.D. and Sterzynski, T. (1994) *Mech. Compos. Mater.*, **30**, 442.
29. Bretas, R.E.S. and Baird, D.G. (1992) *Polymer*, **33**, 5233.
30. De Carvalho, B. and Bretas, R.E.S. (1995) *J. Appl. Polymer Sci.*, **55**, 233.
31. Lee, S., Hong, S.M., Seo, Y. *et al.* (1994) *Polymer*, **35**, 519.
32. Ogata, N., Tanaka, T., Ogihara, T. *et al.* (1993) *J. Appl. Polymer Sci.*, **48**, 383.
33. Kulichikhin, V.G. and Platé, N.A. (1991) *Vysokomol. Soyed.* A, **33**, 3.
34. Engberg, K., Ekblad, M., Werner, P.-E. *et al.* (1994) *Polymer Eng. & Sci.*, **34**, 1346.
35. Jang, S.H. and Kim, B.S. (1994) *Polymer Eng. & Sci.*, **34**, 847.
36. Brostow, W., D'Souza, N.A., Kubát, J. and Maksimov, R.D., in preparation.

37. Ferry, J.D. (1980) *Viscoelastic Properties of Polymers*, Wiley, New York.
38. Brostow, W. and Corneliussen, R.D. (eds) (1986, 1989, 1992) *Failure of Plastics*, Hanser, Munich–Vienna–New York.
39. Brostow, W. and Samatowicz, D. (1993) *Polymer Eng. & Sci.*, **33**, 581.
40. Brostow, W. (1985) *Mater. Chem. & Phys.*, **13**, 47.
41. Hartmann, B. (1983) *Proc. Can. High Polymer Forum*, **22**, 20.
42. Hartmann, B. and Haque, M.A. (1985) *J. Appl. Phys.*, **58**, 2831.
43. Hartmann, B. and Haque, M.A. (1985) *J. Appl. Polymer Sci.*, **30**, 1553.
44. Doolittle, A.K. (1951) *J. Appl. Phys.*, **22**, 1741.
45. Catsiff, E., Alfrey, T. and O'Shaugnessy, M. (1953) *Textile Res. J.*, **23**, 808.
46. Gruntfest, I.J., Young, E.M. and Kooch, W. (1957) *J. Appl. Phys.*, **28**, 1106.
47. Urzhumtsev, Y.S. and Maksimov, R.D. (1968) *Mech. Polymerov*, **4**, 379.
48. Cessna, L.C. (1971) *Polymer Eng. & Sci.*, **11**, 211.
49. Bhuvanesh, Y.C. and Gupta, V.B. (1994) *Polymer*, **35**, 2226.
50. Brostow, W., D'Souza, N.A. and Maksimov, R.D., in preparation.
51. Brostow, W., D'Souza, N.A., Kubát, J. and Maksimov, R.D., in preparation.
52. Kubát, J. (1965) *Nature*, **204**, 378.
53. Kubát, J. and Rigdahl, M. (1976) *Mater. Sci. & Eng.*, **24**, 223.
54. Bohlin, L. and Kubát, J. (1976) *J. Solid State Commun.*, **20**, 211.
55. Högfors, Ch., Kubát, J. and Rigdahl, M. (1981) *Phys. Status Solidi B*, **107**, 147.
56. Kubát, J., Nilsson, L.-Å. and Rychwalski, W. (1982) *Res Mechanica*, **5**, 309.
57. Kubát, J. (1982) *Phys. Status Solidi B*, **111**, 599.
58. Kubát, J. and Rigdahl, M. (1986) in *Failure of Plastics* (eds W. Brostow and R.D. Corneliussen), Hanser, Munich–Vienna–New York, Ch. 4, p. 60.
59. Brostow, W., Kubát, J. and Kubát, M.J. (1994) *Mater. Res. Soc. Symp.*, **321**, 99.
60. Brostow, W. and Kubát, J. (1993) *Phys. Rev. B*, **47**, 7659.
61. Blonski, S., Brostow, W. and Kubát, J. (1994) *Phys. Rev. B*, **49**, 6494.
62. Brostow, W., Kubát, J. and Kubát, M.J. (1995) *Mech. Compos. Mater.*, **31**, 59.
63. Wortmann, F.-J. and de Jong, S. (1985) *Textile Res. J.*, **55**, 750.
64. Wortmann, F.-J. (1985) *Proc. 7th Internat. Wool Textile Res. Conf.*, Tokyo, **1**, 303.
65. Schulz, K., Wortmann, F.-J. and Höcker, H. (1989) *Chemiefasern/Textilindustrie*, **39/91**, T7.
66. Wortmann, F.-J. and Schulz, K. (1991) *Makromol. Chem. Symp.*, **50**, 55.
67. Brostow, W., D'Souza, N.A., Maksimov, R.D. and Sterzynski, T., in preparation.

13

Thermoreversible gelation of rigid rod-like and semirigid polymers

Andreas Greiner and Willie E. Rochefort

13.1 INTRODUCTION

Nestled in the region between the physical states defined as fluid and solid, is that ubiquitous substance called a 'gel'. It is fairly widely accepted that a gel can be defined topologically as three-dimensional network of connected strands swollen by a solvent. A satisfactory 'working' definition of the *physical gel* state was given many years ago by Ferry [1] in which he stated that 'A gel is a substantially diluted system which exhibits no steady state flow'. A further complication of the gel state is that this three-dimensional network can be formed by a number of systems (small molecules, aggregates, biopolymers or synthetic polymers) and in several different ways. Most polymeric gels are broken down into two categories based on how their network strands are connected: chemical or physical gels.

In *'chemical' gels* the network connection (crosslink) is usually a covalent bond, which leads to a thermally 'irreversible' gel. When the crosslinking is purely physical in nature, a 'physical' gel is formed which is thermally reversible. There have been at least two excellent works published in the last few years, *Reversible Polymeric Gels and Related Systems* by Russo [2] and *Thermoreversible Gelation of Polymers and*

Mechanical and Thermophysical Properties of Polymer Liquid Crystals
Edited by W. Brostow
Published in 1998 by Chapman & Hall, London.
ISBN 0 412 60900 2

Biopolymers by Guenet [3], which have been primarily concerned with thermoreversible gels in flexible polymer systems. The present review will focus on thermoreversible physical gels formed from rigid rod-like and semirigid polymers.

As indicated above, reversible gels can be formed by all types of polymers, and are not restricted to crystallizable polymers, polymers with strong hydrogen bonds, or rigid and semirigid polymer liquid crystals (PLC). The only restriction which applies for reversible gelation is that the polymers must be soluble and swellable. Due to the exclusive nature of the gelation process, it is very often difficult to define and identify a reversible gel. Its properties depends on a number of factors: the solvent, structure of the polymer and its molecular weight; polymer concentration in the system; and the thermal history of the sample. Therefore it is not only difficult to determine general structure–property relationships for reversible gels, but it is often non-trivial to identify whether the system in question is truly a gel or, with respect to the working definition given by Ferry above, just a highly viscous liquid with an extremely long relaxation time.

Thermoreversible gels can be described as elastic liquids [4] (the micro-Brownian motion of a polymer in a gel is still possible but the movement of polymer chains past each other is impossible) which can be transformed to viscous liquids by heating, and which will again for a gel upon cooling. Consequently a thermoreversible gel can be best identified by observation of the sol–gel (or gel–sol) transition. This observation can be accomplished by a number of methods which have been nicely outlined by Russo [2]. There are two general classifications of techniques which we will call optical, in which some type of visual observation or scattering of light is used as the tool, and mechanical, in which some method of applying a non-destructive stress and observing the response (or applying a deformation and measuring the stress) is employed. There is merit to each of these two general techniques, but in the spirit of the definition of a gel proposed above, the preference of these authors is to the mechanical response of the supposed gel. Of the mechanical techniques available, the most reliable, non-destructive method is the use of small strain dynamic oscillatory shear testing. This can be done either by imposing a small deformation (strain) and measuring the resulting stress or imposing a small stress and measuring the resulting deformation (strain). In either case, the term 'small' indicates that the intrinsic gel structure is not perturbed, i.e. that the deformation process is completely reversible and non-destructive.

The question of the structure of reversible gels and the mechanism for their formation is not trivial, since it cannot be generalized. Russo [2] has proposed a general classification of 'fishnet gels' for flexible polymer systems and 'lattice gels' for semirigid and rigid polymer systems, which though qualitative, provides a convenient physical picture for the difference in the nature of the gels formed by the two different 'classes' of polymers. Within the flexible polymer systems, the main differentiation is between the non-crystallizable (i.e. atactic polystyrene and polyvinylchloride) and crystallizable (polyethylene) polymers and whether their crosslink points are simple interchain junctions or microcrystalline domains [5]. In rod-like polymer systems, the situation is even more complicated because such issues as the relationship between the phase separation and gelation, the kinetic mechanism of gelation (nucleation and growth vs. spinodal decomposition) and the nature of the crosslinks, or even if the crosslinks exist, are all unresolved [2, 5]. However, regardless of the nature of the interchain interaction, the polymer spanning from branchpoint to branchpoint is not a random coil, i.e. the system cannot be thought of as a traditional Gaussian network. One fact that is known for these rigid polymer systems is that gelation occurs by lowering the solvent quality, most readily by reducing temperature, in even very dilute solutions (below the overlap concentration).

Within all the 'qualifiers' outlined above to determining the nature of the gel, or whether the system is even a gel at all, it was decided that the most logical approach is to discuss the gelation in rod-like polymer systems individually. In this way, the unique gelation characteristics of a particular polymer can be outlined in the context of the general framework we have just discussed. This is manageable because there have actually been relatively few rigid rod polymer + solvent systems studied to date, particularly in any quantitative fashion. In fact, a large majority of the work until recently had been concentrated on the polypeptides, and in particular on poly(γ-benzyl-α, L-glutamate) in various good and less good solvents.

Another qualifying issue which led us to adopt the scheme mentioned above for the discussion of thermoreversible gelation in rigid rod-like polymers is that, as was pointed out by Russo [2], the backgrounds and goals of individuals currently working in the gel field are so diverse, ranging from mathematicians to chemists and chemical engineers, that an attempt to unify this review in terms of structure–property relationships for the various gel systems would surely lead to confusion and misconceptions. In separating the systems, the uniqueness and character of the individual polymers can be emphasized, while still trying to incorporate more global, unifying concepts. The authors also want to draw the attention to amorphous semirigid polymers which form anisotropic, thermoreversible gels from semidilute solutions, which is difficult to understand. Efforts should be made to understand the gelation mechanism on a molecular state.

13.2 POLY(AMINO ACID)S

Synthetic poly(amino acid)s or polypeptides if you prefer, were the first synthetic polymers reported to form a lyotropic (solvent induced) liquid crystalline phase when dispersed in their α-helical conformation in certain solvents [6, 7]. Now they are recognized as just one of a group of rigid backbone polymers with sufficient local stiffness to form lyotropic liquid crystalline solutions. Many of the others, such as the polyaramides and the poly(1, 4-phenylenebenzbisthiazoles (PBT) and -oxazoles (PBO)), can be fabricated into high modulus fibers by wet spinning from a lyotropic solution. However, they are typically soluble only in highly protonated solvents (i.e. strong acids) in which the polymer is charged and often aggregated. On the other hand, the helical poly(amino acid)s may be molecularly dispersed in the uncharged form in a number of solvents, primarily due to the entropy of mixing of their side chains with the solvent [8]. While the presence of the side chain precludes their use for the formation of ultrahigh strength fibers, it makes them ideal candidates for testing the theoretical predictions for rod-like polymers proposed some time ago by Flory [9].

The gel state of rod-like polymers did not receive much attention until the late 1970s and even that was a result of the experimentation directed at testing the predictions of a theoretical Flory phase diagram for rod-like polymers [9]. As indicated above, the solvent plays an important role in both the phase behavior and gelation of rod-like polymers. Very early on Doty *et al.* [10] studied the viscosity–concentration behavior of the poly(amino acid) poly(γ-benzyl-α, L-glutamate) (PBLG; its structure is shown in Figure 13.1) in dilute solution. They studied PBLG in DMF or choroform/dimethyl formamide (DMF) mixtures, solvents with low hydrogen bonding potential which would promote hydrogen bonding stabilizing the α-helix, and dichloroacetic acid, which is itself a hydrogen bonding solvent. They found the following Mark–Houwink–Sakurada relationships:

$$[\eta] = KM^{1.7}\,(\text{DMF})$$

$$[\eta] = KM^{0.87}\,(\text{dichloroacetic acid})$$

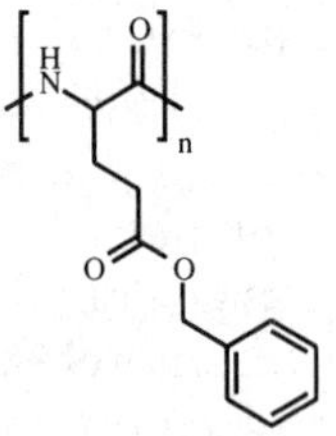

Figure 13.1 Structure of PBLG.

where η is intrinsic viscosity, K is the Mark–Houwink coefficient and M is the molecular weight. The indication from this is that the PBLG behaves essentially as a rigid rod in the solvent DMF, while it possesses a more flexible conformation in dichloroacetic acid. Thus the choice of a low hydrogen bonding potential solvent is key for maintaining PBLG in a rigid rod-like conformation.

Physical gelation of PBLG has to be discussed in connection with association since the associates can be regarded as precursors of physical gels. Early studies on the solution properties of PBLG were limited by the association of PBLG in various solvents. Katchalski noted in his review of association of polypeptide molecules in solvents like benzene and benzylalcohol [11]. Doty *et al.* pointed out that PBLG undergoes molecular association in solvents with very low or zero hydrogen bonding potential such as chloroform, dioxane and benzene [10]. End-to-end association was suggested based on the relationship of the specific viscosity and PBLG molecular weight in solvents promoting association of PBLG. In DMF association of PBLG at low concentration increases with the molecular weight of PBLG and decreases with rising temperature [12]. However at 25°C low molecular weight and high molecular weight PBLG associates in DMF.

An increase of the concentration of PBLG in various solvents leads to a liquid crystalline behavior [6, 7, 13–19] which can result in reversible gelation by lowering of the temperature or by addition of a non-solvent. However thermoreversible gelation is not necessarily related to a liquid crystalline phase. Much of the pioneering work systems of PBLG + solvent in a wide concentration regime was conducted by Miller and coworkers [20–26], with later studies by Russo and coworkers [27–29] and Donald and coworkers [30–33]. Russo [2] and more recently Guenet [3] have published rather extensive review works on reversible polymeric gels which focus primarily on the better studied flexible polymer and biopolymer systems. However, there are sections in each of those works which deal with thermoreversible PBLG gels.

PBLG has been reported to form complex phases with DMF [22, 24, 25, 34–36] which are thermoreversible gels, a process which is obviously accelerated by the presence of water [37]. Thermoreversible gelation of PBLG in benzylalcohol is well described in the literature [30–33, 38, 39]. Isotropic gels as well as anisotropic gels can be formed by system PBLG + benzylalcohol. They are formed from the isotropic state or anisotropic state by temperature decrease. The crucial question is the mechanism of the gelation process and the origin of the anisotropy in the gel state. Phase separation is a possible explanation for gelation. The late stages of phase separation have been documented by video microscopy revealing microscopic phase separation upon cooling of homogeneous mixtures of PBLG + DMF + water (2%) [29]. Formation of crystalline structures has been suggested for the system of PBLG + benzylalcohol. Two crystalline forms have been observed and interpreted as crystal solvate complexes [30, 36]. The gel–sol transition has been assigned to the melting of small crystallites serving as gel junction points [30]. Electron microscopy of vitrified gel samples of the system PBLG + benzylalcohol (from semidilute solutions) visualized a microfibrillar network with microfibrils of about 10 nm [40]. Gels with PBLG at low concentrations (<0.1 wt%) visualized by electron microscopy also show a microfibrillar structure [25]. The nature of the gel junction points of this isotropic dilute systems is not clear.

13.3 POLYDIACETYLENE

Polydiacetylenes are conjugated polymers which can be in a planar or non-planar conformation depending on temperature. The planar–non-planar transition can be regarded as a rod–coil transition. The rod phase of polydiacetylene (4-butoxycarbonyl-methylurethane) (PDA) 4BCMU in dilute solution has been identified as the *trans* isomer and

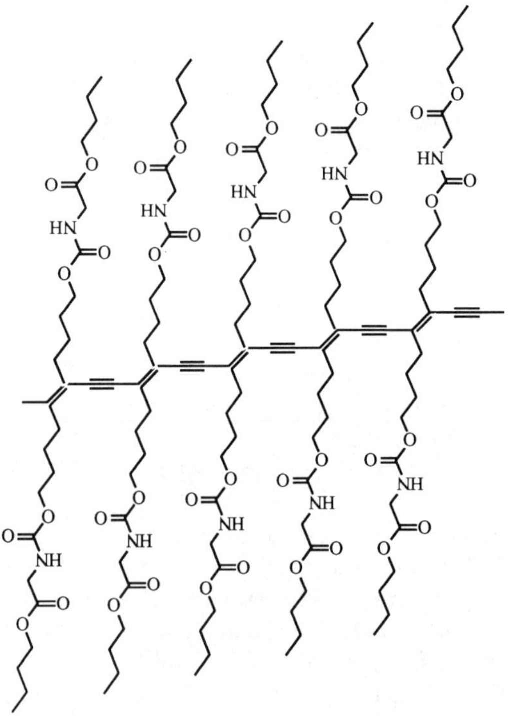

Figure 13.2 Planar structure of PDA 4BCMU.

the coil phase as the *cis* isomer [41]. The planar conformation is stabilized by adjacent side groups via hydrogen bonds [42]. The rod–coil transition is accompanied by a gel–sol transition; in other words an extended chain conformation is essential for gel formation. Gels of (PDA) 4BCMU (Figure 13.2) were prepared by dissolution of (PDA) 4BCMU in toluene at 80°C followed by cooling [42].

The gel–sol transition temperature of (PDA) 4BCMU (~70°C) is independent of polymer concentration. The lowest concentration in toluene where gel formation was observed is 0.06 wt%. Gels of (PDA) 4BCM exhibit optical birefringence which results from a spontaneous nematic alignment in the gel supported by hydrogen bonding [43].

13.4 POLYIMIDE

Thermoreversible gelation was reported for the rigid rod polyimide based on 3,6-diphenylpyromellitic dianhydride (DPPMDA) and 2,2′-(trifluoromethyl)-4,4′-diaminobiphenyl (PFMB) (Figure 13.3) [44]. A solution containing 17 wt% DPPMDA-PFMB was obtained by polymerization of DPPMDA with PFMB in *m*-cresol at high temperatures. The mixture formed a gel upon cooling below 110°C. Mixtures with different concentrations were obtained by dilution with *m*-cresol. Endothermic transitions were observed by differential scanning calorimetry DSC between 70 and 110°C upon heating of the gels depending on the concentration of (1–12 wt%) DPPMDA-PFMB in *m*-cresol. This transition was attributed to an order–disorder transition which was confirmed by polarizing microscopy. Birefringent textures of the gels were observed by polarizing microscopy but disappeared at the temperatures corresponding to the endothermal transition observed in the DSC traces. The gel–sol transition occurred simultaneously with the order–disorder transition as confirmed by the ball-drop experiment. Gel formation was attributed to liquid–liquid phase separation. The

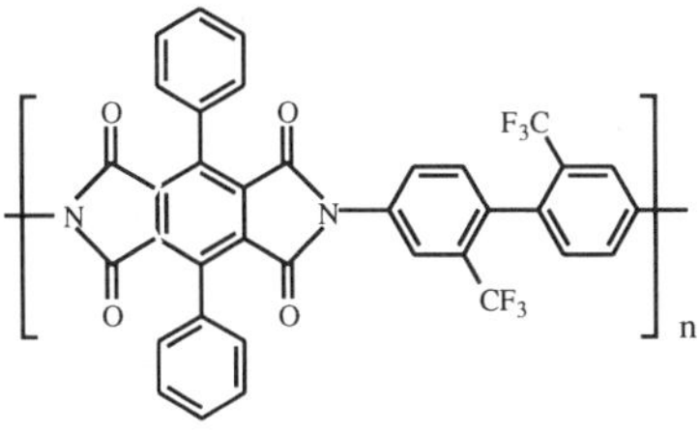

Figure 13.3 Structure of the polyimide DPPMDA-PFMB.

formation of ordered structures was attributed to the local concentration and orientation of the polymer and to the polymer mobility in the gel state.

Formation of thermoreversible anisotropic gels in *m*-cresol was also reported for the segmented rigid rod polyimide BPDA-PFMB synthesized from biphenyltetracarboxylic dianhydride and (perfluoromethyl benzidine) shown in Figure 13.4 [45, 46]. Experiments were performed with mixtures containing 12 wt% polyimide or less. Both the gel–sol and the order–disorder transitions are a function of concentration, temperature and molecular weight. With increasing molecular weight the critical concentration for gelation decreased. With decreasing molecular weight the transition temperatures and heats of transition decreased but increased with concentration. Nematic-like textures were observed in the gel state which disappeared above the transition temperature. The gelation time decreased with increasing concentration but was not linear in the low concentration region. As for the polyimide DPPMDA-PFMB, gel formation has been attributed to a liquid–liquid phase separation forming a solute-rich and a solute poor phase. The anisotropy was attributed to nematic alignment in the solute-rich phase [45].

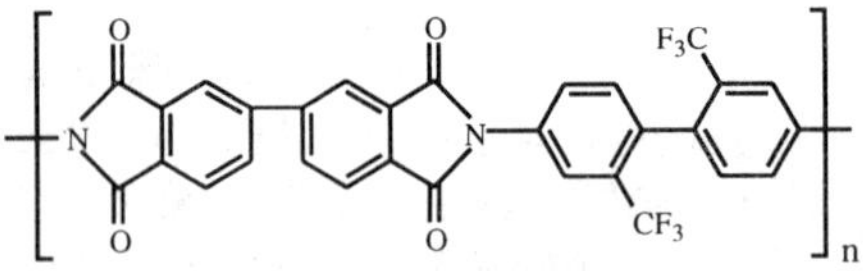

Figure 13.4 Structure of the polyimide BPDA-PFMB.

13.5 POLY(*P*-PHENYLENEBENZOBISTHIAZOLE)

Hot (100°C) 97% sulfuric acid is a good solvent for poly(*p*-phenylenebenzobisthiazole) (PBT) (Figure 13.5). A drastical drop in the solvent power is achieved by a decrease in temperature which causes thermoreversible gelation of the system [47]. The formation of a porous structure from a uniform, homogeneous, isotropic solution has been observed. Indications were found that the gel melting temperature increases with concentration and molecular weight of PBT.

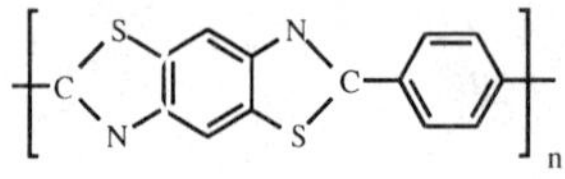

Figure 13.5 Structure of PBT.

13.6 AROMATIC LC POLYESTERS

Para-linked aromatic LC polyesters are semirigid polymers which are typically thermotropic. Although it has only rarely been mentioned in the literature, thermoreversible gelation is a general observation for LC polyesters. An initial study on the thermoreversible gelation has been performed on the polyester shown in Figure 13.6 [48].

PE 1 was obtained with high molecular weight by melt polycondensation of phenylterephthaloyl dichloride and 2,2′-dimethylbiphenyl-4,4′-diol following a published procedure [50]. Molecular weight can be controlled by monomer ratio or by fractionation. PE 1A is a high molecular weight fraction ($M_w = 75000$) of PE 1 with $M_w = 46000$. PE 1B–1E are non-fractionated samples.

PE 1 is fairly soluble in a variety of organic solvents such as chloroform and 1,1,2,2-tetrachloroethane. The Mark–Houwink parameters of PE 1 in chloroform at 20°C are [49]

$$\eta = 4.93 \times 10^{-4} M_w^{0.877}$$

A persistence length of 125 Å was calculated for PE 1 in chloroform at 20°C based on the model of Yamakawa and Fujii [51]. The indication from this is that PE 1 behaves essentially as a semirigid polymer in the solvent chloroform. The glass transition temperature of PE 1 is at 145°C (determined by DSC) and it forms a nematic melt between 180 and 200°C (Figure 13.9a) depending on the molecular weight. Isotropization has not been observed below the range of decomposition (5% weight loss at 450°C, by thermogravimetrical analysis with a heating rate of 20°C min^{-1}) [52]. Crystallinity has never been observed in PE 1, not even with oriented fibers of PE 1 [53], which might be important for the investigation of the gelation mechanism. Thermoreversible gelation of PE 1 has been observed in a variety of solvents (Table 13.1) [54]. In most of these solvents the formation of anisotropic gels depends on the concentration of the polyester, the temperature and on the thermal history of the gels.

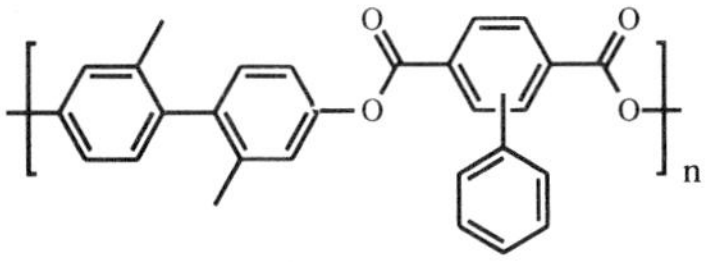

PE 1

PE	M_w
1A	75000
1B	14000
1C	30000
1D	40000
1E	80000

Figure 13.6 Structure and M_w of polyester PE 1 used for gel studies. M_w was determined by static small angle light scattering [49].

Table 13.1 Gel–sol transitions, anisotropy and isotropization temperatures of mixtures of solvents with 20 wt% PE 1D and comparison with solubility parameters

Solvent	Gel–sol transition (°C)[a]	Anisotropy at 20°C[a]	Isotropization temperature (°C)[a]	δ ($J^{0.5}/cm^{1.5}$)[b]
Diphenylether	43	+	51	18.03
3-Phenoxytoluene	96	+	107	18.76
4-Bromanisol	55	+	61	19.60
4-Chlorophenol	87	+	89	24.96
2-Bromo-*p*-xylene	75	+	86	19.28
2-*t*-Butylphenol	36	–	–	21.06

[a] Determined by polarizing microscopy, heating rate 10°C min^{-1}.
[b] δ-values were calculated by the increment system given in [55].

It is not trivial to prepare homogeneous mixtures of polymers and a solvent, especially in a high concentration regime due to the high viscosities. Gels of PE 1A with concentrations of 18 wt% and above were prepared by weighing the desired amount of the components in a DSC pan, followed by several heating and cooling cycles until the glass transition of PE 1A disappeared. Gels with PE 1B–1E were prepared differently in order to provide larger quantities of samples with a higher concentration. The solvent, e.g. 3-phenoxytoluene and the polyester were dissolved in the desired ratio in a low boiling solvent such as chlorofom. Homogeneous samples were obtained after evaporation of the low boiling solvent. The concentration of PE 1 was analyzed by thermogravimetrical analysis. Following this method, 10 g samples of homogeneous gels with $\geqslant$30 wt% PE 1D in different solvents were prepared.

Thermoreversible gels of PE 1 are characterized by a gel–sol transition which is accompanied with a loss of anisotropy at lower concentrations of PE 1 (Table 13.1). At higher concentrations anisotropy is also observed above the gel–sol transition. The gel–sol transitions were observed by optical microscopy, DSC (Figure 13.7), and by dynamic oscillatory shear flow (Figure 13.8). Thermal analysis of the gels was limited by solvent evaporation accompanied by changes in concentration. However, it was possible to run DSC experiments between 20 and 200°C in sealed pans with e.g. 3-phenoxytoluene as solvent. The gel–sol transitions are attributed to endotherms in the DSC traces of the gels, which was verified by dynamic oscillatory shear flow experiments as shown in Figure 13.8. The gel–sol transition is

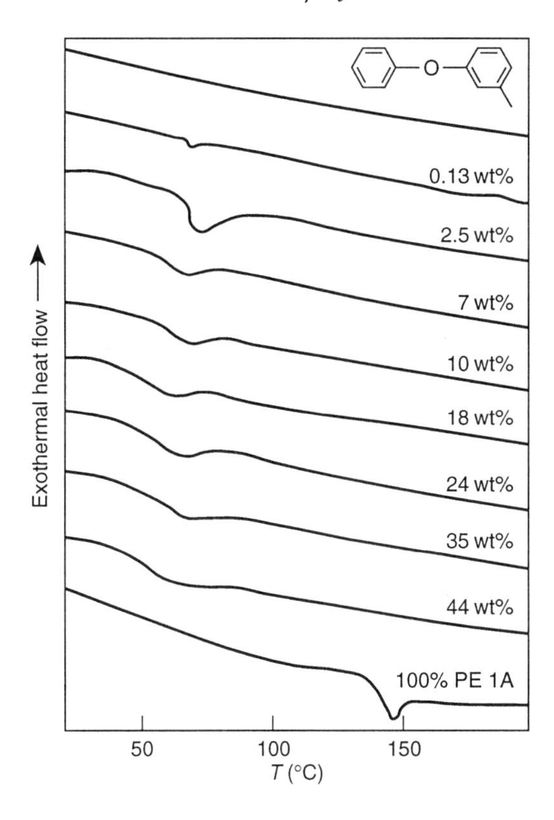

Figure 13.7 DSC curves of mixtures of 3-phenoxytoluene with different concentrations of PE 1A (heating rate $20°C\,min^{-1}$).

accompanied by a dramatic decrease of G' and G''. Formation of isotropic gels was observed with concentrations of PE 1A as low as 0.13 wt% in 3-phenoxytoluene.

PE 1A in bulk is birefringent between crossed polarizers in an optical microscope at 220°C (Figure 13.9a). Gels formed with less than 15 wt% PE 1A in 3-phenoxytoluene exhibit a black and white texture (Figure 13.9b), while those with 15 wt% PE 1A in 3-phenoxytoluene are colored (Figure 13.9c). Their birefringence disappears above the gel–sol transition. In contrast the birefringence is maintained above the gel–sol transition with 35 wt% or more of PE 1A in 3-phenoxytoluene (Figure 13.9d). The observed changes are reversible upon cooling.

The comparison of mixtures with 20 wt% PE 1B and different solvents shows considerable differences in gel–sol transitions and isotropization temperatures. The lowest gel–sol transition temperature was detected with 2-*t*-butylphenol at 36°C and the highest with

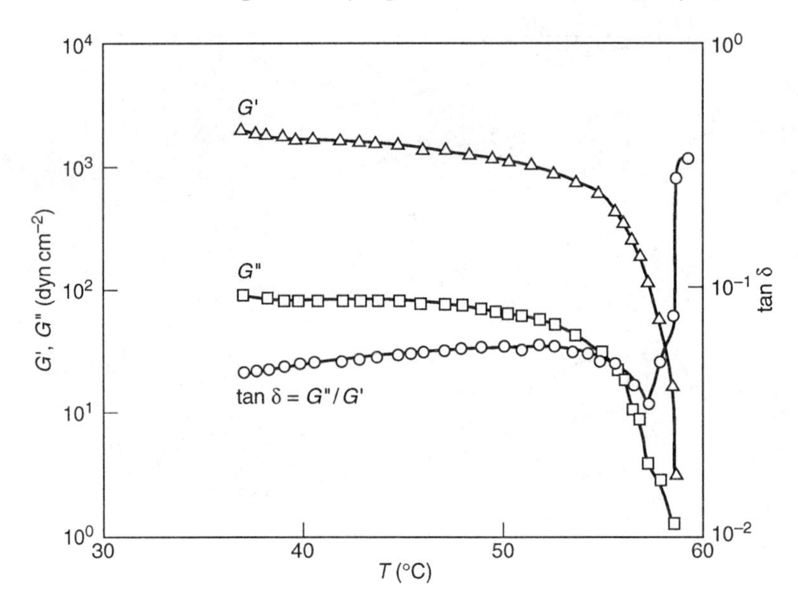

Figure 13.8 Dynamic oscillatory shear moduli as a function of temperature T for a 0.13 wt% mixture of PE 1A in 3-phenoxytoluene. The frequency was 1.0 rad s^{-1} and the strain was 25%.

3-phenoxytoluene at 96°C. Surprisingly, the gel–sol transitions of 3-phenoxytoluene is 53°C higher than the gel–sol transition of the corresponding mixture with diphenylether. A correlation of gel–sol transitions or isotropization temperatures with solubility parameters was not observed.

The gel–sol transitions are of course also sensitive to the molecular weight of the polyester. A plateau region of the gel melting point of gels with 20 wt% of polyesters of different molecular weight in 3-phenoxytoluene is reached at $M_w \sim 40000$ (Figure 13.10).

The mechanism of gel formation as well as the appearance of anisotropy of the non-crystallizable PE 1 is not clear. The anisotropy of polyesters with low amounts of PE 1 (<10 wt%) in the mixture is especially surprising. It could be explained by phase separation forming a polyester-poor phase and a polyester-rich phase and ordering of the polyester in the polyester-rich phase, but the question arises concerning the driving force of the phase separation in the case of the amorphous polyester. Another simple explanation for the anisotropic appearance of the gels which cannot be excluded completely is stress birefringence induced by a volume change upon cooling.

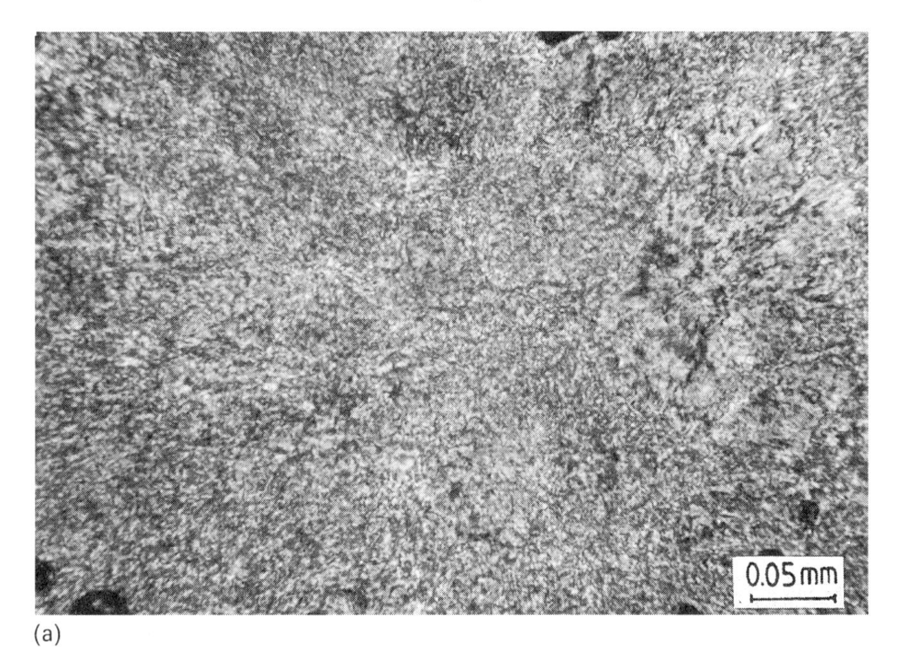

(a)

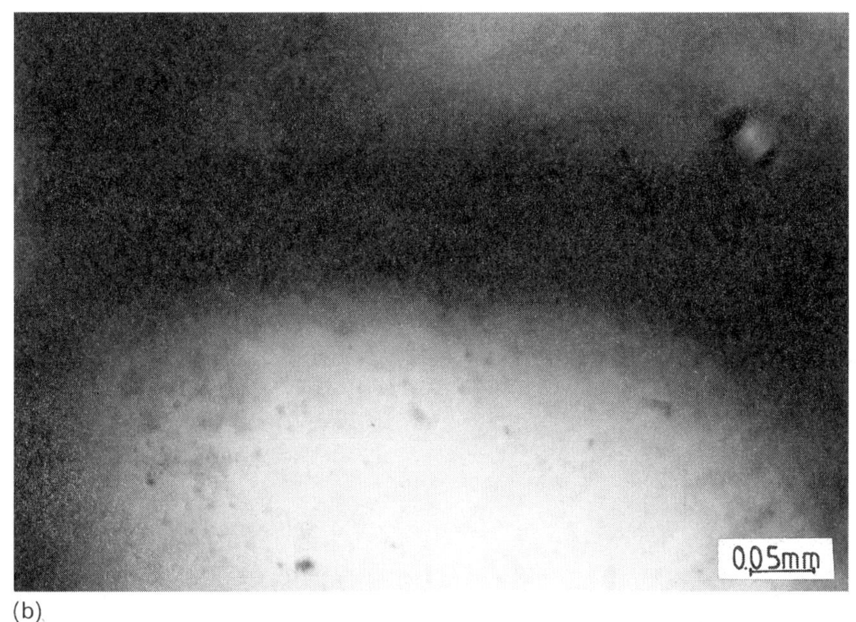

(b)

Figure 13.9 Photographs taken with polarizing microscope of (a) bulk PE 1A at 220°C, (b) a mixture of 12 wt% PE 1A in 3-phenoxytoluene, (c) a mixture of 15 wt% of PE 1A in 3-phenoxytoluene and (d) a mixture of 43 wt% of PE 1A in 3-phenoxytoluene at 25°C.

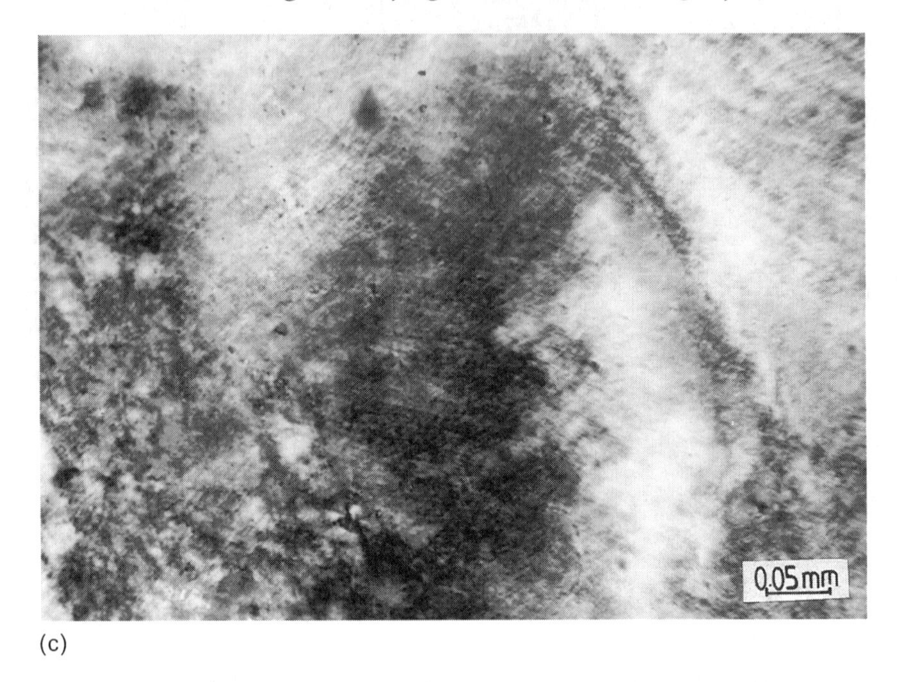

(c)

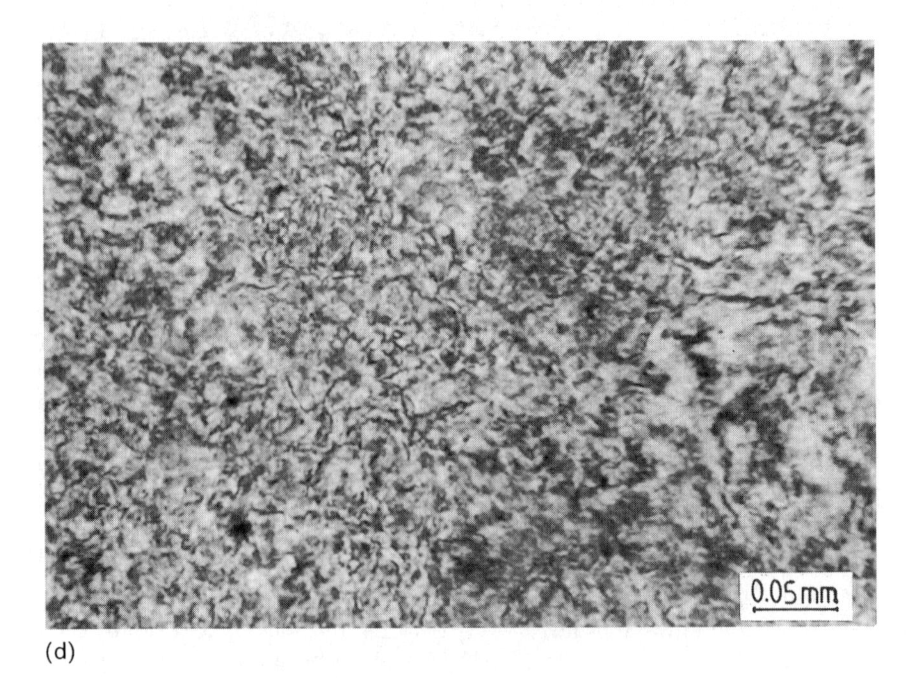

(d)

Figure 13.9 (Continued).

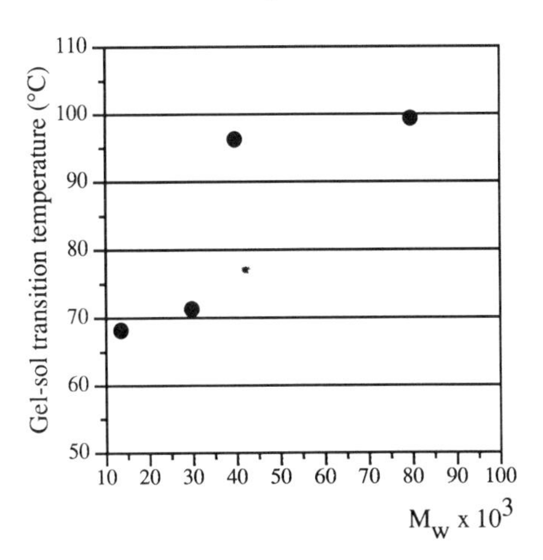

Figure 13.10 Gel–sol transitions of gels composed of 3-phenoxytoluene and 20 wt% of PEs of different molecular weights.

14.7 CONCLUDING REMARKS

This review of thermoreversible gelation of rigid (or semirigid) polymer + solvent systems presented here is probably not a complete description of systems reported in literature but covers a major part of it. Certainly, there are better known rigid or semirigid polymers which have not been investigated with respect to thermoreversible gelation, which marks one of the difficulties in this area. The correlation of polymer structure and thermoreversible gelation is a crucial issue in order to describe, understand and predict the gel properties and gelation mechanisms.

From the data available it can be concluded that crystallization of the bulk polymer is not a necessary prerequisite for thermoreversible gelation. The driving force for gelation by lowering of the temperature even at low concentrations may be phase separation of the polymer–solvent system, but this should be also accompanied by a significant change of the solvent quality. This could be related to a drastic change of the persistence length with variation in temperature. The correlation of the change of persistence length of the polymer and the sol–gel transition could be of interest for understanding the gelation mechanism, in particular for amorphous rigid or semirigid polymers.

Are just simple entanglements responsible for the thermoreversible gelation of rigid or semirigid polymers which becomes predominant at a certain stage of the solvent quality or is it really the formation or ordered structures which drives the formation of thermoreversible gelation? The answer to these questions requires more data on solution properties of the polymers involved here and the examination of more polymer structures.

REFERENCES

1. Ferry, J.D. (1961) *Viscoelastic Properties of Polymers*, Wiley, New York, p. 391.
2. Russo, P.S. (1987) in *Reversible Polymeric Gels and Related Systems* (ed. P.S. Russo) *ACS Symp. Ser.*, **350**, American Chemical Society, Washington, p. 1.
3. Guenet, J.-M. (1992) *Thermoreversible Gelation of Polymers and Biopolymers*, Academic Press, New York.
4. Rehage, G. (1975) *Prog. Colloid. & Polymer Sci.*, **57**, 7.
5. Miller, W.G., Chakrabarti, S. and Seibel, K.M. (1985) in *Microdomains in Polymer Solutions*, (ed. P.L. Dubin), Plenum, New York.
6. Robinson, C. (1956) *Trans. Faraday Soc.*, **52**, 571.
7. Robinson, C., Ward, J.C. and Beevers, R.B. (1958) *Discuss. Faraday Soc.*, **25**, 29.
8. Rai, J.H. and Miller, W.G. (1972) *Macromolecules*, **6**, 257.
9. Flory, P.J. (1956) *Proc. Royal Soc. London*, **A234**, 73.
10. Doty, P., Bradbury, J.H. and Holtzler, A.M. (1956) *J. Am. Chem. Soc.*, **78**, 947.
11. Katchalski, E. (1951) *Adv. Prot. Chem.*, **VI**, 123.
12. Gerber, J. and Elias, H.-G. (1968) *Makromol. Chem.*, **112**, 142.
13. Duke, R.W., DuPré, D.B. (1974) *Macromolecules*, **7**, 374.
14. DuPré, D.B. and Duke, R.W. (1975) *J. Chem. Phys.*, **63**, 143.
15. Murthy, N.S., Knox, J.R. and Samulski, E.T. (1976) *J. Chem. Phys.*, **65**, 4835.
16. Duke, R.W., DuPré, D.B. and Samulski, E.T. (1977) *J. Chem. Phys.*, **66**, 2748.
17. Watanabe, J., Imai, K. and Uematsu, I. (1978) *Polymer Bull.*, **1**, 67.
18. Patel, D. and DuPré, D.B. (1979) *Mol. Cryst. Liq. Cryst.*, **53**, 323.
19. Toriumi, H., Minakuchi, S., Uematsu, I. (1980) *Polymer J.*, **12**, 431.
20. Geobel, K.D. and Miller, W.G. (1970) *Macromolecules*, **3**, 64.
21. Wee, E.L. and Miller, W.G. (1971) *J. Phys. Chem.*, **75**, 1446.
22. Miller, W.G., Wu, C.C., Wee, E.L. *et al.* (1974) *Pure Appl. Chem.*, **38**, 37.
23. Miller, W.G., Rai, J.H. and Wee, E.L. (1974) *Liquid Crystals and Ordered Fluids*, Vol. 2 (eds J.F. Johnson and R.S. Porter), Plenum, New York, p. 243.
24. Miller, W.G., Kou, L., Tohyama, K. and Voltaggio, V. (1978) *J. Polymer Sci. Symp.*, **65**, 91.
25. Tohyama, K. and Miller, W.G. (1981) *Nature*, **289**, 813.
26. Russo, P.S. and Miller, W.G. (1983) *Macromolecules*, **16**, 1690.
27. Russo, P.S., Magestro, P. and Miller, W.G. (1987) in *Reversible Polymeric Gels and Related Systems* (ed. P.S. Russo), *ACS Symp. Ser.*, **350**, American Chemical Society, Washington, p. 152.
28. Russo, P.S., Chowdhury, A.H. and Mustafa, M. (1989) in *Mater. Res. Soc. Symp.*, **134**, (eds W.W. Adams, R.K. Eby and D.R. Maclemore), p. 207.
29. Chowdhury, A.H. and Russo, P.S. (1990) *J. Chem. Phys.*, **92**, 5744.
30. Hill, A. and Donald, A.M. (1988) *Polymer*, **29**, 1426.
31. Horton, J.C., Donald, A.M. and Hill, A. (1990) *Nature*, **346**, 44.
32. Horton, J.C. and Donald, A.M. (1993) *Polymer*, **32**, 2418.
33. Prystupa, D.A. and Donald, A.M. (1993) *Macromolecules*, **26**, 1947.
34. Luzatti, M.C., Spack, G. *et al.* (1962) *J. Mol. Biol.*, **3**, 566.
35. Parry, D.A.D. and Elliot, A. (1967) *J. Mol. Biol.*, **25**, 1.
36. Sasaki, S., Hikata, M., Shiraki, C. and Uematsu, I. (1982) *Polymer J.*, **14**, 205.

37. Russo, P.S. and Miller, W.G. (1984) *Macromolecules*, **17**, 1324.
38. Sasaki, S., Tokuma, K. and Uematsu, I. (1983) *Polymer Bull.*, **10**, 539.
39. Ginzburg, B., Siromyatnikova, T. and Frenkel, S. (1985) *Polymer Bull.*, **13**, 139.
40. Cohen, Y. (1996) *J. Polymer Sci.*, **34**, 57.
41. Lim, K.C. and Heeger, A.J. (1985) *J. Chem. Phys.*, **82**, 522.
42. Sinclair, M., Lim, K.C. and Heeger, A.J. (1983) *Phys. Rev. Lett.*, **51**, 1768.
43. Casalnuovo, S.A. and Heeger, A.J. (1984) *Phys. Rev. Lett.*, **53**, 2254.
44. Cheng, S.Z.D., Lee, S.K., Barley, J.S. *et al.* (1991) *Macromolecules*, **24**, 1883.
45. Lee, S.K., Cheng, S.Z.D., Wu, Z. *et al.* (1993) *Polymer Int.*, **30**, 115.
46. Kyu, T., Yang, J.-C., Cheng, S.Z.D. *et al.* (1994) *Macromolecules*, **27**, 1861.
47. Russo, P.S., Siripanyo, S., Saunders, M.J. and Karasz, F.E. (1986) *Macromolecules*, **19**, 2856.
48. Greiner, A., Rochefort, W.E., Greiner, K. *et al.* (1992) *Makromol. Chem., Rapid Commun.*, **13**, 25.
49. Schmitt, R., Bolle, B., Greiner, A. and Heitz, W. (1992) *Makromol. Chem. Symp.*, **61**, 297.
50. Schmidt, H.W. and Guo, D. (1988) *Makromol. Chem.*, **189**, 2029.
51. Yamakawa, H. and Fujii, M. (1974) *Macromolecules*, **7**, 128.
52. Greiner, A., Rochefort, W.E., Greiner, K. *et al.* (1991) in *Integration of Fundamental Polymer Science and Technology*, Vol. 5, (eds P. Lemstra and L.A. Kleintjens), Elsevier Applied Science, London–New York, p. 258.
53. Motamedi, F., Jonas, U., Greiner, A. and Schmidt, H.-W. (1993) *Liq. Cryst.*, **14**, 959.
54. Bolle, B. (1991) Diploma thesis, Marburg, Germany.
55. van Krevelen, D.W. (1990) *Properties of Polymers*, 3rd edn, Elsevier Sci. Publ. Amsterdam–Oxord–New York.

14

Elastic moduli of polymer liquid crystals

C.L. Choy

14.1 INTRODUCTION

The elastic modulus and acoustic absorption of a polymer are closely related to its molecular structure, so the study of these properties leads to an improved understanding of the polymer. Moreover, these parameters are useful in many practical applications. Flexible chain polymers processed by extrusion or injection molding are normally isotropic, so only two elastic constants are required to describe fully the low strain mechanical behavior. Unlike flexible chain polymers, a high degree of molecular order exists in the nematic melt of a thermotropic polymer liquid crystal (PLC). The rigid or semirigid molecular chains in localized domains have parallel orientation and are readily induced to align in the flow field during processing. Because of their rigid structure, the chains have a long relaxation time, so the high degree of orientation is retained in the solidified material. As a result, an extruded rod of PLC possesses *uniaxial orientation,* thus requiring five independent elastic constants for a full description of its elastic behavior. Similarly, an injection molded bar of PLC has orthorhombic symmetry and needs nine elastic constants for a complete characterization.

Since tensile testing machines are widely available, the elastic constant

Mechanical and Thermophysical Properties of Polymer Liquid Crystals
Edited by W. Brostow
Published in 1998 by Chapman & Hall, London.
ISBN 0 412 60900 2

most commonly determined is Young's modulus. This is usually obtained by stretching a standard tensile bar and measuring the strain along the length of the bar by an extensometer. In some cases the strain in the transverse direction is measured simultaneously, thereby giving the Poisson's ratio. Although other elastic constants may be obtained by quasi-static or dynamic mechanical tests [1, 2], some of the measurements (such as transverse shear modulus) have low accuracy [2]. Moreover, several methods of different precision have to be used in order to obtain the nine independent elastic constants of a sample with orthorhombic symmetry.

To circumvent the experimental difficulties associated with static and low frequency methods, ultrasonic techniques at frequencies ranging from 1 to 10 MHz have been developed [2–6]. It has been found [5] that the complete set of elastic moduli can be obtained even for a sample with dimensions as small as $3 \times 1 \times 1$ mm. Recently, these techniques have been applied to thermotropic main chain PLC [7–10], blends of a PLC and a thermoplastic [10, 11] and a short glass fiber reinforced PLC (C.L. Choy, Y.W. Wong and K.W.E. Lau, unpublished data), and the present review aims to discuss the results of these modulus measurements in terms of structural features, including orientation of PLC chains, aspect ratio of PLC domains and orientation of glass fibers. For side chain PLC, elastic moduli at hypersonic frequency (0.5–4 GHz) have also been determined by phonon spectroscopy [12] and Brillouin scattering [13–15], so we will compare the data at ultrasonic [16] and hypersonic [12–15] frequencies in order to gain some insight into the effect of relaxation.

The acoustic absorption can also be obtained from ultrasonic measurements. However, no work has been done on longitudinal or other main chain PLCs. There is only one study [17] of the ultrasonic absorption of a *comb* PLC, the result of which will be discussed briefly.

14.2 ULTRASONIC TECHNIQUES FOR DETERMINING ELASTIC CONSTANTS AND ACOUSTIC ABSORPTION

14.2.1 General consideration

The elastic behavior of a material is completely characterized by a set of independent stiffness constants C_{ij}. For an orthorhombic material with symmetry axes along the 1, 2 and 3 directions of a Cartesian coordinate system, there are nine independent stiffness constants: C_{11}, C_{22}, C_{33}, C_{12}, C_{13}, C_{23}, C_{44}, C_{55} and C_{66}. For an uniaxial material with the symmetry axis along the 3 direction, $C_{11} = C_{22}$, $C_{44} = C_{55}$, $C_{13} = C_{23}$ and $C_{12} = C_{11} - 2C_{66}$. Therefore, there are only five independent stiffness constants: C_{11}, C_{33}, C_{13}, C_{44} and C_{66}. Once C_{ij} are known, the elastic compliances S_{ij} can be obtained by inverting the stiffness matrix. Then the Young's modulus and Poisson's ratio can be calculated.

For an axially symmetric material such as an extruded PLC rod, the Young's modulus $E(\theta)$ at an angle θ relative to the extrusion direction is given by

$$E(\theta) = [S_{11} \sin^4\theta + (2S_{13} + S_{44}) \sin^2\theta \cos^2\theta + S_{33} \cos^4\theta]^{-1} \quad (14.1)$$

In particular, the axial (E_3) and transverse (E_1) Young's moduli are

$$E_3 = 1/S_{33} \quad (14.2)$$

$$E_1 = 1/S_{11} \quad (14.3)$$

The three Poisson's ratios are given by

$$\nu_{12} = -S_{12}/S_{11} \quad (14.4)$$

$$\nu_{13} = -S_{13}/S_{33} \quad (14.5)$$

$$\nu_{31} = -S_{13}/S_{11} \quad (14.6)$$

where the second subscript in the Poisson's ratios denotes the direction of the applied stress.

14.2.2 Immersion method

The immersion method is commonly employed for the measurements of the elastic constants of polymers [2, 4, 6]. We will briefly describe the experimental procedure and the method of data analysis used by Leung *et al.* [4]. The sample and the transmitting and receiving transducers were suspended from a platform with angle adjusting mechanisms, which in turn was supported by a frame standing at the bottom of a tank filled with silicone oil (Figure 14.1). A pulse generator energized the submerged piezoelectric ceramic transmitting transducer which generated a beam of 10 MHz ultrasonic waves in the silicone oil. The beam impinged on a sample at a set angle and generated both a longitudinal and transverse wave within the sample which were subsequently refracted back into the silicone oil and picked up by the receiving transducer. The received signal and the attenuated driving pulse were observed on an oscilloscope, and the transit time for the ultrasonic pulse to travel from one transducer to the other was measured by a gated time-interval counter with an accuracy of 1 ns.

Consider a sample with orthorhombic symmetry which is cut such that the width, thickness and length directions coincide with the symmetry axes 1, 2 and 3, respectively. When the sample is aligned

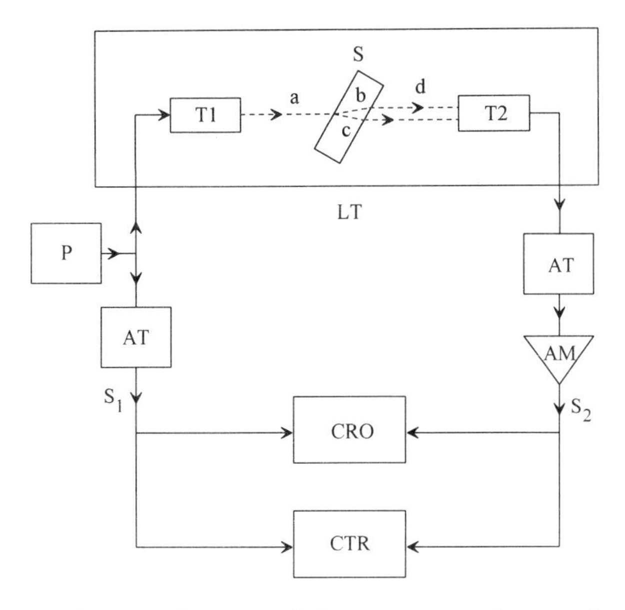

Figure 14.1 Schematic diagram of the experimental setup for ultrasonic measurements by the immersion method. P, pulser; T1, transmitting transducer; T2, receiving transducer; S, sample; a, incident ultrasonic beam; b, refracted beam of longitudinal wave; c, refracted beam of transverse wave; d, transmitted beams. Sample and transducers immersed in silicone oil within the tank (LT). AT, attenuator; AM, wide band amplifier; CRO, oscilloscope; CTR, time interval counter; S1, triggering signal for CRO and start signal for the counter; S2, signal viewed on CRO and stop signal for the counter. (Adapted from [4] by permission of Elsevier Science Ltd.)

so that the incident beam lies in the 2–3 plane (Figure 14.2), the velocities of the two waves (v_L for the quasi-longitudinal wave and v_T for the quasi-transverse wave) generated in the sample are dependent on the incident angle ψ and are related to the stiffness constants C_{22}, C_{33}, C_{23} and C_{44}. The two velocities satisfy the Musgrave equation [18], which can be written in the following form:

$$y^2 + a_1xy + a_2x^2 + a_3y + a_4x = 0 \tag{14.7}$$

where

$$y = \rho v_r^2 - C_{22} \tag{14.8}$$

$$x = \sin^2 \theta_r \tag{14.9}$$

ρ is the density of the sample, θ_r is the angle of refraction and $r = L$ or T. The coefficients a_n are combinations of the stiffnesses: $a_1 = C_{22} - C_{33}$,

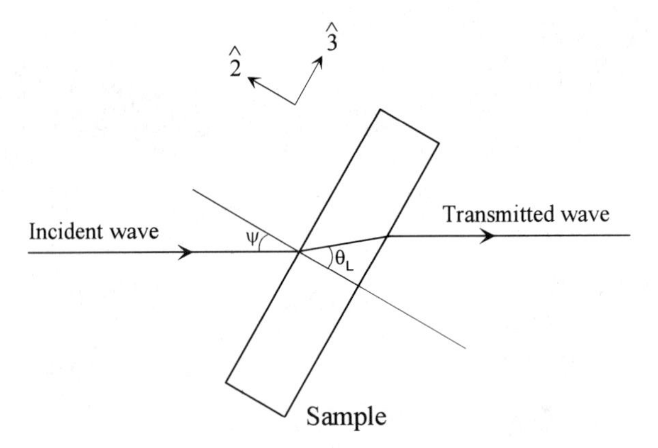

Figure 14.2 Schematic diagram showing incident and transmitted ultrasonic waves. For clarity, only the longitudinal wave in the sample is shown.

$a_2 = (C_{23} + C_{44})^2 - (C_{22} - C_{44})(C_{33} - C_{44})$ and $a_3 = C_{22} - C_{44}$; $a_4 = a_1 a_3 - a_2$ is not independent.

The velocity v_r and the angles of refraction θ_r can be deduced from transit time measurements and Snell's law. The transit time is measured with and without the sample in the path of the ultrasonic beam, and the difference in these transit times is

$$\tau_r = \frac{b}{\cos\theta_r}[v^{-1}\cos(\theta_r - \psi) - v_r^{-1}] \tag{14.10}$$

where b is the thickness of the sample and v is the velocity of sound in silicone oil. Combining this equation with Snell's law:

$$\frac{\sin\psi}{v} = \frac{\sin\theta_r}{v_r} \tag{14.11}$$

the two parameters v_r and θ_r can be calculated.

In principle, only four velocity measurements are required to obtain the four stiffness constants. However, to achieve higher accuracy we adopt the following procedure. First, the sample is set at $\psi = 0$. Only the longitudinal wave is generated in the sample and its velocity v_L is related to C_{22} by

$$C_{22} = \rho v_L^2 \tag{14.12}$$

Once C_{22} has been measured, the variables x and y can be computed from the measured v_r and θ_r at different angles of incidence according to equations (14.8) and (14.9). Then a_n ($n = 1-3$) and hence the stiffnesses C_{23}, C_{33} and C_{44} can be determined by a least squares fit of equation (14.7) to these data. At least ten data points for each of the longitudinal and transverse velocities are used to ensure sufficient accuracy. The overall accuracy of C_{ij} is estimated to be 3–7%, which reflects the uncertainty in the measurements of angle and time, as well as in the thickness of the samples.

Similar measurements for wave propagating in the 1–2 plane and 1–3 plane give the two sets of stiffness constants, (C_{11}, C_{22}, C_{12}, C_{66}) and (C_{11}, C_{33}, C_{13}, C_{55}), respectively. Therefore, the three series of velocity measurements give the nine independent stiffness constants, C_{11}, C_{22}, C_{33}, C_{12}, C_{13}, C_{23}, C_{44}, C_{55} and C_{66}, with C_{11}, C_{22} and C_{33} being determined twice.

We consider now the size requirement for samples in immersion measurements. The ultrasonic pulse typically used consists of only a few periods. At the high frequency of 10 MHz, measurements can be made on samples as thin as 0.7 mm without the problem of pulse overlap arising from multiple reflections in the sample. The ultrasonic attenuation in typical thermotropic PLC at room temperature is not very high, so the upper limit in thickness is about 6 mm. The use of a lower frequency would raise both the upper and lower limits.

A parallel beam of ultrasound may be propagated without divergence only if the beam diameter is much larger than the wavelength. At 10 MHz such a condition holds for a beam diameter of 6 mm. Therefore, a starting material in the form of a cube of side 12 mm is required for cutting the samples for wave velocity measurements in three orthogonal planes. Larger samples are necessary at lower frequencies.

The longitudinal wave absorption coefficient α_L of the sample is determined by comparison of the amplitude of a pulse transmitted through the sample at normal incidence with that of the received pulse when the sample is removed. Reflection losses at the sample surfaces require corrections or may be eliminated by using two samples of different thicknesses [2, 4].

14.2.3. Contact method

The contact method can be used to obtain the complete set of stiffness constants for samples with much smaller dimensions. In this method [3, 5], a piezoelectric ceramic transducer, bonded to one surface of the sample, generated a beam of pulsed 10 MHz elastic waves that was subsequently received by another transducer bonded to the opposite surface. The wave velocity was calculated from the transit time of the ultrasonic pulse measured on a gated time interval counter. Longitudinal and transverse waves were generated using two different types of transducers.

To obtain all the independent stiffnesses of a material with orthorhombic symmetry, we have to measure nine independent velocities v_{ab}, where

a and b refer to the direction of polarization and propagation of the elastic wave, respectively. The three extensional stiffnesses can be deduced from measurements of the longitudinal wave velocities along the three principal axes:

$$C_{11} = \rho v_{11}^2 \tag{14.13}$$

$$C_{22} = \rho v_{22}^2 \tag{14.14}$$

$$C_{33} = \rho v_{33}^2 \tag{14.15}$$

The shear stiffnesses are derived from the transverse wave velocities:

$$C_{44} = \rho v_{32}^2 \tag{14.16}$$

$$C_{55} = \rho v_{31}^2 \tag{14.17}$$

$$C_{66} = \rho v_{12}^2 \tag{14.18}$$

It is noted that $v_{ab} = v_{ba}$, so each shear stiffness can be derived from either v_{ab} or v_{ba}.

The cross-plane stiffnesses C_{12}, C_{13} and C_{23} can be deduced from the velocities of transverse waves propagating at 45° to the principal axes. For example, C_{23} is related to the velocity v_{pq} of the transverse wave propagating in the 2–3 plane and at 45° to the 3 axis:

$$4\rho v_{pq}^2 = [C_{22} + C_{33} + 2C_{44}] - [(C_{22} - C_{33})^2 + 4(C_{23} + C_{44})^2]^{1/2} \tag{14.19}$$

where $p = (1/2^{1/2})\,(0, 1, -1)$ and $q = (1/2^{1/2})\,(0, 1, 1)$. Similar expressions hold for C_{12} and C_{13}. The accuracy of C_{11}, C_{22}, C_{33}, C_{44}, C_{55} and C_{66} is 3%. The cross-plane stiffnesses have a larger error of 8% since the shear wave velocities are not sensitively dependent on these stiffnesses.

A sample with dimensions 3 × 1 × 1 mm is sufficient for obtaining the complete set of C_{ij} by the contact method. A larger dimension of 3 mm is used in the direction where the sample exhibits the highest extensional stiffness (such as the axis of an extruded PLC rod), so that the transit time is sufficiently long to ensure high accuracy.

14.3 ELASTIC MODULI OF THERMOTROPIC MAIN CHAIN POLYMER LIQUID CRYSTALS

To attain a deeper understanding of the elastic properties of thermotropic main chain PLCs, we briefly discuss the chemical composition and physical structure of some commercially available PLCs (Vectra A950, Vectra B950, Vectra B900 and HBA/IA/HQ) for which ultrasonic modulus data are available. Vectra A950 is a copolyester consisting of 73 mol% *p*-hydroxybenzoic acid (HBA) and 27 mol% 2,6-hydroxynaphthoic acid (HNA), and is mainly used for injection molding. Vectra B950 is an extrusion grade copolyesteramide comprising 60 mol% HNA, 20 mol% terephthalic acid (TA) and 20 mol% *p*-aminophenol (AP). Vectra B900 is believed to have a composition almost identical to that of Vectra B950. These three polymers are produced by Hoechst-Celanese Co. HBA/IA/HQ, produced by ICI, is a copolyester comprising 36 mol% HBA, 32 mol% isophthalic acid (IA) and 32 mol% hydroquinone (HQ). Though sounding very different chemically, all these PLCs consist basically of benzene or naphthalene rings linked by ester groups or ester and amide groups (Figure 14.3). We will henceforth refer to Vectra A950 and Vectra B950 by the abbreviated notations, Vectra A and Vectra B, respectively.

The structure of thermotropic liquid crystalline copolyesters and copolyesteramides has been studied by wide angle X-ray diffraction [19, 20], small angle X-ray scattering [21], electron diffraction [21, 22] and transmission [21–25] and scanning electron microscopy [25]. The processing of these PLCs in the elongational flow field during extrusion results in a highly oriented extended chain structure in the solid state. The observed aperiodicity of the meridional X-ray diffraction maxima

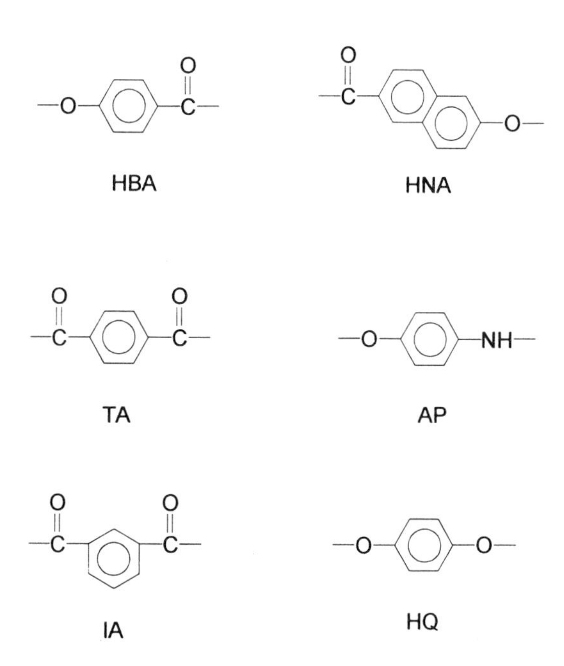

Figure 14.3 Chemical structure of the monomer units of some commercial thermotropic polymer liquid crystals.

implies that the chains consist of a random comonomer sequence. However, the presence of sharp equatorial and off-equatorial Bragg maxima is indicative of three-dimensional order akin to crystallinity. The typical crystallinity of the quenched samples is 20–30%, which remains unchanged after annealing up to 210°C. The crystalline lamellae have a thickness of about 10 nm and a lateral dimension of 100 nm. They are oriented with their thin axis parallel to the molecular chains and are periodically spaced at approximately 34 nm. Since the chains are long (*c.* 200 nm) and extended, there are many tie chains bridging the crystalline lamellae.

Structural features of a larger scale are also observed. Long fibrils, with lateral dimension of 0.3 – 1 μm, are found to lie along the extrusion direction. Closer inspection reveals that the fibrils are composed of finer microfibrils, with the tape-like microfibrils having a minimum width and thickness of 10 and 2 nm, respectively. Although the fibrils and microfibrils are, on average, aligned along the extrusion axis, they are not perfectly oriented on a local scale because they exhibit a worm-like trajectory. It has been conjectured that the microfibril is simply a replication of the molecular chain [25].

14.3.1 Extruded polymer liquid crystals

The elastic moduli of an extruded PLC are largely determined by its molecular orientation, so we first consider the orientation parameters calculated from the observed meridional X-ray reflections. For liquid crystalline copolyesters and copolyesteramides, the sharpness of these reflections gives an indication of the coherence length, a value corresponding to 10 monomers being typical. Therefore, the orientation parameter obtained from the meridional reflection is not specific to the crystalline regions but instead reflects the average orientation of all the chains in the polymer.

From the azimuthal intensity distribution of the strongest meridional reflection ($2\theta = 43°$ for Cu K_α radiation) we have calculated the orientation parameters P_2 and P_4, which are averages of the Legendre polynomials given by

$$P_2 = \langle \tfrac{1}{2}(3\cos^2\alpha - 1) \rangle \tag{14.20}$$

$$P_4 = \langle \tfrac{1}{8}(35\cos^4\alpha - 30\cos^2\alpha + 3) \rangle \tag{14.21}$$

where α is the angle between the axis of the diffracting unit and the draw direction, and $\langle\ \rangle$ denotes average over all units. The Hermans orientation parameter P_2 is commonly used for characterizing molecular orientation in polymers. The parameter P_4 is also required in the analysis of the elastic moduli of extruded PLC, as in our later calculations based on the aggregate model [26].

As shown in Figure 14.4, P_2 and P_4 of Vectra B increase rapidly at low draw ratio (λ) but level off above $\lambda = 4$. The P_2 value of 0.88 at $\lambda > 4$ indicates that very high chain alignment is attained even at moderate draw ratios, and this reflects the high efficiency of the elongational flow field in promoting molecular orientation in the melt.

The first ultrasonic modulus measurement on extruded PLC was performed on Vectra 900 [9]. Since the orientation parameters for this PLC are not available and its moduli are similar to those of Vectra B [10], we will discuss only the elastic moduli of Vectra B. These modulus data were obtained using the contact method because the sample rod at the highest draw ratio ($\lambda = 15$) has a small diameter of 0.8 mm.

As shown in Figures 14.5 and 14.6, the axial extensional stiffness C_{33} and axial Young's modulus E_3 rise sharply with increasing λ but become saturated above $\lambda = 4$, closely following the behavior of P_2 and P_4. The transverse extensional stiffness C_{11}, transverse Young's modulus E_1 and the axial (C_{44}) and transverse (C_{66}) shear moduli show moderate decreases while the cross-plane stiffness C_{13} exhibits a slight increase. The anisotropy patterns of $C_{33} \gg C_{11}$, $E_3 \gg E_1$ and $C_{44} > C_{66}$ arise from the preferential alignment of chains along the draw direction.

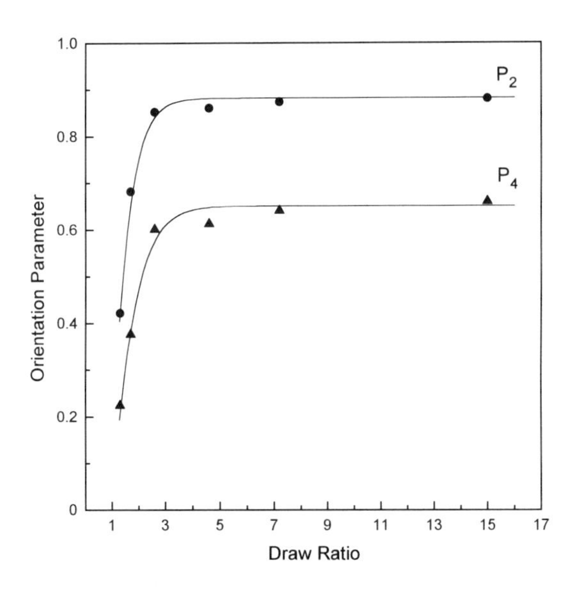

Figure 14.4 Orientation parameters P_2 and P_4 of Vectra B as functions of draw ratio. (Adapted from [10] by permission of the Society of Plastic Engineers.)

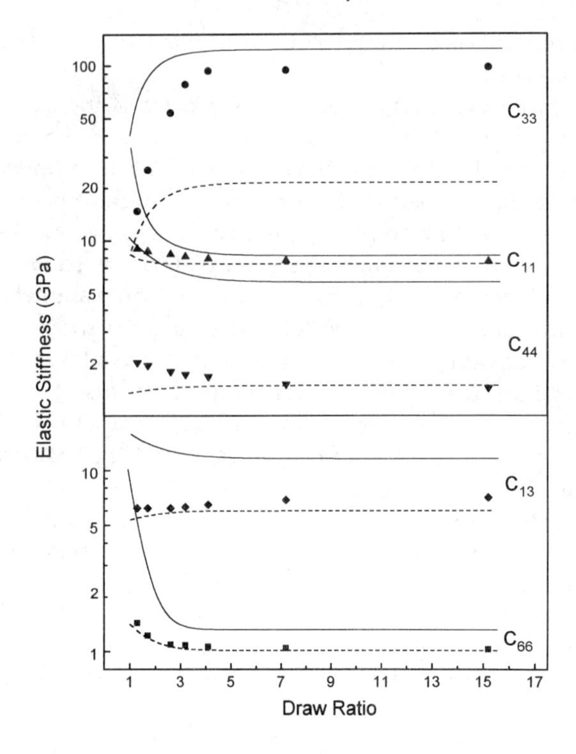

Figure 14.5 Draw ratio dependence of the elastic stiffnesses of Vectra B. The solid and dashed curves denote, respectively, the Voigt and Reuss bounds calculated according to the aggregate model. (Adapted from [10] by permission of the Society of Plastic Engineers.)

In structural applications, engineering parameters such as Poisson's ratios are also used. Figure 14.7 shows that ν_{31} drops drastically while ν_{12} and ν_{13} rise slightly with increasing λ. At $\lambda > 4$, ν_{12} is much higher than ν_{31}, implying that under a transverse stress an extruded rod deforms predominantly in the transverse plane with negligible contraction in the draw direction. The ν_{13} value of about 0.5 at high λ indicates that there is very little change in volume when the rod is subjected to an axial stress.

We have seen that very high molecular orientation and hence high axial moduli C_{33} and E_3 can be attained by drawing thermotropic PLC from the melt. It is also possible to produce highly oriented materials from lyotropic PLC or flexible chain polymers but with greater difficulty. Lyotropic PLC, such as the aramid fiber (Kevlar 49) produced by Du Pont, is spun from a solution followed by drawing at high temperature. Ultra-oriented polyethylene can be prepared from two types of starting material: melt-crystallized material and gel spun from a solution. They

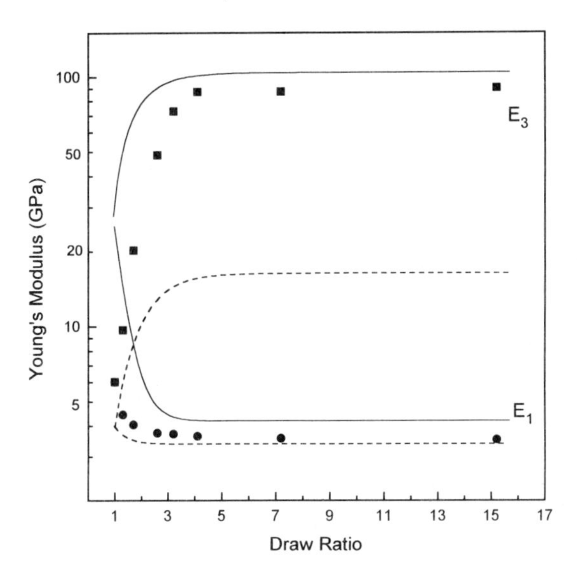

Figure 14.6 Draw ratio dependence of the axial Young's modulus E_3 and transverse Young's modulus E_1 of Vectra B. The solid and dashed curves denote, respectively, the Voigt and Reuss bounds calculated according to the aggregate model. (Adapted from [10] by permission of the Society of Plastic Engineers.)

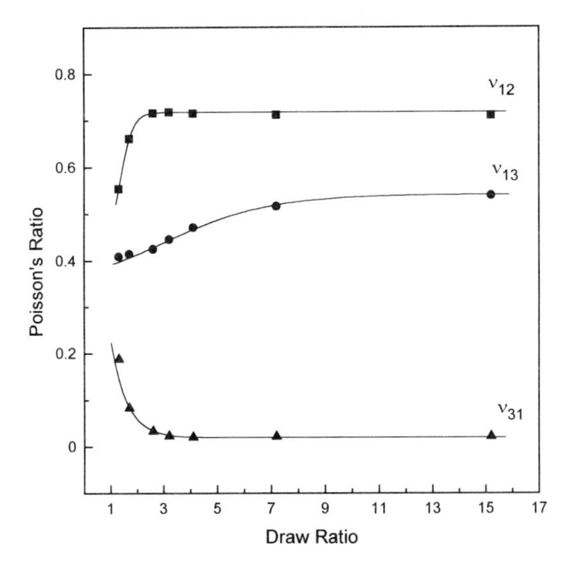

Figure 14.7 Draw ratio dependence of the Poisson's ratios of Vectra B. (Adapted from [10] by permission of the Society of Plastic Engineers.)

are then drawn in the solid state at 70–120°C. As a result of the lower degree of chain entanglement in gel-spun polyethylene, higher draw ratio (and hence higher axial Young's modulus) can be attained, and it is believed that the gel-spun polyethylene fiber (Spectra 1000) produced by Allied-Signal has a draw ratio higher than 100.

Table 14.1 shows a comparison of the elastic properties of Vectra B ($\lambda = 15$), Spectra 1000, polyethylene tape ($\lambda = 27$) prepared from melt crystallized material and Kevlar 49. It is obvious that they all exhibit very high axial Young's modulus and strong anisotropy in the tensile modulus and Poisson's ratio. However, the shear modulus is less anisotropic. The axial Young's modulus of Spectra 1000 is about 40% of the axial modulus of polyethylene crystals [30]. Based on a theoretical estimate [31] and experimental determination using wide angle X-ray diffraction [32], the chain modulus of Vectra B is about 140 GPa, so the observed E_3 at $\lambda = 15$ is approximately 65% of the chain modulus. The ease with which a polymer can be processed to reach a modulus so close to the ultimate limit is quite surprising. Without doubt, this arises from the high degree of molecular alignment in the localized domains of the nematic melt.

We have previously mentioned that available modulus data are usually obtained using a tensile testing machine. To provide a conversion factor between these quasi-static values and the ultrasonic data given in this review, we have measured the axial Young's modulus of extruded Vectra B on an Instron tensile machine at a rate of 2 mm min^{-1}. It is found that E_3 (ultrasonic)/E_3 (quasi-static) is about 1.5, which is the expected effect of frequency.

14.3.2 Aggregate model

A simple theoretical framework for analyzing the elastic moduli of polymers is the aggregate model [26]. In this model, a polymer is

Table 14.1 Comparison of the elastic properties of Vectra B ($\lambda = 15$), Spectra 1000, drawn polyethylene ($\lambda = 27$) and Kevlar 49 (elastic moduli in GPa)

	Vectra B	Spectra 1000 [27]	Drawn polyethylene [28]	Kevlar 49 [29]
E_1	3.50	3.16	4.40	2.49
E_3	90	123	76	113
C_{44}	1.44	1.05	1.44	2.01
C_{66}	1.03	1.00	1.42	–
ν_{12}	0.71	0.59	0.57	0.31
ν_{13}	0.54	0.49	0.44	0.62
ν_{31}	0.018	0.013	0.026	–

assumed to consist of axially symmetric units, whose elastic properties remain unchanged but are increasingly aligned as the polymer is drawn. The aggregate model can be applied in two ways. First, the Voigt average (based on parallel coupling of the units) and Reuss average (based on series coupling of the units) for the Young's modulus E and shear modulus G of the isotropic polymer can be calculated by taking the elastic constants at any draw ratio as the elastic constants of the microscopic unit. Second, the draw ratio dependence of the elastic moduli can be predicted if the orientation parameters and elastic moduli of the microscopic unit are known.

At high orientation, C_{11}, C_{13}, C_{44} and C_{66} of Vectra B are expected to exhibit only slight changes, so we fitted their values at $\lambda > 4$ to a polynomial $a + b/\lambda$ (a, b being constants) and then extrapolated $\lambda = \infty$ to give stiffnesses of the unit: $C^{u}_{11} = 7.5$ GPa, $C^{u}_{13} = 7.2$ GPa, $C^{u}_{44} = 1.35$ GPa and $C^{u}_{66} = 1.00$ GPa. The axial Young's modulus E^{u}_{3} of the unit has already been determined by measurements of the shift of the meridional X-ray reflection resulting from an applied stress [32]. Using the time–temperature equivalence principle, it is estimated that the ultrasonic (10 MHz) E^{u}_{3} at room temperature (23°C) is equivalent to the quasi-static X-ray E^{u}_{3} of 137 GPa at -60°C. Using this E^{u}_{3} value and the above the four C^{u}_{ij} values we obtained $C^{u}_{33} = 145$ GPa. With the five known stiffnesses of the unit and the orientation parameters P_2 and P_4, Voigt and Reuss values for C_{ij} and E_i can be calculated [26].

We first compare the theoretical predictions for the isotropic polymer (Table 14.2) with the values of $E = 6.02$ GPa and $G = 2.20$ GPa observed for isotropic Vectra B900, a PLC with an almost identical composition to Vectra B. It is clear that the observed E and G values generally lie between the Voigt (upper) and Reuss (lower) bounds but are much closer to the Reuss bound.

As shown in Figures 14.5 and 14.6, the magnitude and draw ratio dependence of the moduli C_{11}, C_{13}, C_{44}, C_{66} and E_1 are well predicted by the Reuss model. However, C_{33} or E_3 is close to the Reuss bound

Table 14.2 Elastic moduli (in GPa) of isotropic Vectra B calculated from the elastic constants of the drawn samples

	Draw ratio						
	1.3	1.7	2.6	3.2	4.1	7.2	15.2
E (Reuss)	5.11	4.88	4.57	4.50	4.40	4.19	4.09
E (Voigt)	5.62	7.18	11.8	15.8	18.2	18.0	18.5
G (Reuss)	1.85	1.76	1.64	1.61	1.58	1.50	1.46
G (Voigt)	2.04	2.63	4.41	5.98	6.92	6.83	7.01

only at low draw ratio. As the draw ratio increases, the observed C_{33} or E_3 moves towards the Voigt bound and is only 25% below the Voigt value at $\lambda > 3$. Therefore, we conclude that the parallel and series coupling schemes of the aggregate model can only provide bounds, and it is not justified to use either scheme to obtain precise values of the elastic moduli.

14.3.3 Injection molded polymer liquid crystals

In the injection molding of PLC, the nematic melt is subjected to both shear and elongational stresses during the filling of the mold. Therefore the flow pattern of the melt is more complicated than in extrusion. Because of this, all the ultrasonic modulus measurements have been made only on samples cut from the middle sections of plaques or bars ([7–9], C.L. Choy, Y.W. Wong and K.W.E. Lau, unpublished results). Since this section is at some distance from the gate of the injection mold, the melt has a straight flow front across almost the entire width of the mold. Consequently, this section of the solidified material has orthorhombic symmetry and is characterized by the nine stiffness constants C_{11}, C_{22}, C_{33}, C_{12}, C_{13}, C_{23}, C_{44}, C_{55} and C_{66} if we take the 1, 2 and 3 axes to lie along the width, thickness and length directions of the plaque or bar.

Injection molded plaques or bars of PLC have a skin–core structure [33]. The molecular chains in the skin regions are largely aligned in the mold fill direction while the chain orientation in the core is more or less random. The high molecular alignment in the skin layer is induced by the elongational stress in the fountain flow and is immediately frozen upon contact with the mold surface.

The first ultrasonic modulus measurement on injection molded PLC was carried out by Wedgbury and Read [7] on a copolyester produced by ICI. Only five stiffness constants were obtained because axial symmetry was assumed. More detailed studies were made by Sweeney *et al.* [8] on a PLC of similar composition (HBA/IA/HQ), so we will discuss only the results of this later work.

Measurements were made using an ultrasonic beam of frequency 2.25 MHz and diameter 13 mm. Samples were cut from the long arms of a 6 mm thick picture-frame injection molding. Two samples were tested and the seven stiffness constants for sample thicknesses corresponding to the as-molded plaque and a central region of thickness 2.3 or 2.5 mm are given in Table 14.3. The remaining two stiffness constants, C_{13} and C_{55}, could not be obtained because the fact that the ultrasonic beam diameter is larger than the sample thickness leads to difficulty in generating an ultrasonic beam in the 1–3 plane. Moreover, measurements on samples thinner than 2.3 mm are not

Table 14.3 Stiffness constants (in GPa) of injection molded HBA/IA/HQ

Sample	Thickness (mm)	C_{11}	C_{22}	C_{33}	C_{12}	C_{23}	C_{44}	C_{66}
1	6	7.60	6.99	15.8	5.52	5.53	1.78	0.91
2	6	7.21	6.71	21.6	5.28	5.95	1.43	0.82
1	2.5	8.61	7.10	13.6	5.69	5.51	1.18	1.00
2	2.3	7.80	6.54	21.3	5.10	6.23	1.50	0.68

possible as a result of the low ultrasonic frequency. Because of the lack of data on C_{13}, the assumption of $S_{13} = S_{23}$ was made in order to calculate the Young's moduli and Poisson's ratios shown in Table 14.4.

We first note that the two samples have significantly different stiffnesses. Sample 2 has a C_{33} value about 40% higher, and C_{11} and C_{22} values 5–10% lower than those of sample 1, indicating that sample 2 has a higher degree of molecular orientation. Assuming that these two samples were cut from different regions in the long arm of the picture-frame molding, the result implies that the molecular orientation varies with position. For both samples, C_{33} decreases while C_{11} and C_{22} increase as the thickness is reduced from 6 to 2.5 mm. This is expected because the thinner sample does not contain the skin layer with highly oriented molecular chains.

To study the skin–core effects in more detail, ultrasonic measurements were made on a sample as it was machined progressively thinner from alternate sides, with the removed layer being 0.4 mm thick. Results from successive experiments were compared to obtain the pulse transit times for the layer of material which had been removed. Analyses of these transit times gave the profiles of E_3 and C_{44} for sample 1 shown in Figures 14.8 and 14.9, respectively. Besides the peak in the modulus in the skin region there is another peak about 1.2 mm from the surface of the plaque. This peak has also been observed in static modulus experiments on other injection molded plaques and has been attributed

Table 14.4 Young's moduli (in GPa) and Poisson's ratios of injection molded HBA/IA/HQ

Sample	Thickness (mm)	E_1	E_2	E_3	ν_{21}	ν_{13}	ν_{32}
1	6	3.06	2.85	10.8	0.691	0.442	0.115
2	6	2.92	2.73	15.6	0.705	0.500	0.087
1	2.5	3.72	3.11	8.55	0.658	0.427	0.155
2	2.3	3.53	2.99	14.3	0.654	0.542	0.114

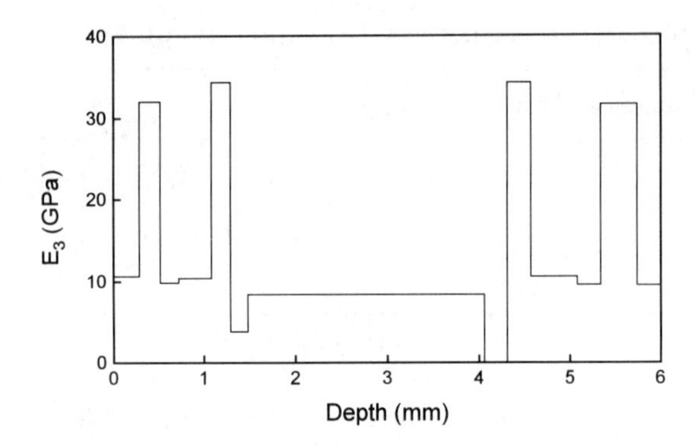

Figure 14.8 Profile of E_3 through plaque thickness for injection molded HBA/IA/HQ. (Reproduced from [8] by permission of Elsevier Science Ltd.)

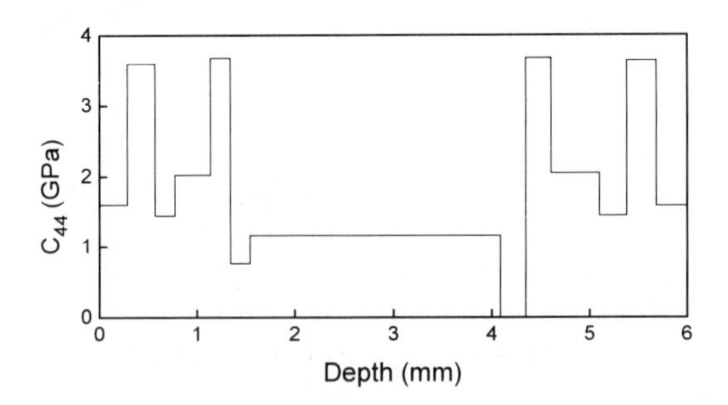

Figure 14.9 Profile of C_{44} through plaque thickness for injection molded HBA/IA/HQ. (Reproduced from [8] by permission of Elsevier Science Ltd.)

to the shearing of the melt against the solidified layer initially formed by shearing [34]. The single value obtained for the central 2.5 mm thick region relates to the testing of the thinnest section and is thus only an average value.

Other approaches have been adopted by Choy *et al.* ([9], C.L. Choy, Y.W. Wong and K.W.E. Lau, unpublished results) to study the effects of the skin–core structure. By using the immersion method at a high frequency of 10 MHz, the stiffness constants of layers as thin as 0.7 mm can be determined. Moreover, the variation of the stiffness constants with the position along the width of each layer can be investigated by the contact method using small samples of dimensions 1.4, 0.7 and 3 mm in the 1, 2 and 3 directions, respectively. We will not distinguish the stiffness data obtained by these two methods since they generally agree to within experimental error.

Injection molded bars of two different geometries have been investigated. For ASTM type I tensile bars with width 12.6 mm and thickness 3.2 mm in the middle section, ultrasonic measurements have been made in the top, middle and bottom layers, each of thickness 0.7 mm. For impact bars of length 126 mm, width 12.6 mm and thickness 6 mm, five layers (0.7 mm thick) equally spaced along the thickness direction have been studied. All the bars were end-gated and samples for measurement were taken from the middle section, within which the stiffness was found to be independent of the position along the length.

Figure 14.10 shows the variation of the extensional stiffnesses with the position along the width in the top and bottom layers of a Vectra A tensile bar. It is seen that the stiffnesses are essentially independent of position and are symmetric with respect to the midplane. The fact that C_{33} is much higher than C_{11} or C_{22} implies that the molecular chains are preferentially aligned in the 3 (length) direction. Since C_{11} is about 10% higher than C_{22}, there is a slight preference for the chains to lie in the 1–3 plane. All these results are consistent with the wide angle X-ray diffraction measurements on another tensile bar of the same

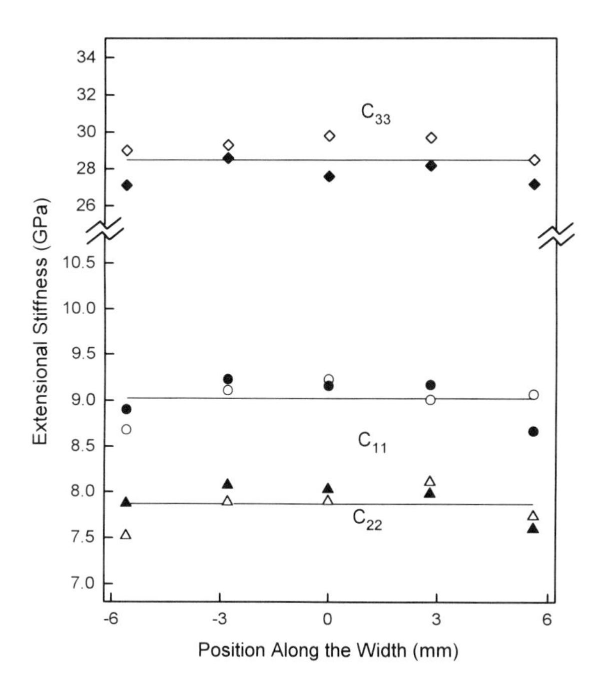

Figure 14.10 Variation of the extensional stiffnesses of injection molded tensile bar of Vectra A with the position along the width. Solid symbols: top layer; open symbols: bottom layer.

material [35]. Unlike the behavior of C_{33} in the skin layers, C_{33} in the middle layer varies greatly with position, decreasing from 29 GPa on the two edges to 12 GPa on the axis (Figure 14.11). C_{11} increases by about 30% as we move from the edge to the axis while C_{22} remains roughly unchanged. The curves for C_{11}, C_{22} and C_{33} are symmetrical about the axis. In the core region near the axis, $C_{11} \sim C_{33}$ and C_{11} exceeds C_{22} by only 20%, indicating approximately random chain orientation. On the two edges, $C_{22} > C_{11}$, implying that the chains lie preferentially in the skin layer (2–3 plane).

The shear moduli also exhibit strong anisotropy in the top or bottom layers (Figure 14.12a). The anisotropy pattern of $C_{55} > C_{44} > C_{66}$ reflects the preferential chain orientation parallel to the 1–3 plane. Figure 14.12b shows that the shear moduli differ by less than 20% in the core, in agreement with the above results on the extensional stiffnesses. The fact that $C_{44} > C_{55}$ on the edges of the middle layer again demonstrates that the chains in this region lie preferentially in the 2–3 plane.

Figure 14.13 shows the variation of C_{33} with the width position in five layers of an impact bar of Vectra A. C_{33} exhibits a symmetric

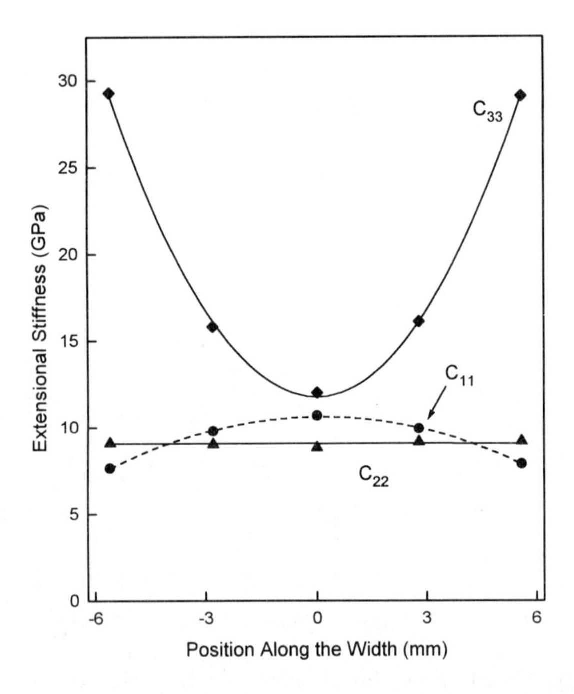

Figure 14.11 Variation of the extensional stiffnesses of injection molded tensile bar of Vectra A with the position along the width of the middle layer.

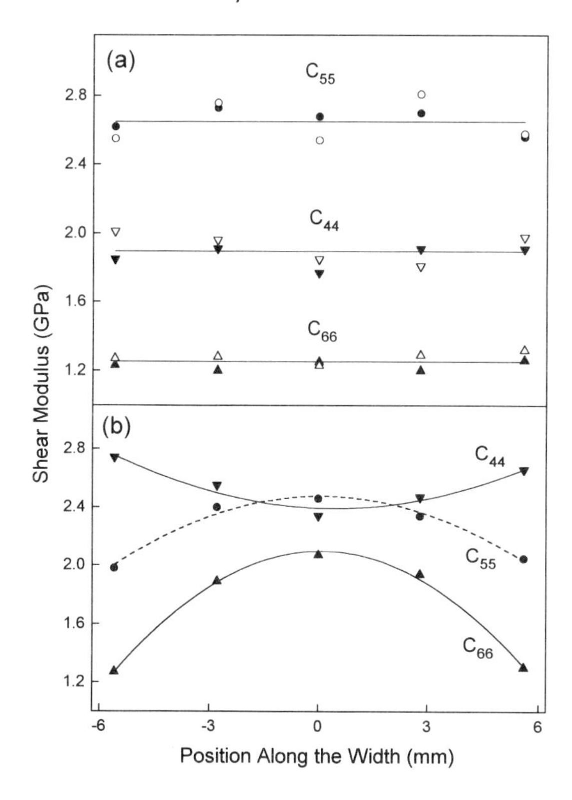

Figure 14.12 Variation of the shear moduli of injection molded tensile bar of Vectra A with the position along the width. (a) Top (solid symbols) and bottom (open symbols) layers; (b) middle layer.

behavior with respect to the axis and the midplane. Its value is independent of position in the top or bottom layer. The values of C_{33} in layers 2 and 4 (at a distance of 1.6 mm from the surface of the bar) are only slightly higher than those in the middle layer.

The stiffness constants for the top and middle layers and the whole of both tensile and impact bars of Vectra A and Vectra B are given in Table 14.5 while the Young's moduli and Poisson's ratios are given in Table 14.6. The top layer has significantly higher C_{33} (or E_3) and slightly lower C_{11} and C_{22} than the middle layer, reflecting the higher molecular orientation in the top layer. C_{55} is higher while C_{44} and C_{66} are lower in the top layer. As expected, the stiffness constants for the whole sample are intermediate between those for the top and middle layers.

The tensile bar of Vectra A has a higher C_{33} but lower C_{11} and C_{22} than the impact bar, which probably arises from the converging flow that develops when the melt flows from the tab of the tensile bar to

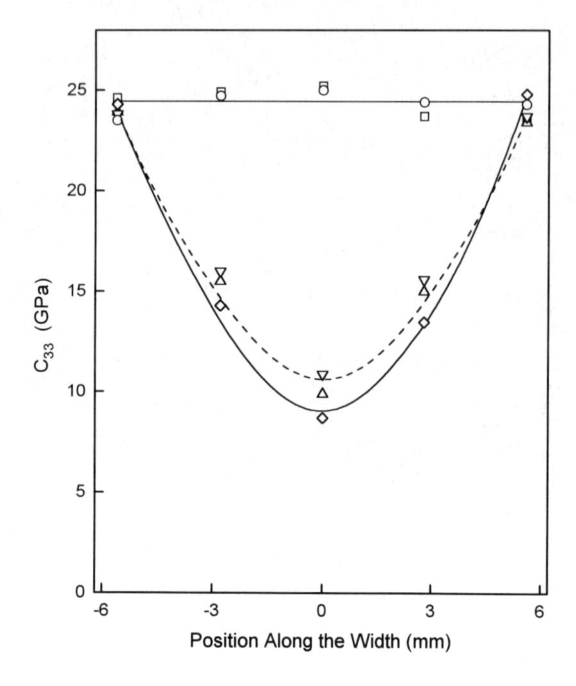

Figure 14.13 Variation of C_{33} of injection molded impact bar of Vectra A with the position along the width. □, △, ◇, ▽, and ○ correspond to data for layer 1 (top), 2, 3, 4 and 5 (bottom).

Table 14.5 Stiffness constants (in GPa) of injection molded Vectra A and Vectra B

		C_{11}	C_{22}	C_{33}	C_{12}	C_{13}	C_{23}	C_{44}	C_{55}	C_{66}
Vectra A	Top	8.86	7.71	29.3	5.38	7.01	6.22	1.87	2.68	1.21
tensile	Middle	9.23	9.10	20.4	6.18	6.62	6.40	2.45	2.26	1.79
bar	Whole	9.07	8.32	23.6	5.72	6.85	6.31	2.04	2.45	1.40
Vectra A	Top	9.00	7.78	24.3	5.49	7.10	6.26	1.80	2.41	1.24
impact	Middle	9.38	9.32	17.1	6.31	6.52	6.38	2.26	1.98	1.67
bar	Whole	9.22	8.48	19.9	5.83	6.88	6.21	1.94	2.33	1.44
Vectra B	Top	8.87	7.42	36.4	5.63	7.21	6.48	1.78	2.77	1.20
impact	Middle	9.35	8.70	23.5	6.76	6.98	6.70	2.32	2.13	1.68
bar	Whole	9.15	7.90	29.7	6.02	7.09	6.53	1.96	2.47	1.34

the narrow neck section. For the impact bars, the substantially higher C_{33} (or E_3) for Vectra B implies that molecular orientation is more readily induced in Vectra B, consistent with the finding of Chung [36].

Table 14.6 Young's moduli (in GPa) and Poisson's ratios of injection molded Vectra A and Vectra B

		E_1	E_2	E_3	ν_{12}	ν_{21}	ν_{13}	ν_{31}	ν_{23}	ν_{32}
Vectra A	Top	4.81	4.28	22.9	0.542	0.609	0.523	0.110	0.442	0.083
tensile	Middle	4.71	4.72	14.9	0.579	0.579	0.452	0.143	0.397	0.126
bar	Whole	4.80	4.50	17.6	0.549	0.586	0.489	0.134	0.422	0.108
Vectra A	Top	4.75	4.23	17.9	0.529	0.594	0.523	0.139	0.435	0.103
impact	Middle	4.73	4.76	11.8	0.562	0.559	0.431	0.173	0.393	0.159
bar	Whole	4.77	4.56	14.0	0.538	0.563	0.501	0.170	0.388	0.126
Vectra B	Top	4.43	3.73	29.6	0.584	0.694	0.499	0.075	0.495	0.062
impact	Middle	3.93	3.67	17.6	0.656	0.703	0.433	0.097	0.434	0.090
bar	Whole	4.38	3.80	23.3	0.598	0.690	0.463	0.087	0.474	0.077

However, the C_{33} value for the injection molded bar corresponds to that for the extruded rod at $\lambda \sim 2$ (Figure 14.5), demonstrating that extrusion followed by drawing is more efficient than injection molding in inducing molecular orientation.

Despite the difference in chemical composition, it is seen from Tables 14.3–14.6 that the stiffness constants of the two copolyesters, HBA/IA/HQ and Vectra A, are not vastly different. The shear moduli C_{44} and C_{66} for Vectra A are 20% and 50% higher than those for HBA/IA/HQ, whereas the transverse Young's moduli E_1 and E_2 for Vectra A are about 50% higher.

14.4 ELASTIC MODULI OF *IN SITU* COMPOSITES CONTAINING POLYMER LIQUID CRYSTALS

It is well known that the addition of filler materials can lead to significant changes in the physical properties of polymers. For example, chopped glass fibers are commonly incorporated into a thermoplastic to increase the stiffness and strength, as well as the heat distortion temperature. However, the processing of chopped glass fiber composites presents some difficulties, the major ones being wear on processing equipment and increased viscosity of the molten polymer.

Because of these reasons, it would be desirable to find an approach in which the reinforcing elements are not present before processing but are formed during the extrusion or injection molding process. To produce these *in situ* composites, a thermotropic PLC is first blended with a thermoplastic in the melt. During the subsequent extrusion or injection molding, the dispersed PLC phase is deformed into the fibrillar domains which then serve as the reinforcing component in the composite [37].

14.4.1 *In situ* composites formed by extrusion

The morphology, molecular orientation and ultrasonic moduli of extruded blends of polycarbonate (PC) and Vectra B have been studied by Choy *et al.* [10]. We will denote these blends by PC + VBX where X is the weight percent of Vectra B in the blend.

Examinations of the fracture surfaces of the extruded rods in a scanning electron microscope reveal that the PLC phase is dispersed in a PC matrix when the volume fraction V_f of PLC is less than 0.3 (the densities of PC and Vectra B are 1.20 and 1.40 g cm^{-3}, respectively, so $V_f = 0.3$ corresponds to 33 wt% of Vectra B). At low draw ratio ($\lambda \leqslant 1.3$), a skin–core morphology is found. The core region, which occupies about 85% of the sample volume, contains mostly spherical PLC domains. In the thin skin layer, however, there are elongated PLC domains aligned along the draw direction. As the draw ratio increases, the aspect ratio of the PLC domains in the core increases, and the distinction between the skin and core diminishes. At $\lambda = 15$, long PLC fibrils are uniformly distributed throughout the rod (Figure 14.14). Examination of these fibrils after extraction of the PC matrix give an average aspect ratio higher than 200. Therefore the PLC fibrils at high draw ratio may be regarded as continuous as far as their effects on the elastic moduli are concerned. At $V_f > 0.55$, phase inversion has occurred and the PLC becomes the continuous phase.

As shown in Figure 14.15, the orientation parameters P_2 and P_4 of the PLC domains in the composites increase more slowly with increasing λ than the bulk PLC. However, as V_f increases, the curve for the composite moves closer to that of the PLC. At high draw ratio ($\lambda = 15$), the composites with $V_f > 0.55$ have essentially the same P_2 and P_4 values as the PLC, but the composites with $V_f < 0.3$ have a significantly lower degree of chain orientation.

The elastic moduli E_1, E_3, C_{44} and C_{66} of the composites are shown as functions of draw ratio in Figures 14.16–14.19. The axial Young's modulus E_3 increases substantially with increasing λ as a result of the higher aspect ratio of the PLC domains and the enhanced molecular orientation within the domains. For E_1 and C_{44}, however, the reinforcement effect is weaker at higher λ because E_1 and C_{44} for the PLC drop slightly with increasing λ. At $\lambda = 15$, E_1 of Vectra B is only 12% higher than that of PC, thereby leading to a very small enhancement in the transverse Young's modulus. Since Vectra B has a transverse shear modulus

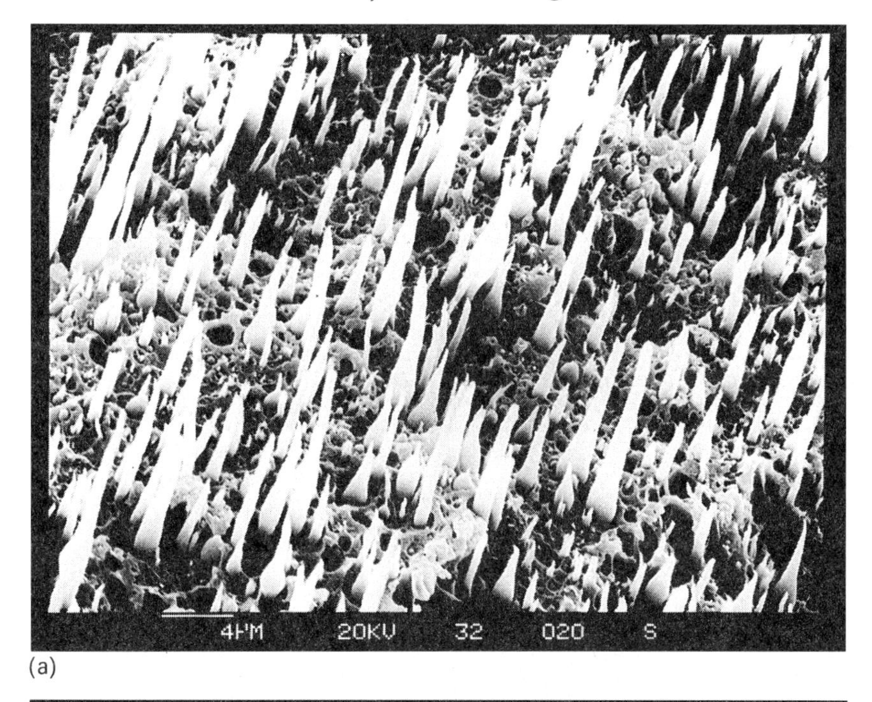

(a)

(b)

Figure 14.14 Fracture surface of an extruded blend (draw ratio $\lambda = 15$) of PC and Vectra B with 30 wt% of Vectra B. (a) Core region, (b) skin region.

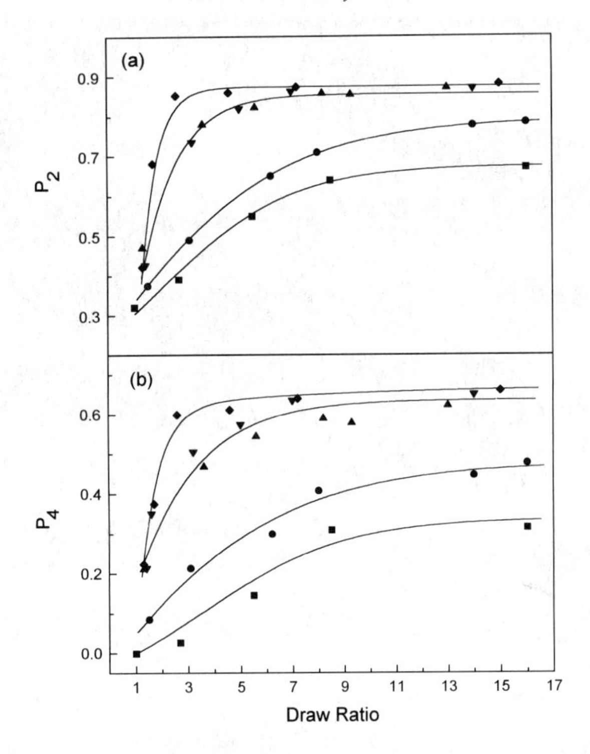

Figure 14.15 Draw ratio dependence of the orientation parameters of PC + Vectra B blends. ■, ●, ▼, ▲, and ◆ refer to blends with 10, 30, 60, 80 and 100 wt% Vectra B. (Adapted from [10] by permission of the Society of Plastic Engineers.)

which decreases to a value below that of PC at $\lambda > 2$, there is a positive reinforcement effect on C_{66} at low λ but a negative effect at high λ.

With all the independent stiffness constants known, it is possible to calculate the Young's modulus at an angle θ relative to the draw axis. This is a valuable parameter which is very difficult to measure on a conventional tensile machine. As shown in the Young's modulus vs. θ plots at $\lambda = 15$ (Figure 14.20), the Young's modulus increases with increasing PLC content at all angles, but the reinforcement effect becomes weaker as θ increases.

As we have seen, the structure of the *in situ* composites with low V_f is akin to that of a short fiber-reinforced composite. Lin and Yee [38] have proposed a two-step procedure for calculating the elastic moduli. First, the axial (E_3^f) and transverse (E_1^f) Young's moduli of the PLC domains in the composites are calculated in terms of the aggregate model. Then the axial Young's modulus E_3 of the composite can be

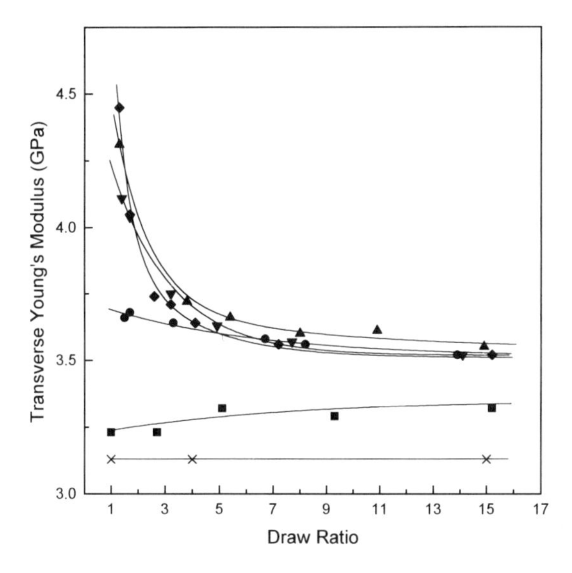

Figure 14.16 Draw ratio dependence of the transverse Young's modulus of PC + Vectra B blends. ■, ●, ▼, ▲, and ◆ refer to blends with 0, 10, 30, 60, 80 and 100 wt% Vectra B. (Adapted from [10] by permission of the Society of Plastic Engineers.)

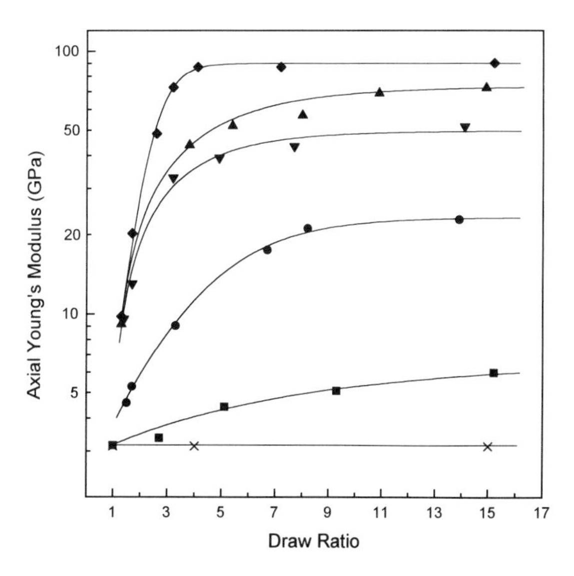

Figure 14.17 Draw ratio dependence of the axial Young's modulus of PC + Vectra B blends, with same legend as Figure 14.16. (Adapted from [10] by permission of the Society of Plastic Engineers.)

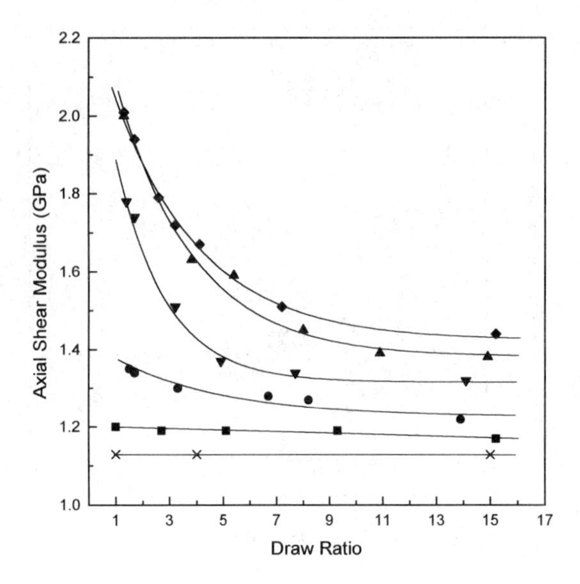

Figure 14.18 Draw ratio dependence of the axial shear modulus of PC + Vectra B blends, with same legend as Figure 14.16. (Adapted from [10] by permission of the Society of Plastic Engineers.)

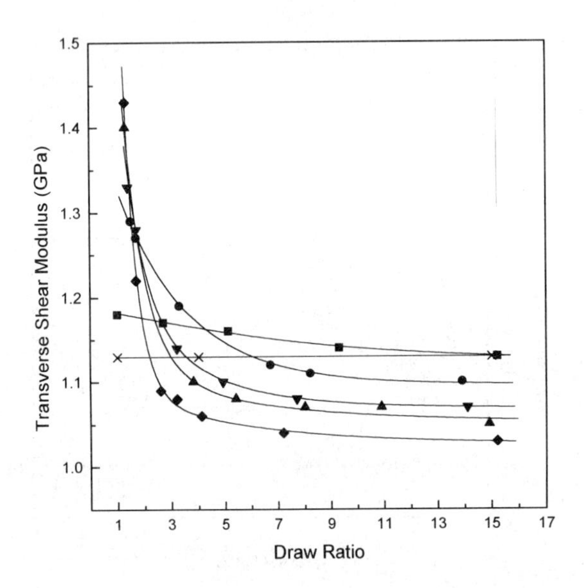

Figure 14.19 Draw ratio dependence of the transverse shear modulus of PC + Vectra B blends, with same legend as Figure 14.16. (Adapted from [10] by permission of the Society of Plastic Engineers.)

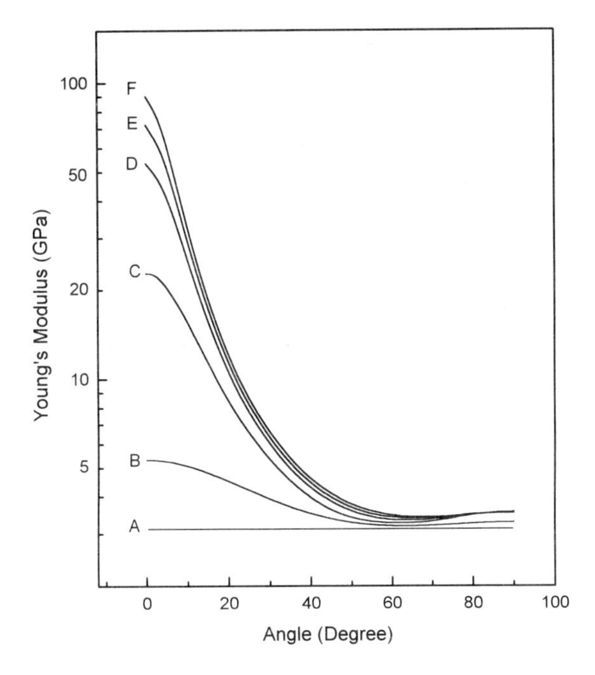

Figure 14.20 Dependence of the Young's modulus of PC + Vectra B blends at $\lambda = 15$ on the angle relative to the draw direction. A, B, C, D, E and F refer to blends with 0, 10, 30, 60, 80 and 100% Vectra B. (Adapted from [10] by permission of the Society of Plastic Engineers.)

obtained from the Halpin–Tsai equation [39]:

$$E_3 = E^m \frac{1 + \zeta \eta V_f}{1 - \eta V_f} \tag{14.22}$$

$$\eta = \frac{E_3^f / E^m - 1}{E_3^f / E^m + \zeta} \tag{14.23}$$

$$\zeta = 2 \frac{L}{D} \tag{14.24}$$

where E^m is the Young's modulus of the matrix and V_f, L and D are the volume fraction, length and diameter of the PLC domains. The transverse modulus E_1 can be calculated by replacing E_3^f by E_1^f and L/D by 1.

Like our earlier calculation for the bulk PLC, E_3^f and E_1^f can be obtained using either the Voigt or Reuss averaging scheme. The aspect ratio L/D can be estimated assuming that the deformation is affine [26, 38] i.e. the deformation of the PLC domains is identical to that of the composite rod, and the volume remains unchanged. It then follows that

$$\zeta = 2\lambda^{3/2} \tag{14.25}$$

Comparisons of the theoretical predictions and experimental data are shown for the composites with 10 and 30 wt% of Vectra B in Figures 14.21 and 14.22, respectively. For PC + VB10, the observed E_3 lies on the Reuss (lower) bound at low λ but is intermediate between the Reuss and Voigt bounds at $\lambda > 4$. For PC + VB30, the observed E_3 is again close to the lower bound at low λ. However, it moves towards the upper bound as λ increases and then follows the upper bound at $\lambda > 4$. For both composites, the transverse modulus E_1 agrees with the Reuss value to within experimental error. The behavior of the composites, particularly that of PC + VB30, is similar to the behavior of Vectra B discussed in an earlier section.

Figures 14.21 and 14.22 also show the predictions obtained by taking the observed moduli of Vectra B as E_3^f and E_1^f. Good agreement with the experimental data is found for E_3 of PC + VB30 and the E_1 values of both composites. The close agreement obtained for E_3 of PC + VB30 may arise from two compensating factors. First, the affine deformation model gives a domain aspect ratio slightly lower than the actual value [40], thereby leading to an underestimate of E_3. However, this is balanced by the effect of taking the observed modulus of Vectra B as E_3^f since the bulk PLC has a higher molecular orientation than the PLC domains in PC + VB30 (Figure 14.15). For PC + VB10, the molecular orientation effect becomes dominant, so the theoretical prediction is significantly higher than the observed E_3.

We have mentioned that the PLC phase is continuous for $V_f > 0.55$. At high draw ratio ($\lambda = 15$), the PLC fibrils in PC + VB10 and PC + VB30 are also essentially continuous, so the axial Young's modulus of all the composites can be analyzed in terms of the rule of mixtures valid for continuous fiber composites:

$$E_3 = V_f E_3^f + (1 - V_f)\, E^m \tag{14.26}$$

As shown in Figure 14.23, the observed E_3 value for $V_f > 0.1$ falls on the straight line predicted by the rule of mixtures. However, the E_3 value for PC + VB10 is below the theoretical prediction, demonstrating that the PLC domains in this composite have a lower degree of chain alignment than the bulk PLC.

According to the theory of continuous fiber composites [40], not only E_3 but also the Poisson's ratio ν_{13} are expected to obey the rule of mixtures. This is verified in Figure 14.24, confirming that highly drawn blends of PC and Vectra B behave like continuous fiber

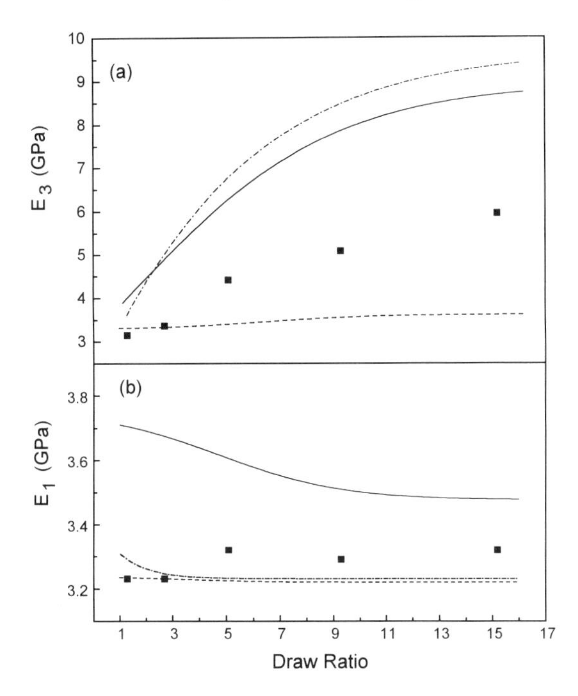

Figure 14.21 Draw ratio dependence of the (a) axial and (b) transverse Young's moduli of PC + VB10. The theoretical predictions based on E_3^f and E_1^f values derived from the Voigt and Reuss model are shown as solid and dashed curves, respectively. The dot-dashed curves are theoretical predictions obtained by taking the observed moduli of Vectra B as E_3^f and E_1^f. (Adapted from [10] by permission of the Society of Plastic Engineers.)

composites. This conclusion is consistent with our thermal conductivity and expansivity data on the same series of blends [40].

14.4.2 *In situ* composites formed by injection molding

Chik *et al.* [11] have determined the ultrasonic stiffnesses of tensile bars of injection molded blends of polycarbonate and Vectra A and have correlated the results with the morphology and molecular orientation. Like the tensile bar of Vectra A, the skin extends around the whole cross-section perpendicular to the length of the bar. In this skin layer, the PLC domains have a high aspect ratio and the molecular chains within the domains are preferentially aligned along the 3 (length) direction. As we move towards the core, the average aspect ratio of the domains decreases and the PLC chains become more randomly oriented. The changes in the domain aspect ratio and molecular

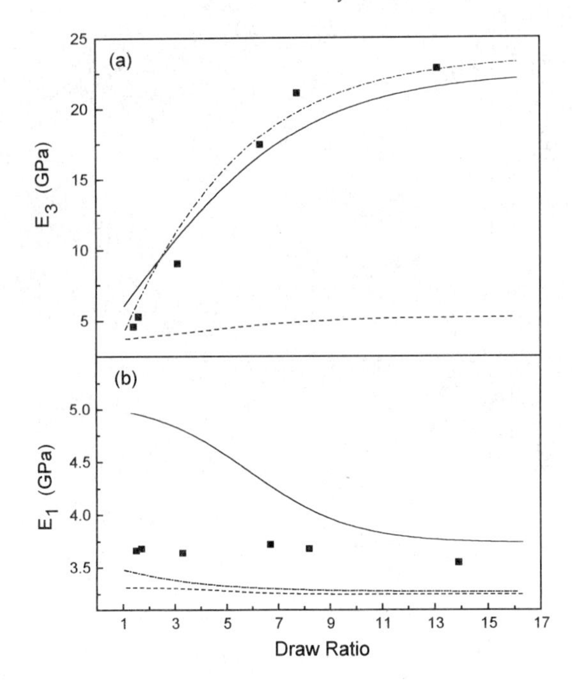

Figure 14.22 Draw ratio dependence of the (a) axial and (b) transverse Young's modulus of PC + VB30, with same legend as Figure 14.21. (Adapted from [10] by permission of the Society of Plastic Engineers.)

orientation are reflected in the behavior of C_{11}, C_{22} and C_{33} for a blend with 20 wt% Vectra A (Figure 14.25). In the top skin layer the stiffnesses are independent of position along the width of the bar, and C_{33} is much higher than C_{11} and C_{22}. On the edges of the middle layer where the skin regions are located, a high C_{33} value of similar magnitude is obtained. However, C_{33} drops substantially as we move towards the axis, and C_{33} is only 10% higher than C_{11} or C_{22} at the core. The pattern of anisotropy is the same as that of Vectra A shown in Figures 14.10 and 14.11, but the magnitudes of C_{11}, C_{22} and C_{33} are lower for the composite since it contains only 20 wt% Vectra A. As the PLC content increases to 60 wt%, the PLC phase becomes continuous, but the position dependencies of C_{11}, C_{22} and C_{33} are similar to those of the composite with 20 wt% PLC.

Figure 14.26 shows the dependence of the Young's moduli on the weight percent of PLC. The Young's modulus in the mold fill direction (E_3) increases substantially while E_1 and E_2 rise slightly with increasing PLC content. However, the increase in E_3 is much less than that exhibited by the extruded blends (Figure 14.23), again demonstrating that injection molding gives rise to a lower degree of chain alignment than drawing.

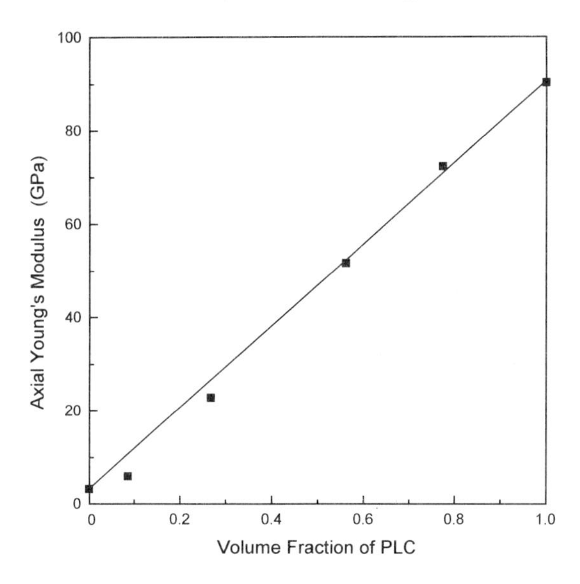

Figure 14.23 Axial Young's moldulus of PC + Vectra B blends at $\lambda = 15$ as a function of the volume fraction of Vectra B. (Adapted from [10] by permission of the Society of Plastic Engineers.)

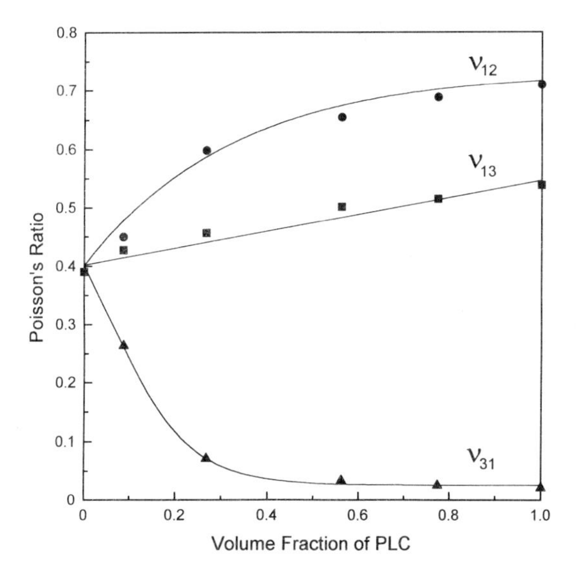

Figure 14.24 Poisson's ratios of PC + Vectra B blends at $\lambda = 15$ as functions of the volume fraction of Vectra B. (Adapted from [10] by permission of the Society of Plastic Engineers.)

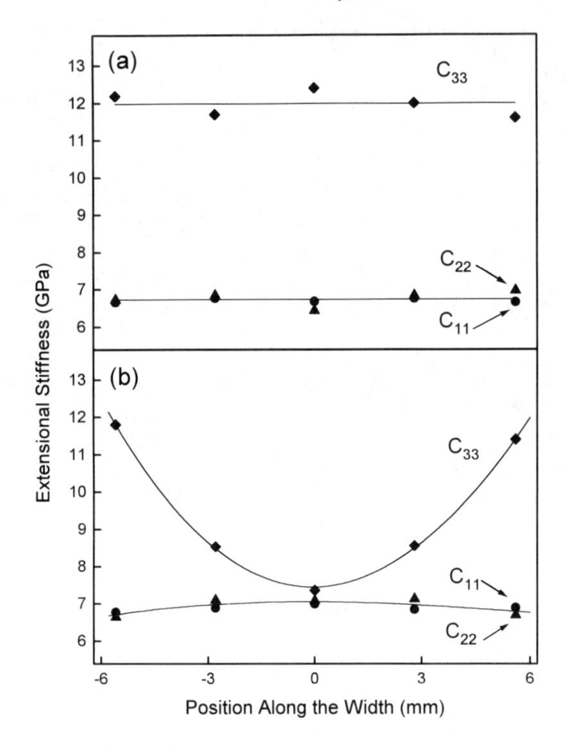

Figure 14.25 Variation of the extensional stiffnesses of injection molded tensile bar of PC + VA30 blend with position along the width. (a) Top layer, (b) middle layer.

As shown in Figure 14.27, all the shear moduli are enhanced by the addition of PLC, with C_{55} showing the greatest reinforcement effect. The high C_{55} value reflects the preferential alignment of the PLC chains parallel to the 1–3 plane. The stiffness constants of the injection molded blends are given in Table 14.7 while the Young's moduli and Poisson's ratios are given in Table 14.8.

14.5 ELASTIC MODULI OF GLASS FIBER-REINFORCED POLYMER LIQUID CRYSTALS

We have seen that composites containing PLC fibrils as reinforcing elements can be readily produced. A more common application of PLC is to serve as the matrix material of short fiber-reinforced composites. Since the current cost of PLC is rather high, the addition of glass fibers results in not only an increase in stiffness but also a reduction in cost. Because of the growing importance of short fiber-reinforced PLC it is

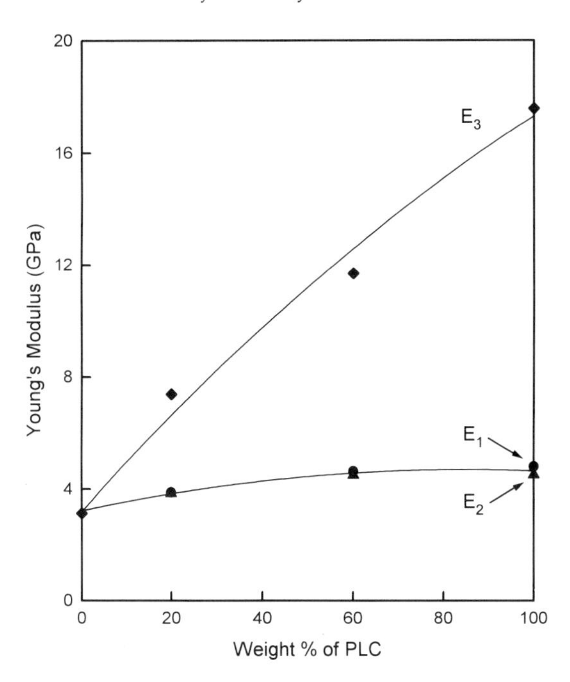

Figure 14.26 Young's moduli of injection molded PC + Vectra A blends as functions of the weight percent of Vectra A.

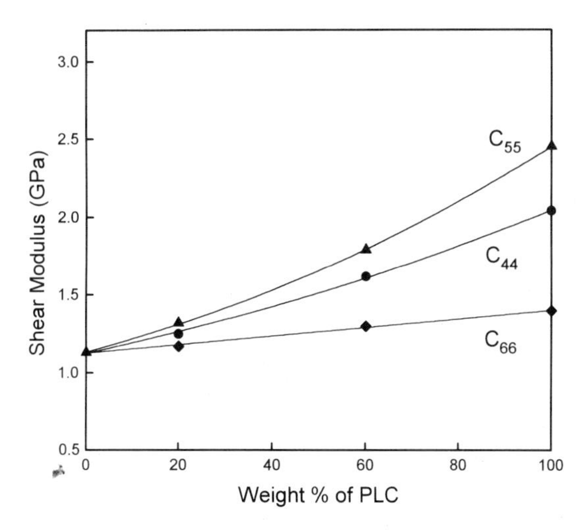

Figure 14.27 Shear moduli of injection molded PC + Vectra A blends as functions of the weight percent of Vectra A.

Table 14.7 Stiffness constants (in GPa) of injection molded blends of polycarbonate and Vectra A

wt% Vectra A	C_{11}	C_{22}	C_{33}	C_{12}	C_{13}	C_{23}	C_{44}	C_{55}	C_{66}
20	6.85	6.72	10.7	4.19	4.30	4.23	1.25	1.32	1.17
60	8.03	7.56	16.2	4.75	5.51	5.08	1.62	1.79	1.30

Table 14.8 Young's moduli (in GPa) and Poisson's ratios of injection molded blends of polycarbonate and Vectra A

wt% Vectra A	E_1	E_2	E_3	ν_{12}	ν_{21}	ν_{13}	ν_{31}	ν_{23}	ν_{32}
20	3.89	3.84	7.39	0.486	0.493	0.392	0.207	0.385	0.200
60	4.63	4.48	11.7	0.491	0.507	0.460	0.181	0.383	0.147

worthwhile to study the stiffness of these materials (C.L. Choy, Y.W. Wong and K.W.E. Lau, unpublished results).

Figures 14.28 and 14.29 show the stiffness constants of 30 wt% glass fiber-reinforced Vectra A as functions of the position along the width. The position dependence and the pattern of anisotropy are similar to those of Vectra A (Figures 14.10–14.12) but the magnitudes of the stiffnesses are higher owing to the reinforcing effects of the fibers. In the top layer the stiffnesses are independent of position, and $C_{33} \gg C_{11} > C_{22}$ and $C_{55} > C_{44} > C_{66}$. These results imply that not only the PLC chains but also the glass fibers lie in the 1–3 plane with preferential alignment along the 3 axis. This preferential orientation of fibers, induced by the flow field during processing, is consistent with the optical microscope observations on injection molded, short fiber-reinforced PLC (C.L. Choy, Y.W. Wong and K.W.E. Lau, unpublished results) and conventional thermoplastics [41, 42].

Like Vectra A, the composite exhibits a high elastic anisotropy on the edges of the middle layer because the PLC chains and glass fibers in this region are preferentially aligned along the 3 direction. Although the degree of anisotropy for both the Vectra A and composite decreases as we move away from the edges, it is still rather high at the core of the composite as compared to almost no anisotropy at the core of Vectra A (Figures 14.10–14.12). This indicates that, even in the core, the fibers show a preference to lie in the 1–3 plane and along the 3 direction.

The stiffness constants and Young's moduli of 30 wt% glass fiber-reinforced Vectra A are given in Tables 14.9 and 14.10, respectively. The elastic moduli of 30 wt% glass fiber-reinforced polyphenylene sulfide (PPS), taken from reference 41, are also shown for comparison. PPS is isotropic and has a Young's modulus and shear modulus of 4.0

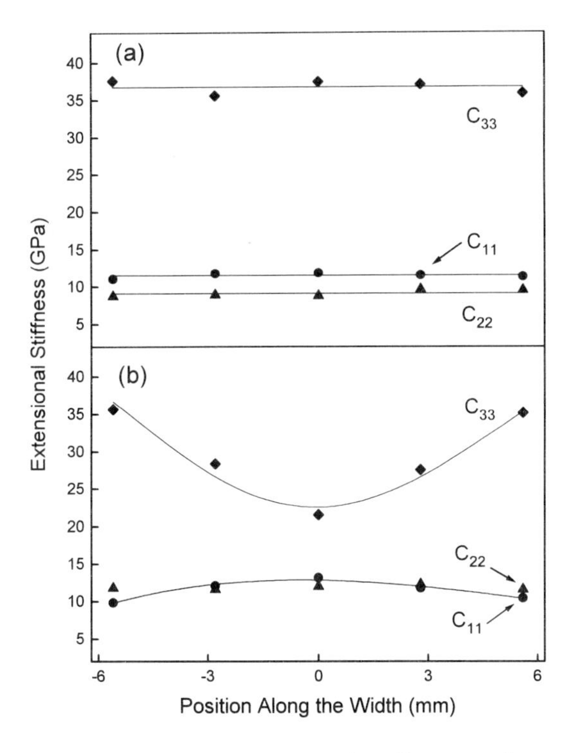

Figure 14.28 Variation of the extensional stiffnesses of injection molded tensile bar of 30 wt% glass fiber-reinforced Vectra A with the position along the width. (a) Top layer, (b) middle layer.

and 1.43 GPa, respectively. As seen from Table 14.6, injection molded Vectra A has a much higher Young's modulus in the 3 direction than PPS, and this leads to the substantially higher E_3 value for the composite with Vectra matrix. In the 1 or 2 direction the Young's modulus of Vectra A is comparable to that of PPS, so the two composites have similar E_1 and E_2 values. For the same reason, C_{44} and C_{55} values for the composite with a Vectra matrix are significantly higher than those for the composite with a PPS matrix.

14.6 ELASTIC MODULI AND ACOUSTIC ABSORPTION OF COMB POLYMER LIQUID CRYSTALS

Unlike main chain PLCs, the mesogenic units in side chain PLCs are located not in the polymer backbone but in the side chains. The dynamics of a side chain PLC is therefore determined by the behavior of both the backbone and the mesogenic groups as well as the coupling between them. Neutron scattering studies [43, 44] reveal that the polymer

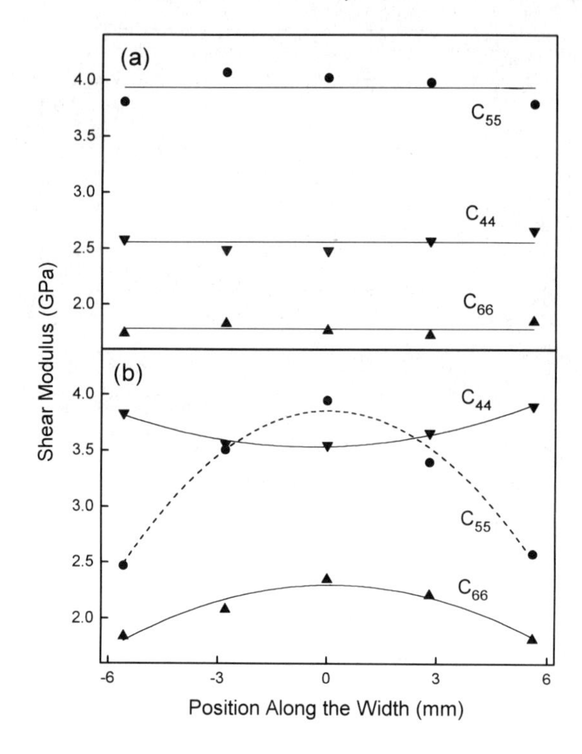

Figure 14.29 Variation of the shear moduli of injection molded tensile bar of 30 wt% glass fiber-reinforced Vectra A with the position along the width. (a) Top layer, (b) middle layer.

Table 14.9 Stiffness constants (in GPa) of injection molded tensile bars of 30 wt% glass fiber-reinforced Vectra A and polyphenylene sulfide (PPS)

		C_{11}	C_{22}	C_{33}	C_{12}	C_{13}	C_{23}	C_{44}	C_{55}	C_{66}
Vectra A	Top	11.6	9.11	36.2	6.31	7.58	7.12	2.56	3.94	1.78
	Middle	11.5	11.9	28.6	6.86	6.93	6.72	3.70	3.18	2.06
	Whole	11.6	10.4	31.1	6.54	7.31	6.84	3.07	3.63	1.88
PPS	Whole	11.7	11.0	18.2	6.30	6.80	6.55	2.28	3.05	2.12

backbone is preferentially aligned perpendicular to the axis of the mesogenic unit. The ratio of the radius of gyration of the backbone perpendicular and parallel to the director is about 1.2 in the nematic phase and 4 in the smectic-A phase. This structural information has been used by several groups of investigators to help interpret the elastic

Table 14.10 Young's moduli (in GPa) and Poisson's ratios of injection molded tensile bars of 30 wt% glass fiber-reinforced Vectra A and polyphenylene sulfide (PPS)

		E_1	E_2	E_3	ν_{12}	ν_{21}	ν_{13}	ν_{31}	ν_{23}	ν_{32}
	Top	7.00	5.39	29.7	0.481	0.625	0.366	0.086	0.528	0.096
Vectra A	Middle	7.17	7.53	23.6	0.533	0.507	0.405	0.123	0.331	0.106
	Whole	7.15	6.43	25.4	0.499	0.555	0.402	0.113	0.405	0.103
PPS	Whole	7.44	7.02	13.2	0.421	0.446	0.377	0.213	0.380	0.203

modulus and acoustic absorption measurements at ultrasonic [16, 17] and hypersonic [12–15] frequencies.

14.6.1 Ultrasonic measurements

Benguigui and coworkers [16, 17] are the only group which has measured the ultrasonic modulus and absorption of side chain PLC. The PLC studied is a polymethylsiloxane with the following chemical structure:

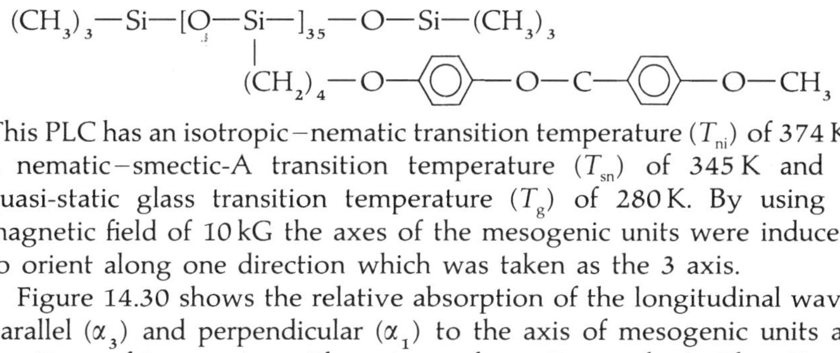

$$(CH_3)_3\text{—Si—[O—Si—]}_{35}\text{—O—Si—}(CH_3)_3$$
$$(CH_2)_4\text{—O—}\langle\bigcirc\rangle\text{—O—C—}\langle\bigcirc\rangle\text{—O—}CH_3$$

This PLC has an isotropic–nematic transition temperature (T_{ni}) of 374 K, a nematic–smectic-A transition temperature (T_{sn}) of 345 K and a quasi-static glass transition temperature (T_g) of 280 K. By using a magnetic field of 10 kG the axes of the mesogenic units were induced to orient along one direction which was taken as the 3 axis.

Figure 14.30 shows the relative absorption of the longitudinal wave parallel (α_3) and perpendicular (α_1) to the axis of mesogenic units as functions of temperature. There is no absorption peak at either T_{ni} or T_{sn}, but a broad peak is observed for both α_1 and α_3 at about 318 K. This is associated with large scale segmental motions of the polymer backbone, so $T_g^{dyn} = 318$ K may be regarded as the dynamic glass transition temperature at the measurement frequency (2 MHz). Since the broad relaxation peak covers a wide range from 280 to 370 K, the dynamic glass transition is superimposed on the mesomorphic transitions, thereby affecting the motions of mesogenic groups in both the nematic and smectic phases.

Figure 14.31 shows the axial (v_{33}) and transverse (v_{11}) velocity of the longitudinal wave as functions of temperature. There is no discontinuity in the velocity at either T_{ni} or T_{sn}. The velocity becomes anisotropic at T_{ni} and the anisotropy, $(v_{33} - v_{11})/v_{11}$, increases from

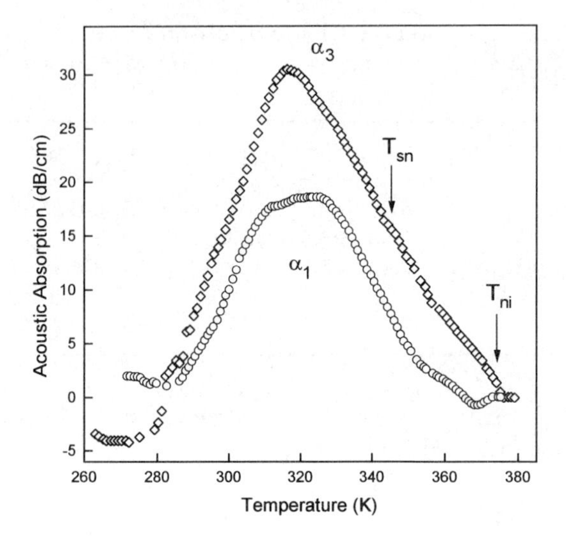

Figure 14.30 Temperature dependence of the relative absorption of the longitudinal wave at 2 MHz parallel (α_3) and perpendicular (α_1) to the axis of the mesogenic units of a side chain polymer liquid crystal with a polymethylsiloxane backbone. (Adapted from [17] by permission of Israel Institute of Technology.)

about 1% in the nematic state to 18% at 270 K. The behavior in the nematic state is different from that of monomer liquid crystals. As a result of their fluidity and continuously broken orientational symmetry, monomer liquid crystals show no elastic anisotropy and no shear stiffness at frequencies from 0 to 10 MHz [45, 46]. Although the nematic range of the PLC is located at 30 K above T_g^{dyn}, it seems that the segmental motion of the backbone has a small but detectable effect on the dynamics of the mesogenic groups, thereby giving rise to the 1% anisotropy in velocity.

Benguigui *et al.* [16] have also measured the longitudinal velocity as a function of angle relative to the 3 axis, from which the stiffness constants C_{11}, C_{13}, C_{33} and C_{44} have been deduced. The axial shear modulus C_{44} was found to be zero in the nematic state (345–374 K). As shown in Figure 14.32, C_{44} is close to zero even at 295 K, but it increases substantially with decreasing temperature, reaching a value of about 0.06 GPa at 270 K. This is more than one order of magnitude lower than the shear modulus of main chain PLC or conventional polymers in the glassy state. In contrast, the values of C_{11} and C_{33} are comparable to those of main chain PLCs or conventional polymers with low degrees of orientation. The stiffness constant C_{13} exhibits the unusual behavior of decreasing with decreasing temperature below a

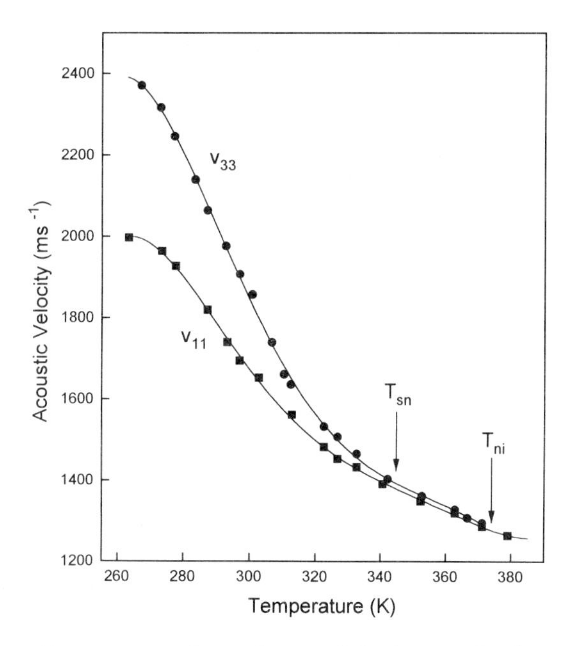

Figure 14.31 Temperature dependence of the longitudinal wave velocity at 2 MHz parallel (v_{33}) and perpendicular (v_{11}) to the axis of the mesogenic units of a side chain polymer liquid crystal with a polymethylsiloxane backbone. (Adapted from [16] by permission of the *J. Physique*.)

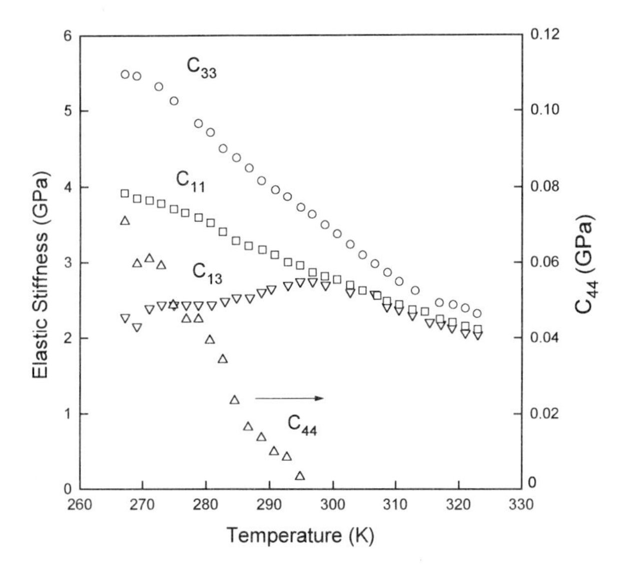

Figure 14.32 Temperature dependence of the elastic stiffnesses of a side-chain polymer liquid crystal with a polymethylsiloxane backbone. (Adapted from [16] by permission of the *J. Physique*.)

temperature of 295 K, which is inconsistent with previous studies on polymers [5, 13, 14, 28]. Since the longitudinal velocity is not sensitively dependent on C_{13}, the unusual temperature dependence of C_{13} may not be real but may arise from the uncertainty of the experimental data.

For side chain PLCs there are two orienting molecular elements which can influence the elastic properties: the mesogenic groups and the monomer units of the polymer backbone. The fact that C_{33} is significantly higher than C_{11} implies that the effect of the preferential orientation of the backbone (orthogonal to the director) is overcompensated by the orientation of the mesogenic side chains.

14.6.2 Brillouin scattering

Kruger and coworkers [13–15] have employed the Brillouin scattering technique to determine the stiffness constants of three side chain PLCs with a polymethacrylate backbone. The chemical structures of these PLCs are shown in Figure 14.33 while the transition temperatures and stiffness constants are given in Table 14.11. For these measurements,

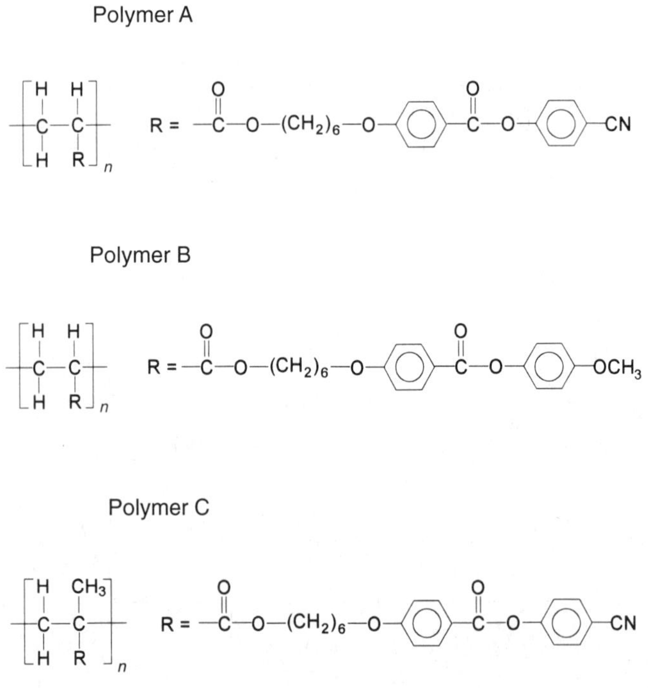

Polymer A

R = —CO—O—$(CH_2)_6$—O—C_6H_4—CO—O—C_6H_4—CN

Polymer B

R = —CO—O—$(CH_2)_6$—O—C_6H_4—CO—O—C_6H_4—OCH_3

Polymer C

R = —CO—O—$(CH_2)_6$—O—C_6H_4—CO—O—C_6H_4—CN

Figure 14.33 Chemical structure of three side chain polymer liquid crystals with a polymethacrylate backbone.

Table 14.11 Transition temperatures and room temperature stiffness constants (in GPa) of three side chain polymer liquid crystals with a polymethacrylate backbone

Polymer	T_g (K)	T_{sn} (K)	T_{ni} (K)	C_{11}	C_{33}	C_{13}	C_{44}	C_{66}
A	295	–	378	5.90	12.9	4.76	1.15	0.95
B	290	357	378	5.49	12.5	4.00	0.64	–
C[a]	274	–	330	6.72	13.7	5.04	1.34	–

[a] The stiffness constants for polymer C refer to a temperature of 220 K.

film samples of thickness about 30 μm were deposited on polyimide coated glass slides which had been rubbed along an arbitrary direction in the film plane to induce the mesogenic units to align in that direction.

Figures 14.34–14.36 show the temperature dependence of the stiffness constants of polymers A, B and C, respectively. As a result of the high frequency (4 GHz) of Brillouin scattering measurements, the sound absorption maximum (and hence the dynamic glass transition temperature T_g^{dyn}) for polymer C was found to be located at 390 K, i.e., at 116 K above T_g. If we assume that T_g^{dyn} for polymers A and B are similar to that of polymer C, then the behavior of the stiffness constants of all three PLCs can be readily understood.

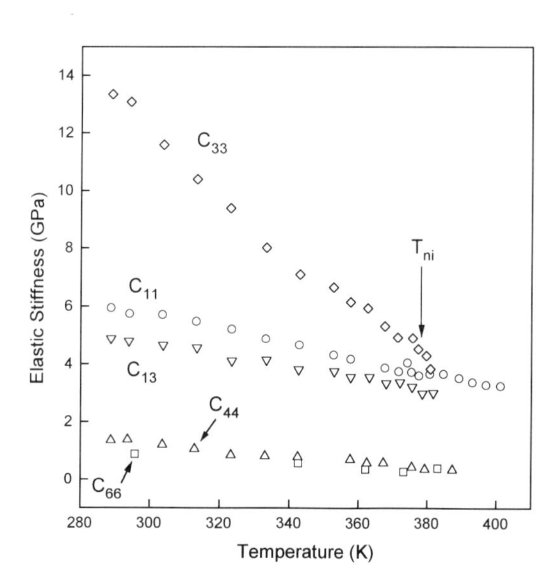

Figure 14.34 Temperature dependence of the elastic stiffnesses of polymer A. (Adapted from [13] by permission of The American Physical Society.)

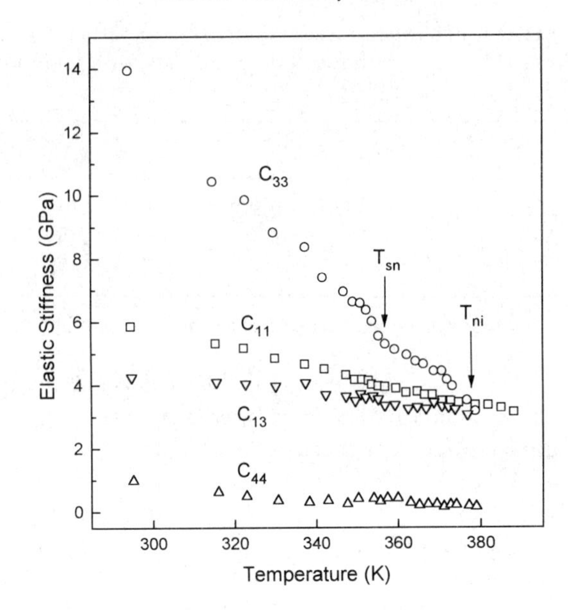

Figure 14.35 Temperature dependence of the elastic stiffnesses of polymer B. (Adapted from [14] by permission of Springer-Verlag.)

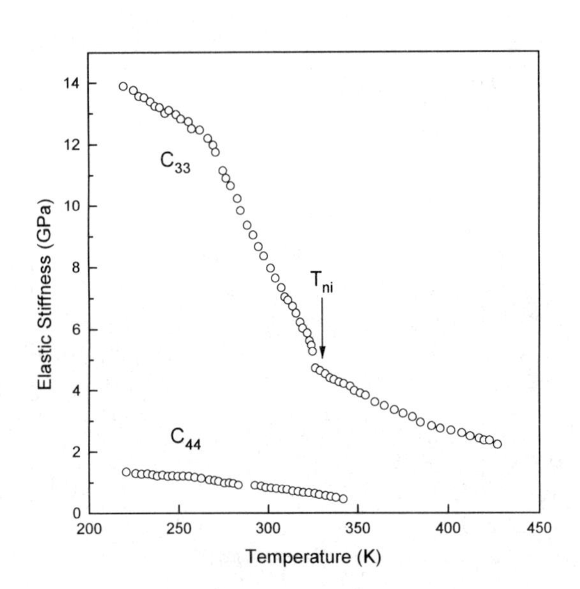

Figure 14.36 Temperature dependence of the elastic stiffnesses of polymer C. (Adapted from [15] by permission of Springer-Verlag.)

Let us first concentrate on polymer A. It is seen from Figure 14.34 that, like the ultrasonic measurements on the PLC with a polymethylsiloxane backbone, the anisotropy in the extensional stiffness first appears at T_{ni}. However, the anisotropy increases sharply with decreasing temperature so that C_{33} exceeds C_{11} by more than a factor of two at 300 K. This is much larger than the anisotropy (a few percent) observed in nematic monomer liquid crystals [47]. The difference in behavior of monomer and polymer liquid crystals indicates that the polymer backbone has a strong influence on the motion of the mesogenic groups. Since $T_{ni} < T_g^{dyn}$, the freezing of large scale segmental motions leads to a severe constraint on the rotation of the mesogenic groups in the nematic state, thereby resulting in a large anisotropy in the extensional stiffness. Probably for the same reason, the shear moduli C_{44} and C_{66} are not zero and have values of about 0.3 GPa even at T_{ni}. At room temperature, both the magnitude and the anisotropy pattern of the stiffness constants resemble those of conventional polymers with a moderate degree of orientation [4]. Like the ultrasonic data on polymethylsiloxane, the anisotropy $C_{33} > C_{11}$ implies that the elastic behavior of side chain PLCs with a polymethacrylate backbone is dominated by the orientation of the mesogenic groups.

Polymer B differs from polymers A and C in that it goes through a nematic–smectic-A transition ($T_{sn} = 357$ K) before reaching room temperature. However, as seen from Table 14.11, the low temperature values of C_{11}, C_{33} and C_{13} are quite similar for all three polymers. The axial shear modulus C_{44} for polymer B is lower by a factor of two. Whether this can be attributed to its smectic state or chemical structure is an open question.

14.6.3 Laser-induced phonon spectroscopy

We have already discussed elastic modulus measurements at ultrasonic (2 MHz) and hypersonic (4 GHz) frequencies. Besides these studies, there is a report on room temperature phonon spectroscopy measurements (at a frequency of several hundred megahertz) on a side chain liquid crystalline elastomer [12]. The mesogenic units are benzoic acid phenyl esters with a methoxy end-group which is attached to the polysiloxane network by a butoxy spacer. The transition temperatures for this polymer are $T_g = 281$ K and $T_{ni} = 329$ K. Mechanical stretching at a temperature a few degrees below T_{ni} to draw ratios $\lambda = 1.05$ and 1.2 induces the mesogenic side groups to orient parallel to the draw direction. The stiffness constants obtained are independent of draw ratio and have the following values: $C_{11} = 5.29$ GPa, $C_{33} = 7.62$ GPa, $C_{13} = 5.80$ GPa and $C_{44} \leqslant 0.3$ GPa.

A legitimate comparison of the elastic anisotropy of different PLCs can be made by considering the behavior at the same relative temperature $T = 0.9\,T_{ni}$. At this temperature $(C_{33} - C_{11})/C_{11}$ is about 2, 44 and 50% at frequencies of 2 MHz, several hundred MHz and 4 GHz, respectively. As explained earlier, the substantially larger anisotropy at the higher frequencies arises from the fact that the dynamic glass transition temperature is higher than T_{ni}. Therefore, at $T = 0.9\,T_{ni}$ the molecular segments in the backbone are immobile and this gives rise to a strong restriction on the motion of the mesogenic groups.

14.7 CONCLUSIONS

It is seen that ultrasonic techniques have two major advantages: high accuracy and the ability to determine the complete set of elastic constants for a sample of small size. As a result of these capabilities, all five independent stiffness constants have been obtained for extruded rods of PLC of high draw ratio ($\lambda = 15$) and small diameter (0.8 mm). Moreover, it has been possible to study the skin–core structure in injection molded PLC by monitoring the variation in stiffness with position. For blends of a PLC and a thermoplastic or glass fiber-reinforced PLC, successful correlation has been obtained between the modulus data and the orientation of the PLC fibrils or glass fibers.

For side chain PLCs, the extensional stiffness in the nematic state is higher along the axis of the mesogenic units than in the perpendicular direction, and the anisotropy increases with increasing frequency. As the frequency increases, the dynamic glass transition is shifted to temperatures higher than T_{ni}, so the effect of the freezing of segmental motions becomes more important.

There have been very few studies of the acoustic absorption of PLCs, probably because of the difficulty in interpreting the data. To date, absorption measurements have been carried out mainly to reveal the relaxation peak associated with the glass transition in side chain PLCs.

ACKNOWLEDGEMENTS

Part of this review was written while the author was on a sabbatical leave at the University of Technology, Sydney, and he wishes to thank Prof. Joe Unsworth, Prof. A.R. Moon and Dr A. Ray for their hospitality. Thanks are also due to Mr H.M. Ma for drawing the figures.

REFERENCES

1. Ward, I.M. (1990) *Mechanical Properties of Solid Polymers*, 2nd ed, Wiley, New York.

2. Read, B.E. and Dean, G.D. (1978) *The Determination of the Dynamical Properties of Polymers and Composites*, Adam Hilger, Bristol.
3. Chan, O.K., Chen, F.C., Choy, C.L. and Ward, I.M. (1978) *J. Phys. D*, **11**, 617.
4. Leung, W.P., Chan, C.C., Chen, F.C. and Choy, C.L. (1980) *Polymer*, **21**, 1148.
5. Choy, C.L., Leung, W.P., Ong, E.L. and Wang, Y. (1988) *J. Polymer Sci. Phys.*, **26**, 1569.
6. Dyer, S.R.A., Lord, D., Hutchinson, I.J. *et al.* (1992) *J. Phys. D*, **25**, 66.
7. Wedgbury, M.K. and Read, B.E. (1986) Report DMA(A) 118, National Physical Laboratory, Teddington, UK.
8. Sweeney, J., Brew, B., Duckett, R.A. and Ward, I.M. (1992) *Polymer*, **33**, 4901.
9. Choy, C.L., Leung, W.P. and Yee, A.F. (1992) *Polymer*, **33**, 1788.
10. Choy, C.L., Lau, K.W.E., Wong, Y.W. and Yee, A.F. (1996) *Polymer Eng. & Sci.*, **36**, 1256.
11. Chik, G.L., Li, R.K.Y. and Choy, C.L. (1997) *J. Mater. Proc. Tech.*, **63**, 488.
12. Degg, F.W., Diercksen, K., Schwalb, G. and Brauchle, C. (1991) *Phys. Rev. E.*, **44**, 2830.
13. Kruger, J.K., Peetz, L., Siems, R. *et al.* (1988) *Phys. Rev. A.*, **37**, 2637.
14. Kruger, J.K., Grammes, C. and Wendorff, J.H. (1989) *Progr. Colloid & Polymer Sci.*, **80**, 45.
15. Kruger, J.K., Grammes, C. and Wendorff, J.H. (1989) *Springer Proc. in Phys.*, **37**, 216.
16. Benguigui, L., Ron, P., Hardouin, F. and Mauzac, M. (1989) *J. Physique*, **50**, 529.
17. Ali, A.H. (1992) PhD Thesis, Technion – Israel Institute of Technology, Haifa, Israel.
18. Musgrave, M.J. (1950) *Rep. Prog. Phys.*, **22**, 77.
19. Gutierrez, G.A., Chivers, R.A., Blackwell, J. *et al.* (1983) *Polymer*, **24**, 937.
20. Biswas, A. and Blackwell, J. (1988) *Macromolecules*, **21**, 3158.
21. Hanna, S., Lemmon, T.J., Spontak, R.J. and Windle, A.H. (1992) *Polymer*, **33**, 3.
22. Spontak, R.J. and Windle, A.H. (1990) *J. Mater. Sci.*, **25**, 2727.
23. Spontak, R.J. and Windle, A.H. (1990) *Polymer*, **31**, 1395.
24. Hudson, S.D. and Lovinger, A.J. (1993) *Polymer*, **34**, 1123.
25. Sawyer, L.C., Chen, R.T., Jamieson, M.G. *et al.* (1993) *J. Mater. Sci.*, **28**, 225.
26. Ward, I.M. (1962) *Proc. Phys. Soc.*, **80**, 1176.
27. Choy, C.L., Kwok, K.W. and Ma, H.M. (1995) *Polymer Compos.*, **16**, 357.
28. Choy, C.L. and Leung, W.P. (1985) *J. Polymer Sci. Phys.*, **23**, 1759.
29. Kawabata, S., Sera, M., Kotani, T. *et al.* (1993) in *Proceedings of the 9th International Conference on Composite Materials*, p. 671.
30. Tashiro, K., Kobayashi, M. and Tadokoro, H. (1978) *Macromolecules*, **11**, 914.
31. Troughton, M.J., Unwin, A.P., Davies, G.R. and Ward, I.M. (1988) *Polymer*, **29**, 1389.
32. Zhang, H., Davies, G.R. and Ward, I.M. (1992) *Polymer*, **33**, 2651.
33. Weng, T., Hiltner, A. and Baer, E. (1986) *J. Mater. Sci.*, **21**, 744.
34. Garg, S.K. and Kenig, S. (1988) in *High Modulus Polymers* (eds A.E. Zachariades and R.S. Porter), Marcel Dekker, New York.
35. Choy, C.L., Leung, W.P. and Kwok, K.W. (1991) *Polymer Commun.*, **32**, 285.
36. Chung. T.S. (1988) *J. Polymer Sci. Phys.*, **26**, 1549.
37. Kiss, G. (1987) *Polymer Eng. & Sci.*, **27**, 410.
38. Lin, Q. and Yee, A.F. (1994) *Polymer Compos.*, **15**, 754.
39. Ashton, J.E., Halpin, J.C. and Petit, P.H. (1969) *Primer on Composite Materials: Analysis*, Technomic, Stamford, Conn.
40. Choy, C.L., Wong, Y.W., Lau, K.W.E. *et al.* (1996) *Polymer Eng. & Sci.*, **36**, 827.

41. Choy, C.L., Kwok, K.W., Leung, W.P. and Lau, F.P. (1992) *Polymer Compos.*, **13**, 69.
42. Choy, C.L., Kwok, K.W., Leung, W.P. and Lau, F.P. (1993) *J. Thermoplast. Compos. Mater.*, **6**, 226.
43. Keller, P., Carvalho, B., Cotton, J.P. *et al.* (1985) *J. Phys. Lett.*, **46**, L1065.
44. Kirste, R.G. and Ohm, H.G. (1985) *Makromol. Chem., Rapid Commun.*, **6**, 179.
45. de Gennes, P.G. (1979) *The Physics of Liquid Crystals*, Clarendon Press, Oxford.
46. Miyano, K. and Ketterson, J.B. (1979) in *Physical Acoustics*, Vol. 14 (eds W.P. Mason and R.N. Thurston), Academic Press, New York.
47. Grammes, C., Kruger, J.K., Bohn, K.P. *et al.* (1995) *Phys. Rev. E*, **51**, 430.

15

Computer simulations

Witold Brostow

15.1 MOTIVATION FOR COMPUTER SIMULATIONS

As we are going to show later in this chapter, computer simulations provide capabilities of solving problems which cannot be solved by theory or experiment, or even a combination of both. Let us start with asking how scientists and engineers approach physical systems. The question is fairly general, not even limited to materials science and engineering (MSE), let alone to polymer liquid crystals (PLCs). The answer is that one first *describes* the system, specifying the parts and their interactions, and then tries to make *predictions* what the system will do in the future under certain imposed conditions.

Formulating a model to describe a physical system involves deciding which known features of the system are important and decisive for its behavior, and which are of secondary or little importance. If we make the model too simple, predictions will be easy to make, but could be wrong. If the model is too complex, the predictions might be more realistic, but they might be hard to make in the first place. Even if one walks the middle ground, it often happens that the model – via a *theory* – leads to conclusions which do not agree with the experiment. There are several possible explanations. Perhaps the experimentalist did not measure what he or she thought is being measured. Or the mathematics used to go from the model to predictions was inappropriate for the

Mechanical and Thermophysical Properties of Polymer Liquid Crystals
Edited by W. Brostow
Published in 1998 by Chapman & Hall, London.
ISBN 0 412 60900 2

problem. However, a more probable reason for the disagreement than the two reasons just mentioned is that at least one important feature of the system was not included in the model.

The missing feature or features can be best found by doing computer simulations. The model implemented on the computer is exactly that assumed for the system. Hidden variables that lurk in nature are eliminated. If there is disagreement now, it cannot be blamed on experimental inaccuracies. If there are problems in the mathematical approximations made in deriving the predictive equations, they will become visible. The theory and the 'computer experiments' are both based on exactly the same model. Parameters of the model can be varied – to see how simulation results depend on them. Assumptions of the computer model can be varied as well – to see which were realistic and lead to sensible results, and which were in fact the root(s) of the problem(s). Improved understanding of the system *and* a better model with enhanced (or possibly even correct) predictive capabilities is a typical result.

While we have noted the serious problems in theory testing related to experimental errors, computer simulations involve possible errors as well. Two questions are important here: (1) is the program really doing what the programmer *thinks* it is doing, and (2) moreover, does the algorithm contain essential features of the simulated system *and nothing else*? These are, respectively, the problems of *verification* and *validation* of simulations. Both problems have been discussed in detail by Bratley, Fox and Schrage [1]. Let me provide here just one example from their book, a verification procedure called *stress testing*. Such testing consists in imposing nonsensical conditions, such as a mole fraction of a component in a mixture which is larger than unity, or a thermodynamic temperature (such as the Kelvin scale) which is negative. The simulated system should then collapse in a predictable way. However, if the simulation produces similar results as before, then something is wrong. Bratley, Fox and Schrage provide several procedures each for verification as well as for validation [1].

15.2 TYPES OF COMPUTER SIMULATIONS OF MATERIALS

We have three basic types of computer simulations which can be applied to polymer systems: Monte Carlo (MC), molecular dynamics (MD) and Brownian dynamics (BD). The last of these is mostly applied to polymer solutions [2, 3] and will not be discussed here further. The first two are of interest in simulations of PLCs.

The MC method has an interesting history. In the late 1930s a group of Polish mathematicians associated with the King John Casimir University in Lviv and which included Stanislaw Ulam spent much time in the Scottish Cafe near the University. Somebody in this group suggested that winning in roulette (taking the bank) in the Casino in Monte Carlo is really a problem in mathematics. Interesting attempts to solve the problem were advanced and argued about – until World War II broke out in 1939. The original problem was never solved – which is why the Casino in Monaco is still in business. However, when Ulam emigrated to the United States, he found one of the results obtained in the Scottish Cafe useful for other purposes. Because of the war and the dispersion of the original participants around the globe, the *Monte Carlo* procedure was published only in 1949 by Metropolis and Ulam [4].

MC relies on making one step at a time, and taking stock of the situation after a certain number of such steps. Applying MC to material systems, we create an object which consists of N particles, such as polymer segments (or atoms, or monoatomic molecules, or ions). One can construct a $6N$-dimensional *phase space*, since each particle has three coordinates of position (for instance the Cartesian ones) plus three coordinates of momentum. One calculates the equilibrium properties of a system by averaging over a large number of states of the system, each state created by the MC procedure. For instance, one can determine experimentally the average radius of gyration $\langle R_g \rangle$ of a polymer. MC gives us two ways of obtaining $\langle R_g \rangle$. We can either obtain a large number of chains and perform averaging over them. Or we can create just one chain, perform, say, 20000 MC steps, and average over all these states of a single chain. Statistical mechanics tells us that the resulting $\langle R_g \rangle$ should be the same. Because of its nature, MC is well suited for the evaluation of equilibrium properties.

MD was developed by Berni J. Alder and collaborators at the Lawrence Livermore Laboratory; the first paper authored by Alder and Wainwright appeared in 1957 [5]. Again consider a system of particles, each at a given time at a location specified, and with velocity (or momentum) also specified. In contrast to MC in which one particle moves at a time, here we let all particles 'run' at once. The velocity $\boldsymbol{v}_i(\boldsymbol{t})$ of the ith particle at time t can be for instance calculated as

$$\boldsymbol{v}_i(t + \Delta t/2) = \boldsymbol{v}_i(t - \Delta t/2) + \boldsymbol{F}_i(t)\,\Delta t/m_i \tag{15.1}$$

Here Δt is the time step assumed for the simulation, $\boldsymbol{F}_i(t)$ the force acting on the ith particle and m_i the mass of the particle. $\boldsymbol{F}_i(t)$ is calculated from potentials which describe interaction between particles (more on this subject below). Since in MD time is an explicit variable, the method can be used to simulate not only equilibrium properties but also time dependent ones. This is particularly important for viscoelastic materials with which we are dealing; MD gives us the capability to investigate processes such as diffussion, flow and fracture.

This is a powerful feature since the motions of polymer chains constitute the key to understanding their properties.

15.3 CONSTRUCTING A PLC SYSTEM ON A COMPUTER

PLCs are relatively complicated; systems of flexible macromolecular chains are simpler. A task simpler still consists of creating a metallic material – that is a system of molecules which are monoatomic. A convenient structure on which one can place the atoms is the two-dimensional hexagonal lattice. The lattice has several advantages [6]: all nearest neighbors are equivalent (compare the square or the simple cubic lattice); the coordination number is $z = 6$, a value which occurs typically in real three-dimensional lattices; at the same time, we retain the perspicuity resulting from the two-dimensionality of the lattice. Thus one can start with a metal-like lattice first (chains of one particle each) and then include the connectedness between the particles. There are several methods of creating chains, and a good one was developed by Mom [7]. The essential step consists of finding two chains such that an end particle of one of them is a nearest neighbor of an end particle of the other chain. When such chains are found, they become connected forming one longer chain. Thus the total number of chains in the material decreases by one. Needless to say, the process is quite easy at the beginning, then more and more difficult, and at some point no more chains can be connected in this way. In the Mom procedure the system is then scanned to see whether there is a particle inside of one of the chains that is a nearest neighbor of one of the end particles of another chain. When such a particle is found, a part of the former chain is connected to the latter. In such a way the number of chains does not decrease, but the lengths of both chains change while new particles appear on their ends. A random choice is made to accept or reject such a move – this to prevent an infinite loop into which the program execution might fall. The Mom method fails when in all chains the end particles are surrounded by particles belonging to the same chain. For such a case a modification of the Mom procedure was developed [8]. A particle is sought that is the nearest neighbor of an end particle of the same chain. Then the chain is rearranged so that the particle becomes connected to the old chain end, and it becomes the new chain end. Such computer generation procedures are relatively time consuming (in one case 18 532 iterations were needed to create a system with 3600 particles [8]). However, polymeric materials with realistic features are created, as confirmed also by MD simulations of stress relaxation [8].

Once a flexible polymeric material is generated, a PLC may be generated by making some sequences in all chains *rigid*, so as to achieve a prescribed concentration θ of LC sequences in the material. The rigidity is achieved by introducing a different interaction potential $u(R)$, where R is the interparticle distance [9] (see Figure 15.1). The double well for the flexible + flexible interactions mimics the *trans* and *gauche* conformations in carbon-like chains. At large intersegmental distances *bond scission* is possible – an important feature when studying behavior in mechanical force fields.

It has been found experimentally that above a certain critical concentration $\theta_{\text{LC limit}}$, the LC sequences form a separate phase [10–15]. That phase constitutes *islands* in the flexible matrix [13–15] (see also Chapter 9 by Hess and López). The next problem is thus creating islands on the computer of appropriate shapes and sizes. Here, once again computer simulations have an advantage over experiments. Experimentally we cannot create islands of the desired size or shape; on a computer it can be done easily, and shapes such as ovals, diamonds or squares have been created (W. Brostow, C. Browning and M. Drewniak, in preparation).

15.4 PERFORMANCE OF SIMULATIONS

The simulation of PLCs can be well performed using the constant temperature MD procedure as described for instance by van Gunsteren

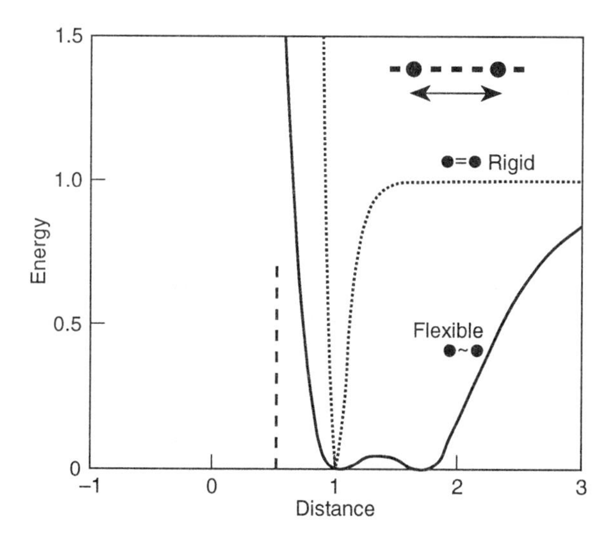

Figure 15.1 Binary interaction potentials used in the simulations. The flexible potential incorporates a double well, making possible conformational transitions (such as *cis* to *trans*).

[16]. As seen in equation (15.1) above, the Newton differential equations of motion are transformed into difference equations employing the finite time step Δt. We also see that the velocities at the time $t + \Delta t/2$ are calculated from those at the time Δt earlier. One usually works for simplicity with reduced (dimensionless) values of energy, distance, time, temperature and mass [17].

To calculate the forces, one has to know which particles (in PLC simulations these are segments, i.e. flexible or rigid) constitute neighbors of a given particle; otherwise the calculation would be longer than necessary while the results are the same. Efficiency in the simulation can be achieved by renewing the list of pairs of nearest neighbors after each time step [9]. Thus dynamic changes in the system are recorded until the end of the simulation, usually defined by the fracture of the material.

It is important to maintain the temperature constant. This can be achieved by multiplying the velocities of the particles at every time step by a factor λ defined as

$$\lambda = \{1 + 0.05(E_{k0}/E_k - 1)\}^{1/2} \tag{15.2}$$

where E_{k0} is a predefined value of the kinetic energy of the systems while E_k is the actual value of that energy. E_{k0} is of course defined in terms of the temperature to be maintained: the multiplying factor 0.05 assures that the coupling is sufficiently weak.

The *periodic boundary conditions* are used to eliminate possible problems resulting from the finite size of the system. This is because we cannot create one mole of particles on a computer; even if we could, the time of a single run would be enormous. If the system is contained in a cubic box, then 26 identical copies ('ghosts') are built on the computer around it. Whenever a segment makes a move such that it leaves the central box, a 'ghost' of that segment automatically enters the box from the opposite side. Thus the number of segments in the box remains constant, and we do not have to worry about surface effects. This is important for systems consisting of a finite number of chains of finite length. We have mentioned in section 15.2 the equivalence of averaging over systems with averaging over time for a single system. Strictly speaking, the equivalence applies to an infinite number of systems on one side and an infinite time on the other. In practice, one generally tries to build a system as large as possible consistent with getting results within a reasonable time.

15.5 RESULTS

Development of an MD code which would create systems of PLC chains on the computer and then subject them to external mechanical forces is apparently not quite easy, in spite of the principles defined above. At the time of writing there was only one paper published with results of such simulations [9], but there is also ongoing continuation of this work (W. Brostow, C. Browning and M. Drewniak, in preparation). We shall report below those results which appear significant within the overall framework of mechanical properties of PLCs.

As discussed in section 15.1, the problem of validation of the simulations is important. Therefore, the first and obvious question is does the system created on a computer as described above and with interaction potentials shown in Figure 15.1 really behave as a real polymer?

To reply to this question, first a simpler system – namely a flexible polymeric material – was simulated and its stress–strain behavior investigated [9]. This was performed for the temperature of 300 K and assuming that the bond dissociation energy is equal to the dissociation energy of a C—C bond in polyethylene. The results are shown in Figure 15.2. We see that the computer-generated diagram of stress σ vs. strain ε has all the features of such diagrams known from experiments; see the regions marked in the figure. Thus the validation of the simulated system has been accomplished.

It has been argued above that simulations can provide us with insights which experiments cannot. The diagram shown in Figure 15.2 lends itself to supporting this statement. Along with stress and strain, in the same simulation the average length of the *cis* (short) bonds, average

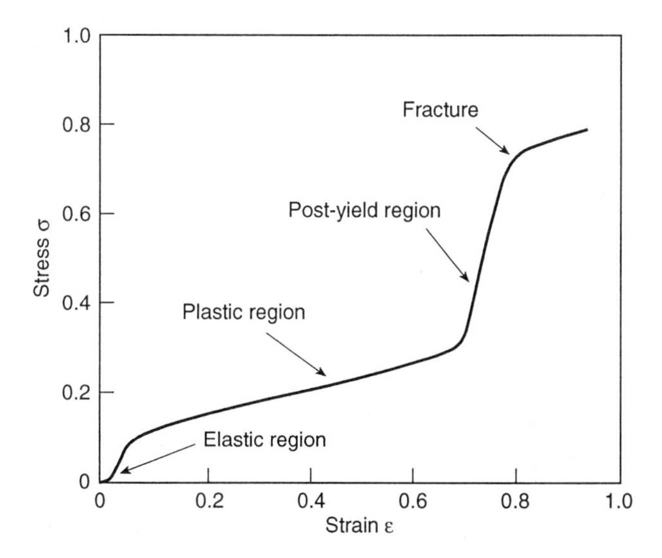

Figure 15.2 Computer-generated diagram of stress σ versus strain ε for polymer consisting of flexible chains only ($\theta = 0$).

length of the *trans* (long) bonds and the fraction of *trans* bonds were followed [9]. The results shown in Figure 15.3 explain the existence of the nearly horizontal plastic region in the stress–strain diagram: it corresponds to the gradual conversion of the bonds from *cis* to *trans*. The molecular-level mechanism of the post-yield region leading to fracture is also explained: this is where the length of the *trans* bond increases. Neither result could be explained on the basis of experimental stress–strain curves alone. Thus, even the validation stage has provided an important result for the understanding of the mechanical behavior of polymers.

At least one more result for a polymeric material consisting of fully flexible chains is worth mentioning here. In Figure 15.4 we show similar stress vs. strain curves as in Figure 15.2, but for several temperatures T. We see clearly, among other things, how in the plastic region increasing T results in lowering of the stress necessary to achieve a given level of strain. This result is expected, but it provides one or more validations of the material we have generated on the computer and of our procedure of imposing stress. At the same time, the results for a flexible polymeric material can serve as a reference for comparison with those for PLCs.

Such a comparison is shown in Figure 15.5 as stress vs. strain curves for random copolymers containing varying concentrations θ of LC (that is rigid) segments. For a given concentration θ one can have a large

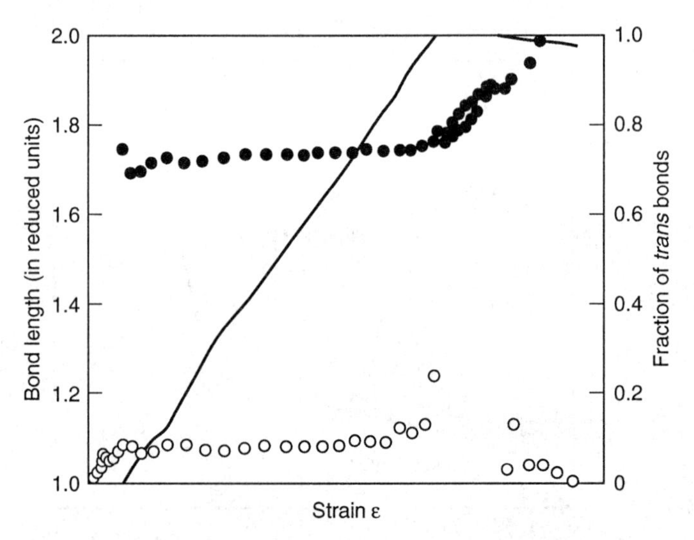

Figure 15.3 Average length of *cis* bonds (lower series of circles), average length of *trans* bonds (upper series of circles) and fraction of *trans* bonds (continuous line), each as a function of the strain ε, for a system of flexible chains ($\theta = 0$). (Redrawn from [9].)

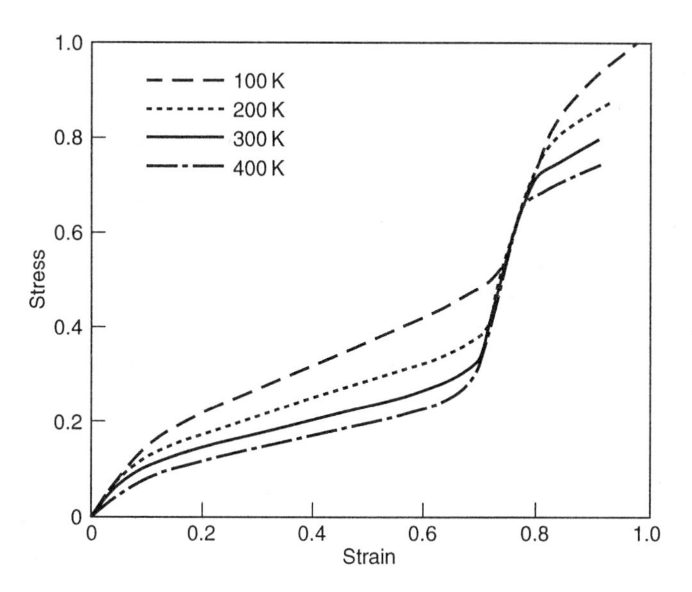

Figure 15.4 Computer-generated diagram of stress σ versus strain ε for $\theta = 0$ at four different temperatures. (Redrawn from [8].)

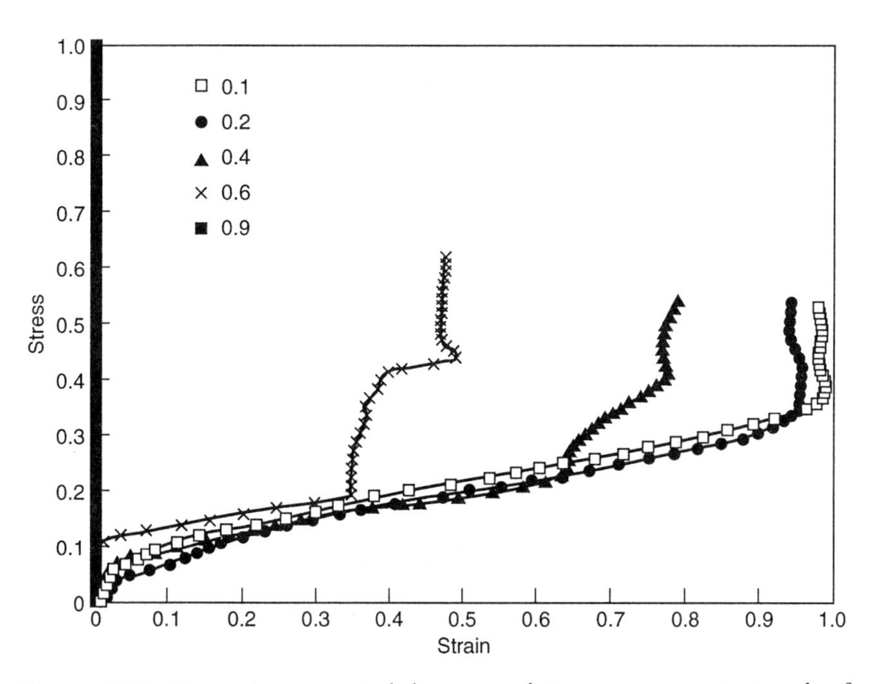

Figure 15.5 Computer-generated diagrams of stress σ versus strain ε for θ values in random copolymers varying from 0 to 0.9.

number of small islands, or a small number of large islands, or something in between. Thus all systems for which we see results in Figure 15.5 contain islands of the same size. As should become clear from other chapters in this volume, or else from Chapter 1 in Volume 1 of this series [18], LC sequences in polymer chains provide a reinforcement, but at the same time they increase the brittleness of the material. This is confirmed by the simulation results shown in Figure 15.5. Consider the rapidly rising parts (the post-yield regions) in the diagrams. When we look at the curves for $\theta = 0$, 0.2 and 0.4, we see how these regions move to *lower* values of the strains. For the highest LC concentration studied, $\theta = 0.9$, the post-yield region appears at a very low value of the strain ε and therefore is hardly visible in Figure 15.5.

Since the rapidly rising post-yield region provides such a useful characteristic of the material, it is shown as a function of θ in Figure 15.6. We see that the curve exhibits a minimum. By starting from $\theta = 0$, we see first the result of introducing brittle LC sequences; then, above $\theta = 0.5$, the strength imparted by the LC sequences makes itself apparent. However, different shapes of the fracture stress vs. θ diagrams are also possible, depending on the size and shape of the islands.

Once again we return to the argument in section 15.1 that simulations can provide us with results which neither experiments nor theory can.

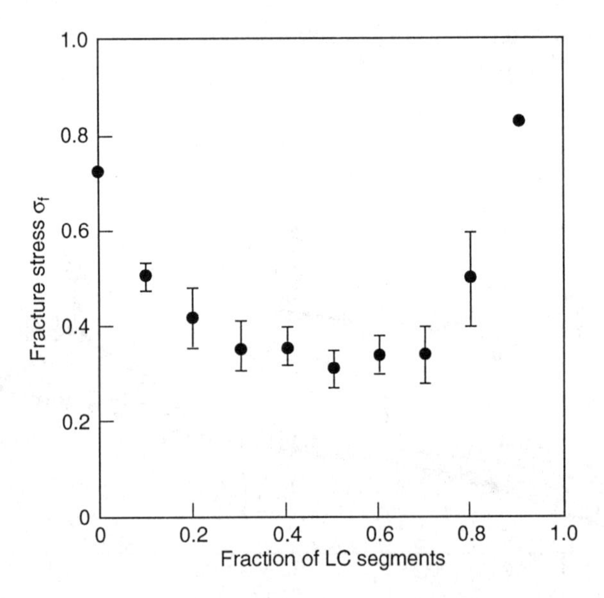

Figure 15.6 The fracture stress (= top of the rapidly rising part of the stress–strain diagram) in reduced units as a function of the LC concentration θ in random PLC copolymers. (Redrawn from [9].)

Among other things, we cannot experimentally create at will copolymers containing blocks of predefined length. The results presented above pertain to random PLCs. In Figure 15.7 we show results for block PLCs with two different lengths of both flexible and LC blocks. The material with shorter blocks here exhibits higher strength. At the same time, simulations show that not only the block length but also the spatial distribution of the blocks affect mechanical behavior. This is the reason why PLCs with the blocks 'organized' together into nearly spherical, square or diamond-shaped islands are being investigated (W. Brostow, C. Browning and M. Drewniak, in preparation). Examples of rectangular or fiber-like islands are shown in Figure 15.8.

Another aspect of mechanical behavior clarified by the simulations is crack formation. When in neighboring and locally aligned chains there is bond scission along a line approximately perpendicular to the alignment, a crack is formed. However, since in the carbon-like chains *cis* and *trans* conformations are allowed (see again the double-well interaction potential in Figure 15.1), the first effect of the application of a tensile force is the *cis* $\mapsto$ *trans* conversion. In Figure 15.9 we see that at the reduced stress levels up to 0.15 the conversion occurs gradually on the right hand side. At the reduced stress equal to 0.20, the entire right half of the material (which is here two-dimensional) underwent this conversion. Gradually, at still higher stress levels, the left hand side undergoes the conformational change *cis* $\mapsto$ *trans* as well.

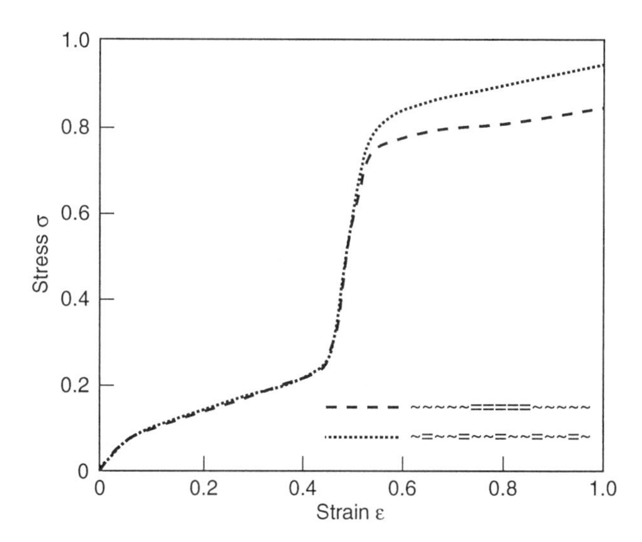

Figure 15.7 Computer-generated diagrams of stress σ versus strain ε for two types of block copolymers. (Redrawn from [9].)

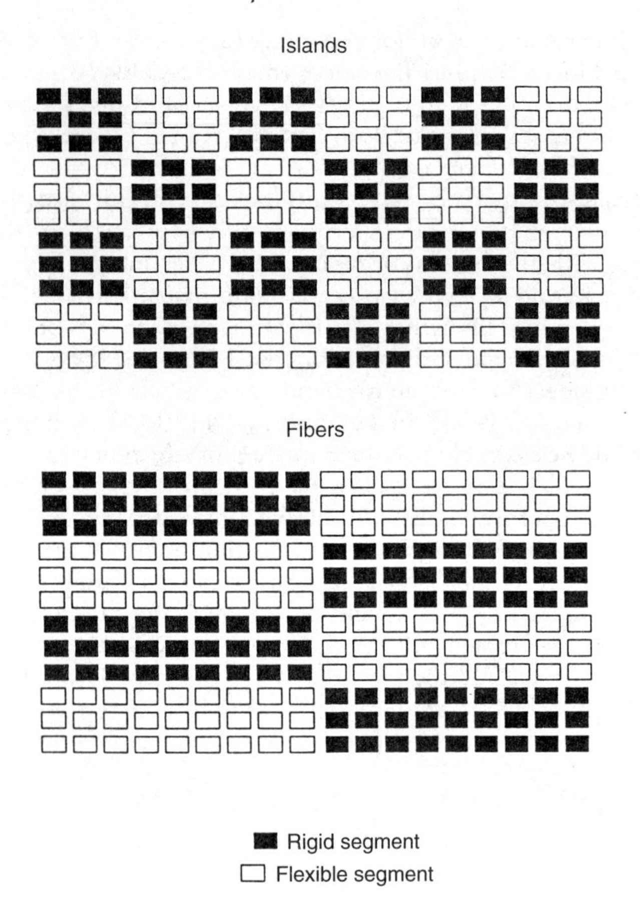

Figure 15.8 Two kinds of LC-rich islands in a PLC material.

At the reduced stress equal to 0.30 the conversion is complete, while the three LC regions in the material (in the middle and on both sides) are easily visible. We have thus also seen the LC reinforcement in action: until the reduced stress of 0.20 was exceeded, it was the LC region in the middle which protected the *cis* conformations on the left side from extension to the *trans* form. The same results confirm also the basic rule of the mechanical behavior of polymeric materials related to the chain relaxation capability (CRC) [19–22]: a polymeric material will relax if it only can. None of the mechanical energy coming from outside has been spent on destructive processes.

As expected, when all *cis* $\mapsto$ *trans* conversions have taken place, and the CRC has thus been exhausted, any further changes in the material must be *destructive* in nature. Now crack formation and then crack

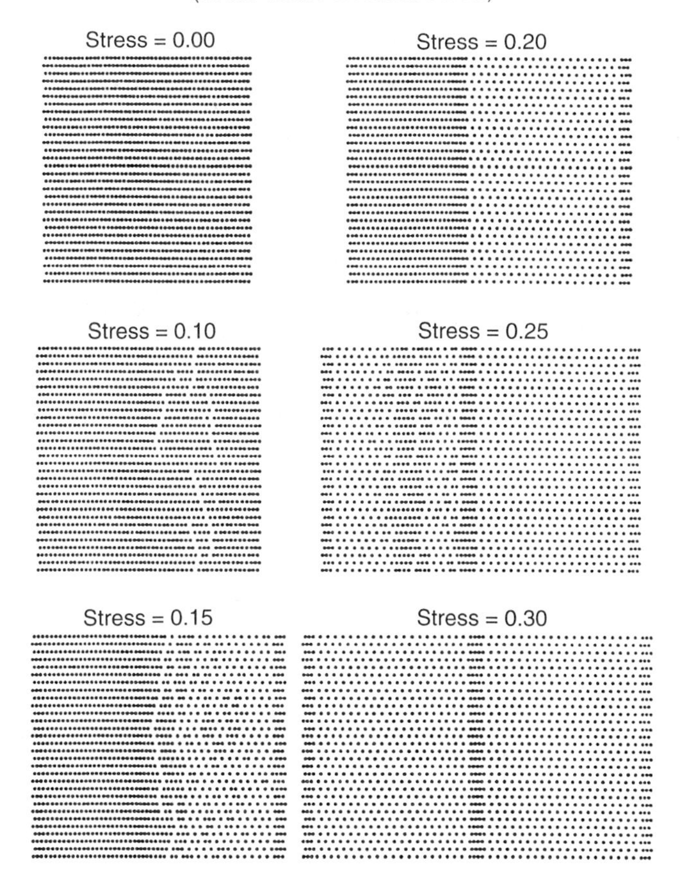

Figure 15.9 *Cis* ↦ *trans* conversions in a two-dimensional PLC material containing three LC regions (in the middle and on the sides) subjected to a horizontal tensile force. The stress values are given in reduced units. (After [9].)

propagation take place. In Figure 15.10 we show the same material as in Figure 15.9 but at the reduced stress level equal to 0.9. Fracture has already taken place; the crack has propagated *approximately along the LC reinforcement* in the middle.

Needless to say, MD simulations of PLCs allow us to investigate also the intermediate stages between conformational conversions shown in Figure 15.9 and the fracture shown in Figure 15.10. In Figure 15.11 an example is provided from a different simulation (W. Brostow, C. Browning and M. Drewniak, in preparation) in which crack propagation has already started but fracture has not yet occurred. It is

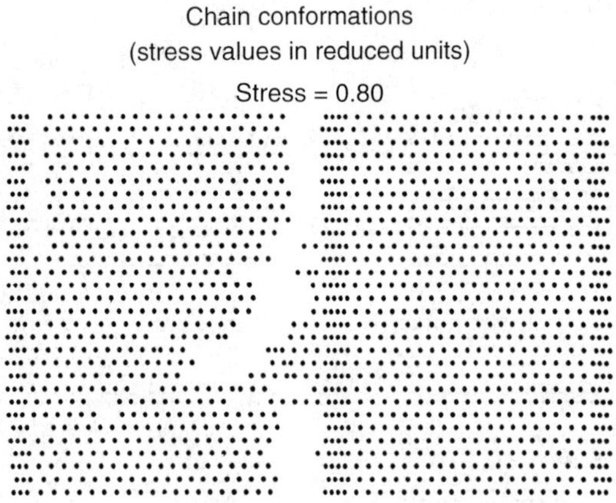

Chain conformations
(stress values in reduced units)
Stress = 0.80

Figure 15.10 Fracture in the material shown in Figure 15.9 but now subjected to the tensile stress = 0.9 (in reduced units). (After [9].)

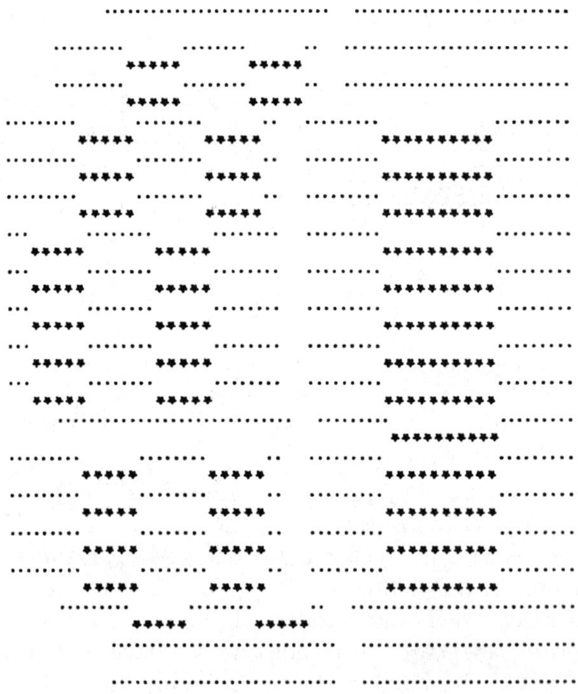

Figure 15.11 Crack propagation stage in a PLC material subjected to a tensile force.

also apparent here that crack propagation depends on the shapes, sizes and spatial locations of the islands.

15.6 CONCLUDING REMARKS

We recall that the first liquid crystals, which were monomer LCs (MLCs), were discovered by Friedrich Reinitzer in 1888 [23] while the first book about them appeared 20 years later [24]. While much has been done in the last century, both MLCs and PLCs remain fascinating object to study and to apply. We have seen several results of MD simulations of PLCs. A tacit assumption was made above all along: the simulated PLCs were all *longitudinal*, that is with LC sequences in the main chain and oriented along the chain backbone. Various other classes of PLCs have been synthesized by chemists [18, 25, 26]. These include *orthogonal*, with the LC sequences also in the backbone but perpendicular to it. There exist as well various kinds of *combs*, and even polymers with three-dimensional LC units. MD simulations should be able to elucidate mechanical behavior of such PLCs as well. This is particularly noteworthy since the promise made in section 15.1 to provide simulation results unobtainable from experiments appears to have been kept.

ACKNOWLEDGEMENTS

The research of the present author referred to herewith has been supported by the Robert A. Welch Foundation, Houston (Grant no. B-1203). Discussions with a number of colleagues and collaborators are appreciated, including Dr Slawomir Blonski (now at Rutgers University, Piscataway, NJ), and Dr Marta Drewniak (now at Solvay Engineered Plastics, Dallas), Prof. Josef Kubát at the Chalmers University of Technology, Gothenburg, Prof. Robert Maksimov at the Institute of Polymer Mechanics of the Latvian Academy of Sciences, Riga, and Prof. Janusz Walasek at the Technical University of Radom.

REFERENCES

1. Bratley, P., Fox, B.L. and Schrage, L.E. (1983) *A Guide to Simulation*, Springer-Verlag, New York–Berlin–Heidelberg–Tokyo.
2. Brostow, W., Drewniak, M. and Medvedev, N.N. (1995) *Macromol. Theory & Simul.*, **4**, 745.
3. Brostow, W. and Drewniak, M. (1996) *J. Chem. Phys.*, **105**, 7135.
4. Metropolis, N. and Ulam, S. (1949) *J. Amer. Statist. Assoc.*, **44**, 335.
5. Alder, B.J. and Wainwright, T.E. (1957) *J. Chem. Phys.*, **27**, 1208.
6. Brostow, W. and Kubát, J. (1993) *Phys. Rev. B*, **47**, 7659.
7. Mom, V. (1981) *J. Comput. Chem.*, **2**, 446.
8. Blonski, S., Brostow, W. and Kubát, J. (1994) *Phys. Rev. B*, **49**, 6494.

9. Blonski, S. and Brostow, W. (1991) *J. Chem. Phys.*, **95**, 2890.
10. Menczel, J. and Wunderlich, B. (1980) *J. Polymer Sci. Phys.*, **18**, 1433.
11. Menczel, J. and Wunderlich, B. (1981) *Polymer*, **22**, 778.
12. Meesiri, W., Menczel, J., Gaur, U. and Wunderlich, B. (1982) *J. Polymer Sci. Phys.*, **20**, 719.
13. Brostow, W., Dziemianowicz, T.S., Romanski, J. and Werber, W. (1988) *Polymer Eng. & Sci.*, **28**, 785.
14. Brostow, W. and Hess, M. (1992) *Mater. Res. Soc. Symp.*, **255**, 57.
15. Brostow, W., Hess, M. and López, B.L. (1994) *Macromolecules*, **27**, 2262.
16. van Gunsteren, W.F. (1988) in *Mathematical Frontiers in Computational Chemical Physics* (ed. D.G. Truhlar), Springer, New York–Berlin.
17. Allen, M.P. and Tildesley, D.J. (1987) *Computer Simulation of Liquids*, Clarendon Press, Oxford.
18. Brostow, W. (1992) in *Liquid Crystal Polymers: From Structures to Applications* (ed. A.A. Collyer), Polymer Liquid Crystals Series, Vol. 1, Elsevier Applied Science, London–New York, Ch. 1.
19. Brostow, W. (1985) *Mater. Chem. & Phys.*, **13**, 47.
20. Brostow, W. (1986) in *Failure of Plastics*, (eds W. Brostow and R.D. Corneliussen), Hanser, Munich–Vienna–New York, Ch. 10.
21. Brostow, W. (1991) *Makromol. Chem. Symp.*, **41**, 119.
22. Brostow, W. (1997) *Reliability and Mechanical Performance of Polymeric Materials*, Springer–AIP Press, Woodbury, NY.
23. Reinitzer, F. (1888) *Monatsch. Chem.*, **9**, 421.
24. Vorländer, D. (1908) *Kristallinisch-flüssige Substanzen*, Enke-Verlag, Stuttgart.
25. Brostow, W. (1988) *Kunststoffe*, **78**, 411.
26. Brostow, W. (1990) *Polymer*, **31**, 979.

Index

Page numbers in **bold** type refer to figures; *italic* type refers to tables.